MULTIPHASE SCIENCE AND TECHNOLOGY

Editorial Advisory Board

MULTIPHASE SCIENCE AND TECHNOLOGY
Volume 1

Edited by

G. F. Hewitt

Engineering Sciences Division and HTFS
U. K. Atomic Energy Authority
Harwell Laboratory, England

J. M. Delhaye

Centre d'Etudes Nucléaires de Grenoble
Service des Transferts Thermiques
Grenoble, France

N. Zuber

Division of Reactor Safety Research
U. S. Nuclear Regulatory Commission
Washington, D.C., U.S.A.

HEMISPHERE PUBLISHING CORPORATION

Washington New York London

DISTRIBUTION OUTSIDE THE UNITED STATES

McGRAW-HILL INTERNATIONAL BOOK COMPANY

Auckland Bogotá Guatemala Hamburg Johannesburg
Lisbon London Madrid Mexico Montreal New Delhi
Panama Paris San Juan São Paulo Singapore Sydney
Tokyo Toronto

MULTIPHASE SCIENCE AND TECHNOLOGY, Volume 1

1 2 3 4 5 6 7 8 9 0 B C B C 8 9 8 7 6 5 4 3 2 1

Library of Congress Cataloging in Publication Data

Main entry under title:

Multiphase science and technology.

Includes bibliographies and index.
Contents: v. 1. Spray cooling of hot surfaces / L. Bolle and J. C. Moureau—The spherical droplet in gaseous carrier streams / Georg Gyarmathy—Boiling in multicomponent fluids / R. A. W. Shock—[etc.]
1. Vapor-liquid equilibrium—Addresses, essays, lectures. 2. Ebullition—Addresses, essays, lectures. 3. Cooling—Addresses, essays, lectures. I. Hewitt, G. F. (Geoffrey Fredrick) II. Delhaye, J. M., date III. Zuber, N.
TP156.E65M84 660.2'96 81-4527
AACR2
ISBN 0-89116-222-4 (Hemisphere)
ISBN 0-07-028428-8 (McGraw-Hill)
ISSN 0276-1459

Contents

Preface

This is the first volume of *Multiphase Science and Technology*, a new international series of books intended to fill an existing gap and bring together materials from different fields such as nuclear energy, chemical processing, petroleum, meteorology, civil engineering, and energetics. The objectives of the series are to provide authoritative overviews of important areas in multiphase systems. The chapters published in the series are systematic and tutorial presentations of the state of knowledge in various areas. The editors hope that the nonspecialist reader can gain an up-to-date idea of the present stage of development in a given area.

Chapter 1 deals with the spray cooling of hot surfaces and is coauthored by Professor Léon Bolle and Doctor Jean-Claude Moureau. Professor L. Bolle has been with the Catholic University of Louvain, Belgium, since 1972 and is currently working on industrial energetics. A former student of Professor L. Bolle, Doctor J. C. Moureau received his doctorate in applied sciences in 1978 and is now in charge of nuclear safety at the Belgian Ministry of Health and Welfare.

The behavior of a droplet in a carrier stream is then examined in Chapter 2, by Doctor George Gyarmathy, Manager of the Turbomachinery R & D laboratories in the Brown Boveri Company, Baden, Switzerland. Doctor G. Gyarmathy has been involved in the theory of wet steam turbine and condensation for the past twenty years and is a recognized expert in these areas.

Chapter 3 is devoted to boiling phenomena in multicomponent fluids. Doctor R. A. W. Shock, the author, works for Heat Transfer and Fluid Flow Service at Harwell, UK. He is a specialist in the design of vertical thermosiphon reboilers and in crystallization.

Wall nucleation depends upon contact angles and wettability phenomena, the subject reviewed by Doctor Jacques Chappuis, Ecole Centrale de Lyon, France, in Chapter 4. Doctor J. Chappuis is currently on a sabbatical year at the Department of Mechanical Engineering of the University of Toronto.

These four chapters constitute the first volume of *Multiphase Science and Technology*. We would like to thank all the authors and reviewers for their outstanding contributions. And we would like also to express our special indebtedness to Mrs. Pauline Wilkes, who did a careful and patient job in typing this book.

The Editors

Chapter 1

Spray Cooling of Hot Surfaces

L. Bolle and J. C. Moureau
Universite Catholique de Louvain, Belgium

1 INTRODUCTION

1.1 Main Uses of Liquid Sprays

A dispersion of small liquid drops in a continuous gaseous phase is generally called a liquid spray. This dispersion has also received various other names according to the dimensions of the droplets produced. Some authors, such as Fortier (1967), distinguish between mist and cloud. A mist is a dispersion including drops smaller than 10^{-1} μm, whereas a cloud refers to a dispersion of larger particles. Others call an aerosol a dispersion of submicrometer particles and a mist much larger drops, for example, drops with a diameter of 150 μm. It is therefore necessary to make clear that, in this chapter, we consider as liquid spray a dispersion of liquid particles with diameters ranging from about 20 μm up to about 1 mm.

A process of disintegration of the liquid phase induces an increase in the interfacial surface area between the liquid and the medium into which it penetrates. This increase can be quite important: in some instances, the initial interfacial surface area can become several hundred times larger. Atomization can thus intensify the physical or chemical processes occurring at the interface, that is, mass, momentum, and energy transfers. This advantage is used in several industrial applications.

Liquid spraying is common in various fields: air conditioning and ventilation, gas absorption, washing and cleaning, fire protection, coating of surfaces, spray drying, combustion, cooling of hot gases, cooling of hot surfaces. This chapter is devoted to the study of this last application.

It is in the steel and metal works industry that one finds the most numerous uses of atomization in order to cool hot walls: cooling of slabs, rolled products, and cylinders in classical mills, and cooling of incandescent metal immediately after its exit from the mould in continuous casting units. In most instances, the amount of heat to be extracted from the metal is large: Fig. 1 shows, for the case of continuous casting, the variation of heat flux density $\dot{q}_w$ in the secondary cooling zone of a rectangular slab.

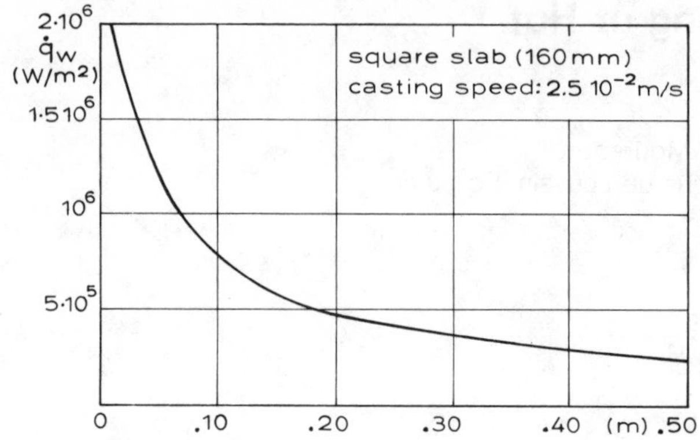

Fig. 1. Variation of the heat flux density as a function of distance to the casting level, according to Weinreich (1969).

Let us consider the typical example of the runout table of a hot strip mill. Between the last finishing stand and the coiling, cooling can be divided into three zones (Fig. 2):

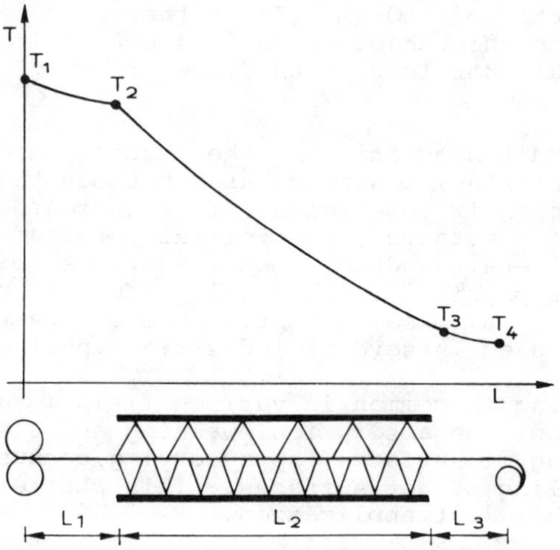

Fig. 2. Cooling zones on the runout table of a hot srip mill.

1. From the last mill to the beginning of the water cooling, the metal is in contact with air. Heat transfer is governed mainly by radiation with the environment, since the strip temperature is high (800 to 900°C).

2. Spray cooling produces an important drop in temperature along stretch L_2. The liquid aspersion is performed differently on the upper and lower sides of the strip. Above, water nozzles are generally mounted along successive rows at a distance of about 2 m from the strip: below, since the water must flow between the supporting rolls, the nozzles are placed nearer and usually have a lower flow rate. What kind of nozzles and how many should be displayed in order to achieve the desired cooling? At what pressure should they work? What is the optimal length L_2? To all these questions, the answer given has sometimes been too empirical and approximate. Indeed, the fundamental problem — determination of the heat transfer coefficient between a hot wall and a water spray — has not been completely solved.

3. Between the end of the spray cooling and the coiling, the strip is again in contact mainly with air: again, radiative heat transfer is predominant.

From a metallurgical viewpoint, the velocity with which heat is extracted from the metal during water cooling is essential in order to obtain a good-quality product. It has been shown that, for low-alloy steels, high strength can be achieved without reducing ductility or weldability provided that the size of the grains can be reduced. One of the best ways to reach this goal is to control cooling during the austenite-ferrite transformation (Morgan et al., 1965, 1966).

For rolled wire rods, a drastic cooling without quenching is desirable. Furthermore, when spraying, the surface temperature risks being considerably lower than that of the center of the rod. Therefore, in order to avoid superficially quenched structures, one has to divide the spray cooling zone into several parts separated by air cooling zones. Indeed, this setup facilitates uniformization of the inside temperature of the rod. Figure 3 illustrates the method for a wire rod with a 5.5 mm diameter. Moreover, Couvreur (1971) insists on the necessity, in design calculations, of taking into account a variation of the surface heat transfer coefficient α with the temperature of the rod. If, instead, one uses an averaged constant heat transfer coefficient, large discrepancies can appear. In Fig. 3, one sees a difference of about 100°C in the rod surface temperature when using constant and temperature-varying coefficients. This proves the usefulness of studies aiming to determine the exact parameters influencing heat transfer in such applications.

In nuclear power plants, spray cooling is one of the safety systems foreseen in case of accident. Schematically, two techniques exist: top spraying, that is, liquid spray onto the core from above; and bottom reflooding, that is, immersion from below. In the latter case, drops can sputter from the upper liquid level and impinge on hot walls. Many related publications — both theoretical and experimental — have described the phenomena occurring in various core configurations and possible incidents.

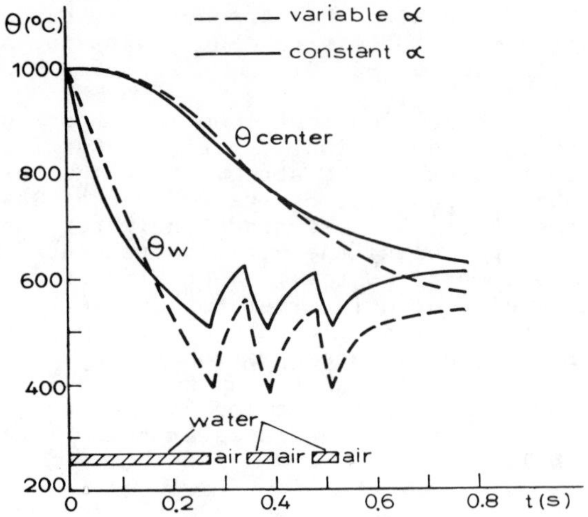

Fig. 3. Time evolution of the wire temperature, according to Couvreur (1971).

Yamanouchi (1968), for instance, has proposed a model for the evaluation of emergency spray cooling for a postulated loss of coolant accident (loca) in a boiling-water reactor. This model has been extended by several authors. Reviews of papers on spray cooling of light-water reactor cores have been published, among others, by Sawan and Carbon (1975). Moore et al. (1973) have studied spray emergency cooling of heavy-water reactors. In many of these texts, the liquid spray is considered only as a means to produce a liquid film flowing over the hot wall. It is the heat transfer between the film and the hot surface that is analyzed in detail and not the transfer due to the impacts of very many individual drops. Therefore, we shall not comment on these studies any further.

Besides the steel-making and nuclear industries, the chemical industry also uses liquid sprays in order to cool hot surfaces, for example, in the cooling of hot vessels and tanks.

Moreover, we think that liquid sprays could also be used in other fields. In 1973, Kawazoe and Kumamaru developed a new technique of spray cooling in order to improve the performance of the extrusion process of plastic-insulated telephone wires. New applications of spray cooling of hot surfaces are to be expected.

1.2 Previous Experimental Studies of Spray Cooling

One can distinguish in the literature two types of experimental work: results obtained in laboratories and measurements performed directly on industrial equipments. Theoretical models will be discussed later.

1.2.1 Laboratory Measurements

The main results published during the last 12 years are

summarized in Fig. 4, depicting the variation of heat flux density at the wall as a function of wall temperature. Later we shall comment on each of the curves appearing in the figure.

The experimental conditions of the tests (nature and surface area of the hot wall, type of spray nozzles, mass flux density of the spray, dimensions of the droplets) are gathered in Table 1.

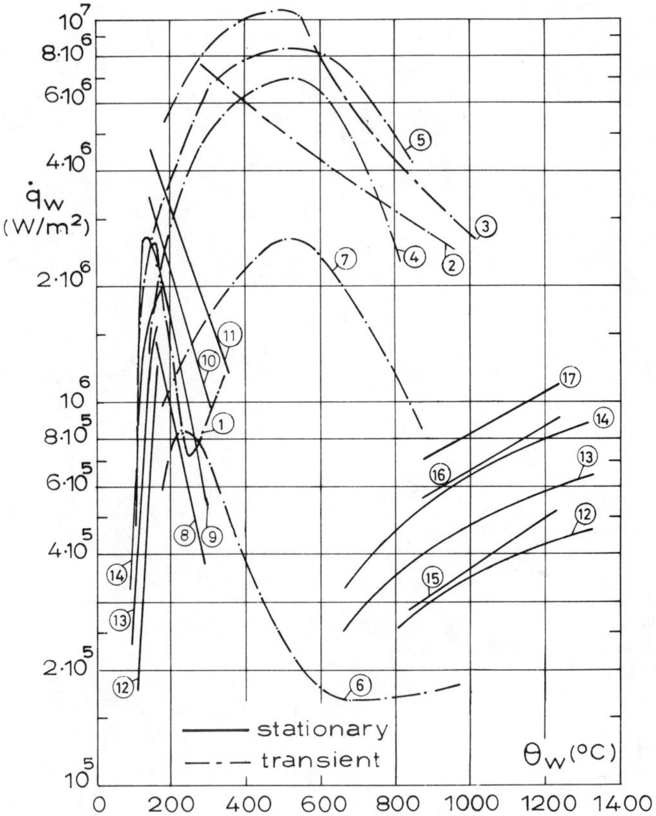

Fig. 4. Heat flux density at the wall as a function of wall temperature (for the nomenclature of the curves, see Table 1).

The experimental procedures can be classified into two categories according to the way they are conducted: (1) nonstationary methods and (2) stationary methods.

Measurements by Nonstationary Methods

Measurements by nonstationary methods involve three steps: first, the metal sample is heated to the desired temperature, then heat is withdrawn and simultaneously the sample is wetted with the

Table 1. *Characteristics of Several Experimental Works*

References	General Characteristics of the Tests: Nature of the Hot Wall	Surface Area of the Hot Wall (cm^2)	Temperature of the Spray (^0C)	Shape of the Spray and Working pressure, Δp (bar)	Mass velocity of the spray, $\dot{m}_\ell$ (kg/m^2 s)	Mean Drop Diameter, d (μm)	Mean Drop Velocity, u_ℓ (m/s)	Characteristics of the curves of Fig. 4 (Number of curve and Main Parameter)
Gaugler (1966)	Horizontal upper face of a copper cylinder plated with chrome	1.2	20	Full cone	0.7–3.7	128–250	16–26	(1) $\dot{m}_\ell = 2\ kg/m^2\ s$
Corman (1966)	Id	Id	Id	Id	57	–	–	(2) $\dot{m}_\ell = 57\ kg/^2\ s$
Auman et al. (1967)	Horizontal plate made of stainless steel (AISI304)	300	21	Fan	–	–	–	(3) –
Lambert and Economopoulos (1970)	Horizontal cylinder made of nickel	3.1	20	1–5	–	–	–	(4) $\Delta p = 1$ bar (5) $\Delta p = 3$ bar
Hoogendoorn and den Hond (1974)	Horizontal disk made of stainless steel (AISI321)	269	20	Full cone, 1–10	0.6–25	200–1000	10–30	(6) $\dot{m}_\ell = 0.6\ kg/m^2\ s$ (7) $\dot{m}_\ell = 25\ kg/m^2\ s$
Toda (1972)	Horizontal disk made of copper plated with gold	1.8	55, 70, 85, 92, 100	Full cone, 40–150	3.8–32	88–146	42–72	(8) $\dot{m}_\ell = 3.8\ kg/m^2\ s$ $\bar{u}_T = 42.3 m/s$ $\bar{d}_T = 88x10^{-6}$ m (9) $\dot{m}_\ell = 4.5\ kg/m^2\ s$ $\bar{u}_T = 48.9 m/s$ $\bar{d}_T = 146x10^{-6}$ m (10) $\dot{m}_\ell = 5.0\ kg/m^2\ s$ $\bar{u}_T = 46.2 m/s$ $\bar{d}_T = 117x10^{-6}$ m (11) $m_\ell = 32\ kg/m^2\ s$ $\bar{u}_T = 72.4 m/s$ $\bar{d}_T = 117x10^{-6}$ m
Junk (1972)	Vertical stainless steel tube	83–207	–	Fan 2–6	–	–	–	(12) $\Delta p = 2$ bar (13) $\Delta p = 4$ bar (14) $\Delta p = 6$ bar
Müller and Jeschar (1973)	Vertical stainless steel plate	20–65	20	Fan Full cone, 2–10	0.3–9	–	11–32	(15) $\Delta p = 2$ bar (16) $\Delta p = 7$ bar (17) $\Delta p = 10$ bar

spray; and finally the sample temperature decay versus time is recorded.

This procedure is relatively simple to apply in practice. It seems, at first sight, close to the conditions existing in most industrial applications, which are all of the transient nature. However, one must not forget that time evolution of the temperature in the sample is a function not only of the metal itself but also of the thermal inertia of all the elements in contact with it (supports, two-phase medium close to the surface, flow configuration of the boundary layer of medium, etc.). Therefore, the correspondance of the laboratory situation to the industrial situation is not made any easier by the transient nature of the procedure chosen.

The disadvantage of the transient method lies in the difficulty of interpreting the experimental results. Indeed, it is difficult to measure with precision the surface temperature of a very rapidly cooled metal sample. The presence of the measuring element on the surface alters the local conditions of the nearby flowing medium and the temperature field inside the sample. In order to avoid this difficulty, the temperature at a certain depth under the surface has to be measured and then the temperature at the wall calculated.

It is necessary to solve what is called the "inverse heat conduction problem," which can be formulated as follows: Let us assume that a metal sample previously heated up to a high temperature (1000°C) is quickly cooled. Knowing the initial conditions (temperature profile in the sample) and the time evolution during cooling of a point near to the surface, find the boundary conditions, that is, the temperature and the heat flux at the surface. Several authors have proposed solutions; including Gaugler (1966), Economopoulos (1968a), and Beck and Chevrier (1971). They require long numerical treatments in which oscillations are often present.

Furthermore, great precision is needed in the positioning of the measuring probe under the surface. Economopoulos (1968a) emphasises that the mean error on the heat transfer coefficient depends strongly on the exact location of the probe measuring the temperature (Fig. 5).

Let us review the main conclusions reached by investigators using the transient method.

In his thesis, Gaugler (1966) suggests the following relation:

$$\dot{q}_w = 4500(\theta_w - \theta_{sat})\dot{m}_\ell^{1/3} \quad \pm 10\% \quad (\mathrm{W/m^2})$$

with $250^\circ\mathrm{C} < \theta_w < 450^\circ\mathrm{C}$

$$0.7\ \mathrm{kg/m^2\ s} < \dot{m}_\ell < 3.7\ \mathrm{kg/m^2\ s}$$

Although Gaugler introduces θ_{sat}, the saturation temperature of water (100°C), he claims that his evolution is valid for water at 25°C.

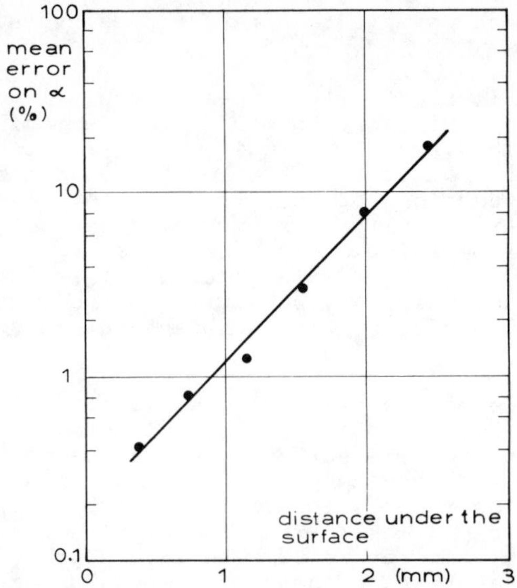

Fig. 5. Influence of the location of the probe on the accuracy.

An example of application of this relation is given in Fig. 4 (curve 1) for a mass flow rate density of atomized water equal to 2 kg/m^2 s. Unfortunately, the domain of validity of this expression is rather limited for industrial spray cooling applications. Besides, for a fixed wall temperature, the heat flux density increases with the momentum of the drops and with the number of drops per unit volume.

Using the same experimental equipment as Gaugler, Corman (1966) has extended the measurements to higher wall temperatures (up to 1000^{0}C) and higher water flow rate densities $\dot{m}_\ell$. In Fig. 4, curve 2, one sees that for $\dot{m}_\ell$ = 57 kg/m^2 s, $\dot{q}_w$ decreases when θ_w increases.

According to Auman et al. (1967), the heat flux density curve (Fig. 4, curve 3) exhibits three zones:

1. For $\theta_w > 650^0$C, the sprayed water is insulated from the wall by a continuous sublayer of vapor; heat transfer is of the film boiling type.

2. For $550 < \theta_w < 650^0$C, the vapor sublayer is not continuous and is unstable; it is characteristic of the well-known transition regime in pool boiling.

3. For $\theta_w < 550^0$C, water drops wet the wall and vaporization is governed by the nucleate boiling regime.

Their results for spray cooling are thus similar to those of the pool boiling regimes. They add a complementary remark: The greater the mass of water per unit surface area and unit time, the greater the heat flux density (Fig. 6.).

One notices that the slope of the curve diminishes when $\dot{m}_\ell$

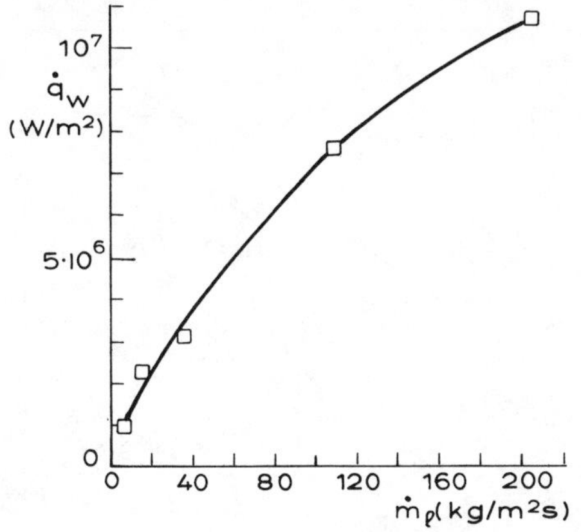

Fig. 6. Heat flux density as a function of mass velocity, for a constant wall temperature.

increases; one could then normally reach conditions such that it would be useless to increase the mass velocity $\dot{m}_\ell$ any further.

The conclusions that can be drawn from the paper by Auman et al. are in fact limited because all necessary information on the test conditions is not given. For instance, these authors do not mention for which $\dot{m}_\ell$ their $\dot{q}_w$ versus θ_w curve has been obtained. At best, one can guess that $\dot{m}_\ell$ is high! Moreover, in Fig. 6, they do not specify the wall temperature. Consequently, their results just exhibit trends.

The heat flux densities obtained by Lambert and Economopoulos (1970) when cooling a nickel cylinder with two different sprays are presented in Fig. 4, curves 4 and 5. They exhibit the same shape as the curve of Auman et al., but it is hazardous to compare them numerically because of a lack of information on the spray characteristics ($\dot{m}_\ell$, drop diameters, etc.).

For Hoogendoorn and den Hond (1974) also, the heat flux density at the wall depends strongly on the wall temperature and the water mass flow rate density. Curves 6 and 7, Fig. 4, are the limits of their range of measurements. It can be seen that:

1. Their heat flux values are lower than those previously presented.

2. Curve 7 decreases for θ_w between 500 and 600^{0}C, whereas curve 6 increases slightly for the same temperatures. Hoogendoorn and den Hond explain this difference: The extremity of curve 6 corresponds to the film boiling regime for each drop, whereas that of curve 7 corresponds to the transition regime.

They present a criterion of existence of calefaction for atomized water (Fig. 7). On the ordinate axis is plotted the temperature above which, according to the authors, the drops do not wet the hot wall but are separated from it by a vapor cushion. This limit temperature (which could also be called a Leidenfrost temperature) varies with $\dot{m}_\ell$ and $\bar{u}_\ell$, a mean velocity of the drops. The criterion is established for a horizontal stainless steel plate and nothing proves that it is still valid for other materials; it should be compared to the criterion introduced by Chevrier et al.

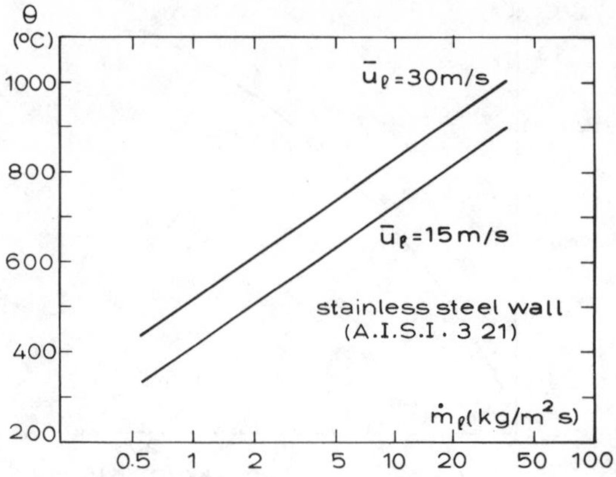

Fig. 7. Limit temperature of calefaction.

(1972) for the quenching of metals in distilled water.

When a hot, semiinfinite solid with thermal effusivity $E_w = \sqrt{\lambda_w \rho_w c_w}$ is suddenly brought in contact with a quiescent, semi-infinite liquid with effusivity $E_\ell = \sqrt{\lambda_\ell \rho_\ell c_\ell}$, the interfacial temperature is instantaneously established and remains equal to

$$\theta_{ci} = \frac{\theta_w E_w + \theta_\ell E_\ell}{E_w + E_\ell}$$

as long as the only controlling phenomenon in the liquid is thermal conduction. Figure 8 shows the contact temperature θ_{ci} as a function of liquid temperature θ_ℓ. The straight line 1 corresponds,

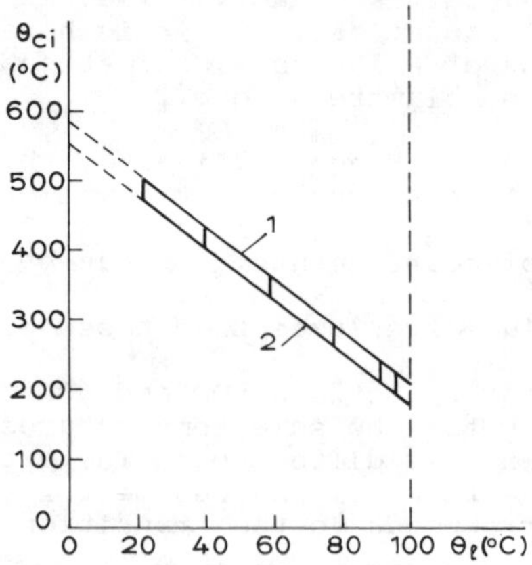

Fig. 8. Calefaction in water quenching.

according to Chevrier et al., to the contact temperature above which calefaction exists. Moreover, line 2 corresponds to the contact temperature below which the nucleate boiling regime exists. This criterion emphasizes the influence of the liquid temperature (which does not appear in the Hoogendoorn and den Hond criterion established for 20°C water) and contains the influence of the nature of the wall and of the liquid (through the effusivities). However, this criterion does not show any effect of the dynamic contact between liquid and solid, whereas Hoogendoorn and den Hond emphasize the importance of this in case of liquid spray.

The experiments performed by Bieth et al. (1976) and Moreaux et al. (1978) consist of cooling a vertical cylinder made of nickel (diameter = 16 mm, height = 48 mm) by means of three full-cone pneumatic atomizers. Although the cooling by pneumatic atomizers is beyond the scope of this chapter, it is interesting to indicate here that these authors consider two cooling regimes: the wetting regime and the nonwetting regime. For the nonwetting regime, which is said to be obtained at high wall temperature, they conclude that the heat extracted from the wall by the drops decreases with the mass velocities of air and water when the drop Weber number

$$\mathrm{We} = \frac{\rho_\ell \, u_\ell^2 \, d}{\sigma}$$

is between 100 and 350, whereas it increases for Weber number between 350 and 5000. The wetting regime obtained at lower wall temperature is characterized by direct contact between the liquid and the hot solid. The heat flux density $\dot{q}_w$ is found to be proportional to the mass velocity $\dot{m}_\ell$: see Fig. 9. The transition

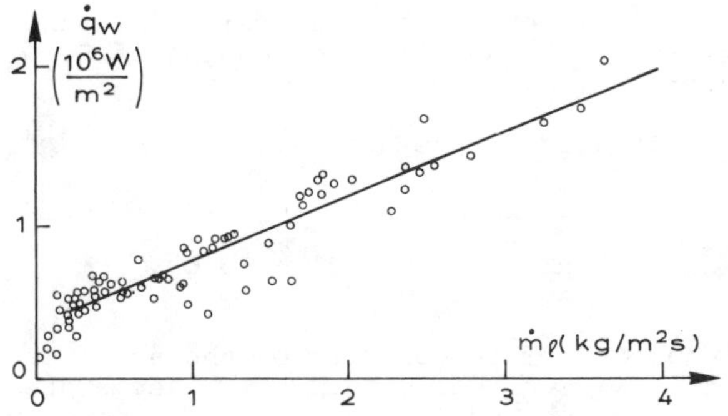

Fig. 9. Heat flux density as a function of mass velocity.

between the two regimes occurs for the transition temperature introduced by Chevrier et al.

This temperature is said to increase when the velocity u_ℓ and the mass velocity $\dot{m}_\ell$ increase. Typical values of the transition temperature are 300 to 450°C. The authors insist also on the influence of salt as NaCl on the domain of existence of the wetting regime. For water with a 20% concentration of NaCl, the wetting regime is obtained at 800°C for $\dot{m}_\ell$ larger than 2 kg/m² s.

Measurements by Stationary Methods

Essentially, measurements by stationary methods consist of

giving the sample a controlled heating energy in order to maintain its surface temperature constant during spraying the liquid. Two ways of determining the heat flux at the wall are possible. On the one hand, heat flux can be established from measurements of the temperature gradient in the sample. Since this method has been used only up to relatively low surface temperatures, we shall not review those types of measurements. On the other hand, heat flux can be deduced from the heating energy given to the sample. Joule effect is then the most common heating means; it requires high electrical currents (several thousands of amperes). The determination of surface temperature is then easier than with the transient procedure.

Let us now comment on the main results obtained by experimenters applying the permanent regime procedure.

Curves 8 to 11 of Fig. 4 can be distinguished by means of the three parameters that, according to Toda (1972), characterize a water spray:

1. The mass flow rate density $\dot{m}_\ell$.

2. A mean velocity of the spray, defined by

$$\bar{u}_T = \frac{1}{2}\frac{\dot{m}_\ell}{\rho_a}\left(\sqrt{1 + \frac{8p_s\rho_a}{\dot{m}_\ell^2}} - 1\right)$$

   where ρ_a is the mass density of the ambient air and p_s is the dynamic pressure of the spray; the meaning of $\bar{u}_T$ will be discussed later.

3. A mean volume diameter of the drops $\bar{d}_T$:

$$\bar{d}_T = 3.7\text{x}10^3 \frac{d_o}{u_{\ell o}}\sqrt{\frac{\sigma_\ell}{\rho_\ell}}\,(1 + 33408\nu_\ell)$$

   with d_o the diameter of the orifice of the nozzle and $u_{\ell o}$ the velocity at the orifice. Note that this last expression, introduced by Tanasawa and Toyoda (1945), is not general and seems suitable only to small sprays with full conical shape.

Unfortunately, the individual influence of these three parameters is not clearly shown.

A criterion of calefaction can be found in Toda's thesis. He considers that the cooling due to each of the drops can be split up into several steps:

1. The drop hits the hot wall and spreads until it forms a liquid film. The thickness of this film depends only on the initial drop diameter.

2. Whatever the wall temperature, the wall is wetted by the liquid: there is thus direct contact between the liquid and the solid. Heat is transported by conduction in the liquid. This phenomenon

is very short.

3. Vaporization then occurs at the liquid — solid interface if the film thickness is larger than a limit value S_F^* given by

$$S_F^* = 1.55 \times 10^{-3} (T_w - T_{sat})^{-1.5} \quad \text{(m)}$$

the liquid film is then insulated from the wall by a vapor sublayer: this is the calefaction under the drop. If, however, the film thickness is smaller than S_F^*, evaporation takes place at the upper surface of the film and there is no calefaction.

We shall discuss this criterion later but must point out here that the model neglects any heat transfer during the period for which the drop is spreading on the wall, which is questionable.

Junk (1972) proposes functional relationships between heat flux density at the wall and $(T_w - T_{sat})$; Δp, the difference of pressures upstream and downstream from the nozzle orifice; p_s, the dynamic pressure of the spray; $\dot{m}_\ell$, the water mass flux density; and ℓ_p, the distance between the nozzle and the wall. Curves 12, 13, and 14 of Fig. 4 correspond to $\ell_p = 0.6$ m and $\Delta p = 2$, 4, and 6 kg/cm^2, respectively.

The following correlation synthesizes the measurements of Müller and Jeschar (1973):

$$\dot{q}_{sr} = [10 u_{\ell o} + (107 + 0.688 u_{\ell o})\, \dot{m}_\ell]\, (\theta_w - \theta_{sat}) \quad \pm 12\% \quad (\text{W/m}^2)$$

with

$$0.3 \leqslant \dot{m}_\ell \leqslant 9 \text{ kg/m}^2\text{ s}$$
$$11 \leqslant u_{\ell o} \leqslant 32 \text{ m/s}$$
$$700 \leqslant \theta_w \leqslant 1200^\circ\text{C}$$

where $\dot{q}_{sr}$ is the heat flux density exchanged between the wall and the spray, not taking into account the radiative heat transfer.

For given θ_w and $\dot{m}_\ell$, the more the water velocity at the orifice $u_{\ell o}$ increases, the more the heat flux density increases. This tendency is illustrated by Fig. 10 for various types of fan spray nozzles (F10.8 — F12.5).

1.2.2 Industrial Measurements

Publications relating measurements on industrial equipment are generally intended less to explain the physics of the phenomena than to give global empirical relations. Their use is often limited because of their specific character.

Bösenberg (1966) discusses correlations concerning cooling on the runout tables of hot strip mills. He shows that the expression of Sibakin and Ikeda,

$$\theta_1 - \theta_4 = -1034 + 0.5\theta_1 + \frac{0.457}{S} - \frac{1728}{t_s} + 234.43\,\frac{\dot{M}_\ell}{L} - 14.01\left(\frac{\dot{M}_\ell}{L}\right)^2$$

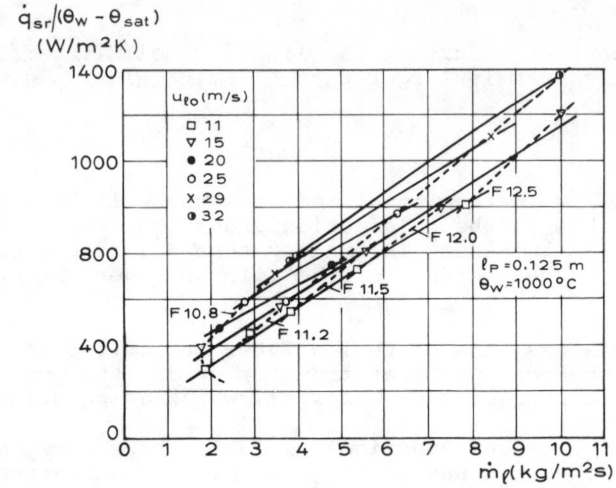

Fig. 10. Heat transfer coefficient as a function of mass velocity, from Müller and Jeschar (1973).

where θ_1 is the temperature of the strip after the last finishing stand, θ_4 is the coiling temperature, S is the strip thickness, t_s is the time duration of spray cooling, and $\dot{M}_\ell/L$ is the water mass flow rate per unit length, and the expression of Lovay and Kreulitsch,

$$\alpha_{LK} = -95.94 + 2.88\text{x}10^4 S + 0.2M_\ell + 10.8SM_\ell$$

with α_{LK} the global heat transfer coefficient over the entire length of the runout table and M_ℓ the total mass of water, are suitable only to the rolling mill considered and are not applicable elsewhere. In order to reach more general results, Bösenberg performed comparative measurements on several hot rolling mills. The resulting heat transfer coefficients that he found are presented in Fig. 11: the slope of the α_w versus $\dot{m}_\ell$ curve increases with $\dot{m}_\ell$, for a constant value of θ_w. This tendency is in opposition to the previously presented results of Auman and Müller.

Economopoulos (1968b) proposes an algebraic relation that should be easily integrated in a general system of automatic control and regulation of the cooling of hot rolled wide strips:

$$\theta_4 = \theta_1 + (\omega_2 - 830) + \frac{1}{U_t}\left[\omega_4 \log\log\left(\frac{S}{S_m}\right) + \omega_3\right]$$

$$+ \omega_1 L + N_p L_p(1 - \omega_1) \quad (^0\text{C})$$

where U_t is the strip rolling speed, N_p is the number of nozzles along spraying length L_p, S_m is the minimal thickness of the strip, and ω_1, ω_2, ω_3, ω_4 are coefficients to be determined in each specific use. Figure 12 shows the application of this formula for a strip thickness of 4 mm.

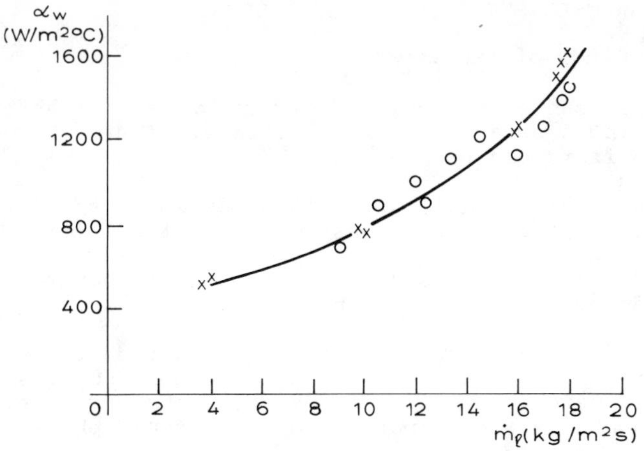

Fig. 11. Heat transfer coefficient as a function of mass velocity.

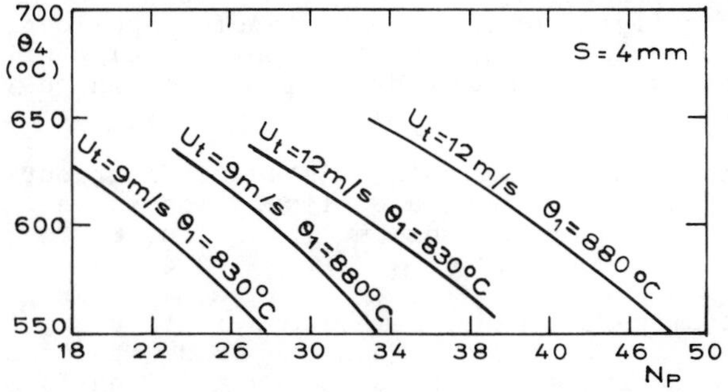

Fig. 12. Variation of coiling temperature with number of spray nozzles.

For Yanagi (1976), the heat flux density evacuated from the strip by means of water sprays follows the experimental relation

$$\dot{q}_w = 1.10\times10^{5}\,\dot{m}_\ell^{0.7}\left(\frac{\theta_\ell}{26}\right)^{-0.54} \quad (W/m^2)$$

Note that, in this expression, the strip temperature θ_w does not appear and that $\dot{q}_w$ decreases when the water temperature θ_ℓ increases.

From this review of the literature, we can conclude that experimental heat transfer results exhibit a large scattering, some of them even being contradictory. These discrepancies are probably due to the numerous parameters that can influence the heat exchange and to very different atomization conditions. Unfortunately, only a few authors take care to mention the characteristics of the spray they used. This is, however, essential for a correct interpretation of the spray cooling heat transfer. Therefore, the following part of this chapter is devoted to a description of the main characteristics of a liquid spray.

2 MAIN CHARACTERISTICS OF A LIQUID SPRAY

2.1 Various Kinds of Atomizers

Atomization can be achieved in several ways. In general, any device generating a spray can be called an atomizer. Three main classes of atomizers can be distinguished:

1. *Pressure nozzles*: Liquid spray is produced by the difference of pressures upstream and downstream from the orifice.

2. *Spinning disks*: Liquid spray is generated by centrifugal forces induced by high rotation speeds.

3. *Twin-fluid or pneumatic atomizers*: In addition to the effect of pressure difference mentioned for pressure nozzles, the mechanical energy content of the gaseous phase causes mixing of the two phases and dispersion of the liquid phase.

A detailed classification of all types of atomizers may be found in Dombrowski and Munday (1968). The subject of this work being spray cooling, however, we shall limit ourselves to atomizers, which are specifically adapted to heat transfer applications. Needs in this field include a rather concentrated spatial configuration of the spray and a given range of drop diameters, which will be discussed later.

The droplets formed by spinning disks are distributed in a plane or a very flat cone. Such a configuration is well adapted to evaporation or aeration operations but is not suited to most applications of wall cooling.

Pneumatic atomizers consume more energy; they are used when very tiny droplets are required. When compared with pressure nozzles, they are characterized by a poor spatial uniformity of cooling, but an increased ability to reach corners or zones behind obstacles. When air velocity is high, the cooling intensity of pneumatic atomizers may be very high.

In the following we shall consider pressure nozzles only. The main types of such nozzles are (1) flat or fan spray nozzles, and (2) swirl spray nozzles.

Flat or fan spray nozzles,: the spray drops are formed by disintegration of a plane liquid sheet and are distributed along a narrow band (Fig. 13) whose section has a very flat elliptical shape. Figure 14 shows projected views of a fan nozzle. Characteristic dimensions of three models of the same type are given. The lengths a_o and b can be larger than those presented, sometimes of the order of 15 to 20 mm for b.

Swirl spray nozzles,: can produce either a hollow- or a full-cone spray. In the hollow-cone spray, the drops are formed by disintegration or a conical liquid sheet and are concentrated in the periphery of a cone. As long as the liquid flow rate is not too large, entrainment of air by the liquid spray is low enough to leave the center of the cone virtually free of drops. In the full-cone spray, the conical liquid sheet oscillates continuously and a central liquid jet is added so that the droplets are distributed

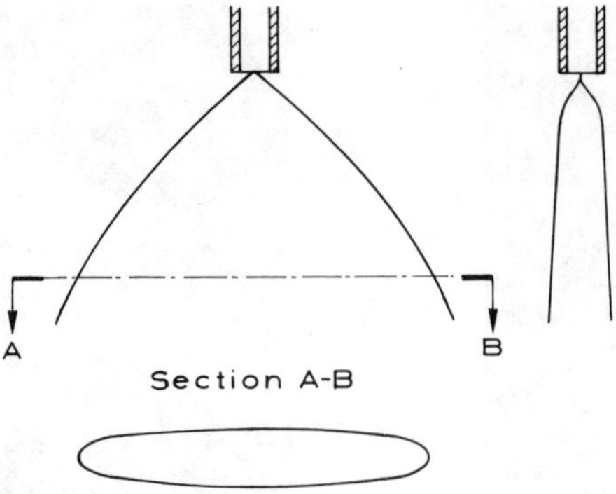

Fig. 13. Spatial configuration of a fan-type spray.

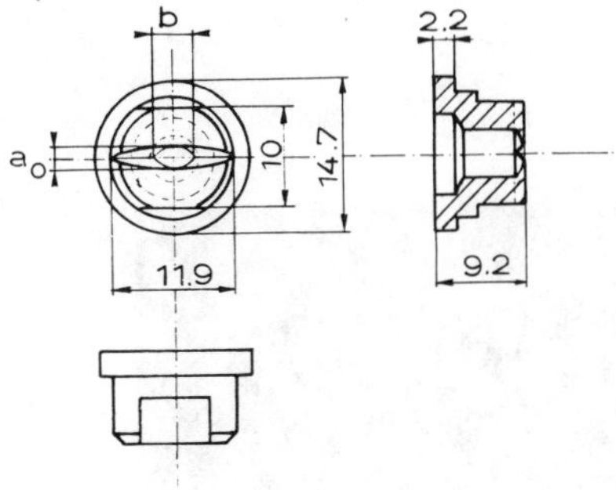

	d_o orifice mm	Spreading angle φ	a_o mm	b mm
F11.2/60	1.2	60°	0.8	1.8
F12/60	2.0	60°	1.4	2.9
F13/60	3.0	60°	2.2	4.6

Fig. 14. Dimensions of three types of fan nozzles.

more or less evenly throughout the whole volume of the cone. Figure 15 shows a characteristic design of a full-cone swirl spray nozzle. Figure 16 presents views of the impingement on a horizontal plate of the three types of sprays described above.

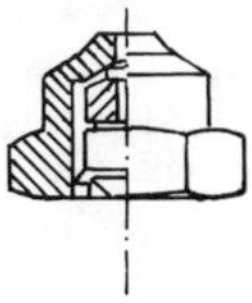

Fig. 15. Full-cone spray nozzle.

Fig. 16. Impact of various types of sprays on a plate, (a) Fan spray, (b) Full-cone spray, (c) Hollow-cone spray. (Courtesy of Lechler Apparatebau KG.)

Since this chapter is devoted mostly to fan-type pressure nozzles, let us examine in greater detail the process of disintegration of the liquid continuum in the vicinity of the nozzle tip and the spatial configuration of the spray below the zone of jet breakup.

Inside the nozzle, just above the orifice, liquid streamlines are impinging one onto the other (Fig. 17), resulting in a local increase of pressure. The design of the V-shaped cut in the nozzle tip is such that the liquid emerges in the form of a plane layer. The spreading angle ϕ of the liquid sheet is a function of the ratio a_0/b, that is, the acuteness of the cut. When this ratio approaches 1, the spreading angle decreases.

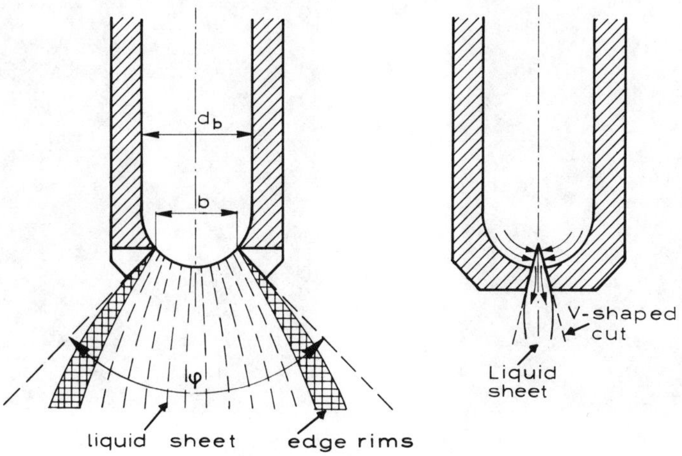

Fig. 17. Development of the liquid sheet for fan spray nozzles.

At the edges of the sheet, two opposite effects occur: on one hand, a contraction of the sheet due to the action of surface tension and, on the other, the tendency for the streamlines to remain nearly rectilinear (except for the rather weak effect of gravity) as a result of the high initial momentum of the liquid in the contracted area of the orifice. Thick rims appear at the boundary, resulting from the effect of surface tension (see Figs. 17 and 18).

The pressure difference at the orifice has a great influence on the extension of the sheet, as can be seen in Fig. 18. It also influences the value of ϕ.

For a very low pressure difference, the sheet is very small and still bounded by thick rims. As the inertia forces are weak, the rims are drawn together and form another narrow sheet in a plane at right angles to the first one.

For low-viscosity fluids, such as water, the velocity in the sheet remains constant; its thickness therefore diminishes.

The disintegration of the sheet is initiated by oscillations in the boundary rims and by oscillations of the sheet itself, which becomes very thin at some distance from the nozzle. A schematic description of this disintegration is represented in Fig. 19. Oscillations of the sheet produce disintegration in two steps: At first, ligaments are formed, reducing the surface energy content of the liquid. Those ligaments, which have very short lifetimes (of the order of 15 μs), then begin to buckle and to oscillate; they finally break into droplets.

The observation of a spray shows that the spatial concentration of drops is not uniform, but varies in the way suggested by

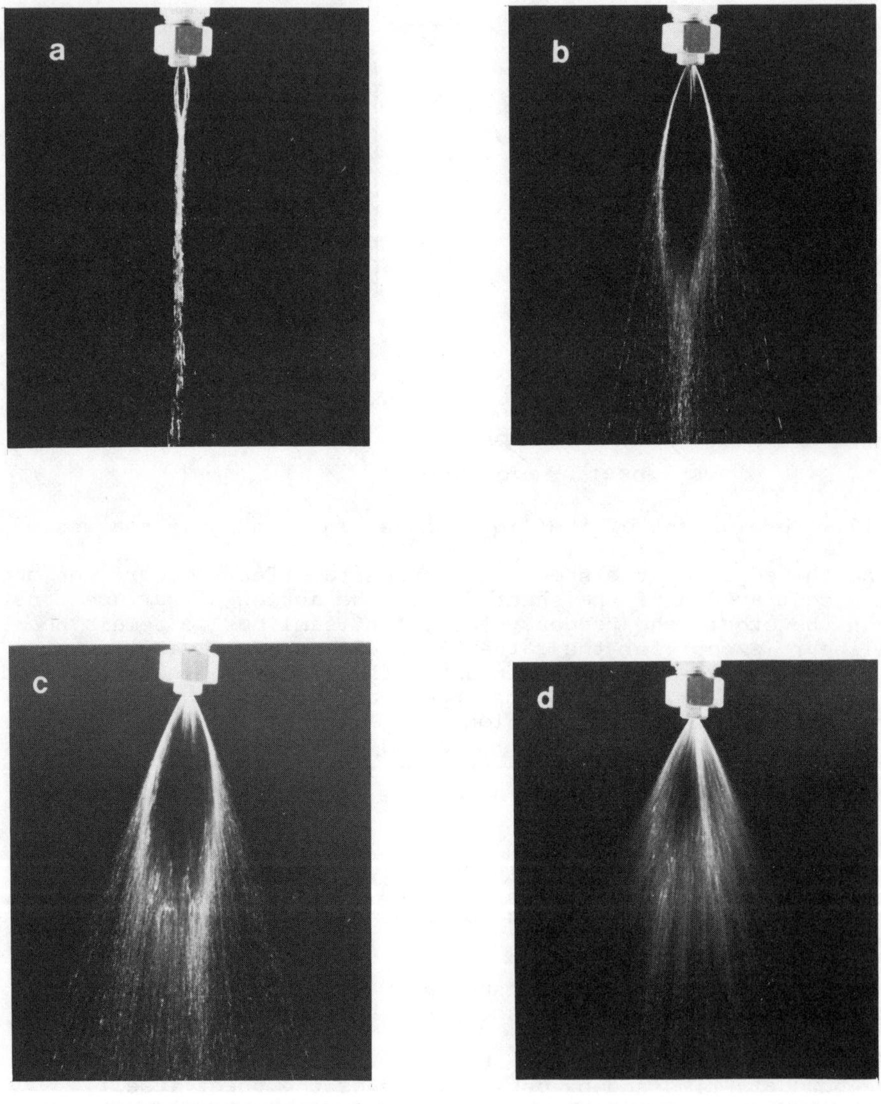

Fig. 18. Views of the liquid sheet for various operating pressures at the orifice of fan spray nozzle. (a) $\Delta p < 0.1$ bar. (b) $\Delta p = 0.4$ bar. (c) $\Delta p = 1$ bar. (d) $\Delta p = 3$ bar.

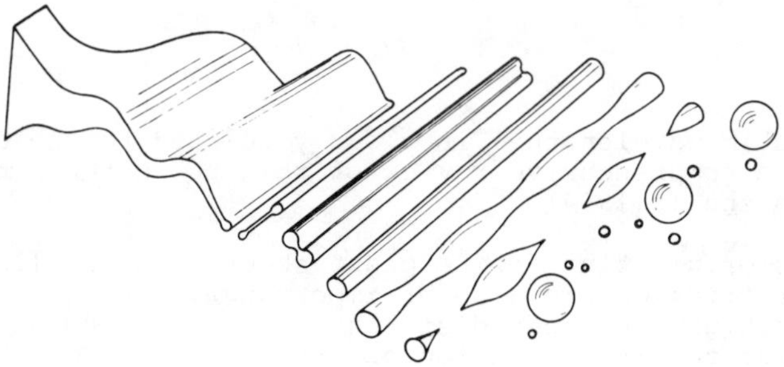

Fig. 19. Breakup of the liquid sheet.

Fig. 20. This is probably related to the oscillatory nature of the disintegration phenomenon of the liquid sheet.

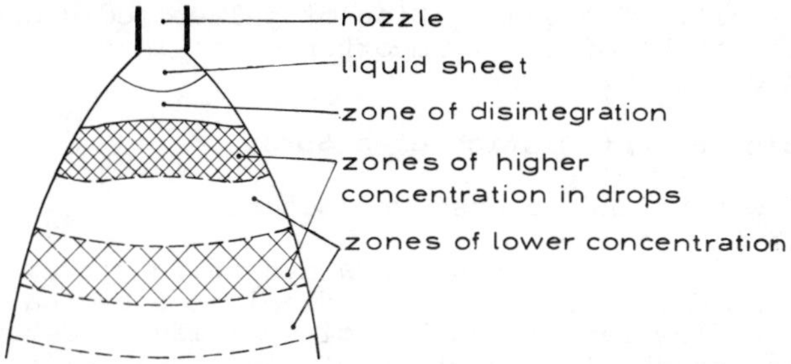

Fig. 20. Variation of drop concentration in a spray.

The diameters of the drops formed are not equal, but distributed over quite a large range (see Sec. 2.3.1 for measured values. The order of magnitude of the mean diameter of the drops increases with surface tension σ_ℓ and mass density ρ_ℓ, but decreases with the pressure difference at the orifice.

Instabilities of liquid jets of circular cross section, falling in still air, were first studied by Rayleigh (1879) for surface tension effects and by Weber (1931) for additional effects of viscosity and inertia forces. Weber showed that the wavelength λ required to produce a maximum instability of a jet of diameter d_o is given by

$$\frac{\lambda}{d_o} = \pi\sqrt{2}(1 + 3Z)^{1/2}$$

where $Z = \frac{\sqrt{\mathrm{We}}}{\mathrm{Re}}$

The Weber and Reynolds numbers are defined respectively as

$$\mathrm{We} = \frac{\rho_\ell d_o u_\ell^2}{\sigma_\ell} \qquad \mathrm{Re} = \frac{\rho_\ell u_\ell d_o}{\mu_\ell} \tag{1}$$

The characteristic wavelength λ is clearly related to the diameter of the drops. A comprehensive review of this subject has been performed by Marshall (1954).

For sprays originating from a plane sheet, the oscillations produced by air friction are also of importance. In this case, one has to distinguish between disintegration of the boundary rims (in a way similar to disintegration of free jets) and of the sheet itself. Some attempt was made to predict theoretically the diameters of the drops; a report on this subject is given by Dombrowski and Munday (1968), but it is not possible to quantify in a model all the secondary effects involved in the mechanism of disintegration.

Once the drops have been formed, they travel through air. Before they arrive at the exact location of use, they can evaporate partially or coalesce. The first phenomenon has been studied by, among others, Ranz and Marshall (1952), and the second by Pasedag and Gallagher (1971). The evaporation of pulverized drops in ambient air seems to be of small importance, at least over short distances (200 to 500 mm).

2.2 Macroscopic Characteristics of a Spray

2.2.1 Total Flow Rate and Related Parameters

Generally, a nozzle is set up in a hydraulic circuit in the way represented by Fig. 21, which also displays a typical laboratory device. The mass flow rate in such a circuit may be determined, in a $(\Delta H, \dot{M}_\ell)$ plane, as the intersection of the characteristic curves of the circuit and the pump, respectively. Actually, the shape of the curve describing the circuit strongly depends on the hydraulic properties of the nozzle.

The mechanical equation of the flow may be written as follows, between sections 1 and O:

$$\frac{p_1}{\rho_\ell} + \frac{u_{\ell 1}^2}{2} = \frac{p_O}{\rho_\ell} + \frac{u_{\ell O}^2}{2} + \Delta W$$

Section 1 is where pressure is measured; it must be as close as possible to the orifice. Section O is the surface normal to the streamlines at the orifice; the area of this surface will be supposed equal to A_O, that of the orifice itself. ΔW is the energy (J/kg) dissipated between sections 1 and O per unit mass, due to viscous losses.

Distributed as well as local head losses are proportional to the initial kinetic energy of the fluid; we therefore have:

$$\Delta W = \xi \frac{u_{\ell 1}^2}{2} \tag{2}$$

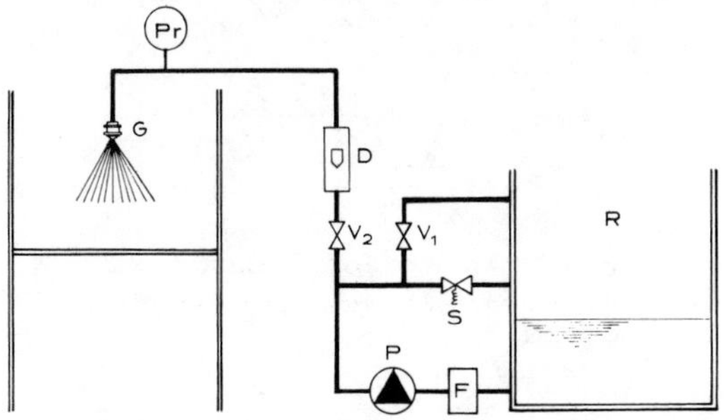

Fig. 21. Experimental apparatus for nozzle testing.

where ξ is a coefficient related to the shape and the roughness of the nozzle near the orifice. The mass flow rate $\dot{M}_\ell$ may be written as

$$\dot{M}_\ell = \rho_\ell \dot{V}_\ell = \rho_\ell A_1 u_{\ell 1} = \rho_\ell A_0 \; u_{\ell 0}$$

Taking into account the preceding relations, it may also be given by

$$\dot{M}_\ell = A_0 \sqrt{\frac{2\rho_\ell \; \Delta p}{1 - (A_1/A_0)^2(1 - \xi)}} \qquad (3)$$

Let us note that p_0 is the ambient pressure, and that

$$\Delta p = p_1 - p_0$$

is the relative pressure of the water just upstream from the orifice.

Identification of Eq. (3) with the following well-known one:

$$\dot{M}_\ell = C_q A_0 \sqrt{2\rho_\ell \; \Delta_p} \qquad (4)$$

yields the phenomenological significance of C_q, which is called the *discharge coefficient*. Typical values of C_q for fan-type pressure nozzles are in the range 0.6 to 0.9.

The measured values of C_q for the three nozzles described in Fig. 14 are listed in Table 2.

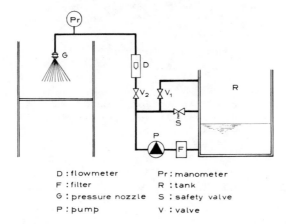

Rather than give the value of the liquid mass flow rate, many authors prefer to characterize the nozzles they use by means of the velocity of the liquid at the orifice. From Eq. (4), this velocity $u_{\ell 0}$ is given by

$$u_{\ell 0} = C_q \sqrt{\frac{2\ \Delta p}{\rho_\ell}} \qquad (5)$$

As an example, for a pressure of 5 bar (5×10^5 N/m^2), the values shown in Table 2 for the total mass flow rate and the velocity at the orifice can be computed.

Table 2. *Values of* C_q, M_ℓ *and* $u_{\ell 0}$

Type of fan nozzle	C_q	Δp = 5 bar: $\dot{M}_\ell$ (kg/s)	Δp = 5 bar: $u_{\ell 0}$ (m/s)
F11.2/60	0.797	2.85×10^{-2}	25.2
F12/60	0.642	6.38×10^{-2}	20.3
F13/60	0.666	14.89×10^{-2}	21.1

2.2.2 Liquid Flow Rate Density ($\dot{m}_\ell$)

The distribution of the drops in the spray is evidently not uniform in all radial directions, especially near the boundaries.

An analysis of the two-dimensional distribution of the liquid flow rate density, measured in a plane normal to the nozzle axis, was published by Müller and Jeschar (1973) (see Fig. 22). It is clear that small differences in the construction of the nozzle in

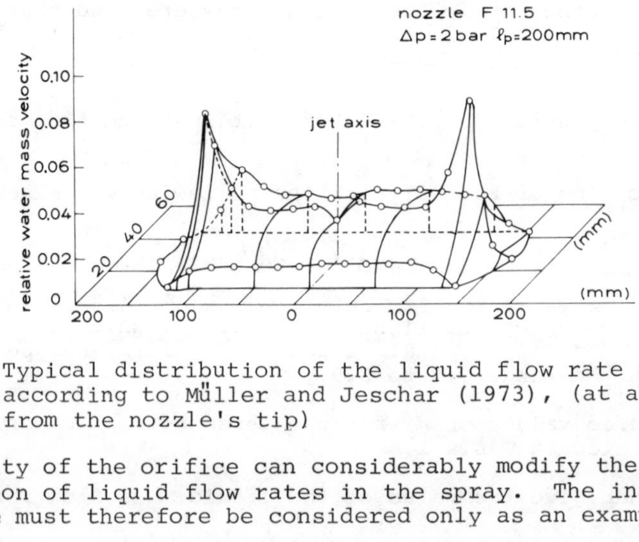

Fig. 22. Typical distribution of the liquid flow rate density, according to Müller and Jeschar (1973), (at a distance ℓp from the nozzle's tip)

the vicinity of the orifice can considerably modify the spatial distribution of liquid flow rates in the spray. The information given here must therefore be considered only as an example.

For the three nozzles used in our heat transfer experiments, we measured a one-dimensional distribution of liquid mass flow rate densities ($\dot{m}_\ell$) level with a plane normal to the axis of the nozzle, and over a region whose dimensions are equal to those of the heated plate (20 X 320 mm) used later. Typical results are given in Fig. 23. Dashed lines show a smoothing, in the sense of a least-squares approximation, of the measured values by a fourth-order polynomial.

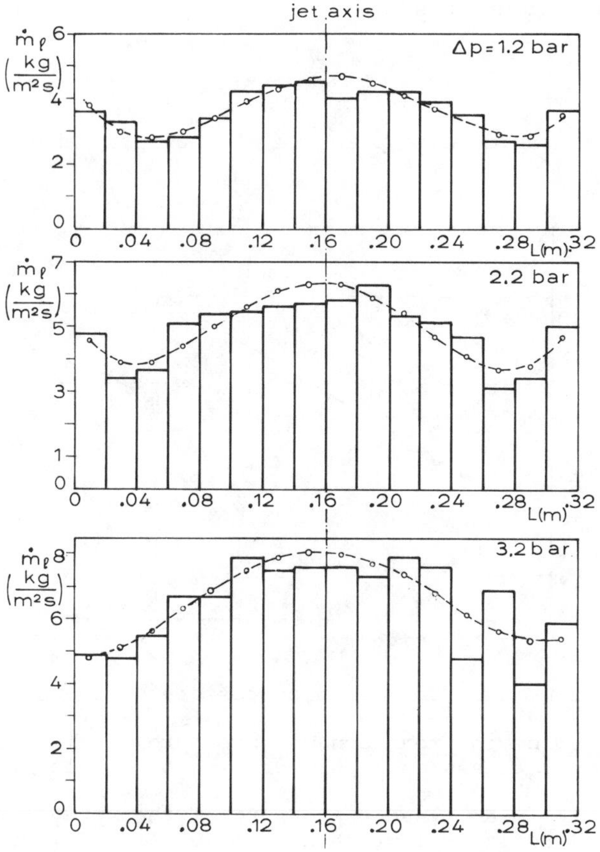

Fig. 23. Distribution of the mass flow rate density (nozzle F13/60; at a distance ℓ_p = 357 mm from the nozzle tip).

In many cases, a higher liquid flow rate density can be observed at both ends of each section normal to the spray axis. This is due to the disintegration of the edge rims of the liquid sheet, which constitute a zone of high liquid flow rate.

2.2.3 Dynamic Pressure of the Spray

It can sometimes be interesting to know the local dynamic

pressure of a spray. Toda (1972) has performed some measurements of this quantity.

We also carried out such experiments, using precision scales supporting a vessel with a rectangular cross section. In order to avoid secondary effects due to the rebound of droplets on a solid surface, we partially filled the vessel with oil. In this way, we obtained records similar to that of Fig. 24. Instants t_1 and t_3 are, respectively, those of start and end of application of spray to the vessel. Weights may be easily converted into pressures by dividing them by the area A_v of the cross section of the vessel. The differences

$$\frac{P_5^* - P_1^*}{A_v} = \frac{P_3^* - P_4^*}{A_v}$$

are equal to the mean dynamic pressure of the spray above the vessel.

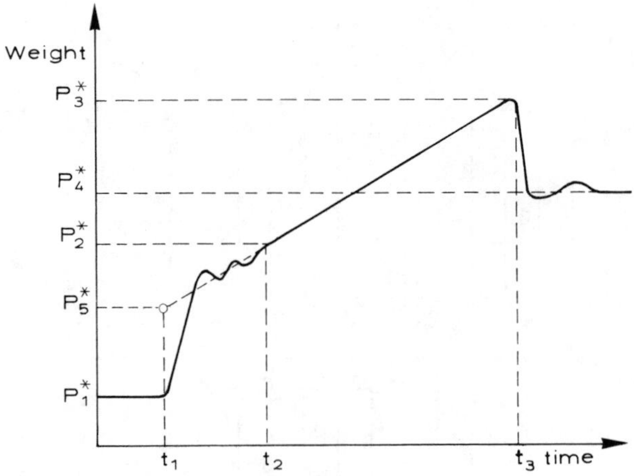

Fig. 24. Measurement of the dynamic pressure of the spray.

The evolution of the mean dynamic pressure is given in Fig. 25 for three different pressures at the orifice of nozzle F13/60.

Calling p_s the mean dynamic pressure of the spray (water plus air) over a total horizontal cross section of area A_s, we can write

$$p_s A_s = \frac{1}{2}\rho_\ell u_\ell^2 A_\ell + \frac{1}{2}\rho_a u_a^2 A_a \qquad (6)$$

where A_ℓ, and A_a are the mean areas of sections crossed by water and air, respectively. The velocities u_ℓ of water and u_a of air are supposed to be uniform for each phase. Obviously,

$$A_s = A_\ell + A_a$$

Equation (6) expresses the energy conservation in case of stagnation of the flow. It is known that the pressure exerted on

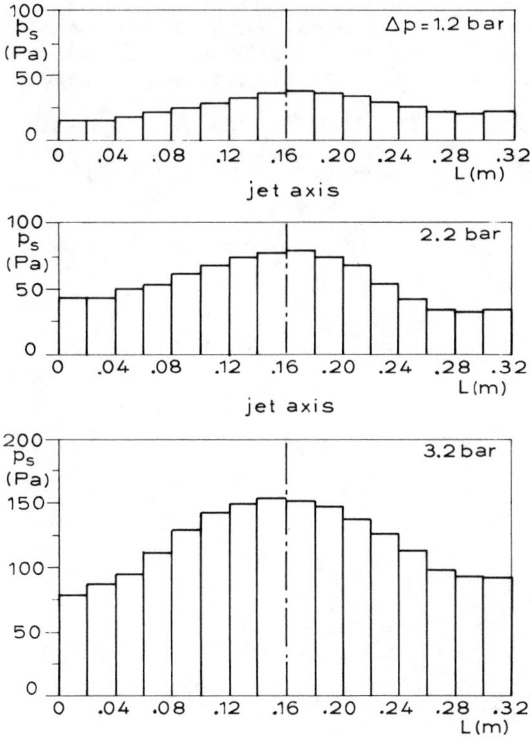

Fig. 25. Mean pressure of impact (nozzle F13/60; ℓ_p = 357 mm).

a plate normal to the streamlines in the spray equals twice the value of p_s (conservation of momentum along the jet axis direction).

On the other hand, momentum of the spray being equal at every distance from the nozzle orifice to the momentum of the liquid at the orifice itself, we have

$$\dot{M}_\ell u_{\ell 0} = \dot{M}_\ell u_\ell + \dot{M}_a u_a \tag{7}$$

Equation (6) can thus be modified, to obtain

$$p_s A_s = \frac{1}{2}\dot{M}_\ell u_\ell + \frac{1}{2}\dot{M}_a u_a$$

$$= \frac{1}{2}\dot{M}_\ell u_{\ell 0} = \frac{1}{2}\frac{\dot{M}_\ell^2}{\rho_\ell A_0}$$

Using relation (4), this yields

$$p_s = C_q^2 \; \Delta p \; \frac{A_0}{A_s} \tag{8}$$

After this relation, the mean dynamic pressure p_s varies linearly with Δp, the relative pressure upstream from the orifice, and inversely with A_s. However the area A_s also depends on Δp, and relation (8) is therefore difficult to apply in practice.

The knowledge of *local values* of the mean dynamic pressure may also be useful in order to evaluate *a mean velocity of pulverization*, in the way suggested by Toda (1972).

Let us consider a small surface of area a_s, normal to the axis of the nozzle. If we know the local liquid flow rate density $\dot{m}_\ell$, Eq. (6) may be rewritten locally as

$$p_s a_s = \frac{1}{2}\dot{m}_\ell u_\ell a_s + \frac{1}{2}\rho_a u_a^2 a_a \tag{9}$$

where $\dot{m}_\ell a_s = \rho_\ell u_\ell a_\ell$

and $a_s = a_\ell + a_a$

We may assume that

$$a_a \cong a_s$$

$$u_a \cong u_\ell \triangleq \bar{u}_T$$

The mean velocity of the spray (assume equal for air and all the droplets) is then given by

$$\bar{u}_T = \frac{m_\ell}{2\rho_a}\left(\sqrt{1 + \frac{8p_s\rho_a}{\dot{m}_\ell^2}} - 1\right) \tag{10}$$

For nozzle F13/60, the mean velocity $\bar{u}_T$ is computed for the central region of the spray, using values from Fig. 23 for $\dot{m}_\ell$ and from Fig. 25 for p_s. The results are compared in Table 3 with the values of the velocity at the orifice $u_{\ell O}$ given by Eq. (5). Actually, the meaning of $\bar{u}_T$ may be emphasized only within the scope of the rough assumptions made above.

Table 3. *Computation of the Mean Velocity $\bar{u}_T$ and the Velocity at the orifice $u_{\ell O}$ (nozzle F13/60)*

Δp (bar)	$\dot{m}_\ell$ (kg/m^2 s)	p_s (Pa)	$\bar{u}_T$ (m/s)	$u_{\ell O}$ (m/s)
1.2	4.25	38	6.38	10.3
2.2	5.75	80	9.40	14.0
3.2	7.65	152	13.04	16.8

2.3 Microscopic Characteristics of a Spray

In the preceding sections we described some global methods of characterization of a spray. In order to describe a pulverization in detail, we must concentrate on the droplets themselves. A statistical description is thus needed. The distributions of diameter and velocity of the drops will be studied successively.

2.3.1 Droplet Diameter Distribution

The disintegration of the liquid sheet results in a spray of very many droplets, whose diameters lie in the range 20 to 1000 μm for fan spray nozzles. The first question is how to measure such small diameters in a flowing medium. The main dependable methods are the impaction method, the molten wax method, thin-wire methods, laser methods, and optical methods.

The *impaction method*, also called the Kühn method, was first used in 1924. Over a short time, a small plane disk is introduced into the spray, normal to the mean trajectory of the drops. The disk is covered by some special substance; for example, soot was used by Oury and Xhrouet (1977), magnesium oxide by Nukiyama and Tanasawa (1938) and by Giffin and Muraszew (1953), and a photographic emulsion by Straus (1949).

Impingement of a droplet leaves a trace or a hole on the disk, with dimensions a function of the diameter of the droplet, but also depending on the velocity of the drop and on the properties of the pulverized liquid and the coating of the disk.

The disadvantages of the impaction method are as follows:

1. It is difficult to find the functional correlation between drop and trace diameters.

2. Two successive drops may leave only one trace.

3. Parasite traces may appear due to the breakup of drops when they impinge.

4. The insertion of the disk into the flow creates a local perturbation; air streamlines are deviated and the smallest drops are carried along the periphery or out of the disk.

As for all other methods, the number of samples measured must be high in order that the statistical evaluation of the parameters of the spray be significant.

The *molten wax method* (Joyce, 1949) consists of pulverizing a molten wax or any other liquid that can be solidified easily at room temperature. The solidified drops may then be classified by current methods used for classification of solids.

In a similar way, Longwell (1943) suggests pulverizing water in a medium at very low temperature in order to freeze the droplets. This method can be applied only to tiny droplets.

Also for pulverizations of very tiny droplets, methods using a *thin wire* at high electric potential or a hot-wire anemometer have been proposed by Geist et al. (1951) and by Goldschmidt and Householder (1969), respectively.

Laser instrumentation systems have been recently developed. A first type of technique is based on the Fraunhofer diffraction of a parallel beam of monochromatic light by the moving droplets (Swithenbank et al., 1976; Weiner, 1979). The diffracted light energy distribution is translated by a minicomputer into the corresponding droplet size distribution. The size range may be varied by changing the focal length of the lens, but there is a lower limit of about 1 μm. Two decades of diameter (5 to 500 μm, for example) can be covered by one lens.

Laser-Doppler systems have also been developed in order to measure the size distribution of droplets. They have been described, among others, by Farmer (1972), Dibelius et al. (1979), and Lee and Srinivasan (1978). A primary light beam illuminates a droplet. The light is then refracted, diffracted, and partially reflected. Based on Mie scattering theory, the angular distribution of light scattered in the forward direction is measured. The diameter of the observed droplet may be deduced after external calibration. This method is especially well suited to tiny droplets (1 to 100 μm in diameter) in dilute flows.

Measurements by laser techniques do not disturb the spray and can be easily automated, but the complexity of the signal processors is such that interpretation of results is very critical.

Optical methods do not introduce any perturbation in the flow. Interpretation of the results is clear in this class of methods, but the photographic equipment required is sophisticated. As an example, in order to photograph a drop with 50 μm diameter traveling at 10 m/s with an unsharp zone of 10% of the diameter, an exposure time of

$$\frac{50 \times 10^{-6} \times 10^{-1}}{2 \times 10} = 0.25 \times 10^{-6} \text{ s}$$

is required. Ultrarapid flashes are needed.

Medenblik (1976) has applied a photographic method of determination of drop diameter distribution to a spray. Furthermore, he has recently developed a fully automatic system for counting drops from a series of photographic proofs and for interpreting the results in terms of distribution curves and mean drop diameters.

Accurate estimations of the drop size parameters or distribution always require a large number of individual measurements, although the total volume of liquid carried by the drops may be very limited. From estimations by Nukiyama and Tanasawa (1938) and Bowen and Davies (1951), a minimum sample population in the range of 500 drops is always needed. In such conditions, these authors estimate that the errors on the mean diameter evaluations are in the range of ±15% and errors on the cumulative distribution curves of ±4%.

We have performed measurements by means of an elaborate

optical method developed by Diamant (1960), who has added to a classical macrophotography device a rotating mirror that permits bringing traveling droplets to a fictitious standstill. A schematic view of this apparatus is shown in Fig. 26. A light beam, originating from (1), illuminates a slit (3) of 43X5300 μm after reflection on a static mirror (2). This beam is then focused on a small zone of the pulverization. The picture of this zone is reflected by the rotating mirror onto the roll film.

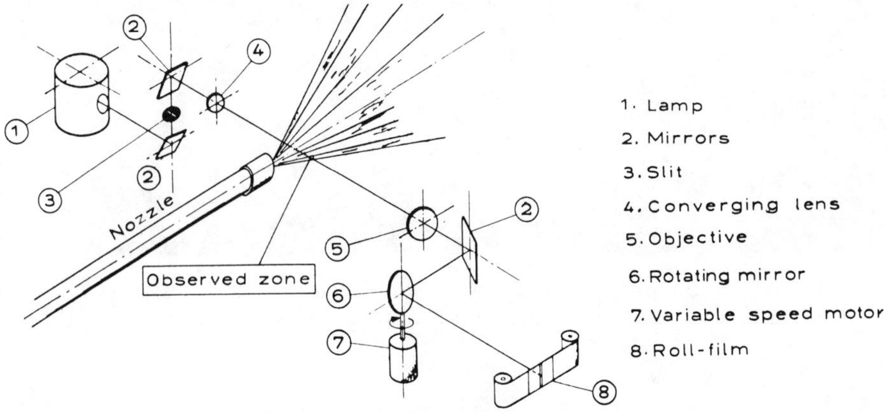

Fig. 26. Schematic view of rotating-mirror microscope.

For a given mirror rotation speed, the picture of a drop will be circular only for one drop velocity, which can be computed (Tolfo and Staudt, 1976), knowing the optical characteristics of the equipment. However, the spray presents a distribution of droplet velocities. The rotation speed of the mirror must therefore be adjusted in order to achieve synchronism with an average droplet velocity, which is easy to find in practice. The picture of a droplet will then be more or less distorted according to its departure from the velocity of synchronism, but the transverse dimension, proportional to the droplet diameter, will be unaffected and thus permits determination of droplet size distribution.

We have used the Diamant microscope placed at our disposal by Le Laboratoire Belge de l'Industrie Electrique (LABORELEC, Brussels). An example of a photograph obtained by this method is given in Fig. 27. Two straight lines, 100 μm distant in reality, can be seen on this photograph. The diameter of the drops is measured normally to those lines.

For given operating points of nozzles F12/60 and F13/60, we have taken a sufficient number of photographs to obtain, in each case, about 300 drop pictures that are clear and sharp enough to measure the diameters. Three representative histograms are shown in Fig. 28.

Drops have been grouped in classes of diameters of 20 μm. This value seems an acceptable compromise: Narrower classes would be a better approach to a continuous distribution function but, on the other hand, it is necessary to have a statistically significant number of drops in each class.

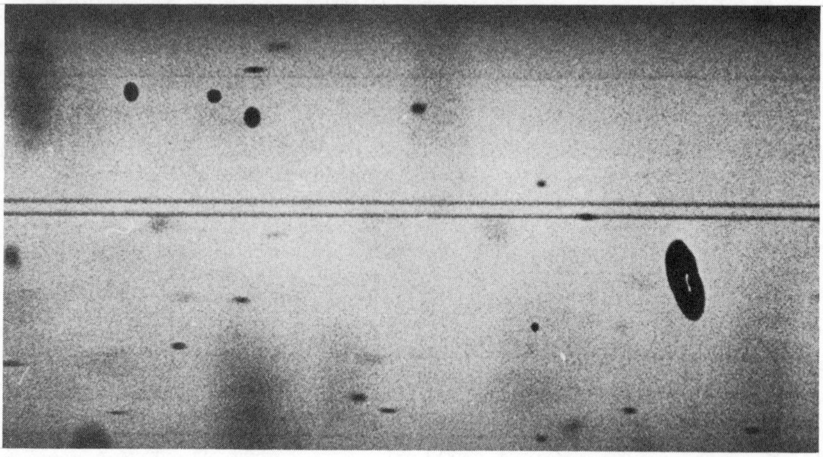

Fig. 27. Example of photograph obtained from the Diamant Microscope.

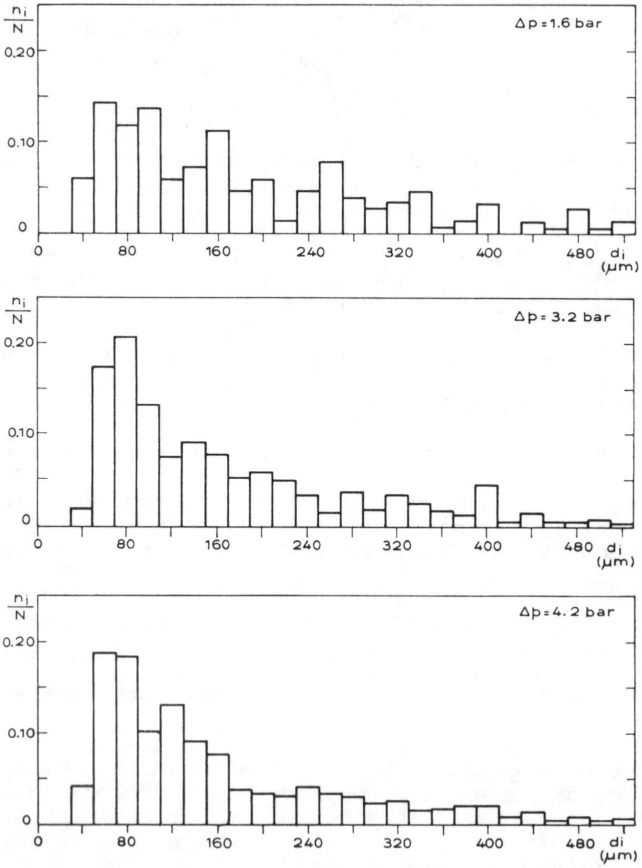

Fig. 28. Examples of droplet diameter distribution (F12/60; ℓ_p = 357 mm).

Each class, referenced with subscript i, is characterized by its average diameter d_i. The number of drops belonging to this class is n_i, whereas N is the total number of counted drops. Evidently,

$$\sum_i n_i = N$$

Drops having a diameter lower than 30 μm cannot be counted or measured; they always appear unsharp, and it is not possible to measure them with any acceptable precision.

The smoothing and systematization of histograms may be done by introducing the *distribution curves of drop diameters*. Several distribution functions $f(d)$ are proposed in the literature. Comprehensive surveys on this subject were performed by Mugele and Evans (1951) and by Lekic et al. (1976).

Nukiyama and Tanasawa's law,

$$f(d) = bd^m \exp(-cd^\delta)$$

where b, m, c, and δ are constants, is frequently used for types of pressure nozzles. Figure 29 shows that this law cannot satisfactorily match the experimental values for a fan-type pressure nozzle (d is expressed in meters; b is such that the law is normalized; the factor 10^6 appears because results are expressed in μm^{-1}).

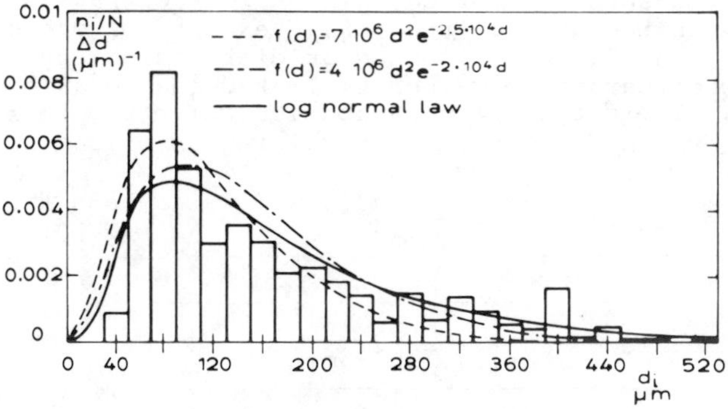

Fig. 29. Appoximation of a given histogram by some distribution curves (F12/60; Δp = 3.2 bar; ℓ_p = 357 mm).

As pointed out by Medenblik (1976), the use of the *log-normal law* gives a good approximation to measured distribution curves for fan-type spray nozzles. Its expression is

$$f(d) = \frac{1}{\sqrt{2\pi}\,\xi d} \exp\left[-\frac{1}{2}\left(\frac{\ln d/d_g}{\xi}\right)^2\right] \tag{11}$$

This looks like a normal distribution law, written in terms of the logarithm of the variable. The two parameters are defined as

$$\ln d_g = \sum_i \frac{n_i \ln d_i}{N} \tag{12}$$

$$\xi^2 = \sum_i \frac{n_i}{N}\left(\ln \frac{d_i}{d_g}\right)^2 \tag{13}$$

It can be seen that d_g is the geometric mean diameter of the drops.

A graphical representation of the log-normal distributed curve has been shown in Fig. 29. The discrepancy between measured and smoothed values is low, except in the vicinity of the peak of the experimental distribution. This peak is always discarded: this is the only disadvantage of the law, but we shall show later that it is of little consequence for heat transfer applications.

Histograms may be recomputed in a smoothed form from the distribution function, noting that

$$f(d) = \lim_{\Delta d \to 0} \frac{n_i/N}{\Delta d} \tag{14}$$

where Δd is the range of each class.

The values of d_g and ξ, computed by means of Eqs. (12) and (13) for several operating points, are reported in Table 4. They are relative to measures performed, like all others, at 357 mm from the nozzle tip, on the axis of the spray, except for one explicitly mentioned point. At 5.2 bar, one cannot distinguish any significant difference between parameter values estimated on the axis or at 0.09 m from it. The drop side distribution curve may thus be considered more or less uniform in the central region of the spray.

Table 4. *Values of Parameters d_g and ξ of the Log-Normal Diameter distribution*

Nozzle	Δp (bar)	d_g (10^{-6} m)	ξ
F12/60	5.2	111	0.709
	5.2[a]	113	0.689
	4.2	140	0.719
	3.2	147	0.703
	1.6	154	0.712
F13/60	3.2	167	0.749
	2.2	163	0.773
	1.2	160	0.783

[a] Measurement at 0.09 m from the axis of the spray.

The size distribution curve gives an accurate description of a spray; it contains more information than when using averaged droplet diameters. However, one finds in the literature characterizations of a spray only in terms of mean diameters. A general formula for computing several kinds of mean diameters was given by Mugele and Evans (1951):

$$(d_{qp})^{q-p} = \frac{\Sigma_i (n_i/N)\, d_i^q}{\Sigma_i (n_i/N)\, d_i^p} \qquad q \neq p \tag{15}$$

where p and q usually have the values 0, 1, 2, or 3.

Some details about the mean diameters of interest in practice are given in Table 5.

Table 5. *Some Mean Diameters of a Spray.*

Name	Mathematical Formula	Physical Significance
Arithmetic mean diameter	$d_{10} = \sum_i \frac{n_i}{N} d_i$	Obvious
Surface mean diameter	$d_{20} = \left(\sum_i \frac{n_i}{N} d_i^2\right)^{1/2}$	Diameter of a drop whose surface times N is equal to the total surface of the spray
Volume mean diameter	$d_{30} = \left(\sum_i \frac{n_i}{N} d_i^3\right)^{1/3}$	Diameter of a drop whose volume times N is equal to the total volume of the spray
Sauter mean diameter	$d_{32} = \frac{d_{30}^3}{d_{20}^2}$	Diameter of a drop for which the volume – surface ratio is equal to that computed for the whole spray

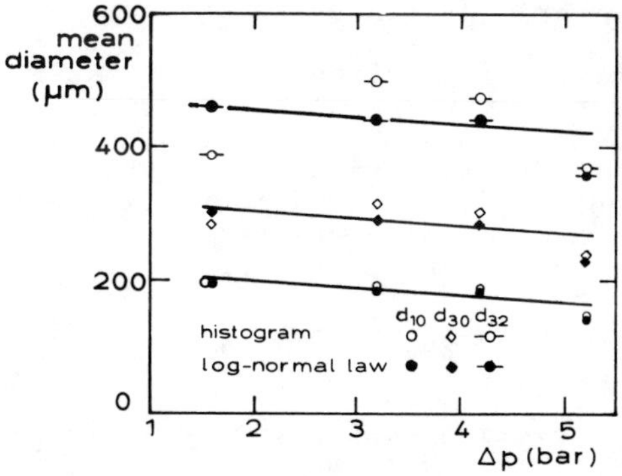

Fig. 30. Variation of mean diameters with pressure (F12/60).

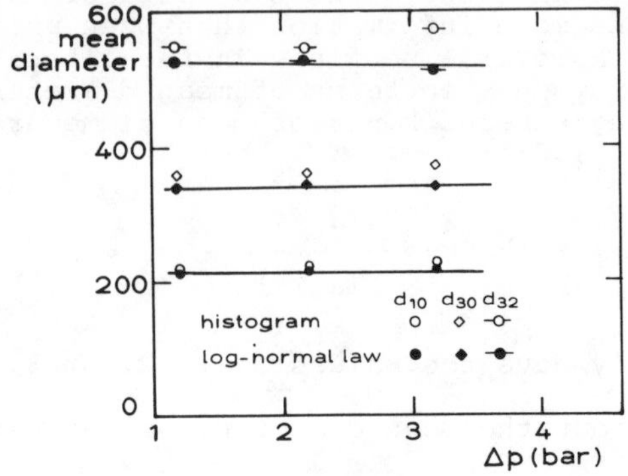

Fig. 31. Variation of mean diameters with pressure (F13/60).

Figures 30 and 31 show the variations of d_{10}, d_{30}, and d_{32} computed from the histograms and the derived log-normal distribution law, for nozzles F12/60 and F13/60. Two facts may be pointed out:

1. The mean diameters decrease very slowly with the pressure at the nozzle.

2. At a given pressure, the mean diameters increase slowly with A_0, the area of the orifice of the nozzle.

These experimental results may be compared with the values obtained from a relation proposed by Dombrowski and Munday (1968):

$$d_{32} = \frac{B}{C_q^{2/3}} \left[\frac{A_0 \sigma}{\Delta p \sin(\phi/2)}\right]^{1/3} \left(\frac{\rho_\ell}{\rho_a}\right)^{1/6} \qquad (16)$$

The constant B is dimensionless, and its value must be determined for each kind of nozzle. A graphical representation of Eq. (16) with the value 0.20 for B is given in Fig. 32. Dombrowski's formula clearly overestimates the influence of A_0 and Δp on the values of d_{32}, in comparison with our experimental results.

2.3.2 Droplet Velocity Distribution

The pictures of the drops obtained by means of the Diamant microscope also permit determination of the velocity of the drops. When the rotation speed ω_m (rpm) is synchronized with the droplet velocity, it is possible to calculate this velocity v_s at synchronism; in the present state of the apparatus, it is:

$$v_s = 1.89\omega_m \times 10^{-2} \quad (\text{m/s})$$

When a drop is traveling faster than v_s, its picture appears as a flatened ellipse (Fig. 33); on the contrary, the pictures of slower drops are elongated ellipses. In Fig. 33, we give the formulas developed by Tolfo and Staudt (1976) in order to calculate

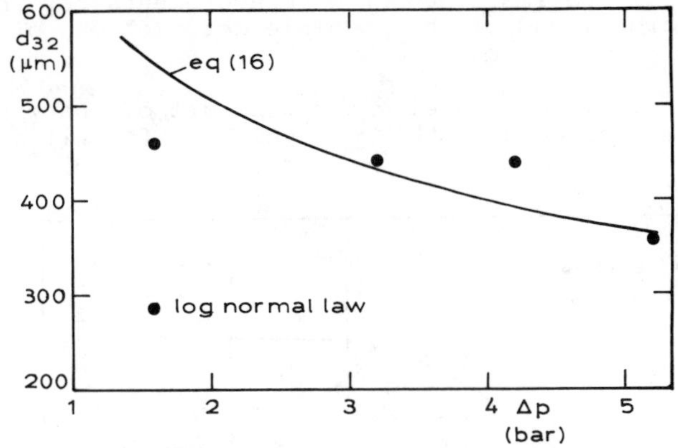

Fig. 32. Variation of Sauter mean diameter with pressure (F12/60).

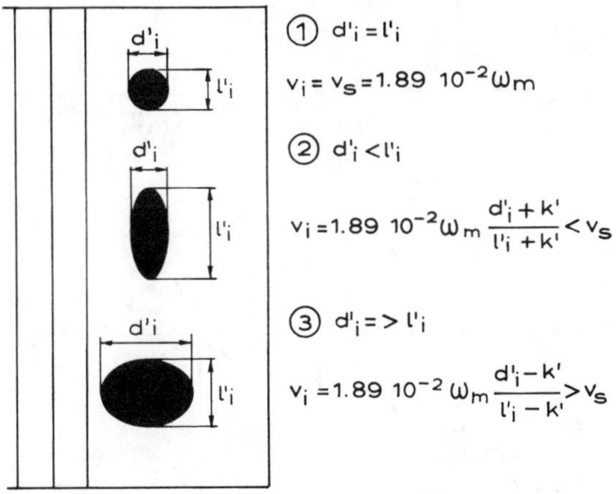

Fig. 33. Drop velocity determination from the photographs.

the droplet velocities, out of synchronism. The constant k' is the product of the slit breadth (43 μm) by the magnification of the microscope (25). These formulas are valid only for spherical drops, which is not strictly the case for the larger ones. The results of measurements will be less accurate for the latter.

The analysis of the same photographs used to draw the histograms of droplet sizes enables us to determine the velocity of each counted drop, and thus to estimate a mean velocity for each class of drops we distinguished earlier. Typical results of such measurements are given in Fig. 34. These diagrams show that:

1. For a given spray, the mean velocity of a class of drops increases with the diameter representative of this class, until an asymptote value is reached.

2. Oscillations in the measured values for large diameters are a

consequence of the small number of measurements for these classes of drops, and of the possible deformations of these large drops.

3. For a given nozzle, increasing the pressure of the water results in an increase of the mean velocity for all classes of drops.

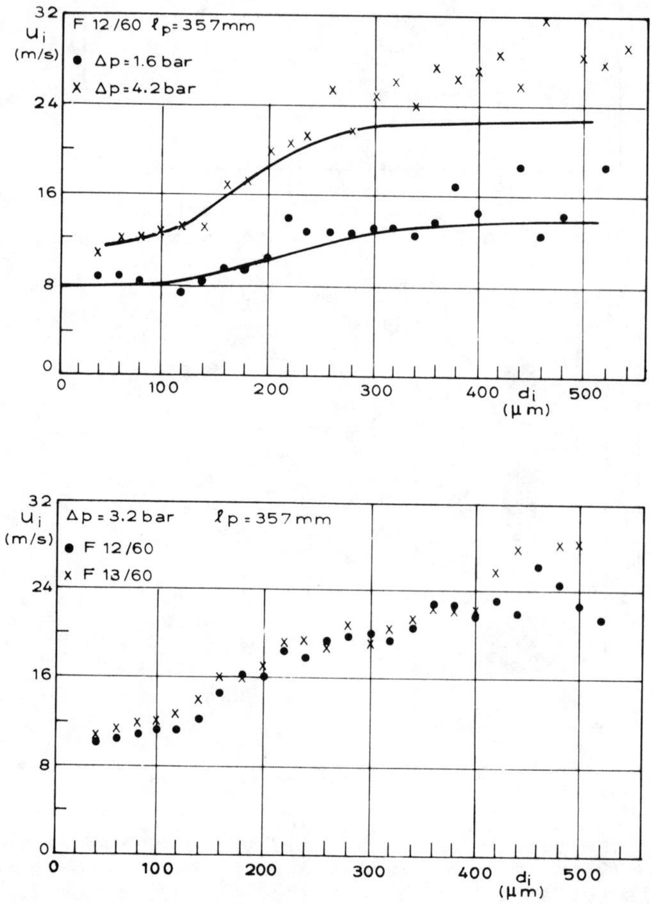

Fig. 34. Velocity of drops as a function of their diameter (ℓ_p = 357 mm).

The velocity of drops belonging to a given class is far from being uniform, but is distributed as shown in Fig. 35. For example, the drops of the 70–90 μm class, for nozzle F12/60 operating at 3.2 bar, have velocities lying in a quite broad range from 6 to 17 m/s. As mentioned, with regard to Fig. 34, the mean velocity of that class is taken as 11.5 m/s.

The asymptotic value of drop velocity is difficult to determine because of the scattering of measurements for large diameters.

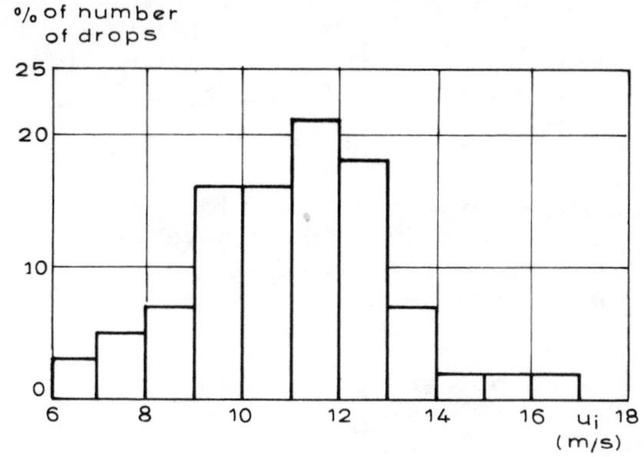

Fig. 35. Drop velocity distribution inside a given class of diameters.

The measured velocities of large drops are generally found to be higher than the velocity $u_{\ell 0}$ at the orifice of the nozzle. As an example, for nozzle F12/60 operating at 1.6 bar, the arithmetic mean velocity of drops belonging to the last 12 classes (d_i > 300 μm), drawn from Fig. 34, is 14.44 m/s. The experimental value of $u_{\ell 0}$ is 11.23 m/s, which is lower.

In fact, this value of $u_{\ell 0}$ at the orifice is an average one. According to Schlichting (1968), the maximum velocity in a turbulent jet may be related to the mean velocity by the approximate relation

$$u_{\ell(max)} = \frac{5}{4} u_{\ell 0} \qquad (17)$$

when $10^4 < Re = \frac{u_\ell d_0}{\nu_\ell} < 10^5$

Applying this relation, we find that

$$u_{\ell(max)} = \frac{5}{4} \times 11.23 = 14.04 \text{ m/s}$$

for the example chosen, which is close to the arithmetic mean value.

Thus, in the present case as in all the others we have encountered, we estimate the asymptotic value of the velocity by using Eq. (17).

We have described above the particular method of drop velocity measurement we used. The same measurements may be performed by means of a laser-Doppler anemometer. The basic principle of this apparatus is the following: A light beam is split into two secondary beams which, by means of mirrors and appropriate lenses, converge onto a small portion of space, the observation zone, where interference occurs. When a small particle moves through this zone, periods of illumination and of relative darkness alternate and modulation of the amplitude of the light emitted by this particle can be related automatically, by means of a sophisticated electronic device, to the particle velocity. The frequency of information treatment can be very high, of the order of 1000 to 10000 measurements a second (Riethmuller, 1977).

2.3.3 Mean Velocity of Drops in a Spray

Histograms of droplet diameter and velocity distributions provide sufficient information to compute several mean droplet velocities. The arithmetic mean velocity is of very important practical use; it is defined as

$$u_{10} = \sum_i \frac{n_i}{N} u_i$$

with summation extended to all the classes of the pulverization.

The relation between u_{10} and the operating pressure Δp is shown in Fig. 36 for nozzles F12/60 and F13/60. Curves for $u_{\ell(max)}$ are also given for comparison.

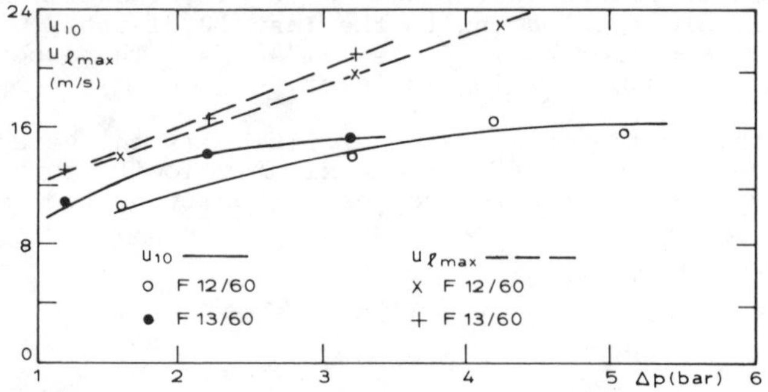

Fig. 36. Variation of arithmetic mean velocity with pressure.

In Sec. 2.2.3 we defined a mean velocity $\bar{u}_T$ of the spray. According to Toda, it is assumed that $\bar{u}_T$ is equal to the mean liquid velocity $\bar{u}_\ell$ as well as the mean air velocity $\bar{u}_a$. The knowledge of the velocity distribution of the drops enables us to reach a better estimation of the air velocity.

Before achieving this, let us first determine *the number X of drops crossing a unit transverse surface area per unit time.*

The liquid mass flow rate density $\dot{m}_\ell$ of the spray is known. Let x_i be the number of drops of the ith class crossing 1 m^2 during 1 s. Evidently,

$$X = \sum_i x_i$$

$$\dot{m}_\ell = \rho_\ell \sum_i x_i V_i$$

$$= \rho_\ell X \sum_i \frac{x_i}{X} V_i \qquad (18)$$

with $V_i = \frac{\pi d_i^3}{6}$

The quantities x_i are proportional to the sample populations n_i of drops of class i measured when establishing the histograms of diameter distribution. We may therefore write

$$\frac{x_i}{X} = \frac{n_i}{N} \qquad (19)$$

ratios n_i/N being known. Equation (18) may now be written in the form

$$X = \frac{\dot{m}_\ell}{\rho_\ell \sum_i (n_i/N)\ V_i} = \frac{6\dot{m}_\ell}{\pi \rho_\ell d_{30}^3} \qquad (20)$$

which provides a way to compute X. Once X is known, every x_i may be calculated from Eq. (19).

Returning to the determinations of mean velocities for air and water in the spray, we first rewrite Eq. (6) in the form

$$p_s = \frac{1}{2}\rho_a u_a^2 + \frac{1}{2}\dot{m}_\ell u_\ell - \frac{1}{2}\frac{A_\ell}{A_s}\ \rho_a u_a^2 \qquad (21)$$

The third term in the right-hand side may be neglected. The second term may be determined: Quantity $\dot{m}_\ell u_\ell$ clearly represents the momentum flux density of the drops. It is equal to

$$\dot{m}_\ell u_\ell = \sum_i x_i \rho_\ell u_i V_i$$

We may thus give a precise definition as well as a means of evaluation of u_ℓ in the form

$$u_\ell = \frac{\sum_i (n_i/N)\ u_i V_i}{\sum_i (n_i/N)\ V_i} = \frac{\sum_i (n_i/N)\ u_i d_i^3}{d_{30}^3}$$

This is a kind of arithmetic mean velocity, the number of drops of each class being weighted by the volume of a drop of this class. This way of weighting the mean value of the velocity clearly gives more importance to the largest drops: The mean value u_ℓ is found to be very close (see Table 6) to the value of $u_{\ell(\max)}$ computed above.

The ratio γ of air and water velocities, defined as

$$\gamma \triangleq \frac{u_a}{u_\ell} \qquad (22)$$

may be determined from Eq. (21):

$$\gamma^2 = \frac{p_s - \frac{1}{2}\dot{m}_\ell u_\ell}{\frac{1}{2}p_a u_\ell^2} \tag{23}$$

Values of γ for nozzle F13/60, computed by means of values of $\dot{m}_\ell$ and p_s mentioned in Table 3, are given in Table 6. From the latter, it may be seen that the mean velocities of air and water differ significantly. The mean velocity $\bar{u}_T$ of Toda for the whole pulverization therefore cannot be used without special care. Our computed values of γ increase less than linearly with Δp, the pressure at the orifice of the nozzle.

Table 6. *Mean Velocities for Nozzle F13/60*

Δp (bar)	X, (10^8 drops/ m^2 s)	u_{10} (m/s)	u_T (m/s)	$u_{\ell(max)}$ (m/s)	u_ℓ (m/s)	γ —	u_a (m/s)
1.2	5.51	10.5	6.0	13.0	12.9	0.325	4.2
2.2	7.35	14.3	9.2	16.6	16.5	0.446	7.4
3.2	9.96	15.4	12.9	21.0	20.6	0.535	11.0

In their one-dimensional analysis of the aerodynamic behavior of liquid sprays, Rothe and Block (1977) predict lower values for both air and water velocities, but some of the required quantities (i.e., a "representative" dimension of the drops) are only very roughly defined. Also, their method of averaging velocities is different from ours.

3 HYDRODYNAMIC BEHAVIOR OF DROPLETS IMPINGING ON A HOT PLATE

3.1 The Single Drop

When a drop impinges on a hot plate, it often breaks into smaller droplets. The occurrence of breakup can be predicted by a simple criterion based on the value of the Weber number of the drop. The Weber number is defined as

$$\mathrm{We} = \frac{u_{\ell n}^2 \rho_\ell d}{\sigma_\ell}$$

where $u_{\ell n}$ is the component of drop velocity normal to the plate. In the cases considered here, the plate is horizontal and $u_{\ell n}$ is thus the vertical component of drop velocity. Since the Weber number compares the drop's kinetic energy to its energy content due to surface tension, a drop will remain whole or not after collision according as its Weber number is small or large. Wachters (1965), Wachters and Westerling (1966) have proposed the following breakup criterion with 400°C plates:

For We < 30, the drop bounces on the surface without breaking up at all.

For 30 ⩽ We ⩽ 80, the drop undergoes considerable deformation when it impinges on the plate but recovers a near-spherical shape when it leaves the plate; sometimes, one or two tiny droplets can leave the bigger one.

For We > 80, the drop spreads on the plate and forms a liquid film that finally breaks up into smaller droplets.

One can question whether this criterion is sufficient: Does the temperature of the plate not influence the behavior of the drop?

Observation by Savic and Boult (1955) and by Toda (1972) relative to the impact of drops on cold plates (20^{0}C) show that, for drops with We > 80, the outer part of the spreading film follows the plate. However, the photographs of Wachters (1965) and high-speed films (5000 frames/s) of Savic and Boult (1955) clearly indicate that, when the plate is hot (400^{0}C), the outer part of the spreading film tends to rise from the plate. Moreover, the same pictures show that the throwing up of liquid in smaller droplets is more explosive when the plate is warmer. In order to acquire a better insight into the possible influence of plate temperature, we shot high-speed films (7000 frames/s) with plate temperatures in the range 800^0 to 1200^{0}C. Our results for 1 < We < 1500 are in agreement with Wachters and Westerling's criterion.

As will be shown later, most of the drops produced by pressure nozzles have a Weber number larger than 80. Therefore, this class of drops will be treated first and in more detail.

3.1.1 Drop with We > 80

High-speed cinematography permits one to follow all the stages of evolution of an impacting drop. We distinguish four phases (Fig. 37):

(a) The drop approaches the plate; its shape is nearly spherical.

(b) A short time after the impact, the drop undergoes large deformation; a continuous film spreads out around a compact central dome.

(c) The dome has disappeared; the film is still more or less continuous.

(d) The film becomes unstable and breaks up into a large number (of the order of 100) of small droplets with small kinetic energy.

During phase (b), the height of the dome decreases linearly with respect to time. This fact has also been pointed out by Harvey. Physically, this linear relation means that the top of the dome does not "feel" the presence of the wall; this concurs with the assumptions of Engel reported by Savic and Boult (1955).

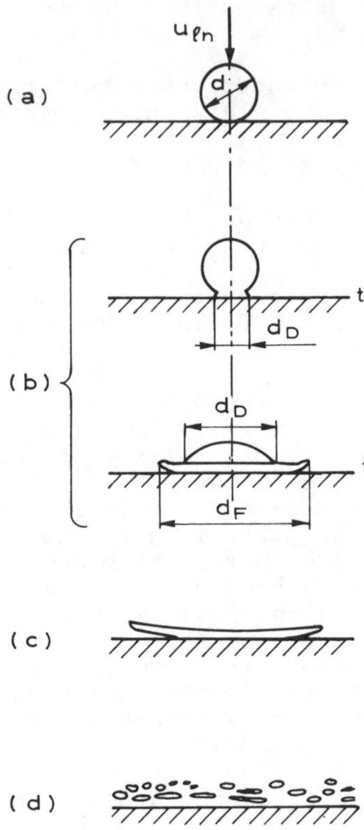

Fig. 37. Breakup of a drop.

We should mention that these authors developed a theoretical analysis of liquid flow inside a drop spreading on a cold wall (assuming an ideal fluid, with negligible viscosity and surface tension).

Their solution for time evolution of dome height does not differ from a straight line during the first instants after impact. The time t^+ after which the dome disappears is given by

$$t+ = \frac{d}{u_{\ell n}} \tag{24}$$

Figure 38 represents time t^+ in a nondimensional form versus the Weber number of the drop. Time t_R is equal to the period of the first natural mode of vibration of a spherical drop, as calculated by Rayleigh:

$$t_R = \pi \sqrt{\frac{\rho_\ell d^3}{16\sigma_\ell}} \tag{25}$$

The experimental points are reasonably well fitted by the correlation

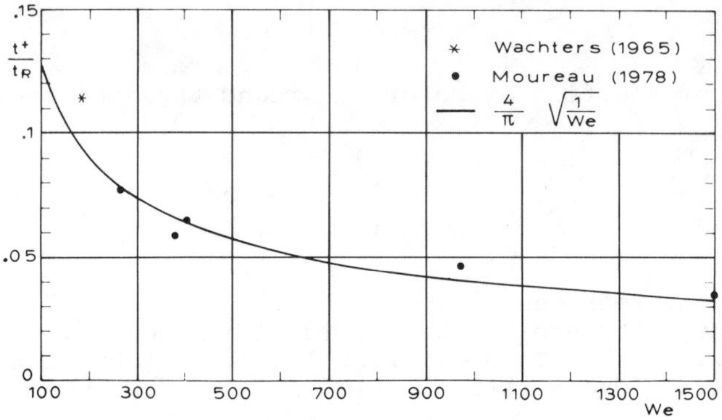

Fig. 38. Variation of time t^+ as a function of We.

$$\frac{t^+}{t_R} = \frac{4}{\pi}\sqrt{\frac{1}{\mathrm{We}}}$$

which is just another way of writing Eqs. (24) and (25).

The dome diameter d_D evolves as shown in Figs. 37 and 39. When $t < 0.2t^+$, the dome diameter is that of the deformed base of the drop; for very short times, indeed, the film does not actually spread out around the base of the drop. When $t \geqslant 0.2\ t^+$, the diameters of film and dome can be distinguished.

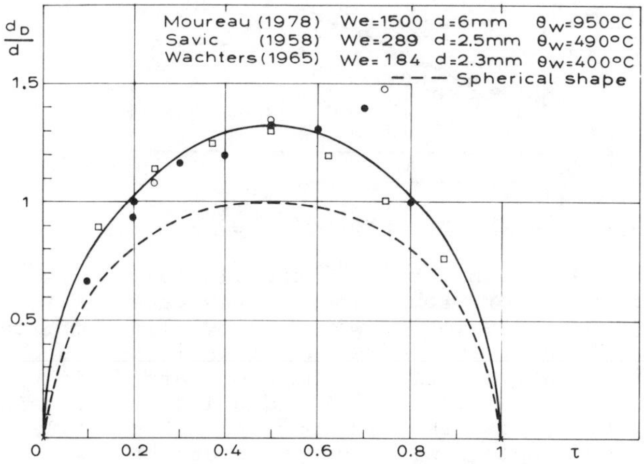

Fig. 39. Evolution of the diameter of the dome.

The experimental points of Fig. 39 are correlated by

$$\left(\frac{d_D}{d}\right)^2 = 6.97(\tau - \tau^2) \qquad (26)$$

with $\tau = \frac{t}{t^+}$

The evolution of the film diameter d_F around the dome may be described by

$$\frac{d_F}{d} = 1.67(3.1\tau - \tau^2) \tag{27}$$

This evolution is represented in Fig. 40. Equation (27) is valid only during the existence of the liquid film, that is, for $0.2t^+ \leqslant t \leqslant 1.2\text{–}1.5t^+$. Indeed, after $1.2\text{–}1.5t^+$, the film becomes unstable and small drops start leaving it.

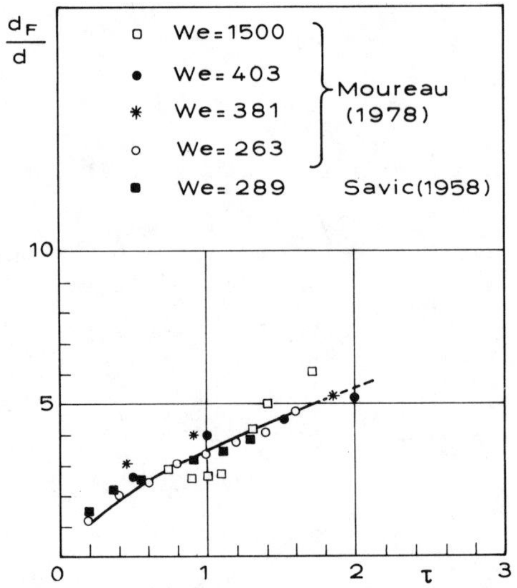

Fig. 40. Evolution of the diameter of the film.

For times larger than 2 to $3t^+$, the liquid film is shredded into several ligaments or drops flowing above the wall. Most of the liquid drops have a velocity with a prevailing horizontal component, so that their influence on the wall is a function of its dimensions. If the wall is small, the drops are rapidly expelled beyond it, although some tiny isolated droplets can subsist near the wall a long time after the disappearance of the film. If the wall is large, the droplets will bounce several times. In both cases, most of these tiny droplets have a Weber number smaller than 80; their behavior will be examined later.

It must be emphasized that the evolutions described above are valid for water and are most likely not applicable to other fluids. Photographs of Savic (1958) show that a drop of molten wax impinging on a plate behaves quite differently. However, results for water drops are the most useful, since water is used in most applications of spray cooling.

3.1.2 Drop with We < 80

As for drops with We > 80, they first flatten out on the hot wall, but afterward their behavior is different: Surface tension is sufficiently important to overcome inertia so that the liquid recovers a spherical shape upon leaving the wall.

The characteristic time is not that after which the dome has disappeared but a time of presence in the vicinity of the wall, which could be defined as the duration between the instant of impact and the instant of separation from the wall due to the rebound. The definition is a macroscopic one and does not consider the possible existence of a vapor layer under the drop. Wachters (1965) has shown that this characteristic time is approximately equal to t_R for 2.3-mm drops impinging on a 400°C plate. Figure 41 shows that our experiments, performed with smaller drop ($d \simeq 1$ mm)

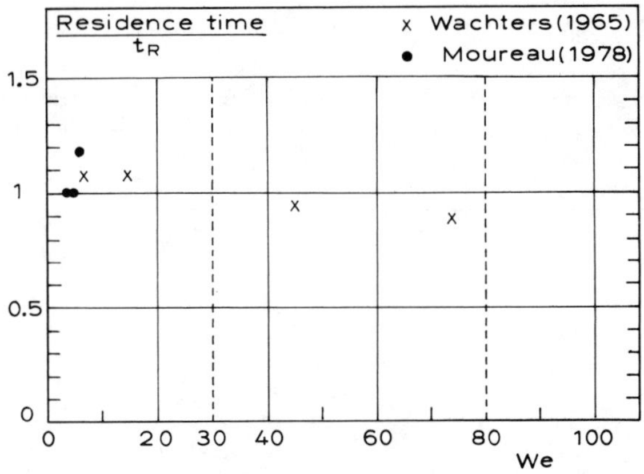

Fig. 41. Residence time of drops with We < 80.

impacting on 800 to 1000°C plates, are in agreement with this result. In Fig. 42 one can see the variation of the Weber number of a rebounding drop $(\mathrm{We})_r$ versus its Weber number before impact $(\mathrm{We})_i$. For very small Weber numbers $(\mathrm{We})_i < 5$, the rebound is nearly elastic. For larger values of $(\mathrm{We})_i$ the energy absorbed by drop deformation increases more and more. For $10 < (\mathrm{We})_i < 40$, the Weber number after rebound is nearly constant, whereas for $40 < (\mathrm{We})_i < 80$, it decreases strongly.

3.2 The Liquid Spray

Knowing the microscopic characteristics of a liquid spray (diameters and velocities of the drops), it is possible to calculate the Weber number of each drop. For liquid sprays produced by pressure nozzles, most of the drops have a Weber number larger than 80. Such a numerical estimation is proposed in Table 7.

We shall assume that the previously described behavior of water drops with diameters ranging from 0.5 to 6 mm can be transposed to smaller drops (20 μm up to 1 mm) making up the liquid spray. One can then calculate for each drop or for each class of drops of a

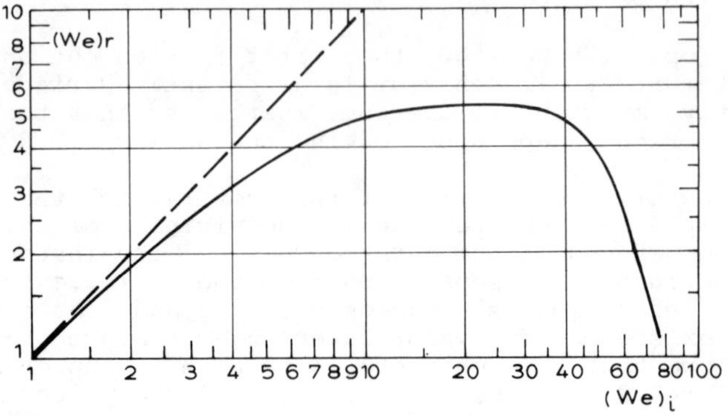

Fig. 42. Weber number of a rebounding drop versus its Weber number before impact.

Table 7. *Percentage of Spray Drops with We > 80*

Type of Pressure Nozzle	Pressure Difference, Δp (bar)	Percentage of Drops with We > 80
F12/60	1.6	76
(d_o = 2 mm,	3.2	91
ϕ = 60°)	4.2	94
	5.2	87
F13/60	1.2	76
(d_o = 3 mm,	2.2	91
ϕ = 60°)	3.2	94

spray the characteristic times (t^+,...) and dimensions (d_D, d_F) as established for single drops. Table 8 gives an example of such a calculation. It must be emphasized that for impacting spray droplets the characteristic times are of the order of 10^{-5} s.

However, an objection can be raised when generalizing the behavior of a single drop to all the drops of a spray: To what extent are the phenomena modified by interactions between the drops? In this matter, two questions arise:

1. Can the liquid films resulting from the impact of each drop not coalesce in such a way so that the hot wall is covered by a continuous liquid film?

2. What happens to the bouncing droplets? Do they not form, in the vicinity of the wall, a zone with high droplet concentration and with a specific influence?

Let us first answer this last question. As soon as pulverization begins, the concentration of droplets increases in the immediate neighborhood of the wall: The dispersed phase forms a

Table 8. *Examples of Characteristic Time and Dimensions of Drops*

Pressure Nozzle F12/60; Δp = 5.2 bar				
d (10^{-6} m)	$u_{\ell n}$ (m/s)	We	t^+ (10^{-6} s)	d_F when $t = t^+$ (10^{-6} m)
60	11.5	109	5.2	210
100	13.0	232	7.7	350
140	16.1	497	8.7	490
200	22.0	1327	9.1	700
300	25.4	2652	11.8	1050
Observed Single Drop				
d (10^{-3} m)	$u_{\ell n}$ (m/s)	We	t^+ (10^{-3} s)	d_F when $t = t^+$ (10^{-3} m)
3.7	2.82	403	1.3	12.95
6	4.27	1500	1.4	21.0

boundary layer. A typical time of formation of this boundary layer is about 0.1 s (Fig. 43), that is, 100 times longer than the duration of breakup of the spray drops. Once the spraying regime

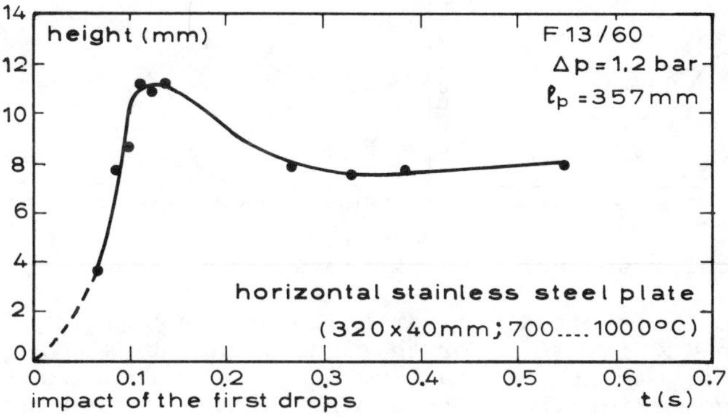

Fig. 43. Evolution of the height of the boundary layer (accuracy: ±0.5 mm).

is established, the height of the boundary layer fluctuates around an average (Fig. 44). This means height has been measured by using a photoelectric cell, which explores the spray picture continuously in the vicinity of the wall. A typical response is

Fig. 44. Boundary layer (the hot plate thickness is 3 mm).

shown in Fig. 45a: An abrupt change in the cell response is observed at the outer edge of the boundary layer. When spraying a cold plate, a boundary layer also exists, but it is different. There is no sudden change in the cell response (Fig. 45b): The gradient of concentration of droplets varies smoothly as the vertical distance is increased.

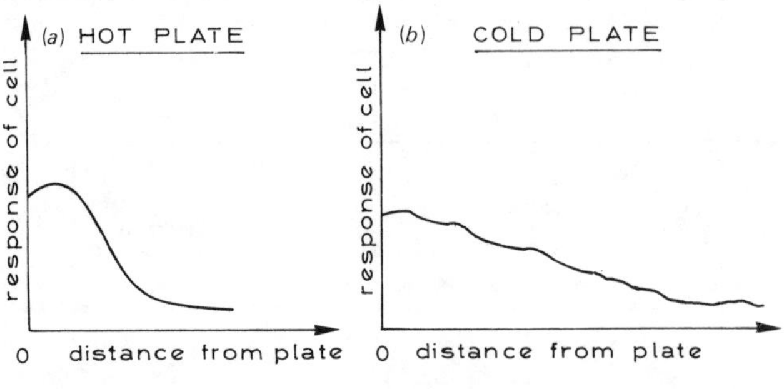

Fig. 45. Typical responses of a photoelectric cell.

An example of longitudinal variation of the boundary layer height is shown in Fig. 46. For that case, the horizontal component of the velocity of bouncing droplets with diameters in the 0.3-mm range varies from 2 to 5 m/s. Their residence time in the boundary layer is thus rather short. Furthermore, the frequency of collisions between vertically moving drops coming from the nozzle and horizontally moving droplets of the boundary layer seems to be low. A detailed analysis should be devoted to this particular point. However, in our experimental conditions with a small plate, we assume as a first approximation that the number of drops with diameter d_i belonging to the ith class crossing the unit surface area per unit time, calculated in sec. 2.3.4, is equal to the

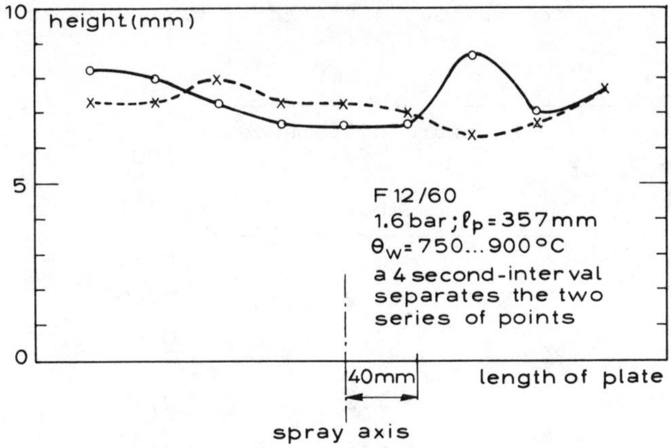

Fig. 46. Height of boundary layer.

number of drops with diameter d_i impinging per unit time on the unit surface area of hot wall.

Let us now answer the first question concerning the existence of a continuous liquid film covering the hot wall.

Consider a drop belonging to the ith class of a liquid spray. After a time t_i^+ following the impact, it will cover a surface area A_i given by

$$A_i = \frac{\pi}{4} d_F^2$$

During that time t_i^+, it covers a mean surface area $\bar{A}_i$:

$$\bar{A}_i = \frac{1}{t_i^+} \int_0^{t_i^+} A_i \, dt$$

If the ith class comprises x_i drops impinging per unit time on the unit surface area, the average fraction of unit surface area covered by the water of the drops of this class is given by

$$x_i \bar{A}_i t_i^+$$

By summation over all the classes of the spray, one finds the average fraction of wall covered by liquid films:

$$\bar{A}_t = \sum_i x_i \bar{A}_i t_i^+$$

Table 9 contains values of $\bar{A}_t$ for the liquid sprays we chose to study in detail; in all these instances, $\bar{A}_t$ is lower than 0.01. These results are to be compared with a value given by Wachters (1965) for the case of a pulverization on a hot metallic disk with

Table 9. *Percentage of Surface Area of Wall Covered by Liquid Films*

Type of Nozzle	Pressure difference, Δp (bar)	$100\bar{A}_t$ (%)
F12/60	1.6	0.39
	3.2	0.40
	4.2	0.45
F13/60	1.2	0.79
	2.2	0.84
	3.2	0.94

a diameter of 20 mm:

$$\bar{A}_t = 0.0003$$

This lower value is probably due to the fact that the liquid spray used by Wachters contains more small drops than ours. Wachters concludes from his observations that on the hot disk the drops do not influence each other. We reach the same conclusion for our rectangular plate: The individual liquid films do not form a continuous film covering the whole plate. Simply, the influences of the drops are very localized.

4 HEAT TRANSFER DURING THE IMPACT OF A DROP

4.1 Existing Theoretical Models

As a first step toward a complete analysis of the heat exchange between a hot wall and a liquid spray, several authors have developed theoretical models concerning the heat transfer caused by a single drop. In this simplified case, two different situations are considered: a single sessile drop lying on the hot surface and a single drop impinging on the hot surface.

4.1.1 The Sessile Drop

Existing theories for a drop on a horizontal plate at high temperature (>300 to 400°C) generally include the following assumptions:

1. The drop is nearly spherical except for the lower part near the wall, which may be more or less flattened.

2. The drop is sessile or nearly sessile until its complete evaporation.

3. A thin sublayer of vapor insulates the drop from the plate.

This state is usually called the "spheroidal state."

Let us immediately note that these conditions are quite far from the behavior of pulverized drops impinging on a hot plate. Indeed, the latter stay near the wall for only a very short time, and undergo large deformations leading, in most instances, to disintegration into many smaller drops. We must, however, mention the studies related to sessile drops, because the models for impinging drops are often rough extrapolations of elaborate models pertaining to sessile drops.

Two shapes of the base of the drop have been considered: spherical and flattened. The studies related to this second category (Wachters, 1965; Gottfried and Bell, 1966; Wachters et al., 1966a) are very much alike; let us sum up the main steps of the last one.

The aim of the model is to determine the vapor film thickness S_v and the evaporation rate of the drop. As far as heat transfer is concerned, Wachters et al. distinguish the base from the rest of the drop: It is the vapor production under the drop that controls the film thickness. Other assumptions are as follows (Fig. 47):

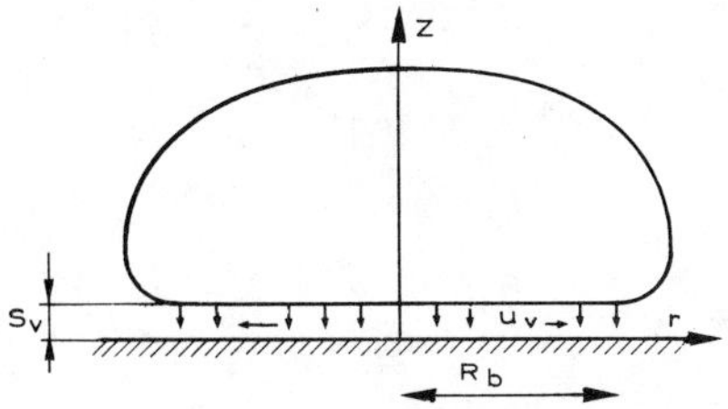

Fig. 47. Single sessile drop with flat basis.

1. The film thickness S_v is constant.

2. Since S_v is much smaller than R_b (the radius of the base of the drop), the vertical component of the vapor velocity can be neglected; there remains only the horizontal component u_v.

3. The vapor flow is laminar and viscous; inertia forces are neglected.

4. The whole drop is at saturation temperature T_{sat}.

5. Radiative heat transfer is not taken into account.

6. The pressure increase in the film does not influence the liquid boiling point; all the heat transferred from the wall is used to vaporize the liquid drop.

On should note that the hypothesis of constant S_v greatly simplifies the calculations. Besides, near the liquid–vapor interface, the vertical component of vapor velocity is certainly not negligible. Finally, one knows that at high wall temperature the radiative heat transfer becomes important.

The governing equations leading to the determination of S_v are the following:

1. In steady state the momentum equation in the vapor film may be written as

$$\frac{dp_v}{dr} = \mu_v \frac{\partial^2 u_v}{\partial z^2} \tag{28}$$

where p_v is the pressure in the film and where the derivatives of u_v with respect to r have been neglected. The boundary conditions are

Zero velocity at the wall: $u_v\ (r, 0) = 0$

Zero velocity at the liquid–vapor interface: $u_v\ (r, S_v) = 0$,

The pressure at the edges of the flattened base is equal to the ambient pressure p_e: $p_v(R_b) = p_e$.

2. The vapor mass flux density $\dot{m}_v$ at the liquid–vapor interface is given by

$$\dot{m}_v = \frac{\lambda_v(T_w - T_{sat})}{S_v} \frac{1}{\Delta h_v} \tag{29}$$

3. A mass balance written for a cylinder with radius r and height S_v leads to

$$\frac{1}{\rho}\pi r^2 \dot{m}_v = \int_0^{S_v} 2\pi r u_v\ dz \tag{30}$$

4. At equilibrium, the pressure integrated over the entire vapor cushion is equal to the weight of the drop of volume V:

$$\int_0^{R_b} 2\pi r(p_v - p_e)\ dr = \rho_{\ell s}\ g\ V \tag{31}$$

with $\rho_{\ell s}$ the mass density of liquid at $T = T_{sat}$. Equation (28) can be integrated in order to find an expression of u_v (with dp_v/dr still unknown) that can be introduced in Eq. (30) combined with (29) to give the pressure distribution in the vapor film:

$$p_v - p_e = \frac{3\mu_v\lambda_v(T_w - T_{sat})}{\rho_v S_v^4\ \Delta h_v} (R_b^2 - r^2) \tag{32}$$

or

$$p_v - p_e = \frac{3\mu_{vs}\ \lambda_{vs}(T_w - T_{sat})(T_w + T_{sat})^3}{8T^3_{sat}\rho_{vs}S^4_v\ \Delta h_v}\ (R^2_b - r^2) \qquad (33)$$

In this last expression viscosity, thermal conductivity, and mass density must be evaluated at the saturation temperature (subscript s) corresponding to pressure p_e. Equation (33) is found from Equation (32) by taking into account laws of variation of μ, λ, and ρ versus temperature.

From relations (31) and (33), one can deduce the vapor film thickness:

$$S_v = \left[\frac{3\pi\mu_{vs}\lambda_{vs}(T_w - T_{sat})(T_w + T_{sat})^3 R^4_b}{16\rho_{\ell s}\rho_{vs}T^3_{sat}Vg\ \Delta h_v}\right]^{1/4} \qquad (34)$$

With this formula, it is possible to evaluate S_v for a drop with given volume V if R_b, the radius of the flattened base, is known. The authors propose a relation between R_b and R_m, the maximum radius of the drop (Fig. 48) and another relation between R_m and V (Fig. 49). The vapor flux density $\dot{m}_v$ may then be found [Eq. (29)]

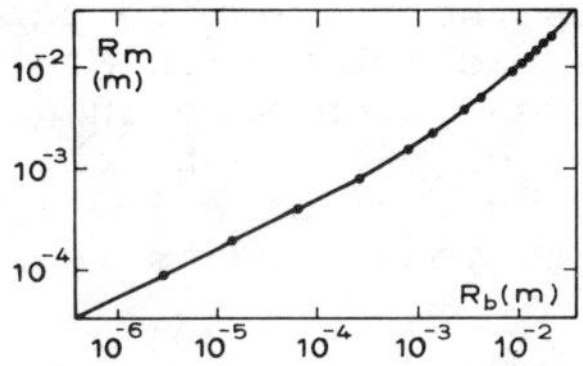

Fig. 48. Variation of R_m versus R_b, according to Wachters (1965).

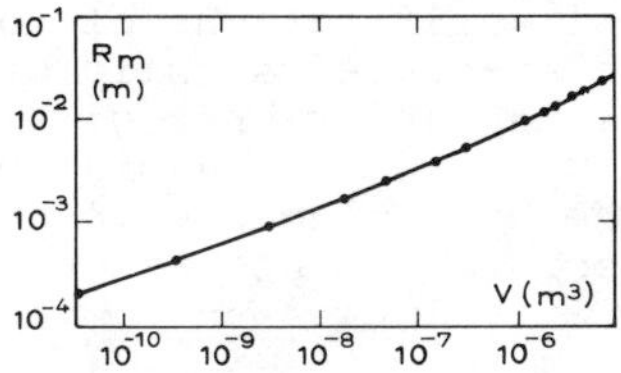

Fig. 49. Variation of R_m versus V, according to Wachters (1965).

and the flow rate $\dot{M}_{\ell 1}$ of liquid evaporated under the drop follows immediately:

$$\dot{M}_{\ell 1} = \int_0^{R_b} 2\pi r\dot{m}_v\ dr$$

$$= \left[\frac{\pi^3\rho_{vs}\lambda_{vs}(T_w - T_{sat})^3(T_w + T_{sat})R^4_b\ V}{3\rho^3_{\ell s}\mu_{vs}T_{sat}\ \Delta h^3_v}\right]^{1/4} \qquad (35)$$

It is interesting to mention numerical values obtained when applying these equations to a few water drops on a plate at 400°C (Table 10). One observes that, for the largest drop (V = 151x10^{-9} m^3, radius = 3.3 mm), the calculated thickness S_v is equal to about 80 μm. Let us note, on the same table, the order of magnitude of the heat flux density extracted from the wall: 10^5 to 6×10^5 W/m^2.

Table 10. *Vapor Film Thickness and Flow Rate of Liquid Evaporated under the Drop*

V (10^{-9} m^3)	Drop Radius if Spherical (10^{-6} m)	R_b 10^{-6} m	S_v 10^{-6} m	$\dot{M}_{\ell 1}$ 10^{-9} m^3	$\rho_{\ell s}\, \Delta h_v \, \frac{\dot{M}_{\ell 1}}{\pi R_b^2}$ (W/m^2)
2.94	889	248	18.7	0.049	5.52×10^5
17.85	1621	775	37.3	0.240	2.75×10^5
44.01	2190	1337	51.5	0.521	2.00×10^5
150.70	3301	2721	76.9	1.450	1.35×10^5

The total flow rate of evaporated liquid is obtained by adding to $\dot{M}_{\ell 1}$, the flow rate $\dot{M}_{\ell 2}$ of liquid evaporated from the whole drop except for the flat base. This last condition could not be evaluated theoretically by the authors. However, they observed experimentally that $M_{\ell 2}$ could be of the same order of magnitude as $\dot{M}_{\ell 1}$.

Experimental verifications of this theory require patient work. In practice, the so-called sessile drop is not motionless on the hot plate but oscillates and vibrates; it can even move in a preferential direction. In order to avoid this displacement, the drop is maintained at the same place by incurving the hot wall. The agreement between theory and evaporation rate measurements is satisfactory.

Wachters (1965) also analyzed the case of a drop with a concave bottom; the vapor film thickness then varies with radial position r. He concludes that the results obtained by this more sophisticated analysis exhibit a greater discrepancy with experiments than results obtained with the simple model (S_v = constant).

Gottfried et al. (1966) assume that the drop keeps its spherical shape during the whole evaporation process. They divide the surface of the drop into two zones:

1. An upper half, A_2, where evaporation if governed by the laws of molecular diffusion; the vapor mass flow rate $\dot{M}_{v2}$ is given by

$$\dot{M}_{v2} = \frac{\delta \tilde{M} \rho_{sat}}{\tilde{R} T_{sat} r} A_2$$

where δ is the molecular diffusion coefficient of vapor into air, $\tilde{M}$ is the molar mass of vapor, and $\tilde{R}$ is the universal gas constant,

2. A lower half, A_1, for which the analysis is similar to that of Wachters et al. described above. From the same momentum equation and similar boundary conditions, they determine the

pressure distribution under the drop. The final expression is, however, much more involved because of the hemispherical geometry.

This theory takes the radiative heat transfer into account. The heat flux transmitted by radiation toward A_1 is given by

$$A_1 F_1 \Gamma_s (T_w^4 - T_{sat}^4)$$

with F_1, a shape factor, equal to

$$F_1 = \left[\left(\frac{1}{\varepsilon_\ell} - 1\right) + \frac{1}{0.682}\right]^{-1}$$

where ε_ℓ is the emissivity of the liquid. Similarly, the radiative heat flux toward A_2 may be written as

$$A_2 F_2 \Gamma_s (T_w^4 - T_{sat}^4)$$

with F_2, the shape factor, given by

$$F_2 = \left[\left(\frac{1}{\varepsilon_\ell} - 1\right) + \frac{1}{0.318}\right]^{-1}$$

One can question whether it is more realistic to consider a flattened or a spherical base for the drop. In practice, the base of the drop is not flat but oscillating. According to Hall (1974), the origin of these oscillations is thermal: For a 400°C wall, they are more pronounced than at 250°C. These oscillations cause the base of the drop to have a shape that is alternately concave and convex; a flat base can thus be considered as an average position. This is probably why the analysis of Wachters et al., based on a constant film thickness, seems in good agreement with experiments.

4.1.2 The Impinging Drop

The heat transfer due to the impact of a single drop on a very hot plate has been treated theoretically by Wachters (1965), Wachters and Westerling (1966), McGinnis and Holman (1969), and Toda (1972).

Wachters and Westerling (1966) have modified their previous model of the sessile drop: When calculating the forces exerted by the drop [as in Eq. (31)], they add to the weight of the drop an inertia term given by

$$\rho_{\ell s} V \frac{d^2 H_m}{dt^2}$$

where H_m is the height of the center of gravity of the drop, measured from the hot wall. Thus, they obtain the following force balance:

$$\int_0^{R_b} 2\pi r(p_v - p_e)\, dr = \rho_{\ell s} V\left(g + \frac{d^2 H_m}{dt^2}\right)$$

Following the same calculation procedure as above, that is, among other things, assuming that the base of the drop is flat, they express the instantaneous thickness of the vapor film as

$$S_v = \left[\frac{3\pi\mu_{vs}\lambda_{vs}(T_w - T_{sat})(T_w + T_{sat})^3 R_b^4}{16\rho_{\ell s}\rho_{vs}T_{sat}^3 V\,\Delta h_v(g + d^2H_m/dt^2)}\right]^{1/4} \qquad (36)$$

This thickness can be evaluated numerically only if the variations of R_b and H_m with respect to time are known. Wachters and Westerling suggest determinging these by high-speed cinematography: they do not propose any general correlation for these time evolutions.

Furthermore, R_b is defined as the radius of the flattened base of the drop. For an impacting drop, which radius will it be? Is it the radius $d_F/2$ of what we called the film around the drop, or is it the radius $d_D/2$ of the dome? Although the texts published by Wachters and Westerling do not permit a definite answer, we believe that R_b is equal to $d_F/2$ when We < 80 and to $d_D/2$ when We > 80.

Wachters and Westerling use Eq. (36) to obtain the vapor film thickness represented in Fig. 50.

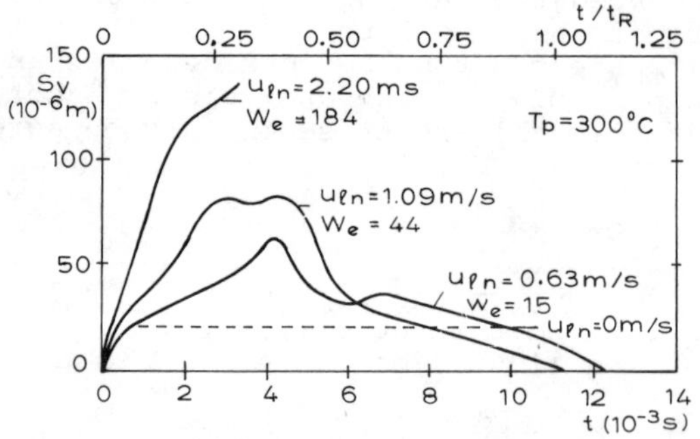

Fig. 50. Time evolution of vapor film thickness, according to Wachters (1965).

The volume change by evaporation at the base of the drop may be written in a similar way:

$$\Delta V_1 = \int_0^t q\left[-\frac{\pi^3\rho_{vs}\lambda_{vs}^3(T_w - T_{sat})^3(T_w + T_{sat})R_b^4 V(g + d^2H_m/dt^2)}{3\rho_{\ell s}^3\mu_{vs}T_{sat}\,\Delta h_v^3}\right]^{1/4} dt$$

where t_q is the period for which the drop remains in the immediate neighborhood of the wall. We have mentioned (cf. sec. 3.1) that Wachters and Westerling suggest equating t_q and t_R:

$$tq = t_R = \pi\sqrt{\frac{\rho_{\ell s} d^3}{16\sigma_{\ell s}}}$$

for bouncing droplets (We < 80). But we also know that for We > 80 this relation is no longer valid because of disintegration of the drop. As a consequence, ΔV_1 cannot be calculated in this latter case, which corresponds to most pulverized drops.

It must be noted that the above relations reply on the same kind of assumptions as those concerning the sessile drop; we have expressed doubts about some of them. Moreover, in the case of an impinging drop, at least two other assumptions are uncertain:

1. Immediate formation of vapor under the drop; for a sessile drop staying several seconds on the wall before collapsing, one can reasonably assume that vapor will be present under the drop, if not immediately, at least during most of the residence time; however, an impinging drop stays in contact with the wall only for a very short time ($\approx 10^{-5}$ s); will vapor be immediately present under the drop? We shall discuss this point in detail later.

2. The hot wall temperature is uniform and equal to T_w. One must admit that the drop induces a local cooling at the surface of the wall. This cooling is difficult to measure. The measurement of a mean wall temperature will certainly not suffice to quantify this very rapid cooling.

Besides, one must not forget that, according to Wachters, a part of the evaporation takes place at the base of the drop and another over the remaining surface of the drop (the lateral evaporation); therefore, a second term ΔV_2 must be added to ΔV_1. Unfortunately, ΔV_2 could not be predicted theoretically by the author. At most, one can believe that the lateral evaporation of an impinging drop will be much smaller than the lateral evaporation of a sessile drop.

Finally, Wachters and Westerling observe that, when θ_w is lower than 400°C, their theoretical analysis differs noticeably from their measurements: Whereas theory predicts an increasing evaporation when θ_w increases, the experiments show the opposite trend. For θ_w higher than 400°C, they do not reach any definite conclusion.

The residence time of the drop in the vicinity of the wall is the main unknown in the analysis proposed by McGinnis and Holman (1969). They define t_q as the time between the instant of liquid-solid contact and the instant corresponding to the end of all heat transfer. Assuming that the impinging drop reaches its saturation temperature instantaneously and that heat is propagating by conduction (no radiation) through an immediately formed vapor film, they state that the heat transfer Q due to the drop is equal to

$$Q = \int_0^{t_q} \frac{\lambda_v A(T_w - T_{sat})}{S_v} dt \tag{37}$$

where λ_v is to be evaluated at $(T_w + T_{sat})/2$ and A designates the drop—solid "contact" surface through the vapor film and is a function of time, as is S_v. Using the works of Savic (1958) concerning the impact of a drop on a cold plate, they claim that A may be written as

$$A = \zeta t d u_{\ell n}^{1.5} \tag{38}$$

where t is the time elapsed since the instant of collision, $u_{\ell n}$ is the component of drop velocity normal to the wall, and ζ a constant depending on the liquid properties. Unfortunately the value of ζ is not given, so that any comparison with our own results for water (see sec. 3.1),

$$d_D^2 = 6.97 d^2 (\tau - \tau^2) \tag{26}$$

is impossible.

If one admits with the authors that (1) all the heat transferred is used to produce vapor, that is, conduction in the liquid is negligible; and (2) all the vapor generated from the instant of "contact" ($t = 0$) remains under the drop, one obtains

$$\rho_v A S_v = \frac{\lambda_v (T_w - T_{sat})}{\Delta h_v + 0.5\, c_v (T_w - T_{sat})} \int_0^{t_q} \frac{A}{S_v} dt \tag{39}$$

This last hypothesis hardly seems admissible. Indeed, how can one admit that vapor cannot be expelled radially from the film? The velocity of the liquid spreading on the wall should then be equal to the radial velocity of the vapor. Indeed, if the spreading velocity of the liquid is greater than the vapor velocity, liquid could touch the wall at the periphery of the drop, which is contrary to the assumptions; if the spreading liquid velocity is smaller than the vapor velocity, all the vapor produced does not remain under the drop.

Continuing their analysis by combining Eqs. (38) and (39) and assuming that thermal properties are a function of temperature only, McGinnis and Holman obtain

$$S_v = \left\{ \frac{2\lambda_v (T_w - T_{sat}) t_q}{3\rho_v [\Delta h_v + c_v (T_w - T_{sat})/2]} \right\}^{1/2} \tag{40}$$

according to which S_v increases with the square root of time. Let us note that the drop diameter and velocity do not appear explicitly in this expression.

Let us consider a typical pulverized drop: d = 200 μm, $u_{\ell n}$ = 15 m/s. The time t^+ after which the dome of the drop has disappeared is given by [cf. (24)]

$$t^+ = \frac{d}{u_{\ell n}} \cong 1.33\text{x}10^{-5} \text{ s}$$

Equating t_q to t^+, one can then calculate thickness S_v by Eq. (40). Table 11 shows that S_v is approximately equal to 25 μm for

$$800 \leqslant \theta_w \leqslant 1000^\circ\text{C}$$

In the last column of the same table, we have recorded the values of S_v obtained when applying Eq. (40) to the drop considered by Wachters: θ_w = 300°C, d = 2.3 mm, $u_{\ell n}$ = 2.2 m/s. Comparing the results with those of Fig. 50, one observes that the thicknesses S_v are quite similar for $t = t^+ = 1.05\text{x}10^{-3}$ s, Wachters finds that S_v is equal to about 70 μm, whereas Eq. (40) leads to a value of more than 65 μm.

Table 11. *Evaluation of Vapor Film Thickness*

$\tau = \frac{t}{t^+}$	Thickness S_v (μm), from Eq. (40)		
	d = 200 μm; $u_{\ell n}$ = 15 m/s		d = 2.3 mm; $u_{\ell n}$ = 2.2 m/s
	θ_w = 800°C	θ_w = 1000°C	θ_w = 300°C
0.1	6.3	7.9	20.7
0.3	11.0	13.6	35.8
0.5	14.2	17.6	46.2
0.8	17.9	22.2	58.5
1	20.0	24.9	65.4
1.2	22.0	27.2	71.6

Finally, the total heat evacuated from the wall by a drop is given by the following expression, which combines relations (37), (38), and (40):

$$Q = \left\{\frac{2}{3}\zeta^2\lambda_v\rho_v\left[\Delta h_v + c_v\frac{(T_w - T_{sat})}{2}\right](T_w - T_{sat})\, d^2u_{\ell n}^3t_q^3\right\}^{1/2} \qquad (41)$$

Besides ζ, whose value is not given, there remains an unknown: time t_q. According to McGinnis and Holman, t_q is a complex function of heat transfer, and drop velocity and diameter. Therefore, one cannot deduce from Eq.(41) any information concerning the functional dependence of Q. From their approach, the authors conclude that additional work is needed.

Toda (1971, 1972) and Toda and Uchida (1973) have proposed a model of heat transfer between a hot wall and a liquid spray. This model first considers the heat exchange caused by the impact of a single drop and then the cooling effect due to all the drops of a spray. In this section, we shall comment only on the first part, that is, that devoted to the single drop.

1. After impact, each sprayed drop is assumed to spread on the wall so that it instantaneously forms a thin liquid film of uniform thickness. This initial uniform thickness S_{Fi} depends on the mean volume diameter d_{30}:

$$S_{Fi} = \zeta_1 d_{30}^{m_1} \tag{42}$$

where ζ_1 is a coefficient and m_1 is a constant to be found experimentally. For water, Toda suggests:

$$\zeta_1 = 0.541\nu_{\ell s}^{0.4}$$

$$m_1 = 0.6$$

with $\nu_{\ell s}$ the kinematic viscosity evaluated at $T = T_{sat}$. The radius of this liquid film R_{Fi} is given by the conservation equation

$$\frac{1}{6}\pi d_{30}^3 = \pi R_{Fi}^2 S_{Fi} \tag{43}$$

Expression (42), with the given numerical values of ζ_1 and m_1, is valid only if

$$890 < \frac{2R_{Fi}\bar{u}_T}{\nu_{\ell s}} < 2\text{x}10^5$$

and if

$$2.15\text{x}10^{-2} < \frac{\mu_{\ell s}\bar{u}_T}{\sigma_{\ell s}} < 1.6$$

The geometric shape of the liquid film as proposed by Toda is rather curious, although it is claimed to be suggested by experiments of drop impingements on a cold plate (20°C). He does not take into account any dome in the spreading liquid as described in sec. 3.1. The time of spreading is neglected! The impacts of a drop on a cold plate and on a hot plate are said to be identical.

2. The liquid film with constant thickness S_{Fi} is supposed to reach saturation temperature T_{sat} instantaneously. Whatever the wall temperature (even very high), the wall is wetted by the liquid: there is direct contact between liquid and solid. Heat transfer is governed by conduction in the liquid. The liquid film is considered as a semiinfinite body at T_{sat}, whose lower face temperature is suddenly increased to T_w and kept at this temperature. The solution of this classical problem of heat conduction (see, for example, Bird et al., 1960) gives as temperature profile in the liquid film:

$$T = T_{sat} + (T_w - T_{sat}) \text{ erfc } \frac{z}{2\sqrt{k_\ell t}} \qquad (44)$$

where z is the coordinate normal to the hot wall and k_ℓ is the liquid thermal diffusivity. The heat flux density evacuated from the wall is given by

$$\dot{q}_1 = -\lambda_\ell \left.\frac{\partial T}{\partial z}\right|_{z=0} = \lambda_\ell (T_w - T_{sat}) \frac{1}{\sqrt{\pi k_\ell t}} \qquad (45)$$

The heat flux density $\dot{q}_1$ decreases when time t increases. Vapor bubbles appear at the liquid–solid interface, and this decreases the liquid film thickness.

If this thickness is larger than a limit value S_F^* given by

$$S_F^* = 1.55\text{x}10^{-3}(T_w - T_{sat})^{1.5} \qquad (46)$$

for water, the liquid film is insulated from the wall by a vapor sublayer of thickness S_v. The times of formation of vapor bubbles and of the vapor sublayer are neglected. However, the decrease of the liquid film thickness is not neglected, so that the heat flux density $\dot{q}_1$ reaches a limit value $\dot{q}_1^*$ corresponding to thickness S_F^*:

$$\dot{q}_1^* = \frac{\lambda_\ell (T_w - T_{sat})}{S_F^*} = \frac{\lambda_\ell (T_w - T_{sat})^{2.5}}{1.55\text{x}10^{-3}} \qquad (\text{W/m}^2)$$

The period t_1 of the exchange by direct liquid–solid contact is evaluated from

$$\lambda_\ell (T_w - T_{sat}) \frac{1}{\sqrt{\pi k_\ell t_1}} = \dot{q}_1^*$$

that is,

$$t_1 = \frac{1}{\pi} \frac{(1.55\text{x}10^{-3})^2}{k_\ell} (T_w - T_{sat})^{-3} \qquad (47)$$

The mean heat flux density $\bar{\dot{q}}_1$ during time t_1 is then equal to

$$\bar{\dot{q}}_1 = \frac{1}{t_1} \int_0^{t_1} \dot{q}_1 \, dt = \frac{2\lambda_\ell}{1.55\text{x}10^{-3}} (T_w - T_{sat})^{2.5}$$

The concept of limit liquid film thickness S_F^* introduced by Toda is original, but its physical basis does not appear very clearly—where did Toda find the numerical values he recommends?

Let us emphasize, however, that Toda admits the existence of direct contact between the hot wall and the liquid, previous to vaporization.

3. Once the vapor sublayer is formed, a second mode of heat transfer begins; it is of the film-boiling type. The theoretical expressions developed by Toda rely on the following assumptions (Fig. 51):

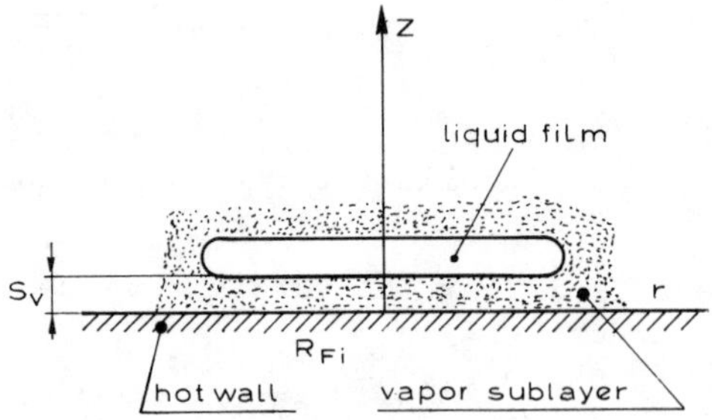

Fig. 51. Model of drop evaporation.

1. The liquid film is shaped as a flat disk supported by a vapor sublayer with constant thickness S_v.

2. The vapor flow is laminar; inertial forces are neglected.

3. The liquid is at saturation temperature T_{sat}.

4. Evaporation above the liquid film is much smaller than evaporation under the liquid film.

5. Since $S_v \ll R_{Fi}$, $\partial T/\partial r = 0$.

Several of these assumptions are similar to those considered in the sessile drop model proposed by Wachters et al. Toda writes similar mass, momentum, and energy conservation equations for the vapor film, but does not neglect the vertical component of vapor velocity. His expressions therefore contain the vertical component u_{vz} and the radial component u_{vr} of vapor velocity: The boundary conditions are

At $z = 0$, $u_{vr} = u_{vz} = 0$ and $T = T_w$.

At $z = S_v$, $u_{vr} = 0$ and $T = T_{sat}$.

At $r = 0$, $u_{vr} = 0$.

At $r = R_{Fi}$ and $z = S_v$, $p_v = p_e$.

He finds that

$$u_{vr} = 3\zeta r(S_v z - z^2) \tag{48}$$

$$u_{vz} = 2\zeta(\frac{3}{2}S_v z^2 - z^3) \tag{49}$$

$$p_v - p_e = -6\rho_v \nu_v \zeta(\frac{1}{2}r^2 - \frac{1}{2}R_{Fi}^2 + S_v\ z - z^2) \tag{50}$$

$$T \simeq T_w + \frac{(T_w - T_{sat})}{S_v - 3\zeta S_v^5/20k_v}(z - \frac{\zeta}{4k_v} S_v z^4 + \frac{\zeta}{10k_v} z^5) \tag{51}$$

where k_v is the thermal diffusivity of vapor and ζ is a coefficient to be determined by means of an equation expressing the balance of forces in the vertical direction due to vapor film and liquid film:

$$\int_0^{R_{Fi}} [p_v(r,S_v) - p_e]2\pi r\ dr = \pi R_{Fi}^2\ \frac{1}{2}\rho_v u_T^2 \tag{52}$$

It should be noted that Toda equates the force induced by the pressure distribution in the vapor film to the force exerted by the flow of air entrained by the drops (at velocity $\bar{u}_T$) on the upper face of the liquid film. In this balance equation, one finds neither the weight of liquid film nor the thrust due to evacuation of vapor under the liquid film.

Combining Eqs. (50) and (52), Toda finds

$$\zeta = \frac{1}{6}\frac{\bar{u}_T^2}{\nu_v R_{Fi}^2} \tag{53}$$

Moreover, if thermal radiation is neglected, a heat balance yields

$$\rho_v u_{vz}(S_v)\ \Delta h_v = \lambda_v\ \frac{\partial T}{\partial z}\bigg|_{z=S_v} \tag{54}$$

Rearranging relation (49) with (51), (53), and (54), the vapor sublayer thickness S_v is found to be

$$S_v = \left(\frac{6k_v \nu_v R_{Fi}^2}{\bar{u}_T^2}\ L\right)^{1/4} \tag{55}$$

with

$$L = \frac{c_{\ell s}(T_w - T_{sat})}{\Delta h_v}\left[1 + \frac{7}{20}\ \frac{c_{\ell s}(T_w - T_{sat})}{\Delta h_v}\right]^{-1}$$

The heat flux density $\dot{q}_2$ during this film-boiling-type heat exchange is given by

$$\dot{q}_2 = -\lambda_v \left.\frac{\partial T}{\partial z}\right|_{z=0} = \frac{\lambda_v (T_w - T_{sat})}{S_v (1 - \frac{3}{20} L)} \qquad (56)$$

which is time independent with S_v.

The heat transfer through the vapor film stops when the liquid film is dried out. The duration t_2 of this heat exchange is said to be found experimentally; for water, Toda suggests

$$t_2 = 6.56 \times 10^6 R_{Fi}^2 . \tau_d$$

where τ_d is the time between two successive drops at the same place on the hot wall. One can show that

$$\tau_d = \frac{S_{Fi} \rho_\ell}{\dot{m}_\ell}$$

This expression contains the mass velocity $\dot{m}_\ell$ of spray drops; indeed, as mentioned above, this model is intended for spray drops.

Figure 52 summarizes the main steps of Toda's model. We have

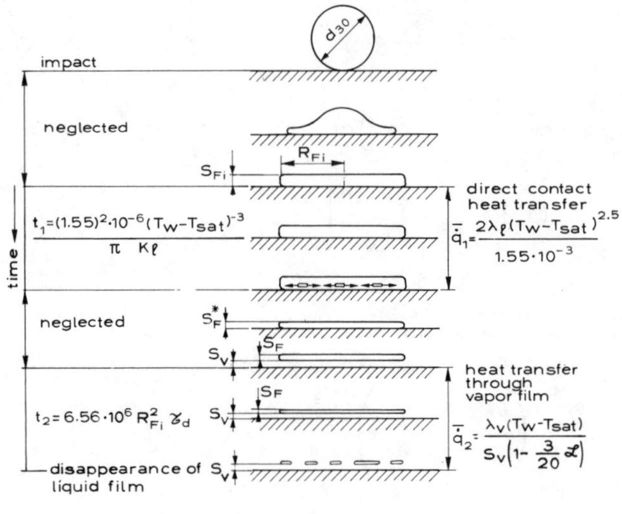

Fig. 52. Main steps of Toda's model.

seen that it contains several questionable assumptions. Let us mention another: The liquid film maintains the same radius R_{Fi} during the whole heat transfer process until it disappears. It is known (cf. Sec. 3.1) that, in practice, an impinging drop spreads out radially on the wall and breaks up into several smaller drops.

In spite of all its unlikelinesses, Toda's work has the merit of systematically investigating phenomena and distinguishing between two modes of heat transfer: by direct contact and through a vapor sublayer. At the end of his thesis, Toda suggests for later investigators to reconsider his starting hypothesis, that is, formation on the wall of a liquid film with constant thickness and at such high speed that no heat can be extracted from the wall. Our own measurements of the behavior of impinging drops urge us to do so. Therefore, the theoretical model we propose explicitly takes into account the heat exchange during the spreading of the drop.

4.2 A New Theoretical Model

A new model of heat exchange between a hot wall and a single drop has been proposed by Moureau (1978). It is based on the above-described behavior of a single drop impinging on a hot horizontal surface (Sec. 3.1).

Two mechanisms of heat transfer are considered:

1. By direct contact between the hot wall and the lower central zone of the liquid dome
2. By conduction through the annular vapor film formed under the periphery of the splashing drop

We shall show that the first type of heat exchange is predominant. The effect of heat radiation will be taken into account when evaluating the heat transfer due to all the drops of a pulverization.

4.2.1 Heat Transfer by Direct Contact

Let us assume that there is direct contact between the liquid drop and the hot wall over a circular surface with diameter d_{DC} given by

$$d_{DC}^2 = \omega^2 d_D^2 = 6.97\omega^2 d^2 \tau(1 - \tau) \tag{57}$$

where d is the initial diameter of the drop (possibly pertaining to the ith class of the spray, and τ is as before:

$$\tau = \frac{t}{t^+}$$

ω is an adjustable parameter whose value can vary between zero and one.

Is this hypothesis credible? Can water at low temperature (20°C) be in direct contact with a wall at high temperature (600 to 1000°C)?

Bradfield (1966) has shown experimentally the existence of contact between a liquid and a hot solid during quenching. Flament (1978) has performed quenching experiments with a silver cylinder where direct contact occurs at temperatures as high as 850°C. Lackmé (1976) has shown that, when a hot metallic pendulum (700°C) hits a cold water surface for a short time (50×10^{-3} s), the duration of insulation by a continuous vapor film is much shorter (5×10^{-3} s), than the duration of direct contact (45×10^{-3} s).

As we have seen, Toda (1972) based his theoretical calculations on a very short period of direct liquid–solid contact.

An attempt to explain this direct contact has been proposed by Lackmé (1976). The wetting could be due to high pressurization of the liquid sheets near the central zone of impact and to the time delay necessary to the phase change.

We therefore admit the assumption of direct contact, seeing that, for impacting spray droplets, the characteristic times are of the order of 10^{-5} s.

In order to evaluate the heat evacuated from the wall by the part of the drop in direct contact (Q_{DC}), we use the well-known error function solution for the one-dimensional conduction problem of an instantaneous temperature change at the face of the wall. Q_{DC} is given by the following expression:

$$Q_{DC} = \int_0^{d_{DC}^+/2} 2\pi r' \frac{\lambda_w (T_w - T_\ell)}{\sqrt{\pi k_w}} \int_{t_0}^{t_1} \frac{dt}{\sqrt{t - t_0}} \, dr' \tag{58}$$

where d_{DC}^+ is the maximum value of d_{DC} during the period of impact; t_0 and t_1 are defined later.

Let us consider a circular crown with mean diameter d' ($d' = 2r'$), drawn on the plate, whose center is the initial point of impact of the drop. This zone is only wetted between times τ_0 and τ_1 (Fig. 53), that is, only when

$$d_{DC} = \omega d_D \geqslant d'$$

Times τ_0 and τ_1 are thus the roots of the following equation, derived from (26) or (57):

$$d'^2 = 6.97 \omega^2 d^2 \tau (1 - \tau) \tag{59}$$

After a first integration, Q_{DC} becomes

$$Q_{DC} = \int_0^{d_{DC}^+/2} 4\sqrt{\pi} r' \frac{\lambda_w (T_w - T_\ell)}{\sqrt{k_w}} \sqrt{t_1 - t_0} \, dr' \tag{60}$$

Since

$$t^+ = \frac{d}{u_{\ell n}} \tag{24}$$

the two roots of Eq. (59), expressed in their dimensional form, are such that

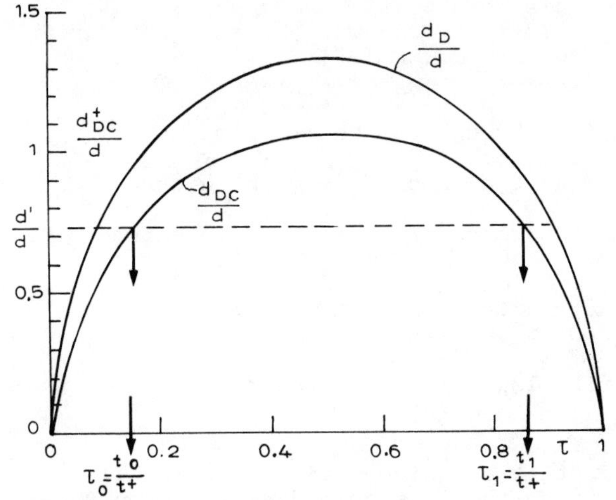

Fig. 53. Evolution of diameter of direct contact when ω is equal to 0.8.

$$(t_1 - t_0)^{1/2} = (t^+)^{1/2}\left(1 - \frac{4r'^2}{1.74\omega^2 d^2}\right)^{1/4}$$

Therefore, one gets

$$Q_{DC} = 4\sqrt{\frac{\pi}{k_w}}\lambda_w(T_w - T_\ell)(t^+)^{1/2}\int_0^{d^+_{DC}/2} r\left(1 - \frac{4r'^2}{1.74\omega^2 d^2}\right)^{1/4} dr'$$

Performing this integration and using the fact that

$$(d^+_{DC})^2 = 1.74\omega^2 d^2$$

one obtains

$$Q_{DC} = 0.70\omega^2\sqrt{\frac{\pi}{k_w}}\ \lambda_w(T_w - T_\ell)\frac{d^{2.5}}{u_{\ell n}^{0.5}} \tag{61}$$

In order to fit theory and experiments, it is appropriate to take for ω^2 a value equal to $\frac{2}{3}$. By choosing this value, we minimize the differences between the theoretical heat flux densities that will be derived below and experimental heat flux densities measured for various liquid sprays on stainless steel.

One finally obtains

$$Q_{DC} = 0.82\sqrt{\lambda_w\rho_w c_w}(T_w - T_\ell)\frac{d^{2.5}}{u_{\ell n}^{0.5}} \tag{62}$$

According to this result, *the exchange of heat is proportional to the thermal effusivity of the wall, the temperature difference, the power 2.5 of the drop diameter, and the inverse of the square root of the velocity before impact.*

4.2.2 Heat Transfer Through the Vapor Film

Let us assume that a vapor film with thickness S_v is present

1. Under the lower surface area of the drop between $(\pi/4)d_D^2$ and $(\pi/4)d_{DC}^2$, when $t \leqslant 0.2t^+$
2. Under the lower surface area of the drop between $(\pi/4)d_F^2$ and $(\pi/4)d_{DC}^2$ when $0.2t^+ \leqslant t \leqslant t^+$
3. Under the entire lower surface area of the drop, that is, $(\pi/4)d_F^2$ when $t^+ \leqslant t \leqslant 1.5t^+$

Time $0.2t^+$ is that at which a film appears all around the drop, with a diameter larger than the dome diameter, and time $1.5t^+$ corresponds to the rupture of the continuous liquid film.

The local heat flux density $\dot{q}_v$ evacuated by conduction through the vapor film is given by

$$\dot{q}_v = \frac{\lambda_v (T_w - T_{sat})}{S_v}$$

where we admit that the liquid is at saturation temperature T_{sat} corresponding to the ambient pressure. The heat exchanged under these circumstances, Q_v, is then equal to ($t_d \triangleq 0.2t^+, t_f \triangleq 1.5t^+$):

$$Q_v = \int_0^{t_d} \int_{d_{DC}/2}^{d_D/2} \dot{q}_v 2\pi r' \, dr' \, dt + \int_{t_d}^{t^+} \int_{d_{DC}/2}^{d_F/2} \dot{q}_v 2\pi r' \, dr' \, dt' +$$

$$\int_{t^+}^{t_f} \int_0^{d_F/2} \dot{q}_v 2\pi r' \, dr' \, dt$$

Cinematographic observations of the impact of drops on hot walls have shown that the front at the periphery of the drop tends to rise from the wall during its spreading movement. The thickness S_v tends to increase with r' and t. If one assigns to S_v a sufficiently small constant value, one can overestimate the value of Q_v.

Applying such a method, the following may be written:

$$Q_v = \frac{\lambda_v (T_w - T_{sat})}{S_v} \left(\int_0^{t_d} \int_{d_{DC}/2}^{d_D/2} 2\pi r' \, dr' \, dt + \int_{t_d}^{t+} \int_{d_{DC}/2}^{d_F/2} 2\pi r' \, dr' \, dt + \int_{t^+}^{t_f} \int_0^{d_F/2} 2\pi r' \, dr' \, dt \right)$$

These integrals are easily performed, since the time evolutions of d_{DC}, d_D, and d_F are known. The final result is

$$Q_v = \frac{\pi \lambda_v (T_w - T_{sat})}{S_v} \left[0.29 d u_{\ell n} t_d^2 - 0.19 u_{\ell n}^2 t_d^3 + 0.14 \frac{u_{\ell n}^4}{d^2} (t_f^5 - t_d^5) - 1.08 \frac{u_{\ell n}^3}{d} (t_f^4 - t_d^4) + 2.61 u_{\ell n}^2 (t^{+3} - t_d^3) + 2.23 u_{\ell n}^2 (t_f^3 - t^{+3}) - 0.58 d u_{\ell n} (t^{+2} - t_d^2) \right]$$

With the help of this expression, for a given value of S_v, it is possible to evaluate Q_v for a drop with known diameter d and velocity $u_{\ell n}$, impinging on a wall at temperature T_w.

4.2.3 Heat Transfer Due to a Single Drop

Numerical evaluations of the energies Q_{DC} and Q_v for two drops typical of those obtained when using pressure nozzles are presented in Table 12.

Table 12. *Calorific Energies Q_{DC} and Q_v for Two Drops Originating from a Fan-type Nozzle*

Case 1: Drop Diameter $d = 500 \times 10^{-6}$ m; Velocity $u_{\ell n} = 20.0$ m/s			
$Q_{DC} = 7.33 \times 10^{-3}$ J	If $S_v = 50 \times 10^{-6}$ m,	$Q_v = 6.19 \times 10^{-5}$ J	$\frac{Q_v}{Q_{DC}} = 0.84\%$
	If $S_v = 25 \times 10^{-6}$ m,	$Q_v = 1.24 \times 10^{-5}$ J	$\frac{Q_v}{Q_{DC}} = 1.69\%$
	If $S_v = 5 \times 10^{-6}$ m,	$Q_v = 6.19 \times 10^{-4}$ J	$\frac{Q_v}{Q_{DC}} = 8.4\%$
Case 2: Drop Diameter $d = 60 \times 10^{-6}$ m; Velocity $u_{\ell n} = 10.5$ m/s			
$Q_{DC} = 5.05 \times 10^{-5}$ J	If $S_v = 6 \times 10^{-6}$ m,	$Q_v = 1.68 \times 10^{-6}$ J	$\frac{Q_v}{Q_{DC}} = 3.33\%$

It should be noted that Q_v is only a small fraction of observed heat-removal rates. Other similar comparisons have led us to the same conclusion. Thus, we admit that the calorific energy Q extracted by a single drop during its impact on a hot wall is fairly well approximated by the energy extracted by direct contact:

$$Q \cong Q_{DC}$$

5 HEAT TRANSFER BETWEEN A HOT PLATE AND A SPRAY: THEORETICAL MODEL AND EXPERIMENTS

5.1 Theoretical Evaluation of the Heat Exchange

The theoretical evaluation of the heat exchange between one single drop and the hot plate, developed in Sec. 4.2, may be extended to the case of the whole spray in the following way.

The heat flux density exchanged between a hot wall and a liquid spray is given by the addition of two terms. The first is equal to the sum of all heat flux densities extracted by the drops of different classes of diameters, whereas the second term is equal to the heat flux density exchanged by radiation between the hot wall and its environment.

In order to evaluate the first contribution, let us remember that the mass flux density $\dot{m}_\ell$ is related to the flux densities x_i of drops with diameter d_i by relation (18). The values of x_i for each class of drops can be determined by some rather sophisticated methods described in Sec. 2.3.1. They lie in the range of 10^6 to 10^7 drops per second and per square meter for the fan-type pressure nozzles we have tested. Once all x_i values are known, the heat flux density evacuated by conduction during direct contact between liquid and plate is given, for drops of the ith class, by

$$Q_i\, x_i$$

where Q_i is equal to the value of Q_{DC} computed for a drop of diameter d_i by means of Eq. (61), the heat transfer through the vapor film being negligible.

By summing over all the classes of the liquid spray, we obtain

$$\sum_i Q_i x_i$$

which is the first term announced.

The heat flux density exchanged by radiation between the hot wall and its environment may be written as

$$(1 - \bar{A}_t)\varepsilon_w\, \Gamma(T^4 - T_e^4)$$

The environment of the plate is large enough to be considered as a perfect absorbant; its temperature is equal to the liquid temperature. The plate emissivity is called ε_w, whereas $\bar{A}_t$ is the average wetted fraction of the plate surface. Obviously, the plate can radiate heat from anywhere except from zones where direct contact occurs. Values of $\bar{A}_t$ are presented in Table 9 and are found, in every case, to be lower than 0.01. We shall thus consider here:

$$1 - \bar{A}_t \cong 1$$

Gathering the two partial results just obtained, we find the final theoretical expression of heat flux density:

$$\dot{q}_w = \varepsilon_w \Gamma (T_w^4 - T_\ell^4) + 0.82 \sqrt{\lambda_w \rho_w c_w} (T_w - T_\ell) \sum_i \frac{d_i^{2.5}}{u_{\ell ni}^{0.5}} x_i \qquad (63)$$

This expression shows the functional dependence of heat flux density on the hot wall characteristics as well as on the spray characteristics. Its determination requires among others, the knowledge of the distribution of droplet diameter and velocity. The coefficient 0.82 is related to the adjustable parameter ω defined above, whose value will be discussed in Sec. 5.3.

5.2 Heat Transfer Measurements

5.2.1 Experimental Device

A schematic view of the apparatus we have used is shown in Fig. 54. A plate P, made of refractory stainless steel (AISI 309 S), 320 x 40 x 3 mm, is brought to high temperature (600 to 1100°C) by

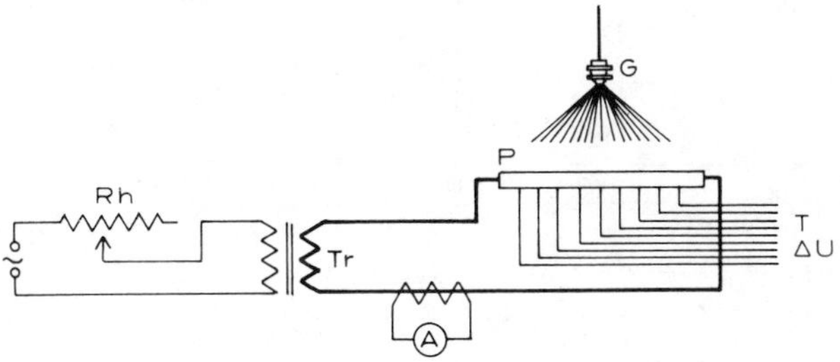

Fig. 54. Schematic view of the spray cooling experimental device.

Joule effect. The ends of this metallic plate are connected to the secondary of a 50-kVA transformer Tr(380 V/15 V). The primary current is adjusted by means of a rheostat Rh designed for this purpose. The plate is fixed at one end and guided at the other so that its longitudinal expansion is free. This device was designed to minimize deformation of the plate.

The liquid aspersion may take place either on the upper face or on the lower of the plate. The temperatures are measured by means of chromel-alumel thermocouples in a 1-mm-diameter stainless steel sheath. Seven or eight thermocouples are inserted or soldered along the center line of the larger dimension of the plate, at a depth of 1.25 mm from the lower face. The sensing head of the thermocouples is electrically insulated from the sheath, and some radiographies show that this sensing head is situated at about 1.2 mm from the end of the sheath. This head is thus at the level of the lower face of the plate.

The differences in electrical potential ΔU existing between two successive thermocouples are also measured. When the plate is kept at high temperature during spraying, the power dissipated in the plate can reach 18 kW (a 2500-A current for a potential difference of about 7 V between the ends of the plate).

All temperatures, potential differences, currents, and water pressures at the nozzle are periodically recorded, and the raw digitalized data are stored on magnetic tapes. All this information is then processed numerically.

5.2.2 Wall Heat Flux Calculations

The sprayed surface temperature distribution is presented in Fig. 55 for a typical experiment. Longitudinal temperature gradients in the central region of the plate are moderate. In the unfavorable situation of the point located at L = 0.20 m in

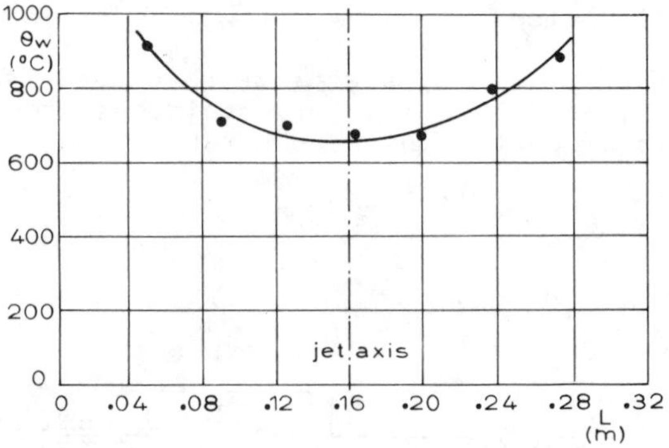

Fig. 55. Longitudinal temperature distribution of the sprayed surface (nozzle F13/60; Δp = 2.2 bar).

Fig. 55, where the temperature gradient is nearly zero on the left but as high as 4 °C/mm on the right-hand side, the longitudinal heat flux due to conduction is approximately 10 W, whereas the heat flux due to spray cooling on the upper surface of 0.04 x 0.04 m^2 near the observed thermocouple is 1200 W. The conduction losses in the longitudinal direction are thus negligible when writing the energy balance near each thermocouple.

The losses by convection and radiation through the lower face of the plate are estimated by means of the following correlation for the heat transfer coefficient α (W/m^2K):

$$\alpha = -4.721 \times 10^{-3} + 1.326 \times 10^{1} T_w - 1.227 \times 10^{-2} T_w^2 + 3.819 \times 10^{-6} T_w^3$$

where T_w, expressed in kelvins, is in the range

$$980\ \text{K} < T_w < 1280\ \text{K}$$

This expression correlates about 40 experimental points obtained from energy balances over the whole plate, evaluated without spraying. The known difference between the free convection intensity from the lower and upper faces of the plate has been taken into account in order to correct experimental results so that they can be applied to the lower face only.

The same experiments have been used to calculate a correlation between the value of the electrical resistance of the plate and the local mean temperature of the metal. In all calculations, the dimensions of the plate are those at ambiant temperature corrected by the thermal expansion coefficient.

We have written a computer program in order to evaluate the correction to be applied to the temperature read on the plate's lower face. This computation takes into account the increased thermal loss from the plate due to the extended surface of the thermocouple and the sink effect in the plate due to this local loss. The local heat exchanges between thermocouple, plate, and ambiency are computed; for further information, the reader is referred to Moureau (1978).

When the lower face surface temperature is known, the heat flux density dissipated by the sprayed face is easily deduced from the local Joule effect intensity. The variation of the inner temperature of the plate can also be calculated by means of the simple one-dimensional solution of the Laplace equation. A typical example of this variation is shown in Fig. 56.

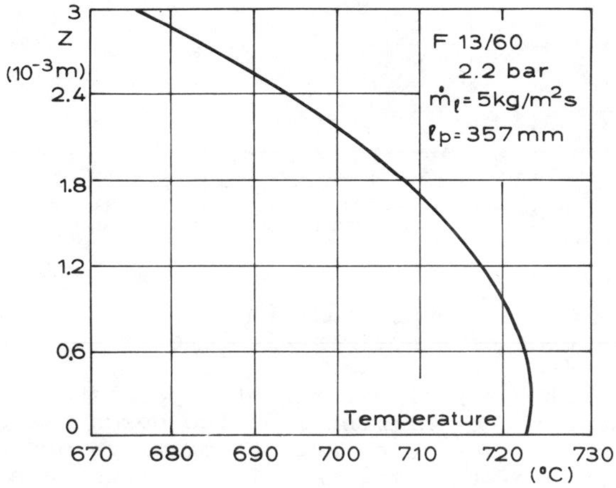

Fig. 56. Variation of the inner temperature of the plate (z = 0: lower dry face; $z = 3\text{x}10^{-3}$ m: upper sprayed face).

It can be seen from Fig. 56 that an uncertainty on the exact position of the sensing head of the thermocouple has only a very slight influence on the measured temperature of the lower face: The temperature profile is extremely flat in the 0–1 mm region near this lower face.

5.2.3 Heat Transfer Results

In each experiment, only those zones located around the three or four central thermocouples are considered for final results, in order that conditions will be as uniform as possible for both temperature and liquid flow rate density. The variations of $\dot{m}_\ell$ from one zone to the other are weak: Each set of points shown in Fig. 57 is thus representative of either a given value of Δp or a narrow range of values of $\dot{m}_\ell$.

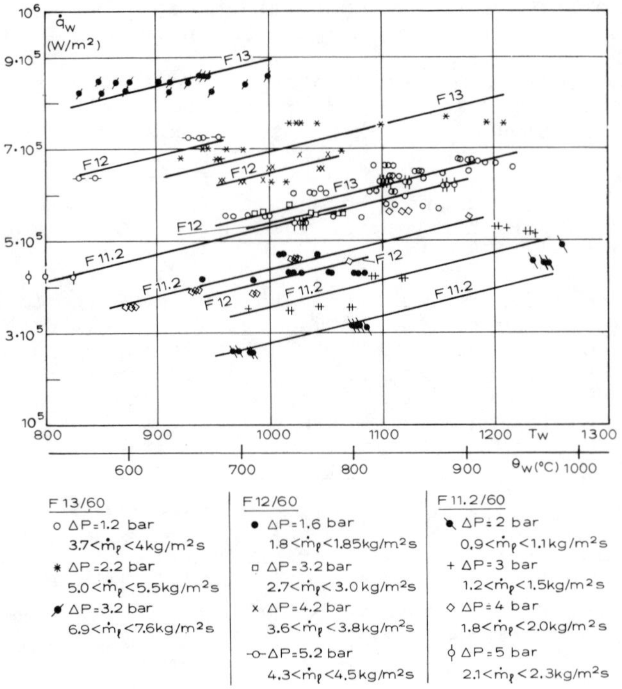

Fig. 57. Heat flux versus temperature for sprays in still air on the upper face of a horizontal plate.

Figure 57 shows 11 sets of experimental points obtained in still air by means of the experimental device described above. A straight line is drawn across each set of points in order to facilitate the reading of this figure. These lines have deliberately been drawn as parallels, whose slope is the average of the slopes of each of the best fit straight lines related to each set of experimental points. Differences between these slopes are not significant.

Figure 57 shows a marked influence of $\dot{m}_\ell$ on the heat flux. More explicitely, for a given nozzle and a given plate temperature, heat flux increases with pressure Δp at the nozzle orifice, and hence with $\dot{m}_\ell$ (Figs. 58 and 59).

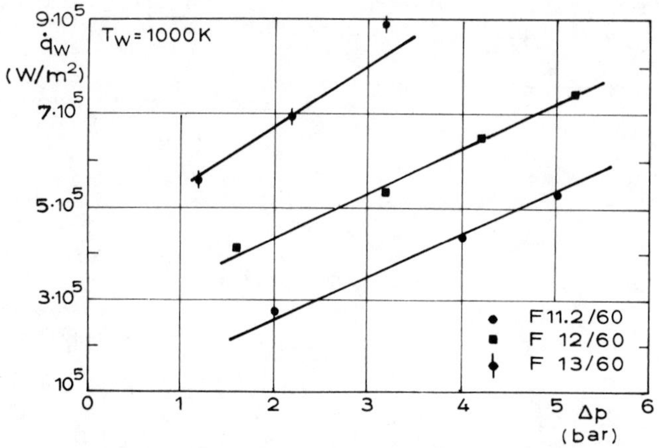

Fig. 58. Variation of heat flux with pressure at the nozzle.

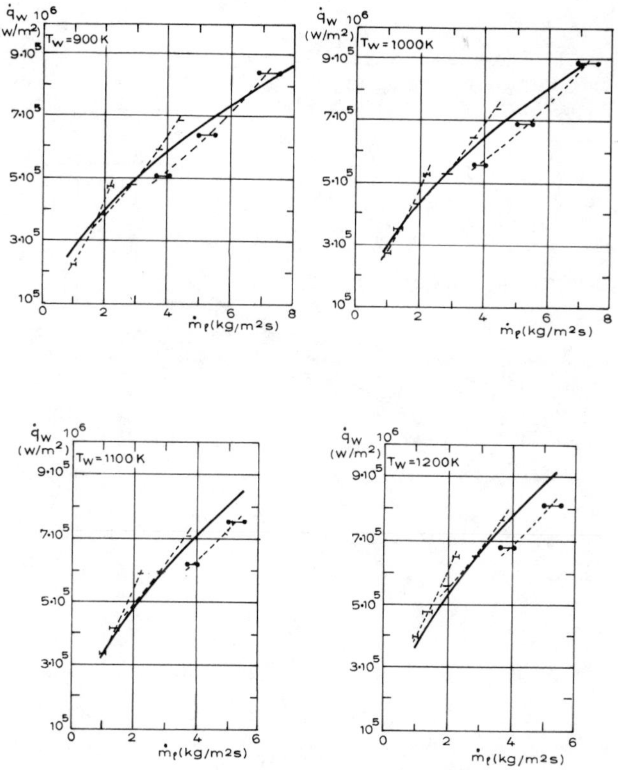

Fig. 59. Variation of heat flux with water flow rate density.

In Fig. 59, individual dotted lines refer to each nozzle

tested. The continuous lines drawn through those dotted lines in the four parts of this figure represent the following correlation:

$$\dot{q}_w = 300T_w(\dot{m}_\ell)^{0.556} \quad W/m^2 \tag{64}$$

where

$$900 < T_w < 1200 \text{ K}$$

$$1 < \dot{m}_\ell < 7 \text{ kg/m}^2 \text{ s}$$

This correlation fits the experimental results with less than 13% error.

Another way of representing the experimental results is shown in Fig. 60. For each temperature considered in Fig. 59, the heat transfer coefficient, defined by

$$\alpha_w = \frac{\dot{q}_w}{T_w - T_\ell} \tag{65}$$

is computed and represented in Fig. 60 by a short trace, corresponding to the range of values of $\dot{m}_\ell$. The temperature T_ℓ of the liquid is 293 K and is equal to the ambient temperature.

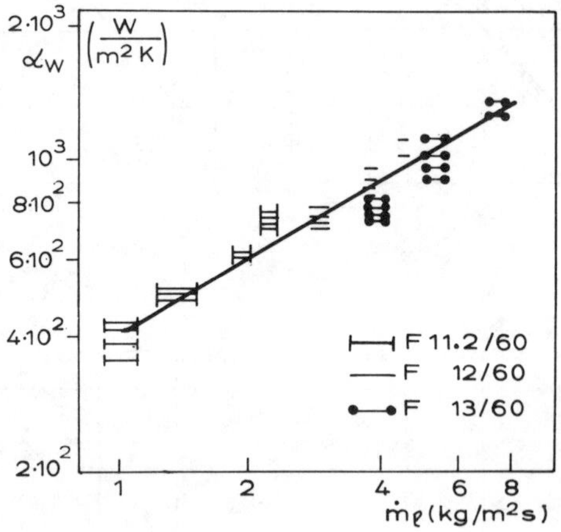

Fig. 60. Variation of heat transfer coefficient with water flow rate density.

The equation of the straight line drawn through Fig. 60 is

$$\alpha_w = 423(\dot{m}_\ell)^{0.556} \quad (W/m^2 \text{ K}) \tag{66}$$

This fits the experimental results with less than 17% error, and can be used whenever the following operating conditions are satisfied:

$$900 < T_w < 1200 \text{ K}$$

$$1 < \dot{m}_\ell < 7 \text{ kg/m}^2 \text{ s}$$

$T_\ell \cong 293$ K, in any case far from the saturation point.

Other experiments have been conducted (Célis and Norhomme, 1977), using nozzles with vertical axis located under the plate, and thus *cooling its lower face*. The plate dimensions are the same as before, and the distance between nozzle tip and sprayed surface is still 357 mm. Results are shown in Fig. 61. The slope of the best fit parallel straight lines for nozzle F11.2/60 is slightly larger than in Fig. 57 (700 instead of 585 W/m^2 K).

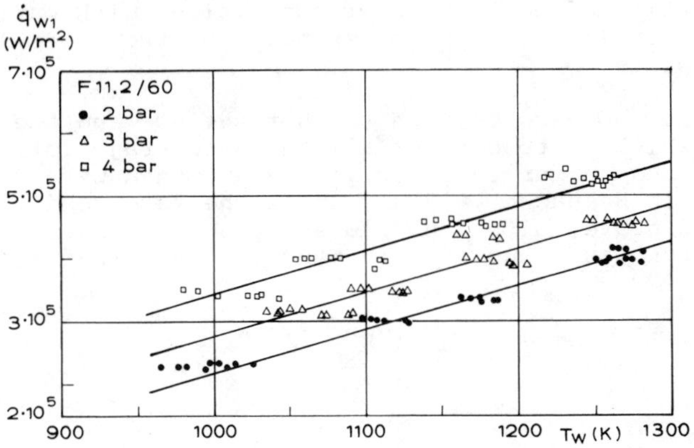

Fig. 61. Heat flux versus temperature for sprays in still air impinging on the lower face of a horizontal plate.

Figure 62 is determined in the same way as Fig. 60. For temperatures equal to 1000, 1100, 1200, and 1300 K, a short trace is drawn, corresponding to the concerned range of values of $\dot{m}_\ell$. The straight line correlating these results is located 15% below the line representing Eq. (66). The heat transfer coefficient α_{w1} corresponding to the lower face of a horizontal plate may thus be written as

$$\alpha_{w1} = 360(\dot{m}_\ell)^{0.556} \quad (\text{W/m}^2 \text{ K})$$

whenever

$$1000 < T_w < 1300 \text{ K}$$

$$0.8 < \dot{m}_\ell < 2.5 \text{ kg/m}^2 \text{ s}$$

Liquid sprays are often used in practice to cool moving metallic strips rapidly. In such applications, a very large horizontal

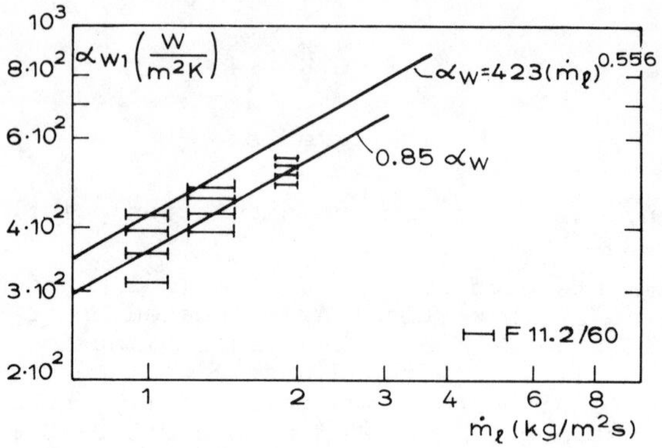

Fig. 62. Variation of heat transfer coefficient with water flow rate density for sprays impinging on the lower face of the plate.

relative velocity appears between the hot surface and the spray, the effect of which is to deflect all the drop trajectories. This has been simulated in our laboratory: Experiments were conducted (Boigelot and Van Hoenacker, 1978), under the same conditions as those described above, with a nozzle spraying the *upper face* of a plate while a fan was inducing a *transverse air velocity* (Fig. 63).

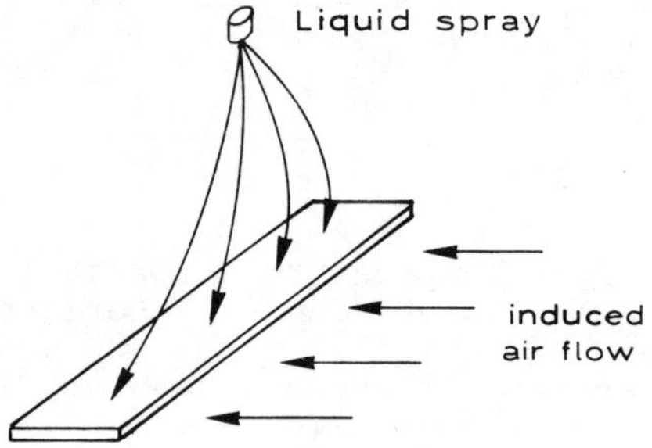

Fig. 63. Configuration of experiments with induced transverse air flow.

Since the transverse air flow influences the local mass flow rate density $\dot{m}_\ell$, new measurements were made. Experimental results are available for nozzle F12/60 spraying at four pressures (1.6 to 5.2 bar) for two different air velocities u_{at} (5.2 and 8.5 m/s). Measured values of $\dot{q}_{wt}$ are shown in Fig. 64 for u_{at} = 5.2 m/s. The slope of the parallel straight lines is equal to 1460 W/m^2 K.

Figure 65 shows, the same way as Fig. 60, the computed values of the heat transfer coefficient α_{wt} versus $\dot{m}_\ell$. Traces correspond to 900, 1000, and 1100 K for u_{at} = 5.2 m/s, 900, 1000, 1100 and 1200 K for u_{at} = 8.5 m/s. In the latter case, values of α_{wt} are very close to one another and traces cannot be distinguished in all cases. The straight line corresponding to results in still air

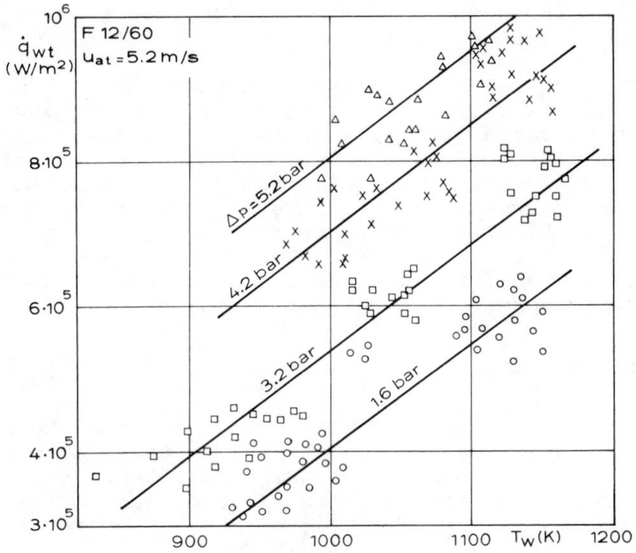

Fig. 64. Heat flux versus temperature for sprays in transverse air flow impinging on the upper face of a horizontal plate.

[α_w from Eq. (66)] is given for comparison.

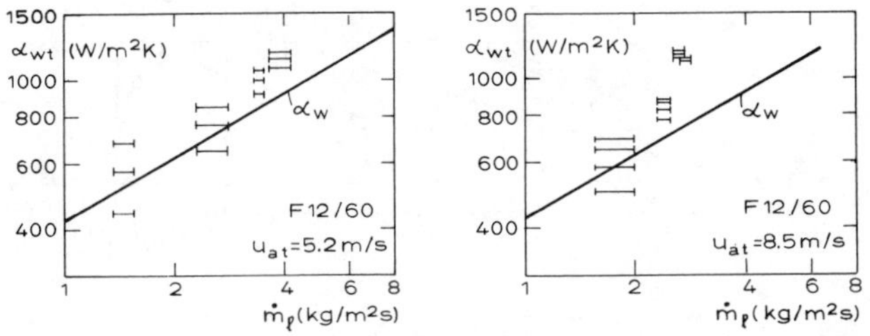

Fig. 65. Variation of heat transfer coefficient with water flow rate density in the case of transverse air flow.

Figure 65 exhibits (1) an increase of heat transfer coefficient due to transverse air flow, and (2) a more pronounced increase of α_{wt} with $\dot{m}_\ell$ than is still air.

5.2.4 Comparison with Other Experimental Results

The above-mentioned results are of the same order of magnitude as those obtained through a stationary experimental procedure, summarized in Fig. 54. In particular, the results of Müller and Jeschar are precisely described and correlated in their publication (1973). A detailed comparison is thus possible. The nozzles used by Müller are of the same type as ours. Their plate is also made of stainless steel, but it is vertical and at a distance of 100 to 200 mm from the nozzle, whereas in our experiments, the plate is horizontal and 357 mm from the nozzle.

We have seen in the preceding section that spraying on the upper or lower face of a plate results in only minor changes for the heat flux. The characteristics of the liquid spray and the effusivity of the plate have a greater influence on heat transfer intensity than the geometric configuration of the experimental device. This face enables us to compare Müller's results with ours, represented in Fig. 57 or by correlation (Eq. 66).

Müller and Jeschar correlate their results by means of the following formula, expressed in watts per square meter:

$$\dot{q}_w = \varepsilon_w \Gamma (T_w^4 - T_e^4) + [10 u_{\ell 0} + (107 + 0.688 u_{\ell 0}) \dot{m}_\ell](T_w - T_{sat}) \quad (67)$$

The first term represents radiative heat exchange between plate and environment. Besides, in the second term for which an uncertainty of ±12% is announced (see Fig. 66), they consider a linear dependance of the heat flux contribution due to the spray on the difference (T_w - T_{sat}); let us note, however, that their experiments were performed with water at 20°C.

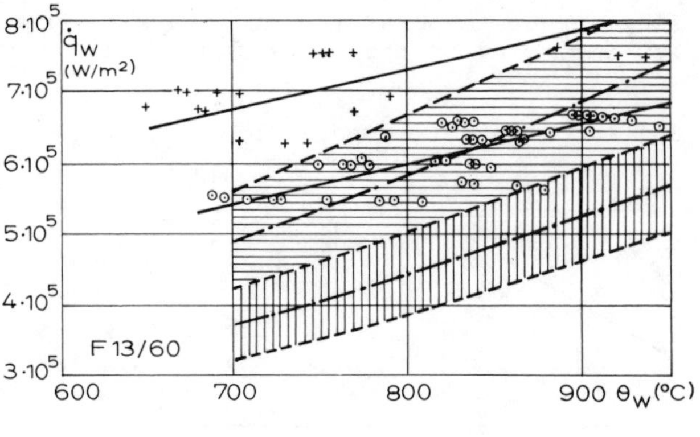

Δp bar	$u_{\ell o}$ m/s	$\dot{m}_\ell$ kg/m²s	our results	Müller & Jeschar
1.2	10.4	3.7....4.0	⊙	(vertical hatching)
2.2	13.3	5.0...5.5	+	(horizontal hatching)

Fig. 66. Comparison between results of Müller and Jeschar (1973) and ours (nozzle F13/60).

For nozzle F13/60, Fig. 66 shows that heat flux values predicted by Müller and Jeschar are, on an average, 20% lower than ours. On the other hand, the increase of heat flux with wall temperature is nearly twice our observed value.

In the case of nozzle F11.2/60 (Fig. 67), the heat flux values

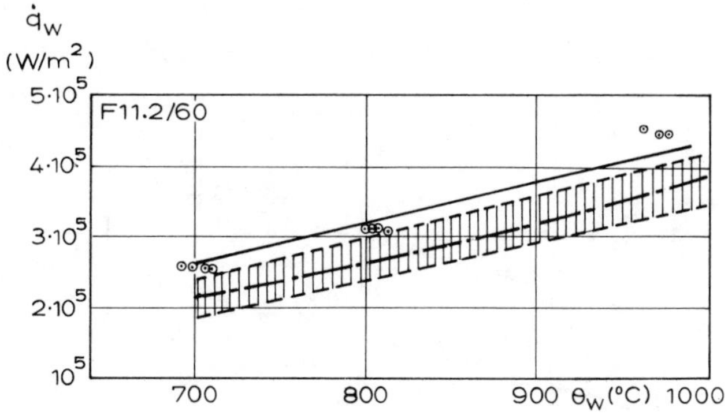

Δp bar	$u_{\ell o}$ m/s	$\dot{m}_\ell$ kg/m²s	our results	Müller & Jeschar
2.0	16.4	0.9...1.1	⊙	⊓⊔⊓⊔⊓⊔

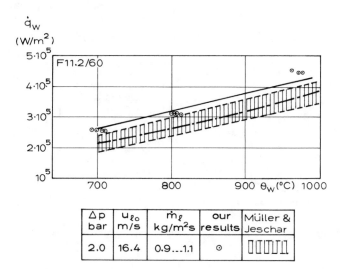

Fig. 67. Comparison between results of Müller and Jeschar (1973) and ours (nozzle F11.2/60).

predicted by Müller are still 20% lower, but the increase with temperature is nearly the same.

Another point of interest is the appreciation of possible effects of the type of experimental procedure upon results. Figure 4 does indeed show that results obtained by means of an unstationary procedure are higher than others: This is probably due mainly to the fact that water mass flow rate densities ($\dot{m}_\ell$) are considerably higher for most nonstationary experiments. A comparison between nonstationary and stationary experimental procedures, applied to the same nozzle in the same spatial configuration, is therefore needed. To this end, we used a method and a sensor designed by CRM (Centre de Recherches Métallurgiques du Bénélux, Liège, Belgium). A 45-mm-diameter nickel cylinder is electrically heated up to the desired temperature. When heating is interrupted, the cylindrical probe is placed in the pulverization at the same distance ℓ_p (357 mm), in a hole bored in a cold plate with dimensions identical to those used in our stationary experiments. Two platinum thermocouples are inserted along the axis of the cylinder. The two recorded temperature profiles are used as inputs in an inverse Fourier problem computation, which yields as output the evolution of heat flux with sprayed face surface temperature.

Figure 68 shows the results obtained by means of this procedure. They agree quite satisfactorily with those obtained under the same conditions by means of a stationary procedure.

5.2.5 Efficiency of Spray Cooling

Many authors (Gaugler, 1966: Pedersen, 1970; Müller and Jeschar, 1973) have introduced the very useful concept of *efficiency of spray cooling*, denoted ε. Quantity ε is related directly to heat flux; it is defined as

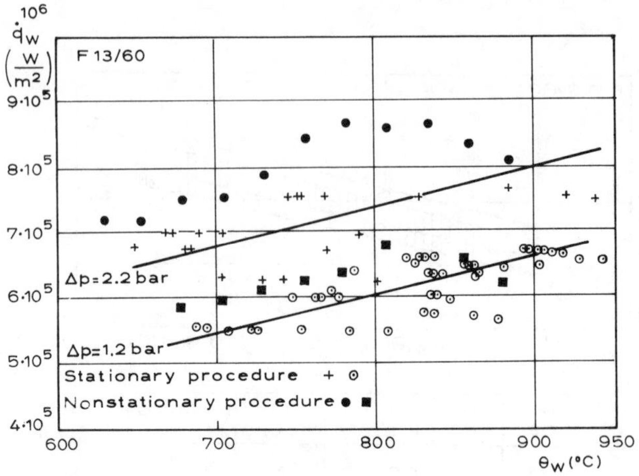

Fig. 68. Comparison of results obtained by means of two different experimental procedures.

$$\varepsilon = \frac{\dot{q}_w}{\dot{m}_\ell[\Delta h_v + c_\ell(T_{sat} - t_\ell)]} \tag{68}$$

The denominator is the thermal power required to heat and vaporize all the water impinging on a unit surface area. Neglecting the superheating of saturated steam and the cooling by entrained air, ε therefore represents the efficiency of use of water in the cooling process, if it is considered that the best use of water is its transformation into saturated vapor.

Figure 69 translates Fig. 56 in terms of efficiency. Points are collected in three ranges of values of $\dot{m}_\ell$ for clarity. It can be seen that efficiency values are rather low for the sprays tested in our experimental work. Another fact of importance is that the lower the mass flow rate density $\dot{m}_\ell$, the higher the efficiency of spray cooling.

In order to have an idea of the order of magnitude of ε, let us suppose that after a given heat transfer process by spray cooling, water was a uniformly saturated liquid. In this case, the efficiency would be

$$\varepsilon = \frac{c_\ell(T_{sat} - T_\ell)}{\Delta h_v + c_\ell(T_{sat} - T_\ell)} = 0.13$$

This value is greater than most of those observed in Fig. 69. Most of our experimental results could thus be explained without emphasizing any prominent effect of change of phase. The reader will see that this is just the main hypothesis of our mathematical model.

The preceding discussion also shows that a current definition of efficiency, that is, the fraction of water that is vaporized during the cooling process, is not adequate.

Another related quantity is often used in practice. It is called the *specific cooling effect of water* η and is defined as the thermal energy extracted from the hot medium by unit mass of water. Clearly, η and ε are proportional:

$$\eta = \varepsilon[\Delta h_v + c_\ell(T_{sat} - T_\ell)] \qquad (69)$$

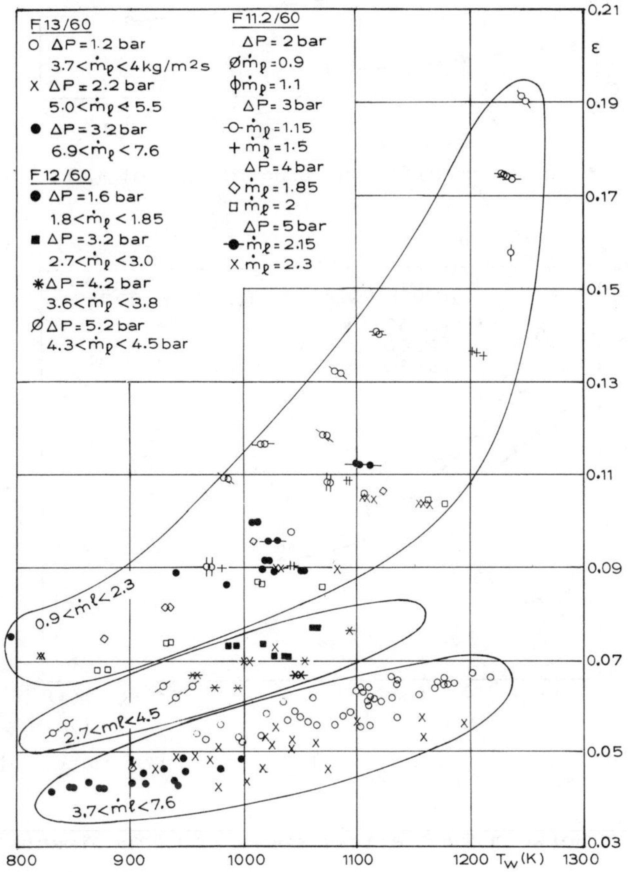

Fig. 69. Variation of efficiency of spray cooling with plate temperature (same conditions as in Fig. 56).

Figure 70 and 71 show, in two different ways, the important trend we have already pointed out: The specific cooling effect of water decreases with water pressure and also with $\dot{m}_\ell$, which is roughly proportional to the square root of Δp, if spreading angle ϕ and distance ℓ_p remain unchanged.

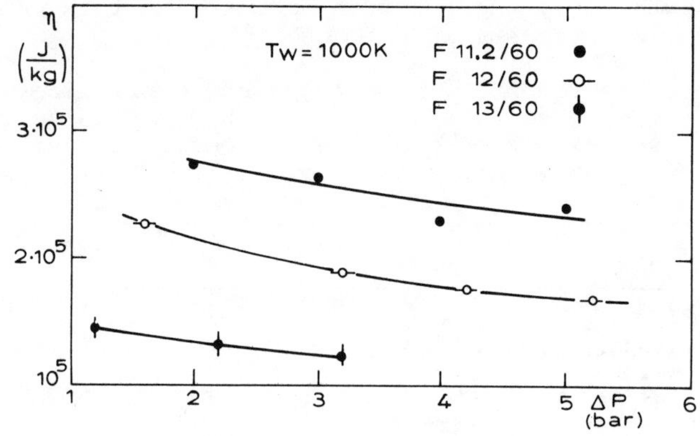

Fig. 70. Variation of specific cooling effect of water with pressure.

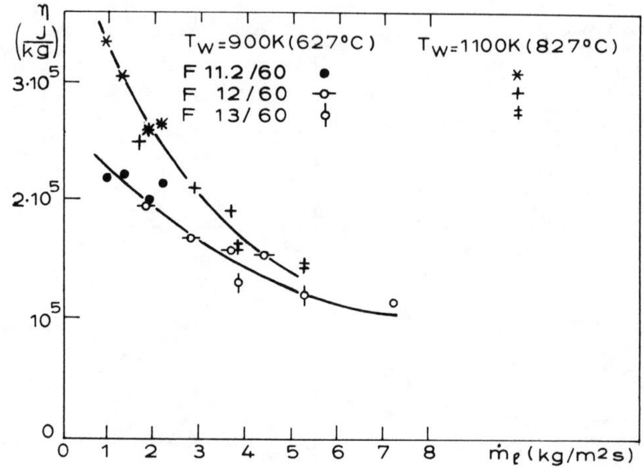

Fig. 71. Variation of specific cooling effect of water with mass flow rate density.

5.3 Comparison between Theoretical Model and Experiments

In Sec. 4.2 we developed a new theoretical model for heat transfer between a hot plate and an impinging drop. The resulting expression (62) of energy transfer for one drop constitutes the basis of the model for heat transfer between a hot plate and a liquid spray, whose final expression is Eq. (63). This relation must still be validated by means of a comparison with experimental results.

Such a comparison is shown in Fig. 72 for results related to nozzles F12/60 and F13/60. Experimental results are summarized by means of the straight lines already drawn in Fig. 57. The curves resulting from Eq. (63) are very nearly straight lines. These theoretical curves are drawn on the basis of distribution curves

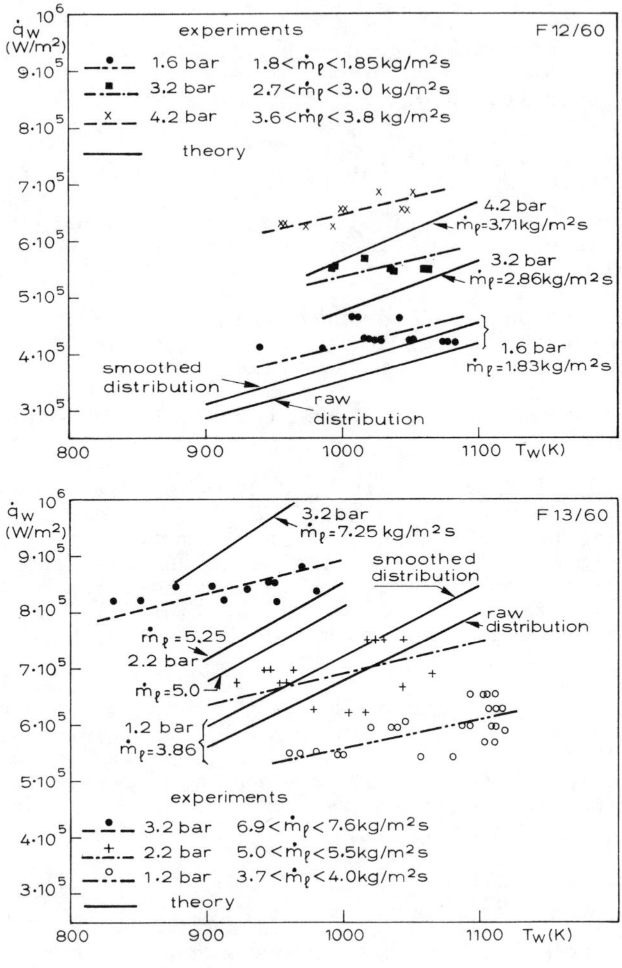

Fig. 72. Comparison between theoretical predictions and experimental results.

of droplet diameter smoothed in the form of log-normal distribution functions and on the basis of mean values of $\dot{m}_\ell$.

In Sec. 4.2, the diameter of the wetted spot of the plate just under an impinging drop was related to the diameter of the dome by the relation

$$d_{DC} = \omega d_D$$

No attempt to quantify ω by means of theoretical arguments has yet been made. We have chosen to evaluate ω by minimizing all the differences between experimental and theoretical curves for the six families of experiments mentioned in Fig. 72. In this way, we found that

$$\omega = 0.82$$

This value has already been included in Eq. (63).

It can be seen from Fig. 72 that:

The increase of $\dot{q}_w$ with T_w predicted by the theoretical model is slightly larger than that resulting from experiments.

The slope of the theoretical curves increases slightly with $\dot{m}_\ell$; this effect is not clear when considering experimental points only.

Theoretical values of $\dot{q}_w$ deduced from Eq. (63) are lower than experimental values for nozzle F12/60 but larger than experimental values for nozzle F13/60.

It is also shown, in Fig. 72, that theoretical results are very sensitive to:

The accuracy of drop velocity and diameter distribution curves. For lower water pressure, theoretical curves are drawn for each nozzle, not only on the basis of smoothed distribution curves but also raw distributions such as those shown in the histograms of Figs. 28 and 34. In one case, the heat flux computed with raw values of d_i and u_i is lower, in the other higher than when computed with smoothed values of diameter and velocity.

The value of liquid mass flow rate density $\dot{m}_\ell$. The nonradiative term of Eq. (63) is related to $\dot{m}_\ell$, along with quantities x_i (see Eq. 18). Values of $\dot{m}_\ell$ for a given set of experimental points lie in a range of $\pm 5\%$ from average value. The same holds for the theoretical curves of $\dot{q}_w$, as shown for nozzle F13/60 at Δp = 2.2 bar.

The mathematical formulation [Eq. (63)] of heat flux clearly indicates that distributions of drop diameter and velocity do have a significant influence. Figure 73 shows, for the two tested nozzles, the distribution of heat fluxes extracted from a plate at 1000 K versus drop class. Curves in Fig. 73 have been drawn on the basis of smoothed distribution curves of d_i and u_i, without taking into account the radiative contribution to heat flux.

In each case represented in Fig. 73, a maximum value of heat flux may be observed in the 300–400 μm zone. Clearly, the contribution of tiny droplets (d_i < 160 mm) to the global transfer of heat is very weak, although these tiny droplets are in greater number.

The asymmetrical shape of the curves giving the contribution of each class of drops to the total heat flux precludes the possibility of replacing real distribution curves of d_i and u_i by mean values of diameter and velocity. For example, if we replace the real spray by droplets of diameter d_{30} only, for which the uniform velocity u^*_{30} is measured on the real distribution curve, the flux density X_{30} of such drops needed to obtain the correct value of $\dot{m}_\ell$ is

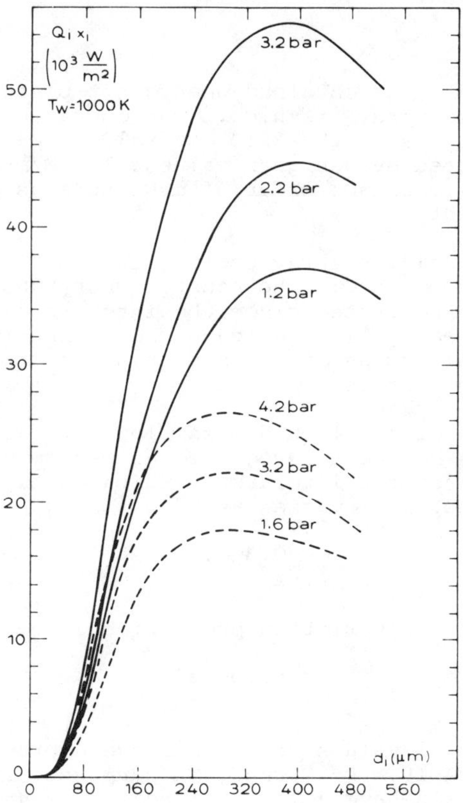

Fig. 73. Heat flux per drop class, without radiation.

$$X_{30} = \frac{6\dot{m}_{\ell}}{\rho_{\ell}\pi d_{30}^{3}}$$

The energy Q_{30} evacuated by a drop with diameter d_{30} is given by

$$Q_{30} = 0.82\sqrt{\lambda_w \rho_w c_w}(T_w - T_{\ell})\frac{(d_{30})^{2.5}}{(u^*_{30})^{0.5}}$$

The heat flux density, without any radiation, is then simply

$$X_{30}Q_{30}$$

When compared with the corresponding heat flux computed on the basis of the real distribution curves, errors as large as 65% to

220% have been found.

6 CONCLUSIONS

Values of heat flux obtained when a hot plate (600 to 1000°C) is cooled by liquid spray are high. Our operating conditions led to heat flux densities in the 2×10^5 to 9×10^5 W/m^2 range. Other authors have reported even higher values: 10^7 W/m^2. These cooling intensities are of the same order of magnitude as those found in quenching experiments.

In order to evaluate spray cooling intensities, two experimental methods can be used: the nonstationary and the stationary method. We chose the latter since the interpretation of experimental results is easier in that case. Our experiments were conducted with fan nozzles of the same type as those of many industrial cooling systems.

The influence of the liquid mass flow rate density $\dot{m}_\ell$ on heat transfer intensity has been recognized as predominant by many authors. Analysis of the literature indicates that heat flux density from the hot wall varies as

$$\dot{m}_\ell^{0.3-0.7}$$

Our experiments give a variation proportional to

$$\dot{m}_\ell^{0.55-0.58}$$

Therefore, when local values of $\dot{m}_\ell$ increase, spray cooling efficiency decreases considerably. For practical purposes, it must be emphasize that the spray should have a $\dot{m}_\ell$ distribution as uniform as possible over the entire cooled surface area.

Most authors consider only mean values of microscopic characteristics of a liquid spray. The new model of heat transfer we present takes into account the distribution curves of drop velocity and diameter. Furthermore, we showed in Sec. 5.3 that the consideration of mean values only can lead to very large errors. The smaller, but very numerous droplets, produced by pressure nozzles, provide only a weak contribution to heat transfer.

Two different basic approaches exist for heat exchange between a drop—or a spray—and a hot plate. Many authors consider that, when a drop reaches the vicinity of a hot plate, a vapor film appears immediately so that heat transfer always occurs—by conduction and radiation—through a vapor layer. Theories of this kind cannot explain the real intensity of heat transfer. Other authors—including ourselves—consider that direct contact between liquid and hot plate can occur to a certain extent, so that during a fraction of the impact duration, heat exchange is quite intense. This fundamental hypothesis enables us to match experimental and theoretical results satisfactorily.

NOMENCLATURE

A	surface area, m^2
c	heat capacity, J/kg K
C_q	discharge coefficient, Eq. (4), (-)
d	diameter, μm
E	thermal efficiency, (-)
F	shape factor, (-)
g	acceleration of gravity, m/s^2
Δh_v	heat of vaporization, J/kg
H	height, m
H	head, J/kg
k	thermal diffusivity, m^2/s
L	distance, m
$\dot{m}$	mass flow rate density, $kg/m^2\ s$
$\dot{M}$	mass flow rate, kg/s
n	number of drops, (-)
N	total number of drops, (-)
p	pressure, (bar)
$\dot{q}$	heat flux density, W/m^2
Q	calorific energy, J
r	coordinate, m
S	thickness, m
t	time, s
T	temperature (K)
u	velocity, m/s
V	volume, m^3
x	number of drops through a unit area per unit time, $m^{-2}\ t^{-1}$
X	total number of drops through a unit area per unit time, $m^{-2}\ t^{-1}$
z	coordinate, m
α	heat transfer coefficient, $W/m^2\ K$
γ	ratio of air and water velocities Eq. (22), (-)
Γ	Stefan-Boltzmann constant for radiation, $W/m^2\ K^4$
ε_w	emissivity, (-)
ε	efficiency, Eq. (68), (-)
η	specific cooling effect, J/kg
λ	thermal conductivity, W/m K
θ	temperature (^0C)
μ	dynamic viscosity, kg/m s
ν	kinematic viscosity, m^2/s

ξ	standard deviation for log-normal distribution law, (-)
ρ	mass per unit volume, kg/m^3
σ	surface tension, m
τ	dimensionless time of disappearance of the dome, (-)
φ	spreading angle of the liquid sheet, (^{0})
ω	constant, (-)

Subscripts

a	air
D	dome
DC	direct contact
e	environment
F	film
g	geometrical mean
i	denotes a class of drops
ℓ	liquid
n	normal
o	nozzle orifice
p	plate
s	spray
sat	at saturation point
T	model of Toda
v	vapor
w	wall

Superscripts

·	quantity per unit time
—	averaged quantity
*	limit value

Dimensionless Groups

Re	Reynolds number, Eq. (1)
We	Weber number, Eq. (1)

REFERENCES

Auman, P. M., D. K. Griffiths, and D. R. Hill 1967, Hot Strip Mill Runout Table Temperature Control *Iron Steel Eng.* vol. 9, pp. 174–181.

Beck, G., and J. C. Chevrier 1971, Comparaison des données de trempe, déterminées à l'aide d'une méthode numérique, à celles du régime permanent *Int. J. Heat Mass Transfer* vol. 14, pp. 1731–1745.

Bieth, M., F. Moreaux, and G. Beck 1976, Caractéristiques du transfert thermique entre un solide à haute température et un jet diphasique *Entropie* vol. 71, pp. 42–48.

Bird, R. B., W. E. Stewart, and E. N. Lightfoot 1960, Transport Phenomena, New York: John Wiley and Sons.

Boigelot, J. B., and L. Van Hoenacker 1978, La transmission de chaleur entre une plaque horizontale et un brouillard d'eau en présence d'un courant d'air transversal. Université Catholique de Louvain.

Bolle, L., and J.C. Moureau 1977, Spray Cooling of Hot Surfaces: A Description of the Dispersed Phase and a Parametric Study of Heat Transfer Results. In *Two-Phase Flows and Heat Transfer*, vol. 3, *Proceedings of NATO Advanced Study Institute*, S. Kakac, F. Mayinger, and T. N. Veziroglu, Eds. Washington, D.C.: Hemisphere pp. 1327–1346.

Bosenberg, K. 1966, Le refroidissement des bandes laminées à chaud sur la table de sortie à rouleaux Thesis, Clausthal Technische Hochschule.

Bowen, I. G., and G. P. Davies 1951, Shell Technical Note, ICT 28.

Bradfield, W. S. 1966, Liquid Solid Contact in Stable Film Boiling, IEC Fundamentals, vol. 5, pp. 200–204.

Celis, M., and H. Noirhomme 1977, Etude expérimentale du transfert de chaleur entre une paroi chaude et une pulvérisation liquide. Université Catholique de Louvain.

Chevrier, J. C., F. Moreaux, and G. Beck 1972, L'effusivité et la résistance thermique des zones superficielles du solide déterminent le processus de vaporisation du liquide en régime de trempe *Int. J. Heat Mass Transfer* vol. 15, pp. 1631–1645.

Corman, J. C. 1966, Water Cooling of a Moving High Temperature Metal Strip Thesis, Carnegie Institute of Technology.

Couvreur, J. 1971, Calculation of the Optimum Geometry of a Spray Cooler for Rolled Wire Rods *C.R.M. Reports* vol. 26, pp. 25–30.

Diamant, S. W. 1960, Etude photomicrographique de la pulvérisation de combustibles liquides, Thesis, Université de Paris.

Dibelius, G., A. Ederhof, and H. Voss 1979, Analysis of Wet Steam Flow in Turbines on the Basis of Measurements with a Light Scattering Probe. In *Two-Phase Momentum, Heat and Mass Transfer in Chemical, Process, and Energy, Engineering Systems*, vol. 2, F. Durst, G. V. Tsiklauri, and N. H. Afgan, Eds. Washington, D.C.: Hemisphere, pp. 619–634.

Dombrowski, N., and G. Munday 1968, Spray Drying *Biochemical and Biological Engineering Science*, Vol. 2 London: Academic Press.

Economopoulos, M. 1968a New Calculation Method of the Heat Transfer Coefficients in Steelmaking Process, *C.N.R.M. Reports* vol. 14 pp 45–58.

Economopoulos, M. 1968b, Study of Cooling on the Runout Tables of Hot Strip Mills, *C.N.R.M. Reports* vol. 17, pp. 33–42.

Farmer, W. M. 1972, Dynamic Particle Size and Number Analysis Using a Laser Doppler Velocimeter *Applied Optics* vol. 11, p. 2603.

Flament, G. 1978, Instabilité de la caléfaction sur un solide à haute température trempé dans l'eau sous-refroidie; étude dynamique et thermique, Thesis, Institut National Polytechnique de Lorraine.

Fortier, A. 1967, *Mécanique des suspensions* Paris: Masson.

Gaugler, R. E. 1966, Experimental Investigation of Spray Cooling of High Temperature Surfaces, Thesis, Carnegie Institute of Technology

Geist, J. M., J. L. York, and G. G. Brown 1951, Electronic Spray Analyzer for Electrically Conducting Particles, *Ind. Eng. Chem.* vol. 43, no. 8, pp. 1371–1377.

Giffen, E., and A. Muraszew 1953, *The Atomization of Liquid Fuels* New York: Wiley.

Goldschmidt, V. W., and M. K. Householder 1969, The Hot Wire Anemometer as an Aerosol Droplet Size Sampler *Atmospheric Environment* vol. 3, pp. 643–651.

Gottfried, B. S., C. J. Lee, and K. J. Bell 1966, The Leidenfrost Phenomenon: Film Boiling of Liquid Droplets on a Flat Plate, *Int. J. Heat Mass Transfer* vol. 9, pp. 1167–1187.

Gottfried, B. S., and K. J. Bell 1966, Film Boiling of Spheroidal Droplets, *I & E.C. Fundamentals* vol. 5, no. 4, pp. 561–568.

Hall, W. B. 1974, The Stability of Leidenfrost drops In *The Tokyo Conference September 1974; Proceedings of the Fifth International Heat Transfer Conference* Washington, D.C.: Hemisphere.

Hoogendoorn, C. J., and R. den Hond 1974, Leidenfrost Temperature and Heat-Transfer Coefficients for Water Sprays Impinging on a Hot Surface In *The Tokyo Conference September 1974; Proceedings of the Fifth International Heat Transfer Conference* Washington, D.C.: Hemisphere.

Joyce, J. R. 1949, *J. Inst. Fuel* vol. 22, p. 150.

Junk, H. 1972, Wärmeübergangsuntersuchungen an einer simulierten Sekundärkühlstrecke für das Stranggiessen von Stahl *Neue Hütte* vol. 1, pp. 13–18

Kawazoe, C., and H. Kumamaru 1973, Some Basic Analysis for the Design of High Speed Tandemized Extrusion Lines for Plastic Insulated Telephone Singles *Sumitomo Elec. Tech. Rev.* vol. 16, pp. 49–55.

Lackmé, G. 1976a, Refroidissement par pulvérisation Note Transferts Thermiques no. 535, C.E.A., Centre d'Etudes Nucléaires de Grenoble.

Lackmé, G. 1976b, Couplage du transfert thermique transitoire et d'une pressurisation inertielle dans l'impact d'un liquide froid sur une plaque chaude, 11–15 octobre Conférence sur les Phénomènes Thermiques et Hydrauliques non Stationnaires, Jouy-en-Josas, France.

Lambert, N., and M. Economopoulos 1970, Measurement of the Heat Transfer Coefficients in Metallurgic Processes *J. Iron Steel Inst.* vol. 10, pp. 917–928.

Lee, S. L., and J. Srinivasan 1978, Measurement of Local Size and Velocity Probability Density Distributions in Two-Phase Flows by Laser-Doppler Technique. *Int. J. Multiphase Flow* vol. 4, pp. 141–155.

Lekic, A., R. Bajramovic, and J. D. Ford 1976, Droplet Size Distribution: An improved Method for Fitting Experimental Data. *Can. J. Chem. Eng.* vol. 54, pp. 399–402.

Longwell, J. P. 1943, Doctoral thesis, Massachusetts Institute of Technology.

McGinnis, F. K., and J. P. Holman 1969, Individual Droplet Heat Transfer Rates for Splattering on Hot Surfaces *Int. J. Heat Mass Transfer* vol. 12, pp. 95–108.

Marshall, W. R. 1954, Atomization and Spray Drying *Chem. Eng. Progress, Monograph series*, vol. 2, no. 50 New-York: American Institute of Chemical Engineers.

Medenblik, H. J. T. 1976, Vergleich von Tropfengrössen-Messverfahren Note AE 01/1-762, Lechler Apparatebau K.G., D-7012 Fellbach, Germany.

Moore, J., D. Hicks, N. Bradley, and I Rowlands 1973 Status of the Steam Generating Heavy Water Reactor *Atom* vol. 195, pp. 7–19.

Moreaux, F., J. C. Chevrier, and G. Beck 1979, Caractéristiques et controle par miniordinateur de refroidissements par pulvérisation In *The Toronto Conference August 1978; Proceedings of the Fifth International Heat Transfer Conference.* Washington, D.C.: Hemisphere.

Morgan, E. R., T. E. Dancy, and M. Korchynski 1965, Improved Steels Through Hot Strip Mill Controlled Cooling *J. Metals* vol. 8, pp. 829–831.

Morgan, E. R., T. E. Dancy, and M. Korchynski 1966, Improving High Strength Low Alloy Steels Through Controlled Cooling *Metal Producing Prog.* vol. 1, pp. 125–130.

Moureau, J. C. 1978, Le refroidissement de parois métalliques très chaudes par pulvérisation d'eau, Thesis, Université Catholoque de Louvain.

Moureau, J. C., L. Bolle, and M. Giot 1979, Influence of the Droplet Size Distribution on the Spray Cooling Efficiency. In *Two-Phase Momentum, Heat and Mass Transfer in Chemical, Process, and Energy Engineering Systems*, vol. 1, F. Durst, G. V. Tsiklauri, and

N. H. Afgan, Eds. Washington, D.C.: Hemisphere, pp. 171–184.

Mugele, R. A., and H. D. Evans 1951, Droplet Size Distribution in Sprays *Ind. Eng. Chem.* vol. 43, no. 6, pp. 1317–1324.

Müller, H., and R. Jeschar 1973, Untersuchung des Wärmeübergangs an einer simulierten Sekundärkühlzone beim Stranggiessverfahrer *Arch. Eisenhüttenwes* vol. 44, no. 8, pp. 589–594.

Nukiyama, S., and I. Tanasawa 1938, *Trans. Soc. Mech. Eng. Japan* vol. 4, no. 14, p. 86.

Oury, C., and C. Xhrouet 1977, Station d'essais de pulvérisation des combustibles liquides, Note de l'Institut Gramme, Liège.

Pasedag, W. F., and J. L. Gallagher 1971, Drop Size Distribution and Spray Effectiveness *Nucl. Technol.* vol. 10.

Pedersen, C. O. 1970, An Experimental Study of the Dynamic Behaviour and Heat Transfer Characteristics of Water Droplets Impinging upon a Heated Surface *Int. J. Heat Mass Transfer* vol. 13, pp. 369–381.

Ranz, W. E., and W. R. Marshall 1952, *Chem. Eng. Prog.* vol. 48, pp. 141–173.

Rayleigh (Lord) 1879, On the Capillary Phenomena of Jets *Proc. Roy. Soc. A* 6801, 29, pp. 71–97.

Riethmuller, M. L. 1977, Laser-Doppler Velocimetry. In *Measurement of Unsteady Fluid Dynamic Phenomena*, B. E. Richards, Ed. Washington, D.C.: Hemisphere.

Rothe, P. H., and J. A. Block 1977, Aerodynamic Behaviour of Liquid Sprays. *Int. J. Multiphase Flow* vol. 3, pp. 263–272.

Savic, P. 1958, The Cooling of a Hot Surface by Drops Boiling in Contact with it Report MT-37, National Research Council of Canada, Ottawa.

Savic, P., and G. T. Boult 1955, The Fluid Flow Associated with the Impact of Liquid Drops with Solid Surfaces Report MT-26, National Research Council of Canada, Ottawa.

Sawan, M. E., and M. W. Carbon 1975, A Review of Spray Cooling and Bottom Flooding Work for LWR Cores *Nucl. Eng. Des.* vol. 32, no. 2, pp. 191–207.

Schlichting, H. 1968, *Boundary-Layer Theory*, 6th ed. New York: McGraw-Hill.

Straus, R. 1949, Thesis, University of London.

Swithenbank, J., J. M. Beer, D. S. Taylor, D. Abbot, and C. C. McGreath 1976, A Laser Diagnostic Technique for the Measurement of Droplet and Particle Size Distribution Department of Chemical Engineering and Fuel Technology, University of Sheffield, Report HIC 245.

Tanasawa, Y., and S. Toyoda 1945, *Trans. Japan Soc. Mech. Eng.* vol. 9, no. 15, p. 52

Toda, S. 1971, A Study of Mist Cooling-Thermal Behaviours of Liquid Films Formed from Mist Drops on a Heated Surface at High Temperatures and High Heat Fluxes *Technol. Rep.* vol. 36, no. 2, Tohoku University, Japan.

Toda, S. 1972, Study of Mist Cooling (1st Report: Investigation of Mist Cooling) *Heat Transfer, Jap. Res.* July—Sept., vol. 1, no. 3, pp. 39—50.

Toda, S., and H. Uchida 1973, Study of Liquid Film Cooling with Evaporation and Boiling *Heat Transfer, Jap. Res.* Jan—Mar. vol. 2, no. 1, pp. 44—62.

Tolfo, F., and P. Staudt 1976, Photomicrographic Analysis of a Spray, *J. Aerosol Sci.* vol. 7, pp. 497—506.

Wachters, L. H. 1965, De warmteoverdracht van een hete wand naar druppels in de sferoidale toestand, Thesis, Technische Hogeschool, Delft.

Wachters, L. H., H. Bonne, and H. J. van Nouhuis 1966, The Heat Transfer from a Hot Horizontal Plate to Sessile Water Drops in the Spheroidal State *Chem. Eng. Sci.* vol. 21, pp. 923—926.

Wachters, L. H., L. Smulders, J. R. Vermeulen, and H. C. Kleiweg, 1966b, The Heat Transfer from a Hot Wall to Impinging Mist Droplets in the Spheroidal State *Chem. Eng. Sci.* vol. 21, pp. 1231—1238.

Wachters, L. H. and N. A. Westerling 1966, The Heat Transfer from a Hot Wall to Impinging Water Drops in the Spheroidal State *Chem. Eng. Sci.* vol. 21, pp. 1047—1056.

Weber, C. 1931, *Z. Angew. Math. Mech.* vol. 11, p. 136

Weiner, B. B. 1979, Particle and Spray Using Laser Diffraction, Soc. of Photo-Optical Instrumentation Engineers, *Optics in Quality Assurance II*, vol. 170, pp. 53—62.

Weinreich, W. 1969, *Klepzig Fachber* vol. 77, pp. 619—622.

Yamanouchi, A. 1968, Effects of Core Spray Cooling *J. Nucl. Sci. Technol.* vol. 5, no. 9, pp. 458—508, and no. 11, pp. 547—558.

Yanagi, K. 1976, Prediction of Strip Temperature for Hot Strip Mills *Trans. I.S.I. Jap.* vol. 16, pp. 11—19.

Chapter 2

The Spherical Droplet In Gaseous Carrier Streams: Review and Synthesis

George Gyarmathy
BBC Brown Boveri & Co., Ltd. Baden, Switzerland

The present chapter is dedicated to the memory of Lang-Shuen Dzung.

1 INTRODUCTION

1.1 Purpose and Scope

Two-phase media consisting of a spatially continuous gaseous phase (the carrier phase) and dispersed liquid droplets are usually termed mists, fogs, aerosols, or sprays. Fog or clouds in the atmosphere, brine above the sea, and the various types of rain are naturally occurring varieties of such two-phase systems.

Technical problems involving droplet dispersions may roughly be classified into two categories: spray-type and mist-type dispersions. Examples of *spray-type* situations include:

Liquid sprays used for the cooling of hot gas streams

Spray columns used in chemical process engineering for mixing or separating vapors and gases

Injection of fuel into combustion air, as in furnaces, in combustion chambers of gas turbines, in the cylinders of diesel engines, and in the carburetor of gasoline engines

Spray-type condensers, e.g., for condensing steam in a steam power plant

The artificial rain created in wet cooling towers in order to dispose of waste heat

---

The author is indebted to J. Smutný for his assistance in carrying out numerical calculations and assembling materials data. Professor G. Dibelius of Aachen (West Germany) is thanked for discussions on dimensional analysis. Numerous valuable comments made by the reviewers of the manuscript, especially by Dr. A. Konorski of Gdansk (Poland), are gratefully acknowledged.

A preliminary version of this chapter was presented orally as an invited review at the Euromech Colloquium 88 "Two-Phase Flow Systems with Condensation Phenomena" (March 30 to April 1, 1977, at Karlsrhue, West Germany) under the title "Growth and Stability of the Liquid Phase in High-Speed Condensing Flow."

In all these cases the liquid is injected into a gaseous stream (or into space filled with gas or vapor) either in the form of individual droplets or in the form of jets and films that eventually break up into drops. The drops are usually large, that is, at least a few micrometers in diameter, but frequently of millimeter size and above. At the point of injection, considerable differences in velocity and temperature of the phases exist in general. In the case of combustible liquids, chemical reactions (burning) may occur in the gaseous phase. In general, spray droplets may be distorted or even shattered into smaller droplets as a result of interaction with the carrier stream. Typically, the spatial distribution of droplets is far from uniform. In some cases droplets may leave the carrier before having a chance to vaporize completely, such as by falling out or by being caught in separating devices.

Examples of the second category, the *mist-type* systems, include:

Atmospheric fogs and clouds

The plume rising from a cooling tower

Two-phase working media used widely in energy conversion, such as wet steam in a steam turbines

Condensing flows in unconventionally operated wind tunnels or in laboratory flow devices such as molecular-beam spectrometers

The fog created in a Wilson cloud chamber for the observation of atomic interaction processes

In contrast to sprays, in these situations there is a great number of droplets dispersed rather uniformly all over the carrier phase, and velocity and temperature differences are moderate or even negligible. The droplets are small, typically of submicrometer size and seldom more than a few micrometers in diameter. As a result, the molecular structure of the carrier may become apparent. Such small droplets are normally formed within the flow itself, rather than being introduced mechanically. (But fine sprays entrained a long way may bear some similarity to mists.) Droplets in mists may be formed by condensation of vapor around tiny, spontaneously generated liquid nuclei, for example, in steam turbines and in wind tunnel flow; or in some cases, as in the atmosphere, foreign particles such as dust or ions play the role of seeding agents around which droplets form. In some applications mist flows are subjected to fast changes of pressure, temperature, and/or flow velocity, an extreme case being the steam turbine. Indeed, two-phase media carrying finely dispersed liquid droplets behave essentially as a gaseous fluid and largely obey the aerodynamic laws governing the flow of compressible media. However, their capability to absorb, carry, and deliver energy is greatly enhanced by the presence of the liquid phase, which can store great amounts of latent heat in little space. The motion of such droplet-laden streams is complicated by internal slip and friction and by sustained deviations from thermodynamic equilibrium during expansion or compression processes.

The key to understanding disperse-liquid systems is the description of heat, mass, and momentum exchange between a single droplet and its gaseous surroundings. This is the topic of this chapter.

As seen from the above variety of practical examples, both the chemical composition and the physical state of the medium may differ greatly. Water droplets have by far the greatest significance for nature and humans; in chemical engineering and combustion, however, other liquids are predominant. The carrier phase may consist of the vapor of the liquid alone (as in steam turbines) or may for its greatest part consist of a noncondensing gas in which small amounts of vapor are diluted (as in all atmospheric situations). In some cases (combustion, spray cooling, etc.), both components have, at least in the neighborhood of the droplets, similar concentrations.

Another widely varying parameter is droplet size. Raindrops are of millimeter size and above. Sprays contain drops of diameters ten or a hundred times smaller. Condensation by nucleation leads to droplets in the micrometer (μm) range or below, but in the very initial stages of their formation these droplets contain only a few dozen molecules, and their size is well below 1 nm. Thus in a comprehensive theory like the one attempted here, one has to consider droplets differring by a factor of 1 million in terms of diameter and by 18 orders of magnitude in terms of mass. Surface effects ("capillarity" effects), which play an important part in the thermodynamics of very small liquid droplets, will have to be accounted for. Also, the state of the carrier phase will show a great diversity. In chemical processes, in heat exchangers, or in the upstream stages of steam turbines, the pressure may be 50 bar or more; in other applications, as in molecular-beam devices or in the condensers of steam turbines, pressures far below atmospheric are common. The carrier is cool in many cases, but high temperatures are by no means rare, an extreme case being combustion, where droplets vaporize in a very hot gaseous environment. The theories concerning droplets are of course applicable, with some simplifications, to spherical solid particles such as occur in pneumatic conveying, in dusty flows, and in rocket exhaust plumes.

It is easy to imagine that the nature of the exchange processes will be quite different under these widely differing conditions. A single, universal, and exact theory describing all situations is certainly far from realization at present. Most existing work is concerned with particular situations in which some or all of the complicating effects may be neglected. For instance, if the carrier is reasonably dense and the droplet is not too small, the gas phase may be assumed to be a continuum. Then the exchange processes may be treated with the usual concepts of heat and mass transfer and of aerodynamics. At the other extreme, molecular kinetics may be used if the droplet is smaller that the mean free path of the gas molecules in the carrier phase. In the domain between the continuum and the free-molecular extremes, however, exact theories are complicated and empirical data are scarce. There is an urgent need for approximate equations covering this range, because droplets grow or shrink into or across this intermediate range during many processes of practical interest.

It is the aim of the present study, first, to review the known (continuum-type and free-molecule type) laws of heat, mass, and momentum exchange between a single drop and its surroundings; second to synthesize them into expressions of more universal (multirange) validity; and third, to derive equations predicting the rate of change of droplet size, temperature, and velocity, and discuss them under rather general conditions.

These latter equations are first-order, ordinary, differential equations and express the rates in terms of the momentary values of the physical parameters describing the droplet (size, temperature, velocity), the carrier phase (pressure, temperature, composition, velocity), and the mass ratio of the two phases. For given initial conditions and a known evolution of the carrier phase, they allow the variation of droplet size, state, and velocity to be calculated by numerical (or in some cases analytic) integration. The integration of the equations for specific situations lies outside the scope of this chapter, but examples will illustrate the steps involved and the simplifications that are possible. Emphasis will be laid on the quasi-steady phenomena that are typical for most sustained growth or vaporization processes. Short-lived transient phenomena, which typically follow sudden environmental changes, will not be analyzed in detail, but their duration and some of their effects will be discussed.

Of course, a number of restrictions will have to be made. The droplet will be assumed to be of spherical shape: this condition is fulfilled in many practical situations, even in sprays. We shall further assume the liquid phase to consist of a single substance only. The carrier phase, however, will be allowed to comprise, besides the vapor of the liquid in question, a second, noncondensible component. Chemical reactions and heat radiation will not be considered, but the applicability to burning droplets will be briefly discussed. Multicomponent non-condensing gases (such as air) mixed with the vapour can be accounted for as a single gas having appropriately averaged properties. The gas-to-vapor dilution ratio in the carrier phase will range from almost infinity (as for "dry" gases, containing no vapor except in the vicinity of the drop) to zero, the latter being the case of pure vapor, for example, steam.

In dealing with substances of various physical natures and composition, it is highly desirable to keep all physical parameters and material properties grouped into nondimensional parameters. Such formulations are in common use in the aerodynamics and heat and mass transfer theory of single-phase, continuum-type fluids. In the present chapter, this dimensional treatment will be systematically extended to two-phase media, including free-molecular transfer processes and nonstationary carrier conditions. A dimensional analysis will lead to the identification of the nondimensional parameter groups required to describe the problem fully. In addition to the well-accepted parameters such as Reynolds, Mach, and Knudsen numbers, Nusselt number, Prandtl and Schmidt numbers, and so on, some new groups will be required to account for the two-phase and the time-rate effects. The application and use of the equations presented for solving various two-phase flow problems will be briefly demonstrated, and critical comparisons will be made with the various growth laws put forward in literature.

1.2 Historical Outline

It is no exaggeration to state that the science of droplets began when Archimedes of Syracuse (died 212 B.C.) established the relationship between the volume and the radius of a sphere. More than 2000 years had to pass, however, before light was shed on the hydrodynamics of flow around droplets. In 1851, Stokes succeeded in calculating the viscous drag of a slow-moving sphere. Soon afterward, the thermodynamic behavior of droplets was analyzed. The theory of capillary equilibrium was elaborated by Sir W. Thomson (later Lord Kelvin) in 1869, J. W. Gibbs in 1878, and R. von. Helmholtz in 1886, leading to a formula for the size of the smallest droplet that is thermodynamically stable in a given environment. This formula was extended to electrically charged drops by J. J. Thomson in 1888. The first treatises bearing on the evaporation of drops were Maxwell's analysis (1877) of diffusion in dense atmospheres and Hertz's experiments (1882) on the evaporation of mercury in vacuum. The latter line of research was later pursued by Langmuir (1913, 1916, 1932) and by Knudsen (1915), who laid down the concepts describing the interaction of single gaseous molecules with liquid or solid surfaces. The drag forces on very small spheres (oil drops) were studied in the famous experiments of Millikan (1911) leading to the determination of the elementary electrical charge.

An important milestone in the development of droplet growth theory was the application of the kinetic theory of gases to the thermal behavior and growth of small droplets (Stodola, 1922). In the 1930s, great advances were made in the theory of heat and mass transfer. The basic analogy between these processes was pointed out by Ernst Schmidt (1929), the pioneer of mass transfer research.

Aerodynamic concepts, like boundary layer and similarity theory, began to be exploited for the prediction of heat and mass transfer to spheres in streaming media, as first done by Frössling (1938). The work done up to the mid-1950s on the evaporation and growth of droplets in hydrodynamic-type (continuum) environments was summed up in a comprehensive, most enlightening treatise by N. A. Fuks (1959).

The application of droplet growth theory to the prediction of the gas dynamic behavior of two-phase flows was pioneered by Oswatitsch (1942), who was the first to combine droplet growth equations with nucleation theory. The awakening of interest toward high-altitude flight in the late 1940s gave new impetus to the application of kinetic theory to the calculation of aerodynamic drag on bodies of various shapes (including the sphere) in low-density atmospheres (Heinemann, 1948; Ashley, 1949), the analysis being soon extended to include heat transfer (Oppenheim, 1953). Buhler and Nagamatsu (1952), in dealing with the condensation of the components of air in hypersonic wind tunnels, extended Stodola's analysis to carrier-gas media, and Schrage (1953) analyzed free-molecular mass transfer with special attention to the interaction of molecules with surfaces. An excellent review of free-molecular flow around spheres was presented by Schaaf and Chambré (1958).

The behavior of droplets of relatively large sizes, as used in chemical engineering, in fuel combustion, and in spray cooling, has been thoroughly analyzed by Rüping (1959) and by Nitsch (1971). An extensive review on the problems of droplet combustion was published by A. Williams in 1973.

Since about 1960, high-speed flows of condensing working media have attracted increased interest with respect to turbines employing steam or other condensible vapors (such as alkali metals) and to hypersonic-flow research facilities. In such flows droplets are produced by spontaneous nucleation of the vapor followed by rapid growth of the initially tiny nuclei. The need for growth laws dealing with droplets of very different sizes have led to expressions covering continuum-type as well as free-molecular carrier media. A review of these developments will be given in Sec. 7. The thermodynamic aspects of wet steam flows have been thoroughly investigated by Konorski (1969, 1976).

A separate line of research, from which droplet growth theory has received important contributions, was that of vacuum physics, especially the branch concerned with the evaporation of liquids in vacuum. Hertz (1882) and Knudsen (1915) were the first to compare the measured evaporation rate of a liquid (mercury) into vacuum with the rate calculated from the kinetic theory of gases. Knudsen postulated that the ratio of these two rates, which he termed "evaporation coefficient," was a statistical but well-defined material property of the substance in question. He realized that experimental values obtained for the evaporation coefficient can be severely falsified by surface contaminations and by temperature measurement errors. Contradictory results ranging between unity and 0.02 obtained by later experimenters have created much confusion in this field. The realization that, for most liquids, the correct evaporation coefficient lies close to unity (Schrage, 1953; Paul, 1962; Mills and Seban, 1967; Hobson, 1973) came too late to prevent the introduction of some incorrect concepts into free-molecular growth laws. More detailed comments on this point will be given in Sec. 3.3.

Similar statistical coefficients, for the so-called thermal accommodation and diffuse reflection, have been introduced to describe the transport of momentum (Maxwell, 1877) and of heat (Knudsen, 1911) by the molecules hitting a surface. Extensive reviews of experimental data have been published by Paul (1962) and by Schaaf and Chambré (1958).

Turning back to the subject of droplet growth in general, it would be desirable to validate existing theories with direct observation of the growth of single drops in a well-controlled environment. The experimental task is a formidable one, especially if the droplets are moving fast, as in high-speed flow. Therefore most existing experimental data are limited either to large drops and spheres (e.g., Frössling, 1938; Ingebo, 1951; Dibelius and Voss, 1979) or to stationary droplets in cloud chambers (Gollub et al., 1974; Chodes et al., 1974).

A second, more feasible experimental method consists of dealing with multitudes of droplets, such as those arising by condensation in supersonic nozzles or in shock-tube experiments. Droplet growth theories can be tested by using them to calculate the evolution of gas dynamic parameters (such as pressure, density, or Mach number) along flow path or time, and comparing the results to measurement (Oswatitsch, 1942; Gyarmathy, 1962, 1976; Wegener and co-workers, 1967, 1969, 1972; Wu, 1972; Barschdorff, 1975). A more direct check of droplet growth rates is possible if the evolution of mean droplet size along the flow has been measured — for example, by optical (light-scattering) methods. Moses and Stein (1978) give a comprehensive review of this subject.

Apart from experimental problems, the reliability of such comparisons is also hampered by the uncertainties in existing nucleation theories. In fact, the influence of competing nucleation rate expressions is of the same order as that of different growth equations. It is to be hoped that further improvements in nucleation theories will benefit growth rate theories, and vice versa.

2 STATEMENT OF PROBLEM, DIMENSIONAL ANALYSIS

2.1 Definition of Sample, Basic Assumptions

In most kinds of mist flows and also in sprays at locations sufficiently far away from the injection point, the distance of droplets from each other and from any walls or obstacles is large enough to render unimportant all direct interaction phenomena, such as impact or coagulation. Therefore these are not considered in the present analysis. However, the indirect interaction of droplets by means of their collective influence on the common carrier is essential for the behavior of most two-phase fluids. This influence is important in all situations where the mass of the carrier cannot be considered to be infinitely large with respect to that of the liquid phase, and will be accounted for in the following.

Fluids carrying a large number of dispersed droplets may be considered as subdivided into cells, each cell containing one droplet and having a shape determined by the momentary pattern of the spatial distribution of the droplets. Such a cell, bounded by a surface $\mathcal{s}$, is shown in Fig. 1a. The cell $\mathcal{s}$ moves along with the droplet, may change its shape and its volume $V_{\mathcal{s}}$, and both energy and matter may flow across its boundary. The total mass contained in each cell shall remain constant in time. *The system considered in the present analysis is the sample contained in one such cell.* The total sample mass $m_{\mathcal{s}}$ consists of the droplet mass m_ℓ and of the mass $(m_v + m_g)$ of the carrier fluid bounded by $\mathcal{s}$, where m_v and m_g are the mass of the vapor and the noncondensible gas components, respectively.

The proper choice of cell size and shape depends on the volume ratio of the phases and on the geometric configuration. In mist flows where the liquid is uniformly distributed in space, cells have more or less regular shapes and their average size follows from the volume ratio of the phases. In sprays, prediction of cell sizes (and shapes) is often much more difficult, and crude assumptions are inevitable.

In this study it will be assumed that the evolution with time

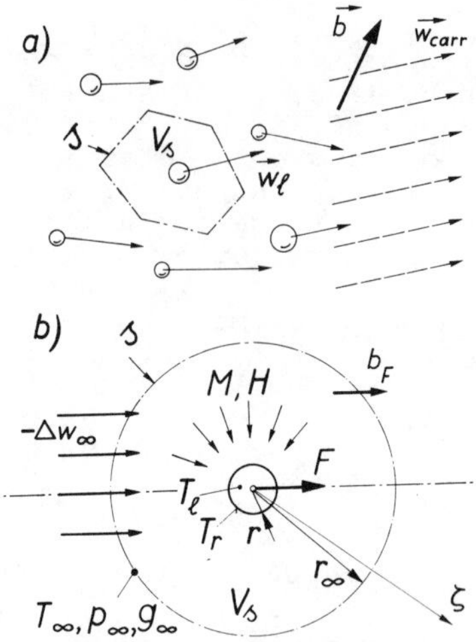

Fig. 1. Definition of the fluid sample considered.
(a) Realistic, (b) Idealized (scale enlarged).

t of pressure, temperature, composition, and velocity of the carrier phase in the far field of the drop, that is, the functions $p_\infty(t)$, $T_\infty(t)$, $x_g(t)$, and $\vec{w}_{carr}(t)$ are known. These functions are of course determined by the external conditions imposed on the fluid as a whole (variation of pressure, flow cross section, body forces, heat transfer, and the like) and on the influence of the liquid phase itself. Their prediction is the task of the fluid dynamics of two-phase media and will not be covered in this study. However, the conservation equations will be formulated for the carrier phase as well as for the droplet. Collision, breakup, coagulation, or formation of droplets will be excluded. The spatial variation of the far-field conditions will be assumed to be smooth, that is, negligible over distances equaling the spacing of droplets. Note that the total mass m_s contained in the sample is constant, even though, in general, the amounts of the gas and the vapor/liquid components may vary in time as a result of convection or diffusion into and out from the cell. The droplet will be spherical. Its radius r, surface temperature T_r, internal mean temperature T_ℓ, and velocity vector $\vec{w}_\ell$ are in general unknown functions of time, with their initial values being specified for some time t_o:

$$\begin{aligned} r(t_o) &= r_o \\ T_r(t_o) &= T_{ro} \\ T_\ell(t_o) &= T_{\ell o} \\ \vec{w}_\ell(t_o) &= \vec{w}_{\ell o} \end{aligned} \tag{1}$$

With ρ_ℓ as the density of the liquid, the mass of the liquid (i.e., the droplet) is

$$m_\ell = \frac{4\pi}{3} \rho_\ell r^3 \tag{2}$$

Introducing x_ℓ, the liquid mass fraction in the sample, as

$$x_\ell \triangleq \frac{m_\ell}{m_s} = \left(\frac{r}{r_o}\right)^3 x_{\ell o} \tag{3}$$

where $x_{\ell o} = x_\ell(t_o)$ means the initial value of x_ℓ, one can express the total mass in the sample as

$$m_s \triangleq m_\ell + m_v + m_g = \frac{m_\ell(t_o)}{x_\ell(t_o)} = \frac{4\pi}{3}\frac{\rho_\ell r_o^3}{x_{\ell o}} \tag{4}$$

The mass fraction of the gas component is

$$x_g \triangleq \frac{m_g}{m_s} \tag{5a}$$

and of the vapor component is

$$x_v \triangleq \frac{m_v}{m_s} = 1 - x_g - x_\ell = 1 - x_g(t) - x_{\ell o}\frac{r(t)^3}{r_o^3} \tag{5b}$$

Assuming that both components behave as thermally perfect gases, the carrier pressure p_∞ is the sum of the partial pressures of the vapor and gas components (Dalton's law):

$$p_\infty = p_{v\infty} + p_{g\infty} \tag{6}$$

Local variations of the overall carrier pressure within the sample are neglected. The partial densities are given as

$$\rho_{v\infty} = \frac{\rho_{v\infty}}{R_v T_\infty} \qquad \rho_{g\infty} = \frac{p_{g\infty}}{R_g T_\infty} \tag{7a,b}$$

with the total mixture density being

$$\rho_\infty = \rho_{v\infty} + \rho_{g\infty} \tag{8}$$

The mass ratio of gas and vapor

$$g_\infty \triangleq \frac{x_g}{x_v} = \frac{\rho_{g\infty}}{\rho_{v\infty}} \tag{9}$$

will be termed the "dilution ratio" of the carrier phase. It is seen from Eq. (5b) that the value of g_∞ is determined by $x_g(t)$ and $x_\ell(t)$ or $r(t)$. By defining the mixture-averaged gas constant $\bar{R}$ as

$$\bar{R} \triangleq \frac{R_v + g_\infty R_g}{1 + g_\infty} \tag{10}$$

the equation of state of the mixture can be written in analogy to

Eqs. (7) as

$$\rho_\infty = \frac{p_\infty}{\bar{R}\ T_\infty} \tag{11}$$

Thus ρ_∞ is seen to be a function of the known variables p_∞, T_∞, x_g and r. Further equations related to the carrier phase and to its averaged properties are given in App. 1.

A fluid domain having a complicated shape, such as the one shown in Fig. 1*a*, is not mathematically tractable within reasonable effort. Therefore we shall assume for all subsequent considerations, as Konorski (1968, 1977) did, that the surface $\mathcal{s}$ is a sphere of radius r_∞ concentric with the drop, see Fig. 1*b*.

Partial pressure and temperature distributions in the vicinity of the drop are illustrated in Fig. 2 over the radial coordinate ζ.

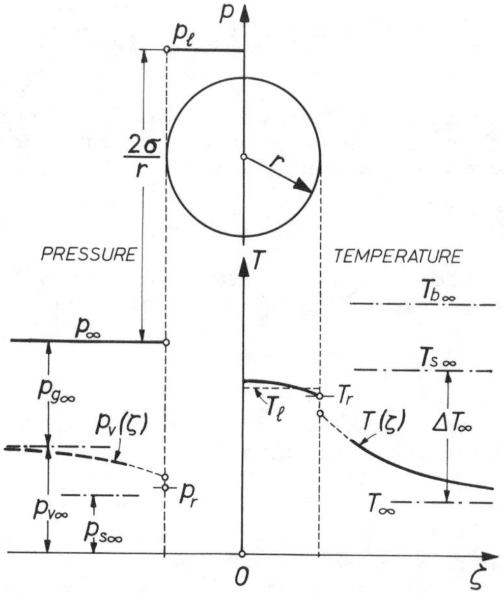

Fig. 2. Pressure and temperature fields in the vicinity of a growing droplet in a mixture carrier.

The variation of $p_v(\zeta)$, $p_g(\zeta) = p_\infty - p_v(\zeta)$ and $T(\zeta)$ as shown is typical for growing drops; in the case of vaporization the slopes would be reversed. The distributions are spherically symmetric in the case of zero relative flow, and cylindrically symmetric if relative flow exists. The far-field values are indicated by dash-and-dot marks. ΔT_∞ is the subcooling of the carrier, as defined by Eq. (180). The overall carrier pressure p_∞ is regarded as uniform, as mentioned above.[1] Within the droplet, due to surface tension, the pressure has a higher value p_ℓ. The temperature T_ℓ means the average within the drop. T_r and p_r are the temperature and vapor pressure of the interface respectively. The discontinuities shown at $\zeta = r$, which cause $p_v(r) \neq p_r$ and $T(r) \neq T_r$, are typical for the so-called slip-flow situation in which the molecular microstructure of the carrier begins to be noticeable. The values of p_v and T are not defined in a zone immediately surrounding the droplet, and $p_v(r)$ and $T(r)$ are extrapolated values. This zone has a thickness of the order of the molecular mean free path in the carrier phase, as discussed later in Secs. 3.3 and 3.4.

The value of r_∞ is defined by the identity

$$\frac{4\pi}{3}\,(r_\infty^3 - r^3)\,\rho_\infty \mathrel{\hat{=}} m_v + m_g \tag{12}$$

from which Eqs. (3) to (5) yield

$$\left(\frac{r_\infty}{r}\right)^3 - 1 = \frac{\rho_\ell}{\rho_\infty}\left(\frac{1}{x_\ell} - 1\right) = \frac{\rho_\ell}{\rho_\infty}\left[\frac{1}{x_{\ell o}}\left(\frac{r_o}{r}\right)^3 - 1\right] \tag{13}$$

As seen, r_∞/r is determined by $\rho_\infty(t)$ and $x_\ell(t)$ or $r(t)$ alone. Figure 3 shows values of r_∞/r as a function of x_ℓ as calculated from Eq. (13), using the values of ρ_ℓ for water, and of ρ_∞ for dry air of T_∞ = 300 K (solid lines), or for saturated steam (dashed lines), at various pressures. For atmospheric pressure (p_∞ = 1 bar), it is seen that $r_\infty/r > 20$ if $x_\ell \approx 0.10$ or less; in such cases the neglect of direct interaction is well justified. It becomes questionable, however, if r_∞/r is 10 or less. This region has been shaded in Fig. 3.

The severity of the influence of the liquid phase on the carrier can be characterized by the change of T_∞ caused by a hypothetical complete evaporation of the droplets, assuming constant values of p_∞ and of the gas content x_g. With $\bar{c}_p$ as the isobaric specific heat of the carrier-phase mixture and L as the latent heat of the liquid per unit mass, we may write for the decrease δT_∞ incurred in T_∞ when the droplet vaporizes completely:

$$m_\ell L = (1-x_\ell)\, m_s \bar{c}_p\, \delta T_\infty \tag{14}$$

Using Eq. (4) and setting $\delta T_\infty / T_\infty < 0.01$ as a rudimentary condition for neglecting the influence of the liquid, one obtains the criterion

$$\frac{x_\ell}{1-x_\ell} < 0.01\,\frac{\bar{c}_p T_\infty}{L} \tag{15}$$

[1]This simplication is permissible, except for fast vaporization in very hot carriers (Kent, 1973).

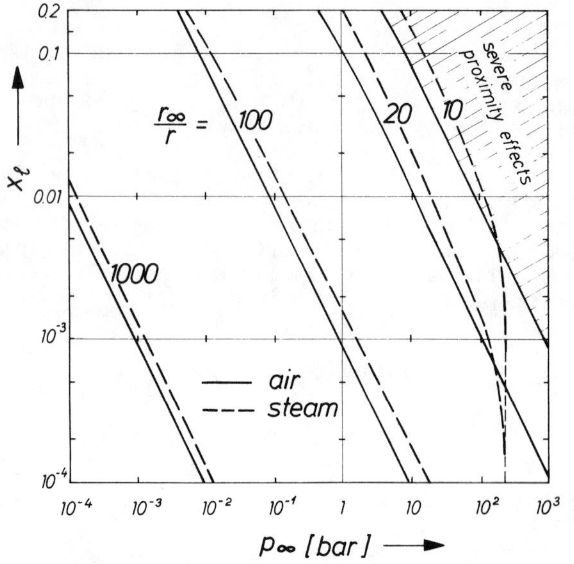

Fig. 3. Relative size of sample domain as a function of carrier pressure and liquid fraction (for water drop). Solid lines: in air, at T_∞ = 300 K. Dashed lines: in saturated steam.

If this criterion is fulfilled, then the carrier-phase conditions are governed solely by the external conditions imposed on the fluid. However, since $\bar{c}_p T_\infty/L$ typically has values of the order of 0.10, the criterion requires x_ℓ to be less than about 10^{-3}, a condition that will be fulfilled only in few cases of practical interest, such as in the early stages of spontaneous condensation or in carriers consisting mainly of a noncondensing gas.

A special class of problems is that of individually burning droplets. Burning drops vaporize due to the heat transfer from the hot environment. Because of the intensity of vaporization, the surrounding air cannot reach the drop surface, and the flame front is kept at some distance from the droplet. Vaporization under such conditions can be treated if r_∞ is chosen to be equal to the flame-front radius and the values of p_∞, T_∞, and x_g are specified on the basis of the combustion process. On the other hand, in the case of "collectively burning" droplets (when the flame front lies outside the cloud of droplets), the situation corresponds to normal vaporization in very hot environments. In this study the radiant transfer of heat to the droplet will not be considered.

In Table 1 the assumptions forming the basis of the present treatment are summarized for sake of later reference.

2.2 Balance Equations

Since the state of the droplet is characterized by four variables, r, T_r, T_ℓ, and $\vec{w}_\ell$, four equations are required to describe its behavior at time t.

Three of these mutually interdependent equations are obtained from

Table 1. *Main Assumptions Regarding the Fluid Sample*

No.	Assumption	Reference to Criteria of Validity and/or to Corrections
	Droplet	
Ia	Spherical shape	Sec. 4.2
Ib	Chemically pure liquid	
Ic	Incompressible liquid	
Id	Droplet interior is a continuum	) Droplet contains at least
Ie	Surface tension is independent of radius	) 100 molecules
If	No internal flow, no rotation	
Ig	Temperature and vapor pressure are uniform over surface	
Ih	Internal temperature field is spherically symmetric and may be sufficiently characterized by one mean temperature value	Sect. 3.6
Ii	There is no electrical charge on the drop	
	Carrier phase	
IIa	Mixture of perfect gases	
IIb	Far-field conditions are spatially uniform but may vary in time	
IIc	Domain of influence is spherical	
IId	Field distributions around the drop are cylindrically symmetric with respect to relative flow direction	
IIe	Constant carrier pressure at the droplet, equaling p_∞	
	General physical situation	
IIIa	No chemical reactions occur within the sample	
IIIb	No thermal radiation	
IIIc	No coagulation or other interference between drops, no formation of new drops	Sec. 2.1

the balances of mass, momentum, and energy, and the fourth equation follows from the balance of the energy fluxes at the drop surface. Because of the presence of mass transfer, the momentum and energy balance equations must account for conductive *and* convective transfer. In App. 2 the three conservation laws are formulated for the entire fluid sample. From these, the balance equations for the droplet and its surface (see below) and for the carrier phase (see App. 2) can be derived. This procedure including the environment of the droplet has the advantage that the definitions of the momentum and energy transfer rates F, H, and H_{int} become apparent. As set out in App. 2, the definitions of these transfer rates are based on the convention that mass transfer means the transition of vapor into liquid, or vice versa, at a temperature equaling that of the droplet surface, T_r, and an overall pressure equaling the carrier pressure p_∞.

The *mass balance* of the droplet is expressed by the differential equation

$$\frac{dm_\ell}{dt} = 4\pi r^2 \rho_\ell \frac{dr}{dt} = M \tag{16}$$

where M is the net rate of vapor mass transfer to the droplet. In the general case of a mixture carrier (vapor and gas), mass transfer involves the superposition of hydrodynamic flow (Stephan flow) of mixture toward the drop and of counterdiffusion within the flowing mixture. In dilute (almost pure gas) carriers the hydrodynamic flow is negligible whereas diffusion is predominant. On the other hand, in pure vapor carriers diffusion has no effect on the mass transfer process because all molecules are of the same substance, and only hydrodynamic flow matters. Because of limitations imposed on droplet growth rate by heat transfer, however, this radial flow is very slow and causes only negligible radial pressure gradients. Therefore, at all carrier compositions, constant pressure p_∞ may be assumed throughout the carrier-phase domain of the sample (assumption IIe in Table 1). Thus M is attributed to diffusion at constant overall pressure.

In order to make this simplified approach formally applicable to nonzero mass transfer in pure vapors, the artifice of an infinitely large diffusion constant has to be introduced (Gyarmathy, 1963). Because of this singular behavior of the mass transfer equation, which will be treated in Sec. 3.1, it is desirable to remove M from Eq. (16). This can be done by using the energy flux balance of the drop surface for expressing M in terms of H and H_{int}. This procedure will lead to the droplet growth laws derived in Sec. 5.

The *momentum balance* (or equation of motion) of the droplet, Eq. (A-30) of App. 2, is written vectorially as

$$m_\ell \frac{d\vec{w}_\ell}{dt} = \vec{F} + m_\ell \vec{b} - \frac{m_\ell}{\rho_\ell} \text{grad } p_\infty \tag{17}$$

where $\vec{w}_\ell$ is the absolute velocity of the droplet, $\vec{F}$ is the drag force, and $\vec{b}$ is the acceleration due to body force fields. The last two terms are usually small.

The relative velocity $\Delta\vec{w}_\infty$ will be defined as that of the droplet with respect to the carrier.[2] The absolute velocity of the carrier far from the drop being $\vec{w}_{\text{carr}}$, we have

[2]This definition is especially convenient in multidroplet flows where droplets move with different absolute velocities in a common carrier. In classical aerodynamics, relative velocity is usually defined the other way, i.e., with respect to the body in question.

$$\Delta\vec{w} \mathrel{\hat{=}} \vec{w}_\ell - \vec{w}_{carr} \qquad\qquad (18)$$
$$\Delta\omega_\infty \mathrel{\hat{=}} |\omega_\ell - \omega_{carr}|$$

The latter equation defines Δw_∞, the scalar of the relative velocity. By symmetry reasons (assumption If and IId), $\vec{F}$ is always parallel to $\Delta\vec{w}_\infty$, although of opposing direction.

The equation of motion will now be formulated in the direction of the momentary drag force $\vec{F}$ (see Fig. 1*b*). This leads to a scalar equation in which $b_F \mathrel{\hat{=}} (\vec{b}\vec{F})/F$ and $w_{carrF} = (\vec{w}_{carr}\vec{F})/F$ are the respective components of $\vec{b}$ and $\vec{w}_{carr}$, and F is the scalar of $\vec{F}$:

$$- m_\ell \frac{d\Delta w_\infty}{dt} = F + m_\ell \left(b_F - \frac{dw_{carrF}}{dt} - \frac{1}{\rho_\ell} \operatorname{grad}_F p_\infty \right) \qquad (19)$$

In the above formulation the values of Δw_∞ and F are inherently positive. The symbol grad_F denotes the gradient in the direction of $\vec{F}$.

Drag is seen to cause a decrease of Δw_∞. The components $\operatorname{grad}_F p_\infty$, b_F, and w_{carrF} are positive if they point in the direction of $\vec{F}$, the drag force acting on the droplet. The magnitude of the last term is negligible with respect to F in many flow situations. Solutions to Eq. (19) for the practically important case of constant values of the last term will be treated in Sec. 3.5.

The *thermal energy balance* of the droplet interior is written as formulated in App. 2, Eq. (A-49), as

$$m_\ell c_\ell \frac{dT_\ell}{dt} = H_{int} - Mc_\ell (T_\ell - T_r) + \frac{m_\ell}{\rho_\ell} \frac{dp_\infty}{dt} \qquad (20)$$

where H_{int} is the heat conduction rate from the droplet surface to the interior. The other terms are usually negligibly small. The second involves the convention, mentioned above, concerning the transfer rates in case of mass transfer.

Defining the *excess* internal average temperature of the droplet as

$$\Delta T_\ell \mathrel{\hat{=}} T_\ell - T_r \qquad (21)$$

the thermal energy balance can also be written as

$$m_\ell c_\ell \frac{d\Delta T_\ell}{dt} = H_{int} - m_\ell c_\ell \frac{dT_r}{dt} - Mc_\ell \Delta T_\ell + \frac{m_\ell}{\rho_\ell} \frac{dp_\infty}{dt} \qquad (22)$$

where dT_r/dt is the rate of change of surface temperature. Usually the last two terms of Eq. (22) are negligibly small.

The fourth equation is given by the *balance of the energy fluxes* at the droplet surface [see Eq. (A-50) of App. 2]:

$$H = - ML_r + H_{\text{int}} \tag{23}$$

Here H is the rate of heat transfer to the droplet and

$$L_r \triangleq L(T_r) + \frac{4\sigma}{3\rho_\ell r} \tag{24}$$

is the latent heat of condensation in case of a curved liquid surface of temperature T_r and radius r. The surface tension σ has been assumed to be independent of droplet radius (assumption Ie). This is permissible except for droplets consisting of very few (<100) molecules, where σ becomes smaller (Tolman, 1949). For a discussion of the variation of σ the reader is referred to Wegener (1969).

The second term in Eq. (24) is usually only a small correction to the latent heat $L(T_r)$ of the bulk liquid and may therefore be neglected; see the discussion of Eq. (A-47) in App. 2.

Since heat transfer is governed by temperature differences, it is evident that all three terms of Eq. (23) are sensitive to the surface temperature T_r. It will be shown in Sec. 5 that for given carrier conditions and for given values of r, Δw_∞, and ΔT_ℓ, Eq. (23) is fulfilled by one and only one value of T_r. Thus the surface temperature is essentially determined by the energy flux balance condition.

Since in most practical situations (except short transients) $H_{\text{int}} \ll H$ is found to hold, Eq. (23) means a very rigid coupling between H and M, in the sense that M is of opposing sign and proportional to H according to

$$M \approx - H/L_r \approx - H/L \tag{25}$$

This means that a droplet is growing if heat can flow *from* its surface to a colder environment ($M > 0$, $H < 0$) and is evaporating if heat flows *toward* its surface from a warmer environment ($M < 0$, $H > 0$).

The driving "forces" for the heat flux H and the vapor mass flux M are the differences of temperature and vapor partial pressure existing between the droplet surface and the carrier phase, that is, the differences $T_r - T_\infty$ and $p_r - p_{v\infty}$ where p_r is the vapor pressure of the droplet surface. As detailed in Sec. 4, p_r is determined by T_r, by the surface tension σ, and by the surface curvature r. For a given droplet p_r is a function of T_r, and increases if T_r increases.

This interdependence of p_r and T_r is of fundamental significance for the thermodynamic behavior of droplets. Let us assume that the conditions are such that $p_r - p_{v\infty} < 0$ holds, which means that the droplet is growing ($M > 0$). This implies (except in some transient situations) that $H < 0$, which in turn means that $T_r - T_\infty > 0$. If now T_r is perturbed by some reason, for example, increased slightly, the heat flux from the drop is increased due to the increase of $T_r - T_\infty$, the mass flux toward the drop is simultaneously decreased because of the rise caused in p_r. With more energy flowing out and less latent heat being liberated, the droplet cools back to the value T_r at which the balance is maintained. Thus it is apparent that Eq. (23) tends to maintain the temperature of the droplet surface at a well-defined value. This surface temperature is established very fast, namely, at the time scale of carrier-molecule collisions with the surface. Therefore T_r may be considered to have at all times the momentary value dictated by Eq. (23) and by the conditions entering the terms of this equation.

In order to solve the conservation equations for the unknown functions $r(t)$, $\vec{w}_\ell(t)$, $T_r(t)$ and $T_\ell(t)$, the transfer rates H, F, M, and H_{int} must be known, and the relationship linking p_r with T_r has to be formulated. The transfer rates will be treated in Sec. 3, p_r specified in Sec. 4, and the methods of solution discussed in Sec. 5. First, however, all factors affecting the behavior of the droplet will be reviewed with the aim of defining the nondimensional parameters of the problem.

2.3 Nondimensional Parameters

2.3.1 Basic Physical Variables

As apparent from the above, the behavior of a droplet is influenced by a great number of parameters including material properties, thermodynamic conditions, geometrical and kinematic data, and so on. In order to reduce the number of such parameters to the minimum required for complete description of the problem, dimensional analysis will be carried through. The resulting nondimensional formulation is formally simpler. It will appear below that properly chosen nondimensional materials data have very similar numerical values for a variety of substances and/or seemingly very different physical conditions. Thus the discussion of results is facilitated. Also it is easier to assess when mathematical simplifications may be made without loss of general validity.

Table 2 lists the physical quantities occurring in the equations given in Sec. 2.1 and 2.2. Together with the equations expressing the transfer rates M, F, H, and H_{int} in terms of the other variables, these equations completely determine the behavior of the droplet. In these latter equations additional material properties will appear (Sec. 3), namely, the dynamic viscosity η, the thermal conductivity λ, the diffusion coefficient D, and the vapor pressures $p_{s\infty}$ and p_r of a plane and curved liquid surface, respectively. In the table the physical quantities are grouped according to their dimension, forming 17 families. The number of independent dimensions is four (length L, mass M, time T, and absolute temperature θ). According to dimensional analysis, the 17 families are interrelated by a minimum of 17-4 = 13 nondimensional groups.

On each line, the quantities belonging to the family in question are classified according to the phase (or component) to which they refer. Quantities characterizing several domains are marked by dotted lines. The coordinates for space and time, ζ and t, are

Table 2. *The Basic Physical and Nondimensional Variables*

	Dimensional Family	Dimension	Physical Quantities, referring to:					Nondimensional Quantities			
			Droplet Interior	Droplet Surface	Carrier Phase Overall	Carrier Phase Vapor	Carrier Phase Gas	Gener.	Variables	Simple Parameters	Groups
A	Density[a]	$L^{-3}M$	ρ_ℓ	—	ρ_∞	$\rho_{v\infty}$	$\rho_{g\infty}$	—	—	g_∞, ρ_∞/ρ_ℓ	—
A	Temperature[a]	θ	T_ℓ	T_r	....	T_∞	....	—	$\boldsymbol{T}_\ell$, $\boldsymbol{T}_r$	T_r/T_∞, Q	—
A	Gas constant, spec. heat[a]	$L^2 T^{-2}\theta^{-1}$	c_ℓ	—	$\bar{R}$, $\bar{c}_p$	R_v, c_{pv}	R_g, c_{pg}	—	—	γ, γ_v, γ_g, $\bar{c}_p/c_\ell$	—
A	Dynamic viscosity[a]	$L^{-1}MT^{-1}$	—	—	η	η_v	η_g	—	—	—	—
A	Thermal conductivity	$LMT^{-3}\theta^{-1}$	λ_ℓ	—	λ	λ_v	λ_g	—	—	λ/λ_ℓ	Pr
A	Diffusion constant	L^2T^{-1}	—	—	D	—	—	—	—	—	Sc
A	Latent heat of vaporization	L^2T^{-2}	—	L	—	—	—	—	—	—	Cl
A	Surface tension	MT^{-2}	—	σ	—	—	—	—	—	—	Eö
A	Pressure	$L^{-1}MT^{-2}$	p_ℓ	p_r	p_∞	$p_{c\infty}$, $p_{v\infty}$	$p_{g\infty}$	—	—	$\Pi_{v\infty}$, $\Pi_{g\infty}$, Π_r, S_∞	Z
B	Mass	M	m_ℓ	—	$(m_v + m_g)$	m_v	m_g	—	—	x_ℓ, x_v, x_g	π
B	Size (radius), space coord.	L	.... r	....	....	r_∞	....	ζ	$\boldsymbol{r}$, $\boldsymbol{r}_\infty$	r_∞/r	Kn
C	Relative velocity	LT^{-1}	...... Δw_∞	....	—	—	—	—	$\boldsymbol{w}$	—	Re
C	Carrier acceler., body force	T^{-2}	—	—	..... dw_{carr}/dt	....	....	b_F	—	—	Fr
C	Time, time constants	T	Δt_{int}	Δt_G, Δt_{vel}	..... Δt_{exp}	....	....	t	τ	—	Sd
D	Mass transfer rate	MT^{-1}	—	M	—	—	—	—	—	α_M	Nu_M
D	Momentum transfer rate	LMT^{-2}	—	F	—	—	—	—	—	α_F	Nu_F
D	Heat transfer rate	L^2MT^{-3}	—	H, H_{int}	—	—	—	—	—	α_H	Nu_H, Bi

[a]Chosen as fundamental families.

common for all domains, and so is the body force field $\vec{b}$, which is represented by its component b_F. On the right hand side of the table, the nondimensionalized quantities are listed. These will be discussed later.

Of the 17 families, a first set A characterizes the thermodynamic and material properties, whereas a second set B is associated with the droplet and fluid sample size. In the simple case of a growing nonmoving droplet in a quiescent carrier of stationary state and zero body force field, sets A and B are sufficient to completely define the physical situation. In more complicated situations one or more families of set C must be included. The velocity family is represented by the relative velocity Δw_∞, because the absolute velocities are not relevant per se.

Of particular interest in our problem is the behavior of the droplet in time: the rate of change of its size, velocity, and internal and surface temperatures. These rates can be conveniently characterized by time constants (Δt_G, Δt_{vel}, and Δt_{int}, respectively), the definitions of which will be given in Secs. 3 and 5. Thus the family "time" is convenient (and sufficient) to express the rates of all physical processes of interest.

Of course, the rates of the resulting processes can optionally be characterized by other means, too. Set D contains the three transfer rates commonly used to characterize phase transition. Although not strictly necessary, these rates will be included in the dimensional analysis. This will enable us to write the expressions of the transfer rates, as given in Sec. 3, in nondimensional form.

2.3.2 Nondimensional Variables

The four variables describing the state of the droplet will be written nondimensionally as

$$\mathbf{r} = r/r_{\text{ref}} \tag{26}$$

$$\mathbf{T}_r = T_r/T_{\text{ref}} \tag{27}$$

$$\mathbf{T}_\ell = T_\ell/T_{\text{ref}} \tag{28}$$

$$\mathbf{w} = \Delta w_\infty/\Delta w_{\text{ref}} \tag{29}$$

where r_{ref}, T_{ref}, and Δw_{ref} mean reference values of drop radius, temperature, and relative velocity.

The spatial coordinates, such as the radial coordinate ζ, can be made dimensionless as ζ/r_{ref}. The above dimensionless variables are listed in the third to last column of Table 2.

2.3.3 Basic Set of Nondimensional Parameters

In order to establish 13 nondimensional groups that can interrelate the 17 dimensional families, it is convenient first to choose four *fundamental dimensions* (i.e., families), and then to use only quantities pertaining to these families to form dimensional groups with each additional dimensional family. The fundamental families will be those of density ($M\,L^{-3}$), temperature (θ), gas constant ($L^2\,T^{-2}\,\theta^{-1}$), and dynamic viscosity ($L^{-1}\,M\,T^{-1}$). These families are shown in the first four lines of Table 2.

Within each family nondimensionalization is achieved by relating the various physical quantities to each other. The most important such nondimensional ratios are listed in the penultimate column of Table 2. In the density family, such ratios are the dilution ratio g_∞ defined by Eq. (9), the ratios defined in Eq. (A-3) (A-4) and ρ_∞/ρ_ℓ. In the temperature family, the thermal effectiveness Q [introduced by Eq. (219)] will later be of significance. The specific heat ratios γ, defined by Eqs. (A-8), as well as $\bar{c}_p/c_\ell$ are nondimensional ratios formed in the gas constant family. The partial pressure ratios $\Pi_{v\infty}$ and $\Pi_{g\infty}$ are defined by Eqs. (A-6), and Π_r expresses the vapor pressure of the droplet surface as

$$\Pi_r \triangleq \frac{p_r}{p_\infty} \tag{30}$$

p_r will be discussed in Sec. 4.1 [cf. Eq. (177)]. The saturation ratio S_∞ is defined by Eq. (179). In the family of mass, the mass fractions x_ℓ, x_v, and x_g were introduced in Sec. 2.1. The nondimensional sample radius $\mathbf{r}_\infty$ will be defined by

$$\mathbf{r}_\infty^3 \triangleq \left(\frac{r_\infty}{r_{\text{ref}}}\right)^3 = \frac{\rho_\ell}{\rho_\infty}\left(\frac{\mathbf{r}_o^3}{x_{\ell o}} - \mathbf{r}^3\right) + \mathbf{r}^3 \tag{31}$$

where the second equality follows from Eq. (13). The ratios α_M, α_F, and α_H listed in the last three families are the "accommodation coefficients" defined in Sec. 3.3.

The four fundamental families and the other 13 families can be connected by the following 13 *nondimensional groups* (cf. the last column of Table 2).

1. *Thermal conductivity* λ is introduced by the *Prandtl number*

$$\text{Pr} \triangleq \frac{\bar{c}_p\,\eta}{\lambda} \tag{32}$$

2. The *diffusion coefficient* D is comprised by the *Schmidt number*

$$\text{Sc} \triangleq \frac{\eta}{\rho_\infty D} \tag{33}$$

No generally accepted named groups seem to exist in the literature that would relate the material properties of the liquid phase to that of the gaseous (or vapor) phase without involving further elements like body forces etc. It is highly desirable to introduce groups expressing L and σ purely in terms of materials and thermal state data, as done below.

3. The *latent heat of vaporization* L could be made dimensionless with $\bar{R}T_\infty$ or $R_v T_\infty$ or $c_{pv} T_\infty$ or any similar product. Such groups occur frequently in two-phase flow literature, seemingly without ever having been named. With a view to the general case of nonperfect vapors we shall nondimensionalize L by using the vapor pressure p_s and setting

$$\mathrm{Cl} \triangleq \frac{L}{p_s\ (v''-v')} = \frac{L}{(p_s/\rho'')(1-\rho''/\rho_\ell)} \tag{34}$$

We refer to this group as the *Clausius-Clapeyron number*[3]. Cl is seen to be a pure material property of the vapor substance, $v'' = 1/\rho''$ and $v' = 1/\rho_\ell$ being the specific volumes of the saturated vapor and liquid, respectively. Obviously, Cl is a function of temperature only. Its values are plotted for a number of substances in Fig. 4.

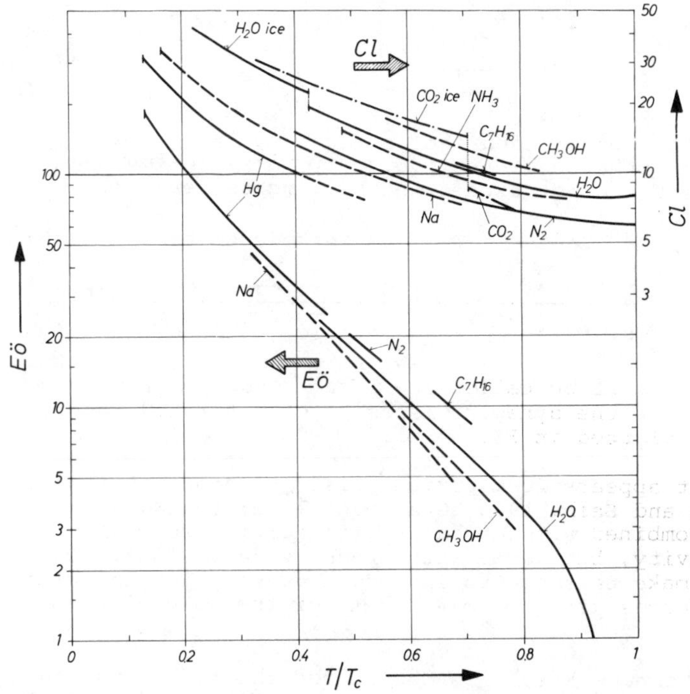

Fig. 4. The nondimensional material property groups Cl and Eö for diverse condensible substances as functions of reduced temperature. (T_c means the critical temperature of the substance.)

[3] See App. 5 for biographical notes on the persons concerned.

It appears that Cl has quite similar numerical values for very different substances. Its gradual decrease with temperature is common to all substances shown.

The reason for the above definition of Cl is the wish to make it be a property of the condensing substance alone and to relate it to the derivative of the vapor pressure function $p_s(T)$. This derivative is given by the Clausius-Clapeyron equation, Eq. (A-13). It appears that Cl is identical to the logarithmic derivative of vapor pressure with temperature:

$$\frac{T}{p_s}\frac{dp_s}{dT} = \frac{d\,\ln\,p_s}{d\,\ln\,T} = \text{Cl} \tag{35}$$

For perfect-gas vapors, $p_s v'' = R_v T$ and $v' \ll v''$, resulting in

$$\text{Cl} \approx \frac{L}{R_v T} \quad \text{(perfect gas vapors)} \tag{36}$$

In calculating droplet growth, Cl may usually be regarded as a constant and evaluated at a reference temperature T_{ref}, provided that T_{ref} is not very different from the droplet surface temperature T_r. Below, the symbol Cl will refer to

$$\text{Cl} = \frac{L(T_{\text{ref}})}{R_v T_{\text{ref}}} \tag{37}$$

where the recommended choice for T_{ref} is $T_{\text{ref}} = T_\infty$. In hot carriers other choices (e.g., $T_{\text{ref}} = T_{b\infty}$) may become necessary, because T_r cannot exceed the boiling temperature $T_{b\infty}$.

4. The *surface tension* σ will be introduced by the new group

$$\text{Eö} = \frac{\sigma}{\eta'' \sqrt{p_s(v''-v')}} = \frac{\sigma}{\eta'' \sqrt{p_s/\rho'')(1-\rho''/\rho_\ell)}} \tag{38}$$

where Eö will be called the *Eötvös number* (pronounced "utvush"). Here η'' is the dynamic viscosity of saturated vapor. Values of Eö are plotted in Fig. 4.

It appears from various lists of nondimensional groups (Weast and Selby 1966, Grassmann 1961) that σ has in the past been combined with mechanical quantities such as the constant of gravity, but never with property data alone. Obviously it would make no sense to introduce gravity into problems where body forces play no role, just for the sake of nondimensionalizing σ.

The name Eötvös is chosen for the following reasons. Eötvös was the first (in 1886) to establish a general law expressing and correlating the surface tension of various liquids.[4] Second, it can be shown that Eötvös' law results in making the function $\text{Eö} = f(T/T_c)$ closely similar for many substances, as evident from the data plotted in Figure 4. Third, in the literature there is a redundancy in the naming of the group $(\rho_\ell - \rho_\infty) b_F r^2/\sigma$, which is variously called the Eötvös number or the Bond number. In order to remove this ambiguity it is suggested that Bond's name be associated with the above group containing the gravity or other body-force field and to use the name Eötvös only for the property data group given by Eq. (38).

Eö is a function of temperature and becomes zero at the critical point. For perfect-gas vapors, $p_s v'' = R_v T$, $\eta'' \approx \eta_v$, and $v' \ll v''$ lead to the approximation

$$\text{Eö} \approx \frac{\sigma}{\eta_v \sqrt{R_v T}} \quad \text{(perfect-gas vapors)} \tag{39}$$

In analyzing droplet growth, Eö may usually be taken to be a constant. In the following, Eö will therefore mean

$$\text{Eö} = \frac{\sigma(T_{\text{ref}})}{\eta_v(T_{\text{ref}}) \sqrt{R_v T_{\text{ref}}}} \tag{40}$$

5. According to the equation of state of the carrier, Eq. (11), *pressure* is related to ρ_∞ and T_∞ by an equation which can be more generally written as

$$\frac{p_\infty}{\rho_\infty \bar{R}\, T_\infty} \mathrel{\hat{=}} Z = 1 \tag{41}$$

It is seen that the pertinent nondimensional group is the *compressibility factor* Z. For perfect gases (assumption IIa), Z has the value of unity. For real gases (especially for vapors at pressures not much lower than critical), Z decreases with increasing pressure, but it seldom becomes less than 0.5. The thermodynamics of real vapors and their classification in terms of the variation of Z with other state parameters has been elaborated by Dzung (1944, 1953, 1955) and Traupel (1952, 1958) and might serve as a starting point for the extension of the present treatment to very high vapor pressures.

6. The *mass* of the droplet m_0 depends on its size, and cannot be expressed by the fundamental families alone. According to Eq. (2),

---

[4] According to Eötvös' law (cf., e.g., Grassmann, 1961), the surface tension of liquids varies with T as $\sigma \approx k_{\text{Eö}} (\rho_\ell/m_{\text{Mol}})^{2/3} (T_c - T)$, where $k_{\text{Eö}} = 2.2 \times 10^{-7}\ \text{m}^2\ \text{kg/s}^2\ \text{K}$ is a constant having a nearly common value for all nonassociating liquids, and m_{Mol} is the molar mass. (Water being of the associating type, its Eö has only half of the above value.) As Einstein (1911) has shown, Eötvös' law is a result of the short-range character of the intermolecular forces in liquids. For biographical notes cf. App. 5.

$$\frac{m_\ell}{\rho_\ell r^3} = \frac{4\pi}{3} = 4.1888 \tag{42}$$

It is seen that nondimensional mass is expressed by a group the value of which is determined by geometric shape. For spheres, $\pi = 3.14159$ (leading to the value given above), and r means simultaneously the radius of surface curvature *and* the measure of size.

7. *Geometric size* or, in general, linear dimension, is represented by the radius r of the droplet. It is related to the fundamental families by the *Knudsen number*,

$$\mathrm{Kn} \triangleq \frac{\eta}{r\ \rho_\infty \sqrt{\bar{R}\ T_\infty}} = \frac{\eta \sqrt{\bar{R}\ T_\infty}}{r\ p_\infty} \tag{43}$$

The second equality follows from Eq. (11). It is seen that the Knudsen number is the reciprocal of nondimensional droplet size. In fact, Kn is often written as $\mathrm{Kn} = \ell/2r$, where

$$\ell \approx \frac{2\ \eta}{\rho_\infty \sqrt{\bar{R}\ T_\infty}} = \frac{2\ \eta \sqrt{\bar{R}\ T_\infty}}{p_\infty} \tag{44}$$

is the mean free path of molecules in the carrier phase far from the droplet.

Since ℓ is unmeasurable, its value is a matter of the theoretical model used. Various authors use various numerical factors in Eq. (44), and their definitions of Kn differ accordingly; cf. App. 1, Eq. (A-12). The *arbitrary* factor 2 in Eq. (44) is an approximation which has the advantage of yielding the formally simplest definition of Kn.

8. *Relative velocity* Δw_∞ is taken into account by the *Reynolds number*,

$$\mathrm{Re} \triangleq \frac{2\ r\ \rho_\infty\ \Delta w_\infty}{\eta} \tag{45}$$

Although Δw_∞ is related here to size rather than to the fundamental families only, the use of Re is preferable to that of the Mach number (cf. Sec. 2.3.3), because its influence on droplet behavior is practically more significant.

9. The *body force field* b_F (e.g., gravity) is usually accounted for by means of the Froude number Fr, defined as

$$\mathrm{Fr} \triangleq \frac{\Delta w_\infty^2}{r b_F} \tag{46}$$

If droplet motion is substantially determined by carrier acceleration rather than by body forces, Fr must be defined by use of dw_{carr}/dt instead of b_F. The influence of Fr on droplet growth is usually small and is neglected in the present analysis (see assumption IVe in Sec. 3.1).

10. *Time* t is made dimensionless by introducing a *nondimensional time* τ, defined as

$$\tau \triangleq \frac{t\ p_\infty}{\eta\ \vartheta} \tag{47}$$

Here ϑ is a nondimensional factor for which a convenient definition will be given in Eq. (233). Setting $\vartheta = 1$ would lead to inconveniently high values of τ in processes of droplet growth, because η/p_∞ is a very short time, typically 10^{-10} s at atmospheric pressure. This time is of the order of Δt_{coll}, the mean time of free flight of a carrier molecule between two collisions with other carrier molecules. For a gas of rigid-sphere molecules, the latter is given (Hirschfelder et al., 1954) by

$$\Delta t_{coll} = \frac{4}{5}\frac{\eta}{p_\infty} \tag{48}$$

If the carrier state is variable, its rate of change is of importance. In literature dealing with high-speed condensing flows, (for example, Gyarmathy and Meyer (1965), Daum et al. (1968), Puzyrewski and Król (1976), Moses and Stein (1978)), the rate of change of environmental conditions is often described by the *expansion rate* $\dot{P}$ of the carrier,

$$\dot{P} \triangleq -\frac{1}{p_\infty}\frac{dp_\infty}{dt} \tag{49}$$

In compression processes $\dot{P}$ has negative values. Equations relating $\dot{P}$ to the rates of change of other parameters will be given at the end of this section. The rate of pressure change can also be characterized in terms of time, for example, by the time Δt_{exp} required for an e-fold change in pressure,

$$\Delta t_{exp} = \frac{1}{|\dot{P}|} = \left| \frac{dt}{d\ \ell n\ p_\infty} \right| \tag{50}$$

To make the process rate nondimensional, Δt_{exp} will be referred to the time η/p_∞, that is, expansion time will be measured in units approximately equaling Δt_{coll}. We propose to call this nondimensional expansion time the *Stodola number*[5]:

$$\mathrm{Sd} \triangleq \frac{\Delta t_{exp}}{\eta/p_\infty} = \frac{p_\infty}{\eta\ |\dot{P}|} \tag{51}$$

thereby honoring Stodola's pioneering work on condensing high-speed flow, on wet steam expansion in turbines, and on the

[5] For biographical notes on Stodola See App. 5.

thermodynamics of very small droplets in steam (Stodola, 1922). The numerical value of Sd indicates the number of collisions suffered by each molecule of the carrier during the time in which an e-fold pressure change occurs. In nucleating (or chemically reacting) flows, Sd is determinant for the character of the phase-transition process and for the magnitude of deviations from equilibrium. The use of Sd to characterize nonequilibrium in *established* two-phase flows will be discussed in Sec. 6.5. Values of Sd will be shown in Fig. 8.

It now remains to nondimensionalize group D in Table 2 that is, the three transfer rates. Since the most common case of droplet growth is associated with laminar relative flow of the carrier fluid, Nusselt-type groups will be used throughout the present discussion. This is in contrast to aerodynamic problems involving predominantly turbulent flows, where Stanton-type groups are more convenient to introduce.

11. Nondimensional *mass transfer rate* M to the drop is expressed by the *Nusselt number for mass transfer*,

$$\mathrm{Nu}_M \triangleq \mathrm{Sh} \triangleq \frac{M}{2\,\pi\,r\,\mathcal{D}(\Pi_{v\infty}-\Pi_r)} \tag{52}$$

where $\mathcal{D}$ is a modified diffusion coefficient defined as

$$\mathcal{D} \triangleq \frac{D\,\rho_\infty\,\bar{R}}{R_v(1-\Pi_{v\infty})} = \frac{\eta\,\bar{R}}{\mathrm{Sc}\,R_v\,\Pi_{g\infty}} \tag{53}$$

Generally Nu_M is termed the *Sherwood number* and abbreviated Sh. The symbol Nu_M is used in the present chapter in order to emphasize the analogy of Sh to the groups Nu_F and Nu_H defined below.

12. The *momentum transfer rate* (or drag force) F will be expressed by the *Nusselt number of drag*,

$$\mathrm{Nu}_F \triangleq \frac{F}{2\,\pi\,r\,\eta\,\Delta w_\infty} \tag{54}$$

Drag being a problem of classical aerodynamics where turbulent flows are of prevalent interest, this laminar-type (viscous) nondimensionalization of F is not used elsewhere. Instead, the "drag coefficient" $c_D \triangleq 2F/\pi\,r^2\,\rho_\infty\,\Delta w_\infty^2$ is commonly used. The relationship between Nu_F and c_D is given by Eq. (76). Nu_F has the advantage of being formally analogous to the other two Nusselt numbers.

13. *The heat transfer rate* H to the drop is nondimensionally represented by the *Nusselt number of heat transfer* or simply Nusselt number:

$$\mathrm{Nu}_H \triangleq \frac{H}{2\,\pi\,r\,\lambda\,(T_\infty^*-T_r)} \tag{55}$$

Here T^*_∞ is a corrected carrier temperature defined in Eq. (71). The nondimensional internal heat flux H_{int} will be expressed by the Biot number, defined in Eq. (66).

2.3.4 Discussion of the Nondimensional Parameters and Their Numerical Values

The following parameters are sufficient for the description of a growing or evaporating droplet in gaseous carrier flow, within the framework of the assumptions listed in Table 1:

Parameters expressing or relating *materials data*

Four "groups": Pr, Sc, Cl, and Eö

Various simple ratios: ρ_∞/ρ_ℓ, R_g/R_v, γ, γ_v, γ_g, c_p/c_ℓ, λ/λ_ℓ, α_M, α_F, α_H

Parameters characterizing *composition* and thermodynamic *state*

Composition: g_∞, or x_v and x_g, or $\Pi_{v\infty}$, or $\Pi_{g\infty}$

Supersaturation: S_∞, or $\Delta T_\infty/T_\infty$ (cf. Sec. 4.1)

Temperatures: $\mathbf{T}_\ell$, $\mathbf{T}_r$, T_∞/T_{ref}, $T_{b\infty}/T_{\text{ref}}$

Parameters expressing *size*

Droplet size: Kn

Sample size: r_∞/r

Two groups specifying *carrier flow* and *rate of carrier change*

Relative velocity: Re

Rate of change of carrier conditions: Sd

In this set there occur seven nondimensional groups, namely, Pr, Sc, Cl, Eö, Kn, Re, and Sd. The groups Z and π have been omitted, because they have fixed values as a result of the assumptions ($\pi = 3.14159$; $Z = 1$). Fr has been omitted because its influence is neglected altogether (assumption IVe in Table 4).

The remaining groups, Nu_M, Nu_F, Nu_H (and Bi) describe the response of the droplet under the conditions thus defined. As mentioned above, however, the droplet response can be alternatively expressed by the time constants Δt_G, Δt_{vel}, and Δt_{int} (cf. Sec. 6.1). These times can be made nondimensional either by referring them to a known time constant of the problem — for example, to Δt_{exp} — or to the time unit ($\eta\, \vartheta/p_\infty$) introduced by Eq. (47). The latter method is more generally applicable, because Δt_{exp} is infinite for constant-pressure carriers.

Numerical values of the parameters are shown in the Figs. 4 through to 8. The values of the Clausius-Clapeyron and Eötvös numbers are shown in Fig. 4 for a few substances as a function of temperature T. Numerical data may be found in Table A-1 (in App. 1). For most liquids (vapors), Cl varies between 5 and 20. Eö is also of the order of 10 at temperatures far below critical, but becomes zero as $T \to T_c$.

The groups Pr and Sc are properties of the carrier mixture only, and strictly speaking depend on the dilution ratio g_∞. For most gases and mixtures their values will be close to unity. For pure air and for pure saturated steam, Pr and Sc are plotted in Fig. 5 over carrier pressure p_∞. Additional data are found in Tables A-1 and A-2, Figure 5 also contains the ratios (carrier/liquid) of density, specific heat, thermal conductivity, and thermal diffusivity, assuming that the liquid is saturated water at pressure p_∞.

Thermal diffusivity is defined as

$$a \triangleq \frac{\lambda}{\bar{c}_p\, \rho_\infty} \quad (56a) \qquad a_\ell \triangleq \frac{\lambda_\ell}{c_\ell\, \rho_\ell} \quad (56b)$$

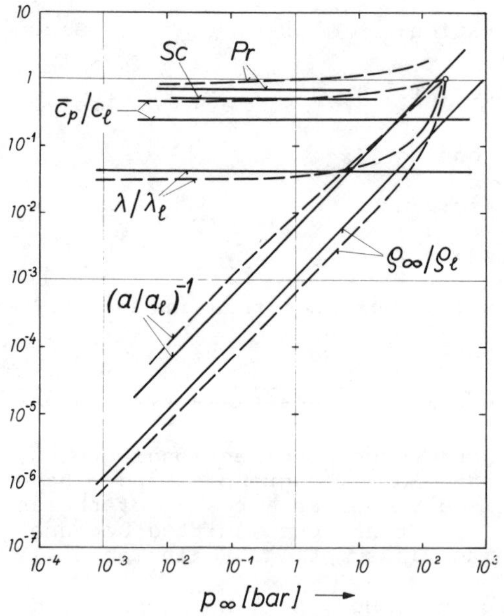

Fig. 5. Nondimensional material properties as functions of carrier pressure (for water drop; Sc refers to the diffusion of water vapor in air). Solid lines: in air of 300 K. Dashed lines: in saturated steam.

The Knudsen number Kn is substantially determined by pressure and droplet size; cf. Eq. (43). Figure 6 shows lines of constant Kn plotted in the p_∞, r plane. It will be seen in Sec. 3 that the value of Kn determines the type of physical behavior of the carrier fluid with respect to the droplet. If Kn $\ll$ 1, the carrier behaves as a continuum, if Kn $\gg$ 1, as a cloud of individual molecules. The reason for this is the following. Kn $\ll$ 1 means that the mean free path in the carrier is much smaller than the droplet size. This means that a molecule hitting and rebounding from the drop surface will, on the average, collide with other carrier molecules in the close vicinity of the droplet. Thus temperature, concentration, and momentum variations within the carrier are continuous. The region Kn $\ll$ 1 is called the *continuum flow* region (upper right in Fig. 6). Here the interaction between the droplet and the carrier is well documented, both theoretically and experimentally. At larger values of Kn there begins to exist a discontinuity between the carrier phase and surface conditions (cf. Fig. 2). This situation, usually referred to in aerodynamics as *slip flow*, is amenable to approximate theoretical analysis. If the discontinuity is large, as typically for $\ell \approx 2r$, the theoretical analysis becomes extremely difficult, and few analyses exist at present (*transition flow*). In the Kn $\to \infty$ limit, however, the situation becomes simple again: All molecules arriving to the surface correspond to far-field conditions, and the interaction can be treated on the basis of the kinetic theory of gases (*free-molecular* range). The limits

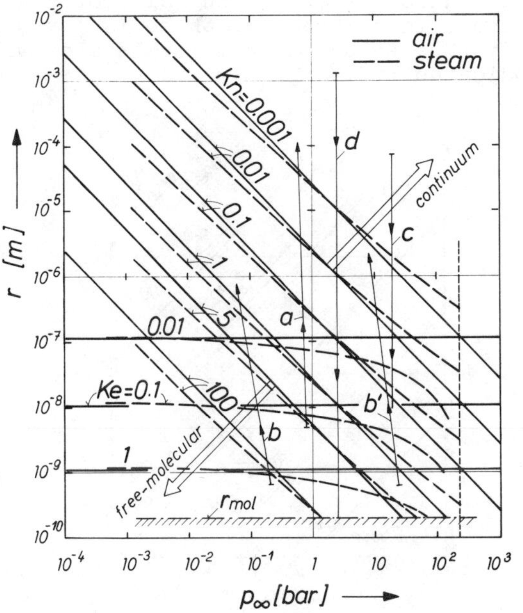

Fig. 6. Plot of the Knudsen and Kelvin numbers of a water drop as functions of carrier pressure and drop size. Solid lines: in dry air at 300 K. Dashed lines: in saturated steam.

of these domains are usually given as follows (Schaaf and Chambré, 1958):

$$\mathrm{Kn} \approx \begin{cases} 0 - 0.01 & \text{continuum flow} \\ 0.01 - 0.2 & \text{slip flow} \\ 0.2 - 5 & \text{transition flow} \\ 5 - \infty & \text{free-molecular flow} \end{cases}$$

Turning back to Fig. 6, it is seen that at atmospheric pressure continuum conditions exist for droplets larger than $r \approx 3.10^{-6}$ m, whereas droplets below $r \approx 10^{-8}$ m are subject to free-molecular conditions. The curves marked Ke will be discussed in Sec. 2.3. The lines a to d indicate typical growth and vaporization processes occurring in practical situations [line a, formation of cloud droplets in the earth's atmosphere; lines b, condensation in low-pressure (b) and high-pressure (b') steam turbines; line c, vaporization of fuel drops in combustion processes; line d, spray droplets used for emergency cooling of water-cooled nuclear power reactors]. The examples a to d show that in many situations droplets grow across wide ranges of the Kn number. This explains the interest in expressions that describe the behavior of droplets in the entire range $0 \leftarrow \mathrm{Kn} \rightarrow \infty$.

The *Reynolds number* Re characterizes the relative velocity Δw_∞. In Fig. 7, lines of constant $\mathrm{Re}/\Delta w_\infty$ are plotted in terms of r and p_∞. Note that for given r and p_∞ the plot yields $\mathrm{Re}/\Delta w_\infty$ from which Re can be calculated for any given value of Δw_∞. For droplets stationary with respect to the carrier phase ($\Delta w_\infty = 0$), Re is zero in the entire field. The shaded boundaries labeled with values of $\mathrm{We}\cdot(1\ \mathrm{m/s})^2/\Delta w_\infty^2$ indicated two

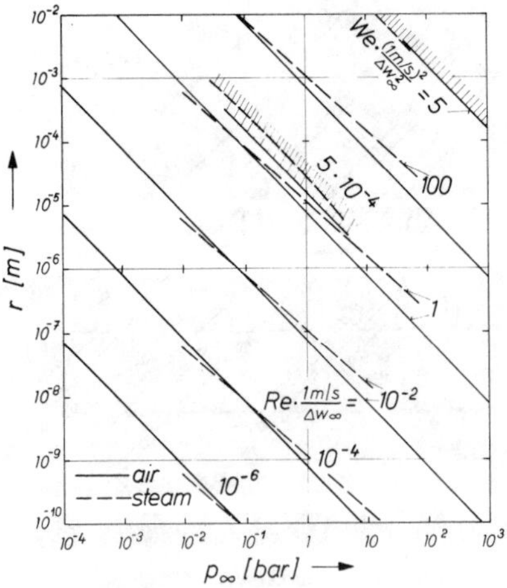

Fig. 7. Plot of the Reynolds and Weber numbers of a water drop as functions of carrier pressure and drop size. (Re and We values are scaled with relative velocity. For $\Delta w_\infty = 0$, Re and We are zero in the entire field.) Solid lines: in dry air at 300 K. Dashed lines: in saturated steam.

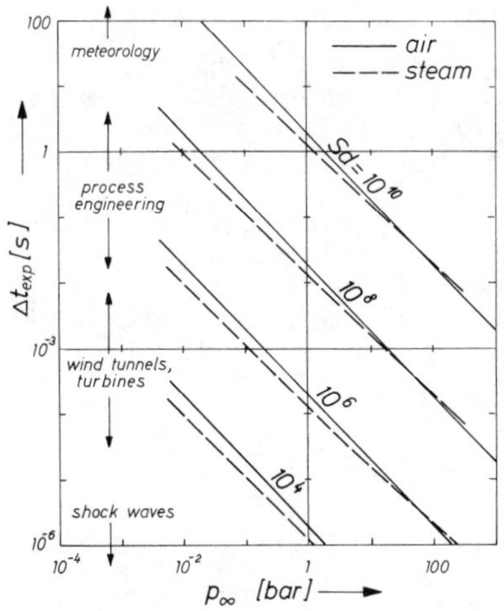

Fig. 8. Plot of the Stodola number for expanding air and steam flows, as a function of carrier pressure and of the time constant $\Delta t_{exp} = 1/\dot{P}$.

examples for the aerodynamic stability limit (We = 5) beyond which no spherical water droplets can exist (cf. Sec. 4.2). The upper limit (with the value 5) holds for Δw_∞ = 1 m/s, the lower (with 5×10^{-4}) for Δw_∞ = 100 m/s. In air and steam the position of the limits is seen to differ only a little.

The *Stodola number* Sd has very high values for most processes of practical interest, except for changes caused by shock waves (where Sd ≈ −10). Figure 8 gives lines of Sd = constant in terms of p_∞ and of the characteristic time Δt_{exp} of the expansion process. In meteorological processes Δt_{exp} is of the order of 100 s or more, and Sd exceeds 10^{12}. In wind tunnels, nozzle and turbine flow Δt_{exp} is of the order of 10^{-4} to 10^{-2} s, and Sd ranges between the values Sd = 10^6 and 10^8 at p_∞ = 1 bar.

2.3.5 Relationships Concerning the Expansion Rate

Some useful relationships relating the *expansion rate* $\dot{P}$ to other time gradients will be given here.[6] In polytropic flow of dry gases, $\dot{P}$ is coupled to the rates of change of T_∞, ρ_∞, and (in rectilinear flow) of $\vec{w}_{carr}$ by the expressions

[6] Strictly speaking, these rates should indicate the change in the environment *of the drop*. If the droplet velocity differs from that of the carrier stream, these rates are different from the ones to which the gas phase is subjected. The difference is usually small, however, and will be neglected.

$$\dot{P} = -\frac{n}{n-1}\frac{1}{T_\infty}\frac{dT_\infty}{dt} = -\frac{n}{\rho_\infty}\frac{d\rho_\infty}{dt} = \frac{|\vec{w}_{carr}|}{\bar{R}\,T_\infty}\frac{d|\vec{w}_{carr}|}{dt} \tag{57}$$

Here n is the polytropic exponent of the flow (e.g., Traupel, 1971). It equals γ in the case of an isentropic expansion. In rectilinear flow $\vec{w}_{carr}$ is usually parallel to $\Delta\vec{w}_\infty$, and therefore its scalar $|\vec{w}_{carr}|$ and its component w_{carrF} are identical.

In the case of two-phase fluids with nonnegligible rates of phase change, the relationship between the rates of T_∞, ρ_∞, and p_∞ is also influenced by the condensation rate $(1/x_\ell)dx_\ell/dt = (3/r)dr/dt$. For frictionless, adiabatic two-phase flow one finds, cf. Eq. (A-51) of App. 2,

$$\frac{1}{T_\infty}\frac{dT_\infty}{dt} = \frac{\gamma-1}{\gamma}\frac{L}{\bar{R}\,T_\infty}\frac{x_\ell}{1-x_\ell}\frac{3}{r}\frac{dr}{dt} - \frac{\gamma-1}{\gamma}\dot{P} \tag{58}$$

$$\frac{1}{\rho_\infty}\frac{d\rho_\infty}{dt} = \frac{\gamma-1}{\gamma}\frac{L}{\bar{R}\,T_\infty}\frac{x_\ell}{1-x_\ell}\frac{3}{r}\frac{dr}{dt} - \frac{1}{\gamma}\dot{P} \tag{59}$$

For fluids comprising a droplet population of nonuniform size, $(3/r)dr/dt$ has to be replaced by the expression

$$\frac{\rho_\ell}{x_\ell}\sum_i 4\,\pi\,r_i^2\frac{dr_i}{dt}$$

which is its mass-weighted average. The summation is to be extended over each individual droplet of the population (index i).

2.3.6 Other Nondimensional Groups

At high relative flow speeds and at high Kn values, the effects of gas compressibility may become important. In such situations the *Mach number* must be taken into account:

$$\mathrm{Ma} \triangleq \frac{\Delta w_\infty}{\sqrt{\gamma\,\bar{R}\,T_\infty}} \tag{60}$$

Here the denominator is the speed of sound in the carrier phase. As seen from Eqs. (43) and (45), Ma follows from Kn and Re as

$$\mathrm{Ma} = \frac{\mathrm{Re\ Kn}}{2\sqrt{\gamma}} \tag{61}$$

As treated in detail in Sec. 4.1, for very small droplets the vapor pressure p_r of the droplet markedly deviates from the corresponding flat-interface vapor pressure $p_s(T_r)$. The criterion for the importance of this so-called *capillarity correction* is the ratio of the group $(2\sigma/\rho_\ell R_v T_\infty)$ to the droplet radius r. In the present study the nondimensional ratio

$$\mathrm{Ke} \triangleq \frac{2\,\sigma(T_{\mathrm{ref}})}{r\,p_{v\infty}}\,\frac{\rho_{v\infty}}{\rho_\ell} = \frac{2\,\sigma(T_{\mathrm{ref}})}{r\,\rho_\ell\,R_v\,T_\infty} \tag{62}$$

will be termed the *Kelvin number*.[7] The second equality follows from Eq. (7a). As indicated, the value of σ at T_{ref} is used in the definition. The deviation between p_r and p_s is significant (more than 1%) if Ke>0.01. The curves of Ke = constant are plotted for water droplets in Fig. 6. It is seen that continuum conditions typically imply negligible capillarity effect, whereas capillarity may become quite significant (Ke$\gg$0.1) in the free-molecular range. Typical values of $\sigma/\rho_\ell\,R_v\,T$ are found in Table A-1 of App. 1.

It should be noted that Ke is related to Kn and to other nondimensional property ratios in the following way:

$$\mathrm{Ke} = 2\,\mathrm{E\ddot{o}}\,\mathrm{Kn}\,\frac{\rho_\infty}{\rho_\ell}\,\frac{\eta_v}{\eta}\,\sqrt{\frac{\bar{R}}{R_v}}\,\sqrt{\frac{T_{\mathrm{ref}}}{T_\infty}} \tag{63}$$

as can be readily verified using Eqs. (62), (43), and (40).

The aerodynamic deformation and breakup of liquid droplets exposed to relative flow is characterized by the ratio of deforming (aerodynamic) forces and restoring (surface tension) forces, which is called the *Weber number*:

$$\mathrm{We} \triangleq \frac{2r\,\rho_\infty\,\Delta w_\infty^2}{\sigma} \tag{64}$$

Droplet deformation and shattering take place beyond We $\approx$ 5 (cf. Sec. 4.2). The position of this limit has been indicated in Fig. 7 for the two relative velocity values Δw_∞ = 1 m/s and 100 m/s. Also, We is related to the parameters given in Sec. 2.3.3, namely,

$$\mathrm{We} = \frac{\mathrm{Re}\;\mathrm{Ma}}{\mathrm{E\ddot{o}}}\,\frac{\eta}{\eta_v}\,\sqrt{\gamma}\,\sqrt{\frac{\bar{R}}{R_v}}\,\sqrt{\frac{T_\infty}{T_{\mathrm{ref}}}} \tag{65}$$

as seen from Eqs. (64), (60), and (40).

Heat conduction within a body, such as the droplet, is commonly described by the *Biot number* Bi, which represents the ratio of the heat flux crossing the boundary of the body to a representative conductive heat flux within its interior. In the case of a droplet one defines

$$\mathrm{Bi} \triangleq \frac{H_{\mathrm{int}}}{2\,\pi\,r\,\lambda_\ell\,(T_r - T_\ell)} \tag{66}$$

[7]For droplets, Ke is a positive number. Negative values of Ke can be used to characterize capillarity effects in bubbles carried in liquids, as evident from the unified formalism introduced by Pattantyús (1977). - For biographical notes on Lord Kelvin see App. 5.

It is seen that Bi is analogous to a Nu_H, except that it refers to the interior of the droplet. The value of Bi depends on the momentary temperature distribution within the droplet (and on internal flow, if the latter is not negligible).

In processes governed by nonsteady heat conduction within bodies (Carslaw and Jaeger, 1959), time is usually nondimensionally expressed by the *Fourier number*,

$$\mathrm{Fo} = \frac{(t-t_o)\, a_\ell}{r^2} \tag{67}$$

in which $(t-t_o)$ is the time elapsed since the beginning of the process. Since heat conduction in the interior of the droplet normally plays a negligible role in droplet growth or vaporization processes, preference will be given to the nondimensional time τ defined by Eq. (47).

2.3.7 Review and Comparison with Literature

In order to facilitate the overview of the numerous nondimensional groups introduced above or occurring in other pertinent literature, a list is given in Table 3. The parameter groups are listed in accordance with increasing complexity of the physical situation described. The four group names marked with an asterisk (*) are the new suggestions made in the present chapter. Their acceptance would formally simplify the nondimensional description of many two-phase phenomena. For each group the table contains the usual symbol found in literature, the symbol (if any) used in the present chapter, and the definition of the group. The last column indicates whether a parameter represents a redundancy, that is, follows from other parameters preceding it in the table.

Some of the groups listed are not used in the present chapter at all. This may be due to simplifications that eliminate certain parameters or to the wish to avoid redundancies. For example, the Laplace number, which characterizes fluid motion within the droplet, is eliminated because of assumption If. The impaction parameter (called K, and in Russian literature sometimes referred to as the Stokes criterion, cf. Golovin and Putnam, 1962) is not used because droplet motion in nonuniform flow fields is not treated in this study. The superheat parameter (in the literature, B) is not expressly used here. The groups Le, Eu, Pe, Bo, and the Stanton numbers are omitted as a result of preference for other numbers from which they follow redundantly. The criterion for free convection, the Grashof number, is eliminated by assumption IVb in Table 4.

It is perhaps useful to restate the physical meaning of these nondimensional parameters (Grassmann, 1961); Weast and Selby, 1966; Zierep, 1972):

Pr is the ratio of momentum diffusivity (i.e., kinematic viscosity η/ρ_∞) to thermal diffusivity $\lambda/\bar{c}_p\,\rho_\infty$ within the carrier.

Table 3. *Review of nondimensional Groups Related to Spherical Particles in Perfect Gas Carriers*

Physical Complexity Involved	Group Name (*suggested)	Symbol: Usual	Symbol: This Chapter	Definition	Interrelationships
Property data of carrier	Prandtl number	Pr	Pr	$\eta \bar{c}_p / \lambda$	—
	Schmidt number	Sc	Sc	$\eta / \rho_\infty D$	—
	Lewis number	Le	—	$a/D = \bar{c}_p \rho_\infty / D$	= Sc/Pr
Property data of condens. substance	*Clausius-Clapeyron number	—	Cl	$L/p_s(v''-v') \approx L/R_v T_{ref}$	—
	*Eötvös number	—	Eö	$\sigma/\eta'' \sqrt{p_s(v''-v')} \approx \sigma/\eta_v \sqrt{R_v T_{ref}}$	—
Carrier state and its rate of change	Superheat parameter	B	—	$c_{pv}(T_\infty - T_{s\infty})/L$	$= -\gamma_v \Delta T_\infty/(\gamma_v - 1) T_s$ Cl
	*Stodola number	—	Sd	$p_\infty^2/(\eta \lvert dp_\infty/dt \rvert) \approx \Delta t_{exp}/\Delta t_{coll}$	—
Droplet size	Knudsen number	Kn	Kn[a]	$\eta \sqrt{\bar{R} T_\infty}/r\, p_\infty$	— a) Note Eq. (A-12)
	*Kelvin number	—	Ke	$2\sigma/r\, \rho_\ell R_v T_\infty$	= Kn Eö $(2 \rho_\infty \eta_v/\rho_\ell \eta) \sqrt{\bar{R} T_{ref}/R_v T_\infty}$
	Laplace number	La	—	$2r\sigma \rho_\ell/\eta_\ell^2$	—
Relative flow	Mach number	M, Ma	Ma	$\Delta w_\infty/\sqrt{\gamma \bar{R} T_\infty}$	—
	Euler number	Eu	—	$p_\infty/\rho_\infty \Delta w_\infty^2$	$= 1/\gamma\, Ma^2$
Relative flow and droplet size	Reynolds number	Re	Re	$2r \rho_\infty \Delta w_\infty/\eta$	$= 2\sqrt{\gamma}$ Ma/Kn
	Weber number	We	We	$2r \rho_\infty \Delta w_\infty^2/\sigma$	= (Re Ma/Eö)$(\eta/\eta_v) \sqrt{\gamma \bar{R} T_\infty/R_v T_{ref}}$
	Péclet number	Pe	—	$2r \rho_\infty \Delta w_\infty \bar{c}_p/\lambda$	= Re Pr
Diverse	Froude number	Fr	Fr	$\Delta w_\infty^2/r b_F$	—
	Bond number	Bo (Eö)	—	$r^2 b_F(\rho_\ell - \rho_\infty)/\sigma \approx r^2 b_F \rho_\ell/\sigma$	= (We/Fr)$(\rho_\ell - \rho_\infty)/2 \rho_\infty$
	Grashof number	Gr	—	$(8r^3 b_F \rho_\infty/\eta^2) \lvert \rho_\infty - \rho_r \rvert$	— Note: ρ_r = carr. dens. at surf.
	Impaction parameter	K	—	$2r^2 \rho_\ell \Delta w_\infty/9 \eta r_T$	— Note: r_T = target radius
	Nondimensional time	—	τ	$t p_\infty/\eta$	—
	Fourier number	Fo	Fo	$(t-t_0) a_\ell/r^2$	—
Transfer process	Nusselt number (heat transfer)	Nu	Nu_H	$H/2 \pi r \lambda (T_\infty^* - T_r)$	—
	Sherwood number	Sh	Nu_M	$M/2 \pi r \mathcal{D}(\Pi_{v\infty} - \Pi_r)$	—
	Nu for momentum transfer	—	Nu_F	$F/2 \pi r \eta \Delta w_\infty$	—
	Stanton number (heat transfer)	St	—	$H/4 \pi r^2 (T_\infty^* - T_r) \Delta w_\infty \rho_\infty \bar{c}_p$	$= Nu_H$/Re Pr
	St for mass transfer	St'	—	$M R_v p_{g\infty}/4 \pi r^2 (p_{v\infty} - p_r) \Delta w_\infty \rho_\infty \bar{R}$	= Sh/Re Sc
	Drag coeff. or Newton number	c_D, Ne	c_D	$F/\pi r^2 \rho_\infty (\Delta w_\infty^2/2)$	$= 8\, Nu_F$/Re
	Biot number	Bi	Bi	$H_{int}/2 \pi r \lambda_\ell (T_r - T_\ell)$	—

[a] Note Eq. (A-12) in App. 1.

Sc is the ratio of momentum diffusivity to mass diffusivity (or diffusion coefficient) D.

Le is the ratio of thermal to mass diffusivity.

Cl expresses the enthalpy of vaporization in terms of the molecular kinetic energy of the carrier.

Eö relates the surface tension forces to the frictional forces created by vapor flowing past the liquid surface at some velocity proportional to $\sqrt{R_v T_{ref}}$.

B relates convected sensible heat to convected latent heat.

Sd relates the time constant of the expansion or compression process, Δt_{exp}, to the collision time interval Δt_{coll} of carrier molecules; if Sd $\gg$ 1, gas kinetic (Maxwellian) equilibrium exists within the carrier itself.[8]

Kn relates molecular mean path to droplet diameter.

Ke compares the diameter of the droplet to the diameter of the critical (smallest thermodynamically stable) droplet under the given carrier conditions.

La relates surface tension forces to forces induced by liquid flow within the drop.

Ma relates relative speed to sonic speed in the carrier.

Eu relates carrier pressure to the aerodynamic pressure variations created by relative flow.

Re is the ratio of inertial and viscous forces (i.e., of convective and conductive momentum transfer) within the carrier.

We relates aerodynamic pressure variations to the overpressure created by surface tension.

Pe is the ratio of convective and conductive heat transfer within the carrier.

Fr relates forces due to aerodynamic pressure variations to forces caused by (gravitational or other) fields.

Bo compares the forces due to fields to those caused by surface tension.

Gr expresses the importance of free convection by relating buoyancy forces caused by density differences to viscous and inertial forces.

---

[8]Note that Sd alone is not decisive for the existence of equilibrium between *phases*, which additionally depends on other factors like droplet size and liquid mass fraction (Sec. 6.5).

K is the ratio of particle stopping distance to target size.

Fo is time expressed in terms of the thermal time lag of the droplet interior.

Nu_H (or Nu) is heat rate measured in terms of conductive flux.

Nu_M (or Sh) is vapor mass flux rate measured in terms of diffusional flux.

Nu_F is momentum flux rate measured in terms of viscous momentum flux.

St, St', and c_D are analogous to Nu_H, Nu_M, and Nu_F, except that convective fluxes within the carrier are used as a reference basis.

Bi is heat flux rate within the drop compared to conductive flux.

Preference for Nu_H, Nu_M, and Nu_F instead of St, St', and c_D is given in this chapter because the behavior of small droplets is predominantly governed by conductive/diffusional/viscous processes rather than by convective transfer.

3 TRANSFER OF HEAT, MASS, AND MOMENTUM TO A SPHERE

3.1 Assumptions and Introductory Remarks

The transfer of heat, mass, and momentum to a sphere under steady conditions (with or without relative flow) is a subject that has been treated extensively in aerodynamics and in heat/mass transfer theory, both under continuum and free-molecular conditions. In two-phase flows, however, the conditions vary in time as a result of two factors: First, the conditions are changing both in the far field and at the droplet surface; second, and less important, the geometric position of the interface is changing due to the change of droplet radius. Little work has been done on this general problem.

We shall therefore follow the approach usually adopted in the literature and shall treat droplet behavior on the basis of the equations giving the heat, mass, and momentum transfer rates under *steady-state conditions* to a sphere of constant radius. In doing so, the assumptions IVa to IVe listed in Table 4 will be made in addition to those already given in Table 1. The limits within which these assumptions are permissible will be discussed in Sec. 3.5. The three transfer processes will be considered to be uninfluenced by each other, even though they may take place simultaneously. This implies, first, the assumptions Va to Vc, see Table 4. These are justified by the fact that the interference of the heat and mass transfer processes with each other, and of both with the momentum transfer, is usually weak. Exceptions and correction formulas will be discussed in Secs. 3.2 and 3.3. Second, it will be assumed here in Sec. 3 that heat and mass transfer are *decoupled* (assumptions VIa, VIb).[9] However, their coupling will be taken into account when

[9] The independence of p_r from T_r is realized, for instance, in the case of nonvolatile solid particles (where p_r=0) or of porous spheres, where p_r can be adjusted by suction or blowing to any desired value.

Table 4. *Assumptions Regarding Heat, Mass, and Momentum Transfer*

No.	Assumption	Remarks
IVa	Droplet radius is constant in time	cf. Ia in Table 1
IVb	Surface temperature and vapor pressure are constant in time	cf. Ig in Table 1
IVc	The far-field conditions are constant in time	cf. IIb in Table 1
IVd	The droplet is in an infinite environment ($r_\infty \to \infty$)	cf. IIc in Table 1
IVe	The transfer process are not influenced by body force fields	
Va	Heat transfer is determined by the temperature and velocity fields	For corrections cf. Secs. 3.2 and 3.3
Vb	Mass transfer is determined by the concentration (partial pressure) and velocity fields	For corrections cf. Secs. 3.2 and 3.3
Vc	Momentum transfer is determined by the velocity field alone	For corrections cf. Secs. 3.2 and 3.3
VIa	Temperature and vapor pressure of the sphere surface are independent of each other	Not used in Sec. 5
VIb	The rates of heat and mass transfer are independent of each other	Not used in Sec. 5

droplet growth is analyzed in Sec. 5. The environment of the droplet will be assumed to be *infinite*, except for some comments made in Sec. 3.5. First, the case of continuum-type and of free-molecular carriers will be treated in Secs. 3.2 and 3.3, respectively. In Sec. 3.4 these results will be synthetized into expressions covering the entire range of Knudsen numbers. In Sec. 3.5 the assumption of quasi-steadiness will be examined. Section 3.6 deals with the heat conduction in the interior of the droplet, whereas Sec. 3.7 is devoted to the relative velocity in accelerated flow systems.

As pointed out in Sec. 2.3.3, it is appropriate to express the transfer rates by the Nusselt numbers defined in Eqs. (55), (52), and (54). The rates of *heat, mass, and momentum transfer* toward a sphere will therefore be set equal to

$$H = 2\pi r \lambda (T^*_\infty - T_r) \, \mathrm{Nu}_H = 2\pi r \lambda T_\infty \mathrm{Nu}_H \left(k_T - \frac{T_r}{T_\infty}\right) \quad (68)$$

$$M = 2\pi r \mathcal{D} (\Pi_{v\infty} - \Pi_r) \, \mathrm{Nu}_M = 2\pi r \mathcal{D} \Pi_{v\infty} \mathrm{Nu}_M \left(1 - \frac{\Pi_r}{\Pi_{v\infty}}\right) \quad (69)$$

$$F = 2\pi r \eta \Delta w_\infty \mathrm{Nu}_F \quad (70)$$

The expressions on the far right will facilitate discussions in Sec. 3.2.5. λ, $\mathcal{D}$, and η are known properties of the carrier phase (mixture or pure vapor). For a given mixture of perfect gases they are a function of temperature alone, and the values used are usually[10] those at T_∞. T_r and Π_r characterize the sphere's surface. T^*_∞ is the *adiabatic wall temperature* of the sphere and follows from T_∞ and the relative Mach number in the form

$$T^*_\infty = T_\infty k_T(\mathrm{Ma}) \quad (71)$$

The nondimensional function k_T will be discussed in Secs. 3.2, 3.3, and 3.4. The deviation of T^*_∞ from T_∞ may normally be neglected if Ma<0.1. The values of r, T_r, Π_r, T_∞, $\Pi_{v\infty}$, Δw_∞, and Ma are considered to be known.

Each of the Nusselt numbers Nu_H, Nu_M, and Nu_F will in general depend on all nondimensional parameters listed in Sec. 2.3.2. However, since heat and mass transfer are decoupled (assumptions Va, Vb), Cl and Eö play no role. If the processes are independent (assumptions Va to Vc), T_r/T_∞ only influences H, and $\Pi_r/\Pi_{v\infty}$ only influences M. In steady-state situations, time τ and the Stodola number are irrelevant. It is found that the Nusselt numbers are usually independent of T_r/T_∞ and $\Pi_r/\Pi_{v\infty}$. As a result, the Nusselt numbers will depend on property ratios such as g_∞ (or $\Pi_{v\infty}$ or $\Pi_{g\infty}$), γ, R_g/R_v, and so on; on two property data groups, Pr and Sc; and on two parameters characterizing the size and relative motion of the droplet, respectively, which are Kn and Re. Of course, the Mach number Ma may be used alternatively instead of Re; see Eq. (63). Since the processes occur outside the droplet, the material properties ρ_ℓ, c_ℓ, λ_ℓ, as well as T_ℓ, p_ℓ, m_ℓ do not enter the problem. The task is to give expressions for the Nusselt numbers in the form

$$\mathrm{Nu}_H = f_1(\mathrm{Kn}, \mathrm{Re}, \mathrm{Pr}, \mathrm{Sc}, g_\infty, \gamma, \ldots) \quad (72)$$

$$\mathrm{Sh} \triangleq \mathrm{Nu}_M = f_2(\mathrm{Kn}, \mathrm{Re}, \mathrm{Pr}, \mathrm{Sc}, g_\infty, \gamma, \ldots) \quad (73)$$

$$\mathrm{Nu}_F = f_3(\mathrm{Kn}, \mathrm{Re}, \mathrm{Pr}, \mathrm{Sc}, g_\infty, \gamma, \ldots) \quad (74)$$

In the following, these expressions will be given first for continuum flow (i.e., in the limit Kn→0), then for free-molecular flow (i.e., Kn→∞), and finally interpolation formulas covering the entire Kn

---

[10] More precise comments are given in Sec. 3.2.4.

range will be given under certain simplifying restrictions.

3.2 Continuum Environment

3.2.1 General

In viscous continuum flow a noncooled, nonheated, nonvaporizing object assumes a temperature T^*_∞ that is higher than the free-stream temperature T_∞, provided that the relative speed is not zero. For a sphere in laminar, compressible flow (Re<1), the solution of the Navier-Stokes equations (Kassoy et al., 1966) indicates the dependence

$$T^*_\infty = T_\infty \, k_T^{ct}(\mathrm{Ma})$$

where (75)

$$k_T^{ct}(\mathrm{Ma}) = 1 + \frac{15}{16}(\gamma - 1)\ \mathrm{Pr}\ \mathrm{Ma}^2$$

and the superscript ct is used for the continuum case. Using the typical values Pr = 0.7 and $\gamma = 1.4$, the correction term is seen to be smaller than 0.003 if Ma<0.1, which at room temperature means that T^*_∞ differs from T_∞ by less than 1 K.

The *heat transfer rate* H is governed by $(T^*_\infty - T_r)$ as seen from Eq. (68). The values of the Nusselt number Nu_H will be discussed further below.

The *mass transfer rate* M depends, as seen from Eq. (69), on the partial pressure ratio difference $(\Pi_{v\infty} - \Pi_r)$ and is proportional to the modified diffusion coefficient $\mathcal{D}$ given by Eq. (53).[11] In the limiting case of pure vapor, $\Pi_{v\infty} \to 1$, $(\Pi_{v\infty} - \Pi_r) \to 0$, and $\mathcal{D} \to \infty$. The values of M remains finite and is a result of hydrodynamic flow of vapor onto the drop rather than of diffusion. Nu_M is usually referred to as the Sherwood number.

The *momentum transfer rate* (or drag force) F is characterized by the Nusselt number Nu_F; see Eq. (70). This latter is related to the drag coefficient c_D commonly used in aerodynamics, as pointed out in Sec. 2.3.3, as

$$\mathrm{Nu}_F = c_D \ \mathrm{Re}/8 \tag{76}$$

In the low-Re limit, sphere drag in an infinite fluid is given by Stokes' law (1845), according to which $c_D = 24/\mathrm{Re}$, giving $\mathrm{Nu}_F = 3$.

[11] Alternative formulations based on the concentration (i.e., partial density) difference $(\rho_{v\infty} - \rho_{vr})$ and on the classical diffusion coefficient D, as encountered sometimes in the literature, are applicable to very dilute carrier mixtures only ($g_\infty \gg 1$) and give erroneous results in pure vapors (Gyarmathy 1963, Nitsch 1971).

In continuum environment (Kn<0.01), the Nusselt numbers are independent of the Knudsen number Kn. Their dependence on the parameters Re, etc., is well known in the entire Re range, from both theoretical and experimental work. The results known from the literature are summarized below.

3.2.2 Nusselt Number Expressions for Quiescent Media

For the case of a spherical droplet in infinite environment, with zero relative flow there exist analytic solutions to the steady-state heat conduction problem (e.g., Carslaw and Jaeger, 1959), the steady-state diffusion problem with arbitrary carrier composition (Gyarmathy, 1963; Nitsch, 1971), and the drag as given by Stokes' equation. The results, expressed by means of Nusselt numbers; are

$$\left.\begin{aligned} Nu_H^{ct} &= 2 \quad &(77)\\ Sh^{ct} \triangleq Nu_M^{ct} &= 2 \quad &(78)\\ Nu_F^{ct} &= 3 \quad &(79)\end{aligned}\right\} \text{for } Re \ll 1$$

The superscript *ct* is used to indicate the continuum case. In presence of relative flow these values are accurate only if the Reynolds number is considerably below unity, cf. below.

3.2.3 Nusselt Number Expressions for Relative Flow

The range of interest for the behavior of small droplets extends from Re = 0 to about Re = 200. In this range many experimental investigations have been conducted, and extensive empirical correlations exist. The general form of the Nu_H^{ct} amd Nu_M^{ct} correlations was first suggested by Frössling (1938).

Heat transfer data, as given by Kramers (1946), Ranz and Marshall (1952), Toei et al. (1966), Lee and Ryley (1968), can be represented for $(Re\, Pr^{2/3}) < 200$ as

$$Nu_H^{ct} = 2\left(1 + 0.30 Re^{1/2}\, Pr^{1/3}\right) \tag{80}$$

where the coefficient 0.30 is taken from Ranz and Marshall and the exponents comply with Frössling's formula.

Mass transfer data were obtained mostly by letting a liquid droplet vaporize in a dry air stream (Frössling, 1938; Ranz and Marshall, 1952; Lee and Barrow, 1968; early work has been critically reviewed by Fuks, 1959). The case of vapor-rich carrier mixtures was investigated by Toei et al. (1966). These results can be summarized as

$$Sh^{ct} \triangleq Nu_M^{ct} = 2\left(1 + 0.30 Re^{1/2}\, Sc^{1/3}\, \Pi_{g\infty}^{1/3}\right) \tag{81}$$

where the constant 0.30 (proposed by Ranz and Marshall) is in the middle of other authors' values (Frössling; 0.276; Toei et al.; 0.325). It is seen that heat and mass transfer obey identical equations, except that Pr is replaced by Sc $\Pi_{g\infty}$. The use of this group has been suggested by Ranz and Marshall on theoretical grounds and empirical support has been found by Toei et al. (1966). It is to be noted that these authors correlated their data in the form $Nu_M^{ct} = 2(1 + 0.30Re^{1/2}\ Sc^{1/3})\Pi_{g\infty}^{0.2}$. However, Eq. (81) represents their data equally well and is moreover theoretically justified.

The inert-gas partial pressure ratio $\Pi_{g\infty}$ in the above expressions is an approximation to the spatial average value proposed by Toei et al.

Numerous *drag coefficient* data compilations exist; see, for instance, Schlichting (1965). A frequently used expression for the range Re<200 is that of Torobin and Gauvin (1959), from which

$$Nu_F^{ct} = 3\left(1 + 0.15Re^{0.687}\right) \tag{82}$$

Figure 9 shows a plot of the three Nusselt numbers as calculated from Eqs. (80) to (82).

3.2.4 Other Effects

The expressions (80) to (82) refer to spheres bounded in an infinite environment, that is, to the case $r_\infty/r \to \infty$. The question of when corrections for *proximity* need to be applied will be dealt with in Sec. 3.5. Here we discuss some further situations in which the above equations need to be corrected.

Strong *free-stream turbulence* has an enhancing effect on the transfer rates. However, since this effect becomes noticeable only above Re ≈ 10^3 (Hayward and Pei, 1978), it will not be considered

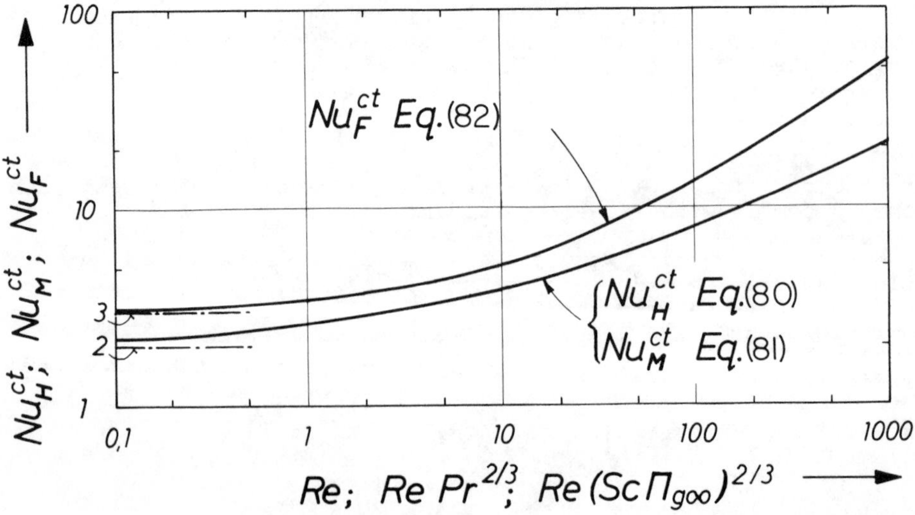

Fig. 9. Empirical dependence of continuum-flow Nusselt numbers on Re, Pr, and Sc $\Pi_{g\infty}$.

here. Ohta et al. (1975) have considered the case of *fluctuating flows*.

The influence of the *material properties* is taken into account in Eqs. (68) to (70) by using the data η, λ and $\mathcal{D}$ of the carrier *mixture*. Usually η and λ are not simple linear combinations of the component properties, and $\mathcal{D}$ is a mixture property anyway, being infinity for pure vapors. These property data should be evaluated at a temperature lying between T_∞ and T_r, but usually data referring to T_∞ give sufficiently good results. In the case of very *high temperature differences*, as occur in hot gases, the *one-third rule* has proven to be quite satisfactory (Kassoy and Williams, 1968; Hubbard et al., 1975; Yuen and Chen, 1976). According to this rule, the material properties have to be provided for the temperature $T_{1/3} = T_r + \frac{1}{3}(T_\infty - T_r)$. A simple estimate will illustrate the magnitude of errors that can be incurred. For gases the transport coefficients η, λ, and $\mathcal{D}$ increases proportionally to T^i, where the exponent i has some value usually between 0.7 and 1.5. Taking as an example a hot environment with $T_\infty/T_r = 2$, the transport property values are found to vary in the near field by some factor between 1.6 and 2.8 depending on i. Use of values pertaining to a false temperature (e.g., to T_∞) may then cause errors up to 80%. In the case of great *compositional variations* (i.e., $|\Pi_r - \Pi_{v\infty}| > 0.5$) *and* very different *molecular masses* of the two components, the carrier properties should be specified for an intermediate composition corresponding to $\Pi_{1/3} = \Pi_r + \frac{1}{3}(\Pi_{v\infty} - \Pi_r)$.

A particular correction is needed for the mass transfer rate in the case of high vapor partial pressure. In *rich mixtures* this rate is no longer a linear function of the difference $(\Pi_{v\infty} - \Pi_r)$ as suggested by Eq. (69), but rather a logarithmic one. This nonlinear law is given, for example, by Nitsch (1971) for $Re \ll 1$. It amounts to setting (with superscript nl for nonlinear)

$$Sh^{ct,nl} = Nu_M^{ct,nl} = Nu_M^{ct} \frac{1-\Pi_{v\infty}}{\Pi_{v\infty} - \Pi_r} \ln \frac{1-\Pi_r}{1-\Pi_{v\infty}} \tag{83}$$

This nonlinearity leads in the case of evaporation ($\Pi_r > \Pi_{v\infty}$) to corrections of 10% and more if the vapor partial pressure are such that $(\Pi_r - \Pi_{v\infty})/(1-\Pi_{v\infty}) > 0.2$. For a given value of Π_r, the correction is the greatest in pure-gas carriers ($\Pi_{v\infty} = 0$), being, for example, a factor of 1.4 for $\Pi_r = 0.5$. In the case where $\Pi_r = 1$, diffusion changes to hydrodynamic outflow of pure vapor, and $Nu_M^{ct,nl} \to \infty$ regardless of the value of $\Pi_{v\infty}$. In the case of condensation ($\Pi_r < \Pi_{v\infty}$), the correction has little practical significance, because it implies the accumulation of a gas cushion around the growing droplet, a phenomenon that could occur only as a result of very rapid, sustained growth in a motionless environment.

Cross-influence effects: The expressions (80) to (82) give heat and mass transfer rates without or with relative flow, that is, without or with simultaneous *momentum transfer*. The cross-influence of *heat transfer* on mass transfer and on drag is mostly overshadowed by the above-mentioned variation of the transport properties with temperature, and is automatically taken care of, in an approximate way, when the one-third rule is applied.

The cross-influence of *mass transfer* on heat transfer and on drag stems from the change of the radial temperature and velocity profiles caused by the radial mass flow. The effect on heat transfer can be simply obtained for the case of quiescent environment by solving the heat conduction problem in a slow radial flow field (Rüping, 1959). The Nusselt number of heat transfer, Nu_H^{ct}, is modified by simultaneous mass transfer into

$$\mathrm{Nu}_{H,M}^{ct} = \mathrm{Nu}_H^{ct} \frac{M\ c_{pv}}{4\ \pi\ r\ \lambda}\left[1-\exp\left(-\frac{M\ c_{pv}}{4\ \pi\ r\ \lambda}\right)\right]^{-1} \quad (84)$$

It is seen that mass flow toward the drop ($M>0$) enhances the heat transfer whereas mass outflow ($M<0$) reduces it, regardless of the direction of heat transfer. The above derivation is made for no transverse flow (Re = 0, that is, $\mathrm{Nu}_H^{ct} = 2$). The much more complicated problem of Re>0 has been solved by Montluçon (1975) for pure vapor carriers and under the significant restriction that η and λ are constant.

In the case of *droplet* growth, heat and mass transfer are related according to Eq. (25). If the expression $M \approx -H/L$ is inserted into Eq. (84), and H is expressed according to $H = 2\ \pi\ r\ \lambda(T_\infty - T_r)\mathrm{Nu}_{H,M}^{ct}$, one obtains

$$\mathrm{Nu}_{H,M}^{ct} \approx \mathrm{Nu}_H^{ct} \frac{L}{c_{pv}(T_\infty - T_r)} \ln\left[1 + \frac{c_{pv}(T_\infty - T_r)}{L}\right] \quad (85)$$

It is apparent that, for a given substance, the correction is the more important the greater the difference of T_∞ and T_r. Evaporation is slowed down and condensation is enhanced. An extreme case is evaporation in a very hot environment, where, for example, $T_\infty/T_r = 3$. With the typical value $L/c_{pv}T_r = 2$, one finds that Nu_H^{ctr} is reduced by 31%, showing that the correction may be quite significant in such cases. If, however, $T_\infty/T_r \approx 1$ (say, 0.8–1.2) the correction may be calculated from the linearized formula (Gardner, 1964)

$$\mathrm{Nu}_{H,M}^{ct} \approx \mathrm{Nu}_H^{ct}\left[1 - \frac{1}{2}\frac{c_{pv}(T_\infty - T_r)}{L}\right] \quad (86)$$

Typically, this gives 5% correction if T_∞ and T_r differ by 20%. Thus it is seen that large effects of mass transfer are inevitably coupled to large temperature differences and should not be quantitatively analyzed without taking the variation of the transport properties into account.

The influence of mass transfer on drag has been investigated by Hamielec and Hoffmann (1967) and by Montluçon (1975). The corrections are found to be smaller than for heat transfer, and are therefore always overshadowed by the variation of the transport properties (Yuen and Chen, 1976).

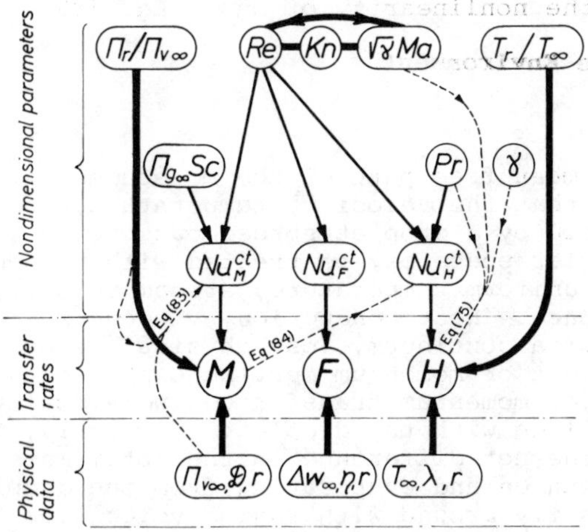

Fig. 10. Factors determining the transfer rates in continuum flow. ▬ decisive influence; —— moderate influence; ----- correction.

3.2.5 Review of the Crucial Parameters

The functional dependences expressed by the above equations are illustrated in Fig. 10. Here influences are shown by arrows, the thickness of which is an indication of the strength of the influence. As seen from the far right side of Eqs. (68) to (70), the transfer rates H, M, and F are determined as follows:

By physical data that are mostly dimensional: the sphere radius, a transport property and the far-field term of the appropriate driving force[12]

By the appropriate Nusselt number

By a nondimensional ratio describing the magnitude of the driving force (i.e., $\Pi_r/\Pi_{v\infty}$, 1, and T_r/T_∞, respectively).

The physical data are shown at the bottom of the figure, the nondimensional parameters in the upper part. Of these, Kn, Re (or Ma), and two ratios characterize the droplet size and the driving forces. Kn exerts no direct influence, and Ma appears only by its influence on the recovery temperature coefficient k_T^{ct}, Eq. (75). In anticipation of Sec. 3.3, $\sqrt{\gamma}$ Ma is used instead of Ma. The property data are represented by the product $\Pi_{g\infty}$ Sc and by Pr and γ, the latter having but a weak effect by way of the recovery temperature. The resulting three Nusselt numbers are moderately influenced by Re.

---

[12]Except in the case of Δw_∞, which is the driving force itself.

Usually weak influences are those of mass transfer on heat transfer, Eq. (84), and of the nonlinearity of Nu_M^{ct}, Eq. (83).

3.3 Free-Molecule Environment

3.3.1 General

If Kn>5, the mean free path of the molecules in the carrier phase is so large that the velocity, temperature, and compositional disturbances created by a droplet spread out over a large domain. Therefore the carrier phase may be treated with respect to the droplet as having uniform temperature, stream velocity, and composition throughout space. Heat conduction, viscous shear, and diffusion being properties of a continuum, lose physical relevance. The transfer rate of heat or momentum must be expressed as the sum of the amounts of energy or momentum transferred individually by the molecules that collide with the droplet. Similarly, the mass transfer rate is the net difference in the total mass of vapor molecules condensing on and evaporating from the droplet. Since molecules of gases fly around with random velocities that statistically obey a (Maxwellian) probability distribution, the mathematical prediction of the transfer rates involves integration over the surface of the sphere and over the range (0 to ∞) of molecular velocities. As far as the molecules *arriving* at the surface are concerned, the problem is straightforward.

It is noted that the *number* of molecules impinging on unit area of a stationary surface, for example, a drop[13] in unit time is given by the kinetic theory of gases (e.g., Hirschfelder et al., 1954) as $p_\infty/\sqrt{2\pi m_{mol} k_B T_\infty}$, where p_∞ is the pressure and T_∞ the temperature of the gas, m_{mol} the mass of one molecule, and $k_B = m_{mol}R = 1.38\times10^{-23}$ J/K, the Boltzmann constant. From this one finds the total *mass* of vapor and gas molecules impinging on unit surface in unit time to be

$$\beta_{v\infty} = \frac{p_{v\infty}}{\sqrt{2\pi R_v T_\infty}} \quad (87a) \qquad \beta_{g\infty} = \frac{p_{g\infty}}{\sqrt{2\pi R_g T_\infty}} \quad (87b)$$

For the combined mass flux of both carrier-phase components we shall set

$$\beta_\infty = \beta_{v\infty} + \beta_{g\infty} \triangleq \frac{p_\infty}{\sqrt{2\pi \hat{R} T_\infty}} \quad (87c)$$

where a mean gas constant $\hat{R}$ has been introduced (and defined) by the last equality. To relate $\hat{R}$ to R_v and R_g, see Eq. (A-52) in App. 3. Typical numerical values of β_∞ at normal pressures and temperatures are of the order of 100 kg/m^2 s.

---

[13] The Brownian motion of droplets containing many molecules (assumptions Id, Ie) may be neglected.

3.3.2 Accommodation Coefficients

Great difficulties arise in calculating the velocities and directions of flight of molecules *rebounding* from a surface. These are determined not only by the impingement velocity and direction, but also by the details of the interaction process with the molecules constituting the surface. Extremes such as immediate specular reflection (as of a ball from a smooth elastic wall) on the one hand, and perfect accommodation (i.e., complete effacing of the incoming molecule's previous history) on the other, are conceivable and certainly do occur occasionally. The essential question is the statistically relevant mean degree of accommodation. A global measure of this are the so-called *accommodation coefficients* conceived by Maxwell (1877), Hertz (1882), and Knudsen (1911). Comprehensive reviews on the subject have been published by Schaaf and Chambré (1958) and Paul (1962).

Thermal accommodation: Since accommodation coefficients at the same surface may be different for different gases, we have to distinguish between the vapor and gas components. The average thermal energy of the rebounding vapor and gas molecules can be characterized by temperatures, $T_{\text{reb},v}$ and $T_{\text{reb},g}$, if the velocity distribution of these molecules is assumed to be Maxwellian. Following Knudsen, one sets

$$T_{\text{reb},v} \mathrel{\hat{=}} T_\infty + \alpha_{Hv}\,(T_r - T_\infty) \tag{88a}$$

$$T_{\text{reb},g} \mathrel{\hat{=}} T_\infty + \alpha_{Hg}\,(T_r - T_\infty) \tag{88b}$$

where T_r and T_∞ are the temperatures of the surface and of the gas phase respectively, and α_{Hv}, α_{Hg} are the *thermal accommodation coefficients*. The values of these are obviously between zero and unity. No theories being available, they have to be determined experimentally.

For most gas/solid surface combinations α_H lies above 0.8 and for liquid surfaces it is nearly unity (Schaaf and Chambré, 1958; Paul, 1962). In dealing with droplets, one may therefore set

$$\alpha_{Hv} \approx \alpha_{Hg} \approx 1 \tag{89}$$

However, to maintain generality in the expressions given below, α_{Hv} and α_{Hg} will be retained.

In treating *mass transfer*, a *mass accommodation coefficient* α_{Mv} is introduced to express the ratio $\alpha_{Mv} \mathrel{\hat{=}} \beta_{v,\text{adh}}/\beta_{v\infty}$ of the number of molecules that at least temporarily adhere (condense) to the droplet surface, to the total number of impinging vapor molecules. (Gas molecules, even if present, do not contribute to net mass transfer.) Langmuir (1913, 1916) and Knudsen (1915) have analogously defined an *evaporation coefficient* $\alpha'_{Mv} \mathrel{\hat{=}} \beta_{\text{meas}}/\beta_r$ relating the measured mass flux of actually emitted molecules, β_{meas}, to the theoretically possible maximum flux β_r from a droplet of temperature T_r. The value of β_r is determined mainly by the vapor pressure

$p_r(T_r)$ of the droplet, that is, the pressure at which pure vapor would be in thermodynamic equilibrium with the drop when both have temperature T_r. As is obvious from Eq. (87a),

$$\beta_r = \frac{p_r(T_r)}{\sqrt{2\pi R_v T_r}} \tag{90}$$

As seen β_r is determined by the properties of the substance and by the surface temperature. For extremely nonvolatile substances, such as solids, p_r and therefore β_r are zero.

In the general case of volatile substances ($0<p_r \lesseqgtr p_{v\infty}$) in nonvacuum environment ($p_{v\infty}>0$), mass transfer is the net difference between the adhering (condensing) vapor mass flux $\beta_{v,\text{adh}}$ and the emitted mass flux $\beta_{v,\text{emi}}$. Using the mass accommodation coefficient α_{Mv} and Knudsen's evaporation coefficient α'_{Mv} as defined above, we may write

$$\beta_{v,\text{adh}} \triangleq \alpha_{Mv}\,\beta_{v\infty} \tag{91a}$$

$$\beta_{v,\text{emi}} \triangleq \alpha'_{Mv}\,\beta_r \approx \alpha_{Mv}\,\beta_r \tag{91b}$$

The values of α_{Mv} and α'_{Mv} are not necessarily equal, because $T_r \neq T_\infty$. Usually, however, this distinction is not made, and the last equality above is used. From the total flux $\beta_{v\infty}$ of vapor molecules, the fraction $(1-\alpha_{Mv})\beta_{v\infty}$ rebounds, as does the entire molecule flux $\beta_{g\infty}$.

The values of α_{Mv} have to be obtained by measurement. In such measurements β_r is calculated from Eq. (90) using measured values of T_r. In order to obtain β_r with not more than a few percent uncertainty, temperature measurement has to be extremely accurate and the surface has to be very clean. Since Knudsen's time, because of these difficulties, experiments produced a wild scatter with values ranging between 0.01 and 1. The controversy about the correct values of α_{Mv} is not yet completely settled. According to recent critical evaluations and newer measurements (Schrage, 1953; Paul, 1962; Mills and Seban, 1967; Hobson, 1973; Bryson et al., 1974; Fukuta and Armstrong, 1974; Narusawa and Springer, 1975), one may assume that α_{Mv} of most liquids, including water, lies fairly close to unity:

$$\alpha_{Mv} \approx 0.7 - 1 \tag{91c}$$

Some substances, the molecules of which assume different states in the vapor and liquid phase, are likely to have lower values (Paul, 1962). However, the very low values of 0.02 to 0.2 given in some earlier publications are likely to be due to measurement error.[14]

[14] Values of α_{Mv} close to unity in no way contradict the observation that the *net* flux of definitively condensing molecules on a growing drop may be much (e.g., 10 to 50 times) smaller than any of the incoming vapor molecule flux $\beta_{v\infty}$ or the theoretically maximum emitted flux β_r. Some of the literature on condensing flow is haunted by the confusion of α_{Mv} with the resulting net flux fraction.

Considering these uncertitudes, typically the value $\alpha_{Mv} = 0.9$ is recommended for use in the equations given below.

For the *transfer of momentum* it is assumed, following Maxwell (1877), that the rebounding flux of molecules is subdivided into two parts: one part α_F completely accommodated to T_r and emitted in random directions (so-called diffuse reflection) and another part $1 - \alpha_F$ which has maintained its original temperature T_∞ and is specularly reflected. [More sophisticated assumptions, such as the decoupling of directional and temperature behavior or the distinction between the accommodation of tangential and normal components of momentum, as proposed by Schaaf and Chambré (1958), will not be considered here.]

Since this *diffuse reflection coefficient* α_F is not necessarily equal for the vapor and carrier-gas molecules, we express the diffusely reflected molecular mass flux densities by defining separate coefficients according to

$$\beta_{v,\mathrm{diff}} \triangleq \alpha_{Fv}\, \beta_{v\infty} \quad (92a) \qquad \beta_{g,\mathrm{diff}} \triangleq \alpha_{Fg}\, \beta_{g\infty} \quad (92b)$$

The remaining part of the incoming molecular fluxes, $(1-\alpha_{Fv})\beta_{v\infty}$ and $(1-\alpha_{Fg})\beta_{g\infty}$, are the specularly reflected ones. The experimental values of α_F are close to unity (Millikan, 1923b; Schaaf and Chambré 1958; Paul, 1962). Therefore one may usually set

$$\alpha_{Fv} \approx \alpha_{Fg} \approx 1 \quad (93)$$

As Paul (1962) notes, for given substances (of surface and gas) and given temperatures, the numerical values of α_H, α_M, and α_F must be expected to follow the rule $1 > \alpha_F > \alpha_H > \alpha_M > 0$. There is no existing theory, however, to predict or quantitatively interrelate the accommodation coefficients.

3.3.3 Transfer Rates

Taking the empirical values of the accommodation coefficients and applying kinetic theory to the impinging molecules as outlined above, the free-molecular transfer rates to a sphere of radius r can be formulated as follows. All conditions, including r, are taken to be constant in time, and T_r and p_r are taken to be uniform over the entire surface. Each flux is the sum of the contributions of the vapor and the gas molecules. Since the molecular masses of these are different in general, we distinguish between Mach numbers referred to $\sqrt{\gamma\, R\, T_\infty}$ of the vapor and of the gas, setting

$$\mathrm{Ma}_v \triangleq \frac{\Delta w_\infty}{\sqrt{\gamma_v\, R_v\, T_\infty}} \quad (94a) \qquad \mathrm{Ma}_g \triangleq \frac{\Delta w_\infty}{\sqrt{\gamma_g\, R_g\, T_\infty}} \quad (94b)$$

and we introduce, following Hill et al. (1963), the abbreviations

$$K_v \triangleq \frac{\gamma_v+1}{2(\gamma_v-1)} \quad (95a) \qquad K_g \triangleq \frac{\gamma_g+1}{2(\gamma_g-1)} \quad (95b)$$

Typically K has the value of 3 (for $\gamma = 1.4$). Analogously to the two Mach numbers, two different values will arise for the adiabatic wall temperature (or "recovery temperature") T^*_∞.

Heat Transfer:

The molecular transfer of energy to nonabsorbing bodies has been analyzed by, for example, Oppenheim (1953). The results were summarized by Schaaf and Chambré (1958). For the sphere the heat transfer rate is given as

$$H = 4\pi r^2 (T^*_{\infty v} - T_r)\, \beta_{v\infty}\, K_v\, R_v\, \alpha_{Hv}\, k_{Hv}$$

$$+ 4\pi r^2 (T^*_{\infty g} - T_r)\, \beta_{g\infty}\, K_g\, R_g\, \alpha_{Hg}\, k_{Hg} \qquad (96)$$

where

$$T^*_{\infty v} = T_\infty \left(1 + \frac{2\gamma_v}{3K_v} \mathrm{Ma}_v^2 + \ldots\right) \triangleq T_\infty\, k^{fm}_{Tv}(\mathrm{Ma}_v) \qquad (97a)$$

$$T^*_{\infty g} = T_\infty \left(1 + \frac{2\gamma_g}{3K_g} \mathrm{Ma}_g^2 + \ldots\right) \triangleq T_\infty\, k^{fm}_{Tg}(\mathrm{Ma}_g) \qquad (97b)$$

$$k_{Hv} = 1 + \frac{\gamma_v}{6} \mathrm{Ma}_v^2 + \ldots \qquad (98a)$$

$$k_{Hg} = 1 + \frac{\gamma_g}{6} \mathrm{Ma}_g^2 + \ldots \qquad (98b)$$

The above expression for H results from kinetic theory as the *difference* of an incoming energy flux (which is in the general case of $\Delta_{w\infty} \neq 0$ nonuniform over the sphere) and of an outgoing flux that is uniform as a result of assumption Ig in Table 1.

The influence of relative speed is seen to be felt in two ways: by raising the carrier temperature "seen" by the droplet and by direct enhancement of the impinging vapor and gas molecule fluxes as represented by the factors k_{Hv} and k_{Hg}. The above expressions are Taylor series approximations to the exact, but mathematically complicated, functions given in the above references.[15] The truncated terms amount to a maximum of 5% of the Ma^2 term if $\mathrm{Ma}<1$.

Mass Transfer:

From the kinetic theory developed by Oppenheim (1953), it follows that the net mass of vapor absorbed by the sphere in unit time is

$$M = 4\pi r^2 (\alpha_{Mv}\, \beta_{v\infty}\, k_{Mv} - \alpha'_{Mv}\, \beta_r) = 4\pi r^2 \alpha_{Mv} (\beta_{v\infty}\, k_{Mv} - \beta_r) \qquad (99)$$

[15] The recovery temperature expressions given in some of the literature contain an error, as pointed out by Kemp (1979).

where $\alpha_{Mv} = \alpha'_{Mv}$ has been set according to Eq. (91*b*). The factor

$$k_{Mv} = 1 + \frac{\gamma_v}{6} \mathrm{Ma}_v^2 + \ldots \tag{100}$$

represents the influence of the relative speed, and is found to be identical to k_{Hv}.

Momentum Transfer (drag):

Expressions for the drag force acting on various bodies in free-molecular flow were first derived by Ashley (1949). His result for the sphere, if written for the two kinds of molecules (with T_r and p_r again uniform over the surface) is

$$F = 4\pi r^2 \Delta w_\infty \beta_{v\infty} \left(\frac{4}{3} k_{Fv} + \frac{\pi}{6} \alpha_{Fv} \sqrt{\frac{T_r}{T_\infty}} \right) + 4\pi r^2 \Delta w_\infty \beta_{g\infty} \left(\frac{4}{3} k_{Fg} + \frac{\pi}{6} \alpha_{Fg} \sqrt{\frac{T_r}{T_\infty}} \right) \tag{101}$$

where the relative speed influence is represented by the factors

$$k_{Fv} = 1 + \frac{\gamma_v}{10} \mathrm{Ma}_v^2 + \ldots \tag{102a}$$

$$k_{Fg} = 1 + \frac{\gamma_g}{10} \mathrm{Ma}_g^2 + \ldots \tag{102b}$$

which are again first-order series approximations to the rigorous functions. Ignoring terms beyond the second leads to errors of less than 5% of the Ma^2 terms at subsonic speeds.

In Eq. (101) the sphere is assumed to absorb none of the molecules (M = 0) and to have a temperature T_r different from T_∞. Usually, however, very little error is incurred by setting $T_r/T_\infty = 1$ in Eq. (101). For low Mach numbers ($k_F \approx 1$) and for $\alpha_F \approx 1$, the expression in parentheses has the numerical value (1.33 + 0.52) = 1.85. Cross effects arising from absorption or emission have been analyzed by Brock (1964a,b). They are negligible under the conditions pertaining to droplets.

3.3.4 Nusselt Number Expressions

The Nusselt numbers are obtained by inserting the free-molecular expressions for the three transfer rates, Eqs. (96), (99), and (101), respectively, into Eqs. (68) to (70). By defining suitable average material properties, the contribution of the vapor and gas molecules to the transfer rates can be represented by a common term. The derivation is outlined in App. 3. The overbar means mass-weighted averaging as defined by Eq. (A-7) in App. 1. According to Eqs. (A-67), (A-75), and (A-83) in App. 3, one finds

$$\mathrm{Nu}_H^{fm} = \frac{B_H}{\mathrm{Kn}} \tag{103}$$

$$\mathrm{Sh}^{fm} \mathrel{\widehat{=}} \mathrm{Nu}_M^{fm} = \frac{B_M \Pi_{g\infty}}{\mathrm{Kn}} \tag{104}$$

$$\mathrm{Nu}_F^{fm} = \frac{B_F}{\mathrm{Kn}} \tag{105}$$

where the coefficients B_H, B_M, and B_F are independent of droplet size and usually have values around unity. The exact values depend on the properties and composition of the carrier phase and on the accommodation coefficient values. They slowly increase with relative Mach number Ma. B_M and B_F are also influenced by the temperature ratio T_r/T_∞, and B_M is affected by the partial pressure ratios Π_r and $\Pi_{v\infty}$. According to Eqs. (A-68), (A-76), and (A-84) in App. 3, their expressions are

$$B_H = \sqrt{\frac{2}{\pi}}\,\frac{\gamma-1}{\gamma}\,\frac{\overline{(R^{3/2}\,K\,\alpha_H)}}{\bar{R}^{3/2}}\,\mathrm{Pr}\,k_H(\mathrm{Ma}) \tag{106}$$

$$B_M = \sqrt{\frac{2}{\pi}}\,\sqrt{\frac{R_v}{\bar{R}}}\,\alpha_{Mv}\,\mathrm{Sc}\left[k_M(\mathrm{Ma}) + \left(1 - \sqrt{\frac{T_\infty}{T_r}}\right)\frac{\Pi_r}{\Pi_{v\infty}-\Pi_r}\right] \tag{107}$$

$$B_F = \sqrt{\frac{2}{\pi}}\left[\frac{4}{3}\,\frac{\overline{(R^{1/2})}}{\bar{R}^{1/2}}\,k_F(\mathrm{Ma}) + \frac{\pi}{6}\,\frac{\overline{(R^{1/2}\,\alpha_F)}}{\bar{R}^{1/2}}\sqrt{\frac{T_r}{T_\infty}}\right] \tag{108}$$

Eqs. (A-69), (A-71), and (A-82) in App. 3 give the pertinent Mach number functions as

$$k_H(\mathrm{Ma}) = 1 + \frac{\gamma}{6}\,\mathrm{Ma}^2\,\frac{\bar{R}\,\overline{(R^{1/2}\,K\,\alpha_H)}}{\overline{(R^{3/2}\,K\,\alpha_H)}} \tag{109}$$

$$k_M(\mathrm{Ma}) = 1 + \frac{\gamma}{6}\,\mathrm{Ma}^2\,\frac{\bar{R}/R_v}{\Pi_{v\infty}-\Pi_r}\,\Pi_{v\infty} \tag{110}$$

$$k_F(\mathrm{Ma}) = 1 + \frac{\gamma}{10}\,\mathrm{Ma}^2\,\frac{\bar{R}\,\overline{(R^{-1/2})}}{\overline{(R^{1/2})}} \tag{111}$$

These functions are approximations having errors under 5% of the Ma^2 term up to Mach numbers close to unity. At Ma = 1, B_H and Nu_H^{fm} are typically enhanced with respect to their values at Ma = 0 by 20 to 30%, whereas B_F and Nu_F^{fm} are increased by 10 to 20%. The value of B_M (and Nu_M^{fm}) may rapidly increase with Ma, provided that $\Pi_{v\infty}-\Pi_r \ll 1$; as will become evident below, this sensitivity to Ma is merely a matter of definitions and does not affect the mass transfer rate M itself.

In the free-molecular case, the adiabatic wall temperature T_∞^* occurring in Eq. (68) has the value

$$T_\infty^* = T_\infty\,k_T^{fm}(\mathrm{Ma}) \tag{112}$$

where

$$k_T^{fm}(\text{Ma}) = 1 + \frac{2\gamma}{3}\,\text{Ma}^2\,\frac{\bar{R}\,\overline{(R^{1/2}\,\alpha_H)}}{\overline{(R^{3/2}\,K\,\alpha_H)}} \tag{113}$$

cf. Eqs. (A-55), (A-56), and (A-65) in App. 3.

In the simple but practically important case of Ma = 0 and $T_r = T_\infty$, the B coefficients have the following values, with the subscript 0 referring to this special case (i.e., the *stationary, isothermal sphere*):

$$B_{H0} = \sqrt{\frac{2}{\pi}}\,\frac{\gamma-1}{\gamma}\,\text{Pr}\,\frac{\overline{(R^{3/2}\,K\,\alpha_H)}}{\bar{R}^{3/2}} \approx \sqrt{\frac{2}{\pi}}\,\frac{\gamma-1}{\gamma}\,\text{Pr}\,K\,\bar{\alpha}_H \tag{114}$$

$$B_{M0} = \sqrt{\frac{2}{\pi}}\,\sqrt{\frac{R_v}{\bar{R}}}\,\text{Sc}\,\alpha_{Mv} \approx \sqrt{\frac{2}{\pi}}\,\text{Sc}\,\alpha_{Mv} \tag{115}$$

$$B_{F0} = \sqrt{\frac{2}{\pi}}\left[\frac{4}{3}\,\frac{\overline{(R^{1/2})}}{\bar{R}^{1/2}} + \frac{\pi}{6}\,\frac{\overline{(R^{1/2}\,\alpha_F)}}{\bar{R}^{1/2}}\right] \approx \sqrt{\frac{2}{\pi}}\left(\frac{4}{3} + \frac{\pi}{6}\,\bar{\alpha}_F\right) \tag{116}$$

where the last equalities apply if $R_v \approx R_g$ and $K_v \approx K_g$. For $R_v = R_g$ and $K_v = K_g$, and for the case of pure-vapor or no-vapor carriers, these approximate equalities become accurate.

Assuming that $\alpha_{Hv} = \alpha_{Hg} = 1$, $\alpha_{Mv} = 0.9$, and $\alpha_{Fv} = \alpha_{Fg} = 1$, which is, as pointed out above, a fair estimate for numerous substances, Eqs. (114) to (116) yield the following numerical values for the two most common carrier fluids:

$$\text{Dry air:}\;\begin{cases} B_{H0} = 0.48 \\ B_{M0} = 0.36 \\ B_{F0} = 1.48 \end{cases} \qquad \text{Steam:}\;\begin{cases} B_{H0} = 0.69 \\ B_{M0} = 0.57 \\ B_{F0} = 1.48 \end{cases}$$

Here $\gamma = \gamma_g = 1.40$, $K = K_g = 3$, Pr = 0.70, Sc = 0.5, $\Pi_{g\infty} = 1$ have been assumed for air, and $\gamma = \gamma_v = 1.30$, $K = K_v = 3.83$, Pr = 0.98, and (based on self-diffusion) Sc = 0.8 for steam. In pure vapors, diffusion is not relevant for the mass transfer process, and the specification of B_{M0} using the self-diffusion value of Sc has no significance. This choice has no influence on the mass transfer rate, as will be evident from Eq. (131*b*).

The free-molecular mass transfer rate can be expressed from Eq. (99) in terms of the partial pressure ratios $\Pi_{v\infty}$ and Π_r as

$$M = 4 \pi r^2 \alpha_{Mv} \frac{p_\infty}{\sqrt{2 \pi R_v T_\infty}} \left[\hat{\Pi}_{v\infty} k_{Mv} (\text{Ma}) - \Pi_r \sqrt{\frac{T_\infty}{T_r}} \right] \quad (117)$$

It is apparent that M is a function of p_∞, T_∞, T_r, $\Pi_{v\infty}$, and Π_r. M is never infinite (the cases $T_\infty = 0$ or $T_r = 0$ being excluded and becomes zero only if

$$\frac{\Pi r}{\Pi_{v\infty}} = \frac{p_r}{p_{v\infty}} = k_{Mv}(\text{Ma}_v) \sqrt{\frac{T_r}{T_\infty}} \quad (118)$$

This behavior of M is in contrast to continuum carriers where the mass transfer rate is zero for $\Pi_{v\infty} = \Pi_r$ (as evident from Eq. (69) and from the Nu_M^{ct} values given in Sec. 3.2), and moreover is essentially proportional to $(\Pi_{v\infty} - \Pi_r)$. The free-molecular rate behaves in the same manner only if $\text{Ma}_v = 0$ and $T_\infty = T_r$. This difference in the behavior of M, and the fact that both Nu_M^{ct} and Nu_M^{fm} are defined by one and the same equation, Eq. (52), is the cause for the anomalous behavior of B_M for $(\Pi_{v\infty} - \Pi_r) \to 0$, as apparent from Eq. (107). The free-molecular value of M for the case of $\Pi_{v\infty} = \Pi_r$ is given by Eq. (A-72) in App. 3.

3.3.5 Review of the Crucial Parameters

Figure 11 shows the structure of the functional relationships

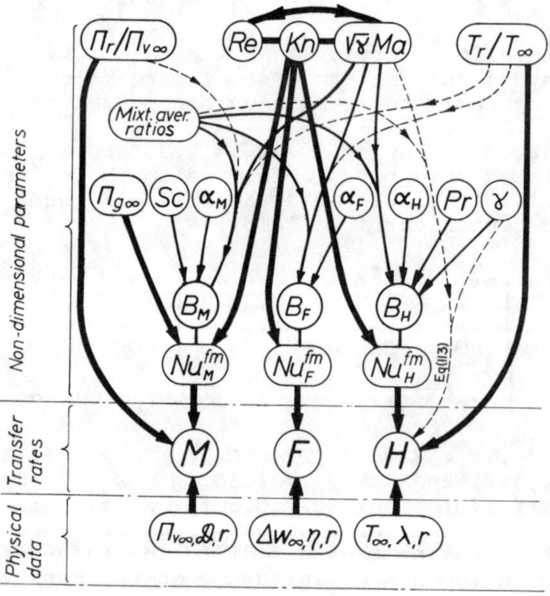

Fig. 11. Factors determining the transfer rates in free-molecule flow. ▬▬ decisive influence; —— moderate influence; ----- correction.

etermining the values of the transfer rates M, F, and H. As in ig. 10, the thickness of the arrows indicates the strength of each nfluence. By definition of the Nusselt numbers, the dependence n the dimensional physical quantities shown on the bottom of the igure and on the driving-force ratios $\Pi_r/\Pi_{v\infty}$ and T_r/T_∞ is the ame as in the continuum case.

The main differences to Fig. 10 consist of the presence of he Kn influence (which is a very strong one) and the replacement f the Re effect by the influence of Ma only (which is, by the way, eaker than that of Re). There appear some new nondimensional aterials properties, namely, the molecular accommodation coeffici-nts α_M, α_F, α_H and the ratios of various mixture-average properties. he factors B_M, B_F, and B_H are used to express the combined influence f Ma and of the materials data.

Among the weak influences we see the dependence of B on the atios $\Pi_r/\Pi_{v\infty}$ and T_r/T_∞, and the recovery temperature effect n H.

3.4 Multirange Expressions

3.4.1 General

In many two-phase flow situations of practical interest, grow-ng or evaporating droplets pass across a variety of Knudsen number anges during their lifetime; cf. Fig. 6. Even though in the ntermediate (slip and transitional) ranges both theoretical and mpirical information (Burrows, 1957; Rose, 1964) are scarce, there s a great practical interest for multirange correlations from which he three Nusselt numbers Nu_H, Nu_M, and Nu_F can be calculated for any alue of Kn. It is desirable that such expressions be valid for eynolds and Mach numbers up to at least Re $\approx$ 100 and Ma $\approx$ 1.

For the drag coefficient c_D (or Nu_F) of the sphere, where mpirical data are abundant, such correlations have been made over very broad range (Re<1000, Ma<6), and remarkable accuracies have een achieved (Carlson and Hoglund, 1964; Crowe et al., 1971; alsh, 1975; and Henderson, 1976). The mathematical structure of he functions is rather complicated, however, and in some instances he correlations are "patched" together by use of different expres-ions in different domains.

No comparably exhaustive correlations exist for the heat and ass transfer rates. The very limited available data on the heat ransfer in subsonic transition flow, for example, by Takao (1963), ave been correlated by Sherman (1963).

The aim of the following treatment is to use the theoretical nd empirical information existing about continuum and free-olecular flow conditions, as summarized in Sec. 3.2 and 3.3, for he construction of multirange expressions giving Nu_H, Nu_M, and Nu_F all in analogous form as functions of Kn, Re (and Ma). Of these, he Nu_F expression will be checked against the existing extensive orrelations and against Millikan's classical (1923a) measurements, nd Nu_H will be compared to one experimental finding. It is elieved that once the expression of Nu_F is empirically verified, he analogous expressions of Nu_H and Nu_M become trustworthy, too.

The multirange expressions given below are generalizations

of similar expressions derived earlier by Gyarmathy (1962, 1963), Carlson and Hoglund (1964), Kang (1967), Fukuta and Walter (1970), and May (1974).

3.4.2 The Cunningham Correction

As first pointed out by Cunningham (1910) from theoretical considerations concerning the slip range, the influence of Kn on c_D can be represented by dividing the continuum value by a correction term of the form(1 + constant x Kn). Millikan (1923b) has measured (as a by-product of his famous experiments aimed at the determination of the elementary electrical charge) the drag of small, slow-moving oil droplets in air over a very wide range of Knudsen numbers. His result, when written in terms of Nu_F is

$$Nu_F^{Mill} = \frac{3}{1 + f(Kn)Kn} \tag{119}$$

with

$$f(Kn) = 1.547 + 0.519 \exp\left(- \frac{0.699}{Kn}\right) \tag{120}$$

wherein the constants have been adapted to the definition of Kn stated in Eq. (43). It is seen that the Cunningham constant varies from 1.547 to 2.066 as Kn goes from 0 to ∞. For Kn = 0, the continuum result expressed by Eq. (79), that is, $Nu_F^{Mill} = Nu_F^{ct} = 3$, is obtained. In the free-molecular limit, Millikan's result is $Nu_F^{Mill} = 1.452/Kn$, which compares very well with the result $B_F = 1.48$ obtained theoretically in Sec. 3.3.4 for any single-component carrier in the case of completely diffuse reflection ($\alpha_F = 1$). (The value 1.452 could be obtained exactly by setting $\alpha_F = 0.94$.)

A simple rational approach to derive the Cunningham constant theoretically consists of subdividing the environment of the droplet into two concentric zones and applying free-molecular transfer laws in the inner zone and continuum laws in the outer zone. This zonal model is of course applicable to all three transfer processes, and has been employed by Kang (1967) and Fukuta and Walter (1970).[16] The thickness of the inner zone is usually taken to be of the order of the mean free path ℓ and expressed as $c \cdot \ell$, with the constant c of the order of 1.

This approach leads to expressions of the form

$$Nu = \frac{Nu^{ct}}{1 + [(Nu^{ct}/B) - 2c/(1 + 2c\ Kn)]\ Kn} \tag{121}$$

[16] Due to the neglect of Stefan flow in the diffusion equation, both authors' treatment is limited to dilute ($\Pi_{g\infty} \approx 1$) carriers. Their "condensation coefficient" is identical to α_{Mv} and should correctl be assigned a value close to unity.

where Nu, Nu^{ct}, and B mean any of Nu_H, Nu_H^{ct}, B_H, or Nu_M, . . ., or Nu_F, . . ., respectively. Here the Cunningham "constant" is seen to be a function of Kn. With the choice $c = 0.25$, particularly close agreement with the function f(Kn) of Millikan can be obtained.

By setting $c = 0$, Eq. (121) goes over into the Cunningham-type expression

$$Nu = \frac{Nu^{ct}}{1 + (Nu^{ct}/B)\ Kn} \qquad (122)$$

where the factor of Kn is a constant. In order to show the influence of c, Nu/Nu^{ct} as calculated from Eq. (121) has been plotted in Fig. 12 for various values of Nu^{ct}/B and for the two values $c = 0$ and $c = 0.25$. It is seen that the influence of c is moderate (15% at most) and limited to the range Kn ≈ 0.1 to 2.

Sherman (1963) has shown that sphere heat transfer data can be satisfactorily correlated by Eq. (122) over the entire Kn range. Two further arguments are in favor of setting $c = 0$ and using Eq. (122) instead of (121). Willis (1966), in calculating sphere drag

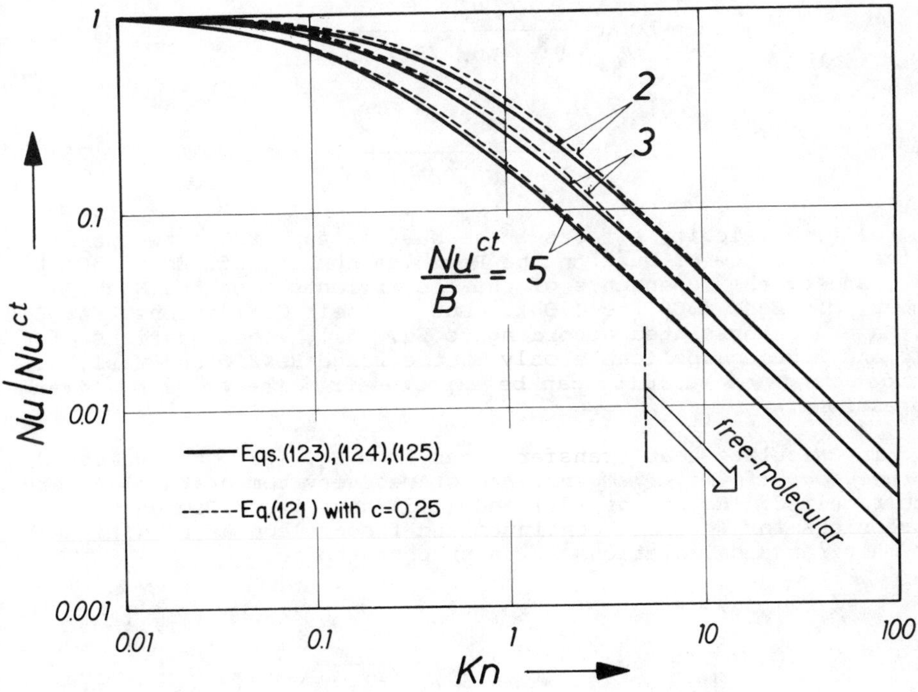

Fig. 12. Variation of the Nusselt numbers with the Knudsen number according to the multirange expressions Eq. (121) and (123)-(125), for various constant values of Nu^{ct} and B. (Nu means any of Nu_H, Nu_M, Nu_F; and B any of B_H, $B_M \cdot \Pi_{g\infty}$, B_F.)

in the transition range, has found that molecular theory and Millikan's data agree with respect to the behavior of Nu when $Kn \to \infty$ and $Nu \to Nu^{fm}$, showing that the difference $Nu - Nu^{fm}$ goes toward zero as 1/Kn. This contradicts Eq. (121), which for $c>0$ predicts proportionality to $1/Kn^2$. Second, Eq. (122) was found to predict correctly the drag coefficient at the transition from continuum to slip flow (Gyarmathy, 1962). Thus Eq. (122) is found to perform satisfactorily on both margins of the intermediate domain. Since, in addition, it is algebraically simpler, it will be given preference over equations of the type of Eq. (121), despite the latter equation's close similarity to Millikan's empirical data.

3.4.3 Multirange Expressions for the Nusselt Numbers

Thus following multirange expressions are found for the Nusselt numbers of heat, mass, and momentum transfer to the sphere:

$$Nu_H = \frac{Nu_H^{ct}}{1 + (Nu_H^{ct}/B_H)\ Kn} \tag{123}$$

$$Sh \mathrel{\widehat{=}} Nu_M = \frac{Nu_M^{ct}}{1 + (Nu_M^{ct}/B_M)\ Kn} \tag{124}$$

$$Nu_F = \frac{Nu_F^{ct}}{1 + (Nu_F^{ct}/B_F)\ Kn} \tag{125}$$

The relative velocity affects these Nusselt numbers in two ways: by the dependence of Nu^{ct} on the Reynolds number, cf. Eqs. (80) to (82), and by the dependence of the coefficients B on the Mach number, cf. Eqs. (106) to (108). Due to their definitions, Kn, Re, and Ma are interrelated according to Eq. (61). The equations of Nu^{ct} and B being applicable only in the range $Re<200$ and $Ma<1$, no broader range of validity can be expected from the above multirange expressions.

To calculate heat transfer rates from Eq. (68), the values of the *adiabatic wall temperature* (or recovery temperature) T_∞^* are also required. Equations (75) and (113) give two different expressions for T_∞^* under continuum and free-molecular conditions. For multirange calculations it is proposed to set

$$T_\infty^* = T_\infty\, k_T(Ma)$$

with

$$k_T(Ma) = 1 + \left[\frac{15}{16}(\gamma-1)\ Pr + \frac{2\,\bar{R}\ \overline{(R^{1/2}\ \alpha_H)}}{3\ \overline{(R^{3/2}\ K\ \alpha_H)}}\ \gamma\ Kn\right]\frac{Ma^2}{1+Kn} \tag{126}$$

This function becomes identical to k_T^{ct} and k_T^{fm} for $Kn \to 0$ and $Kn \to \infty$, respectively. The dependence on Kn is such that transition between the two extremes occurs in the range Kn = 0.1 to 10. The values of k_T^{ct} and k_T^{fm} being never very different, any errors of the above interpolation are expected to be negligibly small. Of course, $k_T = 1$ may be set for Ma<0.1.

3.4.4 The Stationary, Isothermal Sphere

For the practically important case of *negligible relative velocity* (Re = Ma = 0) and *isothermal sphere* ($T_r = T_\infty$), the values of Nu^{ct} and B are given by Eqs. (77) to (79) and (114) to (116). Inserting these into Eqs. (123) to (125), one obtains (with the subscript 0 referring to zero relative flow and $T_r = T_\infty$)

$$Nu_{H0} = \frac{2}{1 + (2/B_{H0})\ Kn} \tag{127}$$

$$Sh_0 \triangleq Nu_{M0} = \frac{2}{1 + (2/B_{M0})\ Kn} \tag{128}$$

$$Nu_{F0} = \frac{3}{1 + (3/B_{F0})\ Kn} \tag{129}$$

where B_{H0}, B_{M0}, and B_{F0} are given by Eqs. (114) to (116).

By using the material properties of dry air and of steam, as in Sec. 3.3, and assuming that $\alpha_H = \alpha_F = 1$, $\alpha_{Mv} = 0.9$, the following expressions are obtained

Air:

$$Nu_{H0} = \frac{2}{1 + 4.18\ Kn} \tag{130a}$$

$$Nu_{M0} = \frac{2}{1 + 5.63\ Kn} \tag{131a}$$

$$Nu_{F0} = \frac{3}{1 + 2.02\ Kn} \tag{132a}$$

Steam:

$$Nu_{H0} = \frac{2}{1 + 2.89\ Kn} \tag{130b}$$

$$Nu_{M0} = 0 \tag{131b}$$

$$Nu_{F0} = \frac{3}{1 + 2.02\ Kn} \tag{132b}$$

In pure vapor carriers, Nu_{M0} is seen to become zero. However the mass transfer rate M obtained from Eq. (69) remains finite, as is apparent from the following.

Using the Nusselt numbers obtained above for the stationary isothermal sphere, Eqs. (127) to (129), and inserting them into Eqs. (68) to (70), the transfer rates of heat, mass, and momentum can be written for this special case as follows:

$$H_0 = \frac{4\ \pi\ r\ \lambda\ (T^* - T_r)}{1 + (\sqrt{2\pi}/Pr)[\gamma/(\gamma-1)][\bar{R}^{3/2}/(R^{3/2}\ K\ \alpha_H)]\ Kn} \tag{133}$$

$$M_0 = \frac{4\ \pi\ r\ (D/R_v\ T_\infty)(p_{v\infty} - p_r)}{\Pi_{g\infty} + (\sqrt{2\pi}/Sc)(1/\alpha_{Mv})\ Kn} \tag{134}$$

$$F_O = \frac{6\pi\, r\, \eta\, \Delta w_\infty}{1 + [9\sqrt{2\pi}/(8 + \pi\, \bar{\alpha}_F)]\ \mathrm{Kn}} \qquad (135)$$

It is seen that all three expressions are of identical structure, except for M in the case of pure-vapor carriers ($\Pi_{g\infty} = 0$). The transfer rates are proportional to a driving difference and to the droplet radius r. The latter also enters implicitly by way of Kn. In the case of pure vapors, M is strictly proportional to 1/Kn. The presence of a noncondensing gas ($\Pi_{g\infty} > 0$) is seen to reduce the mass transfer rate. The expressions of heat and momentum transfer are influenced by the mixture composition only through the carrier properties λ, η, γ, Pr, and the mass averages designated by overbars and calculated in accordance with Eq. (A-7).

3.4.5 Relative Flow and Other Corrections

More general expressions for H, M, and F, which include the effects of *relative velocity* and of $T_r \neq T_\infty$, are not explicitly presented here. They are of similar general behavior and can easily be derived by inserting the general expressions giving Nu^{ct} and B, that is, Eqs. (80) to (82) and Eqs. (106) to (108), into the multirange relationships Eqs. (123) to (125), and then using the Nu values thus obtained to calculate H, M, and F from Eqs. (68) to (70). It is noted that any two of the values Kn, Re, Ma determines the value of the third, by way of Eq. (51).

Although the above procedure is adequate in most cases of practical interest, further refinements may be appropriate in some special situations. For example, in the case of *vapor-rich carriers*, Nu_M^{ct} should be replaced by the corrected (nonlinear) values $\mathrm{Nu}_{M,nl}^{ct}$ obtained from Eq. (83). (See Sec. 3.2 for comments on the need for this correction.) Another correction may arise if the *spatial density* of droplets is very high (i.e., $r_\infty/r<20$, see Fig. 3). A correction to Nu_H^{ct} is given by Eq. (153) for the simple case of spheres subject to nonsteady cooling (or heating) in a stagnant, constant-temperature environment. No proximity corrections are available for Nu_M^{ct}. Regarding Nu_F^{ct}, corrections may be taken from the literature dealing with particle drag; cf. Rudinger (1965).

Effects arising from the *simultaneous presence of mass transfer* can be accounted for by using in Eq. (123) the corrected parameter $\mathrm{Nu}_{H,M}^{ct}$ as given by Eqs. (84) or (85).

3.4.6 Comparison with Experimental Data

Comparison with empirical results can readily be made for Nu_F only. The most recent empirical correlations for the sphere's aerodynamic drag are those of Walsh (1975) and of Henderson (1976) which have been critically discussed by these authors (1977). In the subsonic range these correlations reproduce the available empirical data to within 7% and 16%, respectively. Henderson's correlation, which has the convenience of using a single analytic expression for the entire subsonic range, can be written, if c_D and Ma are expressed by Eqs. (76) and (61) and $\gamma = 1.4$ is set, as

$$Nu_F^{Hend} = \frac{3}{1 + f_1(Kn)} + \left[f_2(Re) + \frac{Re^3\ Kn^2}{488} + \frac{Re^9\ Kn^8}{39340} \right] e^{-0.2113\ Kn\ \sqrt{Re}}$$

$$+ \left(1 - e^{-0.4226\ Kn}\right) \frac{Re^2\ Kn}{53.33} \qquad (136)$$

where

$$f_1(Kn) = \left(1.531 + 0.5544 e^{-0.699/Kn}\right) Kn \qquad (137)$$

$$f_2(Re) = 0.5625\ Re\ \frac{1 + 0.0844(0.03\ Re + 0.48\ \sqrt{Re})}{1 + 0.03\ Re + 0.48\ \sqrt{Re}} \qquad (138)$$

This equation is valid for Ma<1, that is, for $Kn < 2\sqrt{\gamma}/Re = 2.366Re$. It has to be compared to Eq. (125) combined with Eqs. (82), (108), and (111), which give for a single-component carrier (air), using $\gamma = 1.4$, $\alpha_F = 1$, and $T_r = T_\infty$,

$$Nu_F^{air} = \frac{3(1 + 0.15\ Re^{0.687})}{1 + 18(\sqrt{\pi}/2)[(1 + 0.15\ Re^{0.687})/(8 + \pi + Kn^2\ Re^2/5)]Kn} \qquad (139)$$

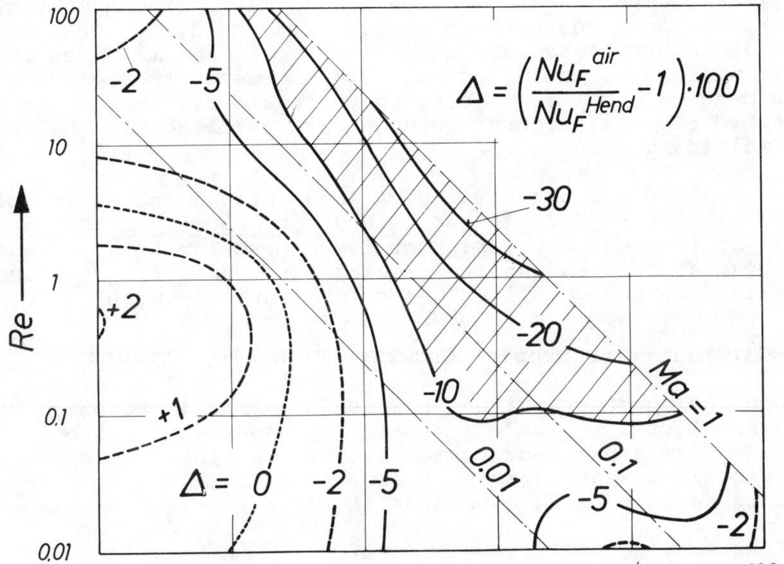

Fig. 13. Percentage deviation Δ between Nu_F^{air} as obtained from Eq. (139) and Henderson's empirical drag correlation, Eq. (136), as a function of Kn, Re, and Ma.

The percentage deviation Δ of Nu_F^{air} from Nu_F^{Hend} is shown in Fig. 13 over the Kn, Re field up to the limit Ma = 1. It is seen that Δ is small in the continuum (Kn $\to$ 0) and free-molecular extremes (Kn $\to \infty$). Deviations in excess of 10% (shaded area) occur, however, in the intermediate range for Mach numbers in excess of typically 0.1. A comparison with the correlation by Crowe et al. (1971) has lead to deviations of similar order. One may conclude that Eq. (125) is numerically less accurate than the best c_D correlations found in literature. Because of its partially theoretical origin, however, Eq. (125) can be more readily generalized to carrier fluids other than pure air. An important benefit from the reasonable agreement found above is that the analogous expressions for Nu_H and Nu_M, Eqs. (123) and (124), may be assumed to represent satisfactorily the dependence of these coefficients on Re, Kn (and Ma).

For Nu_M no empirical data, and for Nu_H only few empirical data are available outside the continuum range. Kavanau's (1955) semiempirical equation for air, as cited by Schaaf and Chambré (1958), is equivalent to setting Kn/B_H = 3.42Ma/Re Pr, which gives $B_H = 2\sqrt{\gamma}$ Pr/3.42. Using the data γ = 1.4 and Pr = 0.70, one obtains B_H = 0.484, a value that is in perfect agreement with the value B_H = 0.48 calculated for this case from Eq. (114).

3.4.7 Review of the Crucial Parameters

For the continuum and free-molecular cases, the dependence of the Nusselt numbers and of M, F, and H on the pertinent nondimensional parameters has been reviewed in Secs. 3.2.5 and 3.3.6, respectively. The dependence structures are shown in Fig. 10 and 11. The multirange expressions given in Sec. 3.4.3 being mere interpolation formulas between these two cases, they contain all influences shown in these two figures. Therefore no separate scheme will be drawn for the multirange case.

The relative strength of the various influences will comply more closely with Fig. 10 if Kn<0.1 and more closely with Fig. 11 if Kn>1. It is seen, however, that the dependence of M, F, and H on the physical data and on the driving-force ratios $\Pi_r/\Pi_{v\infty}$ and T_r/T_∞ will be strong at all values of the Knudsen number.

3.5 Applicability to Droplet Growth; Promixity Effects

Since the equations given in Secs. 3.2 to 3.3 are based on the premises listed in Table 4, their applicability to droplets of changing size and to environments that are finite in space and changing in time is not granted automatically. Therefore these assumptions have to be discussed critically.

3.5.1 Constancy of r (Assumption IVa).

With the exception of the free-molecular case, the carrier-phase conditions (T, Π_v, Δw) locally vary in the vicinity of the sphere. Obviously, the spatial scale of these near-field variations is dictated by the size of the sphere. If r changes, the distributions must change their scale too. The near-field distributions have been analyzed in detail only for continuum conditions, and the various cases treated range from spherical symmetry at no flow (Re = 0) (e.g., Carslaw and Jaeger, 1959; Gyarmathy, 1963), to the much more complicated laminar or boundary-layer/wake-type flow situations at higher Re (e.g., Stokes, 1851; Ranz and Marshall, 1952; Hamielec and Hoffmann, 1967; Hayward and Pei, 1978).

However, a change of r is significant only if the near-field distributions cannot adapt themselves fast enough. Let Δt_G be the characteristic time constant of radius change,

$$\Delta t_G \triangleq \frac{r}{|dr/dt|} \tag{140}$$

We now estimate the adaptation time of the environment for the temperature near field under continuum conditions in the absence of relative flow. If a constant positive temperature difference $T_\infty - T_r$ is maintained, a heat flux

$$H = 4\,\pi\, r\, \lambda (T_\infty - T_r) \tag{141}$$

is flowing toward the drop, as follows from Eq. (68) by setting $T^*_\infty = T_\infty$ and $\mathrm{Nu}_H = 2$. The excess thermal energy stored in the near field can be estimated as

$$E \approx \bar{c}_p\, \rho_\infty\, (T_\infty - T_r)\, \frac{4\,\pi \left[(r + 2r)^3\right] - r^3}{3} \tag{142}$$

where the "thermal displacement thickness" of the near field has been estimated to be of the order of $2r$.

The criterion for sufficiently fast adaptation is that dE/dt is only a small fraction of H. For example, we say arbitrarily that

$$\left| \frac{1}{H} \frac{dE}{dt} \right| < 0.1 \tag{143}$$

With the expressions given above, this criterion becomes

$$\Delta t_G > \frac{3^3 - 1}{0.1} \frac{\bar{c}_p\, \rho_\infty}{\lambda} r^2 = 260 \frac{r^2}{a} \tag{144}$$

where a is the thermal diffusivity of the carrier medium; cf. Eq. (56a).

Since in practical problems Δt_G is usually not known a priori, it must be estimated. A good estimate to M is given by Eq. (25). This gives with $\mathrm{Nu}_H = 2$ for continuum, no-flow conditions,

$$4\,\pi\, r^2\, \rho_\ell \frac{dr}{dt} \approx \frac{4\,\pi\, r\, \lambda (T_r - T_\infty)}{L} \tag{145}$$

which with Eq. (140) leads to

$$\Delta t_G \approx \frac{r^2 \rho_\ell L}{\lambda \, |T_\infty - T_r|} \tag{146}$$

Upon insertion of this into Eq. (144) it appears that r may be treated as quasi-constant if

$$|T_\infty - T_r| < \frac{\rho_\ell/\rho_\infty}{260} \frac{L}{\bar{c}_p} \tag{147}$$

This condition is seen to be independent of the value of r (for continuum-type carriers). For example, in case of water droplets suspended in steam of $p_\infty = 10$ bar, we have $\rho_\ell/\rho_\infty = 172$ and $L/\bar{c}_p = 776$ K, giving $|T_\infty - T_r| < 510$ K, a condition more than fulfilled in most practical situations. For $p_\infty = 1$ bar the limit would be about 5100 K. It is thus apparent that the heat transfer equations for a constant-size sphere are applicable to growing droplets except in cases of high-density, highly superheated carriers. Such extreme situations may occur, for example, in combustion chambers and rocket nozzles, and have been analyzed by Torda (1973). In these cases typically $T_\infty > T_{b\infty} \approx T_r$, and the criterion for the quasi-constancy of r is, from Eq. (147),

$$\frac{T_\infty - T_{b\infty}}{T_{b\infty}} < \frac{\rho_\ell}{\rho_\infty} \frac{\gamma - 1}{260\,\gamma} \frac{L}{\bar{R}\, T_{b\infty}} \approx 10^{-3}\ \mathrm{Cl}\ \frac{\rho_\ell}{\rho_\infty} \tag{148}$$

The limit set for T_∞ is seen to be the more stringent the higher the carrier gas density ρ_∞.

3.5.2 Constancy of T_r and p_r (Assumption IVb).

Because of the intense bombardment by the gas-phase molecules and the simultaneous reemission of molecules from the surface, as discussed in Sec. 3.3.1, the surface layer of droplets very quickly assumes a temperature dictated by dynamic equilibrium. Therefore the constancy of T_r and p_r is linked in the case of droplets growth to the constancy of the far-field environmental conditions.

3.5.3 Constancy of Far-Field Conditions (Assumption IVc).

Again the temperature field will be chosen as example. The radius will be constant. Nonsteady conditions now imply that the temperatures in the near field do not follow the rate of change of T_∞. If the near field retained its temperature at T_r while T_∞ was changing, its excess thermal energy given by Eq. (142) would change at the rate

$$\left|\frac{dE}{dt}\right| \approx \bar{c}_p\, \rho_\infty \left|\frac{d\,T_\infty}{dt}\right| \frac{4\,\pi\,(r + 2r)^3}{3} \tag{149}$$

Quasi-steady conditions prevail if this energy change is only a small fraction of H, as required by Eq. (143). Using Eqs. (141) and (149), the condition (143) is found to be fulfilled if

$$\left| \frac{d\,T_\infty}{dt} \right| < \frac{0.1}{3^2} \frac{\lambda}{\bar{c}_p\,\rho_\infty} \frac{|T_\infty - T_r|}{r^2} \tag{150}$$

Here $|dT_\infty/dt|$ is given by the rate of the process to which the carrier phase is subjected. Introducing the Stodola number from Eq. (51) and using Eqs. (43), (41), (32), and (57), the above condition assumes the form

$$\mathrm{Sd} > 90\,\frac{n-1}{n}\,\frac{\mathrm{Pr}}{\mathrm{Kn}^2}\,\frac{T_\infty}{|T_\infty - T_r|} \tag{151}$$

The right side is seen to be essentially determined by the value of Kn. Taking as example a droplet of $r = 1$ µm in air of $p_\infty = 1$ bar, Fig. 6 indicates that $\mathrm{Kn} \approx 0.03$. Setting $n = 1.4$, $\mathrm{Pr} = 0.7$, and assuming that $T_\infty/|T_\infty - T_r| \approx 30$, which is a typical value for far-from-equilibrium conditions, the condition is fulfilled if $\mathrm{Sd} > 6\mathrm{x}10^5$. As seen form Fig. 8, at $p_\infty = 1$ bar this is found to hold even in the fast expansions occurring in wind tunnels and steam turbines. For larger droplets, however, the condition would not be fulfilled, unless the process were correspondingly slower.

It follows from the close analogy of the three transfer processes that the conditions imposed on the rate of environmental change by the concentration and velocity fields are of the same order of magnitude. The conclusion is that constancy of far-field conditions may be assumed in most practical situations. Caution has to be exercised only if the environmental change is very fast ($\mathrm{Sd}<10^5$) or if the droplet is relatively large ($\mathrm{Kn}<0.01$).

If the condition (151) is not fulfilled, the transfer processes have to be treated as nonsteady. The case of thermal transients during droplet evaporation in a hot carrier has been numerically analyzed by Hubbard et al. (1975). The effects of the carrier-side transients were found negligible beside the transients within the droplet.

3.5.4 Infinite Environment (Assumption IVd).

As seen from Fig. 3, the assumption of an infinite environment ($r_\infty/r \to \infty$) is a severe idealization for fluids having pressures above atmospheric and carrying significant liquid mass fraction. Only the continuum case needs to be considered. If r_∞ is finite, each droplet has its own "yard" with which it exchanges heat, mass, and momentum. At boundary with neighboring drops (i.e., at $\zeta = r_\infty$), the radial gradients of the T, Π_v, and Δw fields are zero by symmetry reasons. This case is illustrated in Fig. 14, using the spherically symmetric temperature fields of a stationary sphere as an example. Unsteady heat conduction in this situation has been analyzed by Konorski (1968).

There are two errors incurred when the transfer-rate expres-

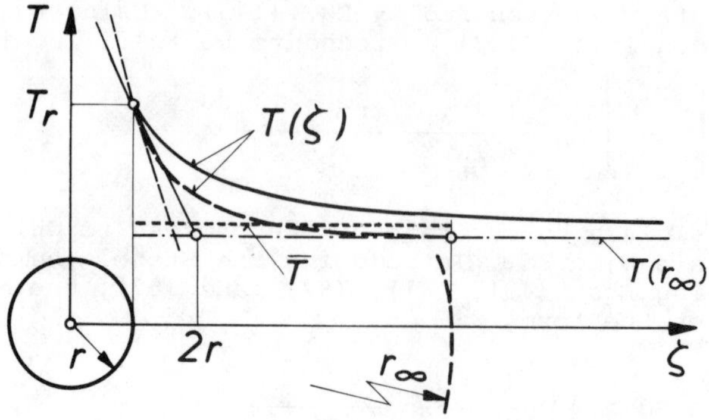

Fig. 14. Shape of the radial temperature profile for infinite and finite spacing of droplets, at given values of T_r and $T(r_\infty)$; for quiescent, growing droplet, under continuum conditions. Solid line: infinite environment. Dashed line: finite environment (radius r_∞) having averaged temperature $\bar{T}$ different from $T(r_\infty)$.

sions derived for the infinite environment are used for finite-sized fluid samples. First, for a given difference $T_r - T(r_\infty)$, the temperature gradient at the surface is underestimated. Second, a distinction has to be made between the sample-boundary temperature $T(r_\infty)$ and the spatially averaged carrier temperature $\bar{T}$. In the energy balance of the carrier (App. 3), T_∞ means $\bar{T}$, whereas in all heat transfer and droplet growth equations T_∞ should obtain the value $T(r_\infty)$. Not making this distinction means setting $T(r_\infty) = \bar{T}$, by which the temperature gradient is further underestimated. If the temperature distribution $T(\zeta)$ over $r < \zeta < r_\infty$ is known, the gradient $dT/d\zeta$ at $\zeta = r$ and the spatial mean temperature

$$\bar{T} \triangleq \frac{3}{r_\infty^3 - r^3} \int_r^{r_\infty} T(\zeta)\, \zeta^2\, d\zeta \tag{152}$$

can be calculated for given values of $T(r_\infty)$ and $T(r) = T_r$.

In estimating the magnitude of the correction for a given value of r_∞/r, a principal difficulty arises in that the transfer of heat (or mass or momentum) in a finite, adiabatically or near adiabatically bounded sample is an inherently nonsteady process. The solution of the time-dependent conduction problem would be required for a domain the radius r_∞ of which is in general changing with time. For spherical symmetry, analytic solutions should be possible, although no published results seem to exist.

An estimation of this "proximity correction" was carried out by calculating the nonsteady temperature profiles existing outside the sphere as its surface temperature T_r is approaching $T(r_\infty)$. This latter, as well as r and r_∞, were assumed constant. By neglecting the higher-order eigenfunctions (the influence of which decay rapdily with time) – that is, by retaining the basic time-dependent solution only – the correction was found to be of the order of 10% for $r_\infty/r = 20$, and 20% for $r_\infty/r = 10$. About four-fifths of these corrections are due to the gradient steepening and one-fifth to the difference between $\bar{T}$ and $T(r_\infty)$. This correction leads to a Nusselt number Nu_H^{prx} (the superscript indicates the proximity effect). In Eq. (68) the temperature T_∞^* should then be calculated by setting $T_\infty = \bar{T}$, that is, using the value given by the conservation equations. The dependence of Nu_H^{prx} on r_∞/r is found to be correlated in the range $r_\infty/r > 3$ by the expression

$$\mathrm{Nu}_H^{prx} = \mathrm{Nu}_H^{ct} \left[1 + 3.8\ (r_\infty/r)^{-1.24}\right] \tag{153}$$

Here Nu_H^{ct} is the infinite-environment steady-state Nusselt number given by Eq. (77). Equation (153) is plotted in Fig. 15.

No analysis is available for the practically more important case of expanding carrier flows. There the change of r_∞ and the distributed heat sink effect of expansion work make analytic solutions more difficult. Such solutions would be of interest for high-pressure steam flow in nozzles and turbine blade passages.

Proximity effects in particle drag have been subject to detailed investigations; cf. Rudinger (1965). At high Reynolds numbers (Re>100), Nu_H and Nu_M are not affected by neighboring droplets unless the drop happens to be in another drop's wake and their distance is less than five diameters (Miura et al., 1977). Proximity effects between two stationary drops were recently analyzed by A. L. Williams et al. (1978).

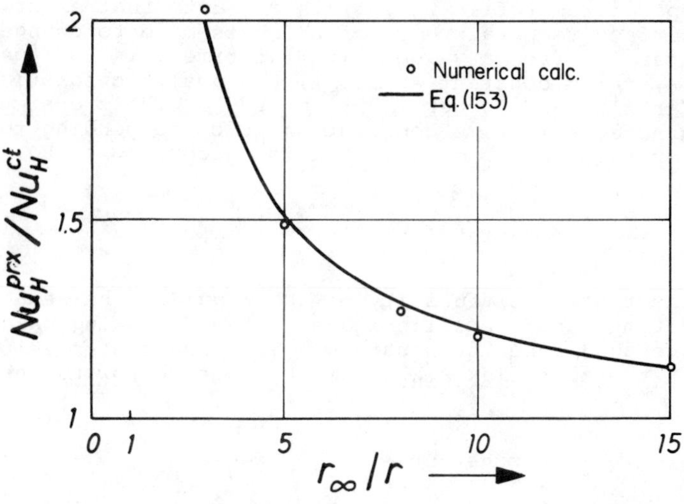

Fig. 15. Estimated magnitude of the proximity correction for heat transfer.

3.5.5 Body-Force Effects (Assumption IVe).

The primary effect of body forces is to cause relative motion, which is accounted for by the Re or Ma dependence of the Nusselt numbers. The secondary effect is that of "free convection." This is small in the case of small droplets (unless extreme temperature fields and very large body forces exist simultaneously) and will be neglected throughout this study.

3.6 Heat Conduction in the Interior of the Droplets

The inner temperature of small droplets differs in most situations so little from the surface temperature that the two may be set equal ($T_\ell = T_r$). Important differences may arise, however, when the environmental conditions are changing quickly. An extreme case, for example, is that of spray cooling. Here relatively large and cold droplets are suddenly brought into a vapor-rich environment that is much hotter.

Transient heat conduction in a sphere is exhaustively analyzed in the book of Carslaw and Jaeger (1959). Application to spray cooling has been treated in detail by Rüping (1959), Nitsch (1971) and Kulić et al. (1975, 1978). If the droplet is large, internal circulation may be significant. This case is not considered here (assumption If). If Re>20, this effect may lead to an accelerated thermal adaptation of the droplet (Dibelius and Voss, 1979).

The concern of the present investigation is focused on situations in which any transient phenomena resulting from a sudden exposure of the droplet to a new environment (as in the case of spray injection or shock wave passage) have already disappeared, but the *environmental conditions keep changing continually*. Then the internal temperature profile $T(\zeta, t)$ is close to that existing in a sphere the surface temperature T_r of which is made to change linearly with time, starting from a value $T_r(t_0)$ at time t_0. For the initial condition $T(\zeta, t_0)$ = constant = $T_r(t_0)$, the analytic solution is known (see Carlslaw and Jaeger, 1959, p. 235). The transients decay within a characteristic time constant Δt_{int} corresponding to a Fourier number value of Fo = 0.3, that is, according to Eq. (67), within

$$\Delta t_{int} = 0.3\,\frac{r^2}{a_\ell} \tag{154}$$

Here a_ℓ is the thermal diffusivity of the liquid. The remaining quasi-steady temperature profile leads to the following values of the volume-averaged inner temperature T_ℓ, of the center temperature T_{cent}, and of the heat flux H_{int} from the surface toward the interior:

$$T_\ell - T_r \approx \text{constant} = -\frac{1}{15}\,\frac{r^2}{a_\ell}\,\frac{d\,T_r}{dt} \tag{155a}$$

$$T_{cent} - T_r \approx \text{constant} = -\frac{1}{6}\,\frac{r^2}{a_\ell}\,\frac{d\,T_r}{dt} \tag{155b}$$

$$H_{\text{int}} \approx \text{constant} = \frac{4\,\pi\,r^3}{3}\,\rho_\ell\,c_\ell\,\frac{d\,T_r}{dt} \qquad (155c)$$

If Eqs. (155*a*) and (155*c*) are inserted into Eq. (66), it appears that for the quasi-steady case the Biot number has the value Bi = 10. Higher Bi values may occur during the transients taking place during the time Δt_{int} following a sudden change of the environment.

Except during such transients, the surface temperature T_r of a droplet is dictated by its size and by the environmental conditions, as will be shown in Sec. 5. In most cases, especially in pure-vapor ($g_\infty = 0$), vapor-rich ($g_\infty < 1$), and hot gas-rich ($g_\infty >> 1$, $T_\infty > T_{b\infty}$) carriers, one may set

$$\frac{d\,T_r}{dt} \approx \frac{d\,T_{b\infty}}{dt} = -\,\dot{P}\,\frac{T_{b\infty}}{\text{Cl}} \qquad (156)$$

whereas in cold gas-rich carriers ($g_\infty >> 1$, $T_\infty < T_{b\infty}$)

$$\frac{d\,T_r}{dt} \approx \frac{d\,T_\infty}{dt} = -\,\dot{P}\,T_\infty\,\frac{n-1}{n} \qquad (157)$$

is the better approximation. The errors incurred in using the wrong estimate are rarely significant, though. The last equalities in Eqs. (156) and (157) follow from Eq. (A-166) in App. 1 and Eq. (57), respectively. The latter is valid for polytropic carrier expansion processes having exponent n.

If no other comment is made, the approximation expressed by Eq. (156) will be adopted in the following. Equation (155*c*) is then written as

$$H_{\text{int}} \approx \frac{4\,\pi\,r^3}{3}\,\rho_\ell\,c_\ell\,\frac{d\,T_{b\infty}}{dt} = m_\ell\,c_\ell\,\frac{d\,T_{b\infty}}{dt} = -\,m_\ell\,c_\ell\,T_{b\infty}\,\frac{\dot{P}}{\text{Cl}} \qquad (158)$$

This quasi-steady expression for H_{int} will be used in the analysis of droplet growth in Sec. 5.1.4 and Secs. 5.2 to 5.3. The temperature differences within the droplet follow from

$$\frac{T_{\text{cent}} - T_r}{T_{b\infty}} \approx \frac{1}{6}\,\frac{r^2}{a_\ell}\,\frac{\dot{P}}{\text{Cl}} = \pm\,\frac{1}{6}\,\frac{a}{a_\ell}\,\frac{\text{Pr}}{\text{Cl Sd Kn}^2} \qquad (159)$$

$$\frac{T_\ell - T_r}{T_{b\infty}} \approx \frac{1}{15}\,\frac{r^2}{a_\ell}\,\frac{\dot{P}}{\text{Cl}} = \pm\,\frac{1}{15}\,\frac{a}{a_\ell}\,\frac{\text{Pr}}{\text{Cl Sd Kn}^2} \qquad (160)$$

It is seen that *the smaller is the droplet size* — that is, the higher the value of Kn — *the smaller are the temperature differences*. In a constant environment (Sd $\to \infty$), the quasi-steady case of course gives uniform temperature $T_{\text{cent}} = T_\ell = T_r$. For the last equality, Eqs. (32), (33), (51) and (56*a*) were used. In Eqs. (159), (160) the positive sign applies to carrier expansion (cooling), the negative sign to compression (heating).

Typical values of a/a_ℓ, Pr, and Cl can be taken from Figs. 4 and 5. It is seen that the decisive factor is the product $Sd \cdot Kn^2$, which may be very different according to droplet size, carrier pressure, and rate of the process; cf. Fig. 6 and 8. For a droplet of $r = 1$ μm in saturated steam of $p_\infty = 1$ bar (giving Kn = 0.03) and expanding at a rate corresponding to $Sd = 10^7$ (which is a fast rate, typical of steam turbines and supersonic nozzle flow, cf. Fig. 8), one obtains

$$T_{\text{cent}} - T_r \approx 6\text{x}10^{-5}\, T_{b\infty} \approx 0.02 \text{ K}$$

a negligibly small value. Important values of the order of 10 K would be found in droplets of $r = 20$ μm and larger. With a_ℓ being (for water) of the order of 10^{-7} m^2/s, the adjustment time Δt_{int} given by Eq. (154) is of the order of a few microseconds for droplets of 1 μm size. See Fig. 20 for further numerical data.

The above quasi-steady model of heat conduction within the droplet, as formulated by Eqs. (155) to (160), applies only to times $t > (t_0 + \Delta t_{\text{int}})$, where t_0 denotes the time at which the environmental change has been effected. Processes involving considerable changes in droplet size have a time duration of the order of Δt_G or longer. If $\Delta t_{\text{int}} \ll \Delta t_G$, then droplet growth or vaporization may be analyzed on the basis of Eqs. (155) to (160). This condition can be written by using Eqs. (154) and (160) as

$$0.3\, \frac{c_\ell}{L}\, |T_\infty - T_r| \ll \frac{\lambda_\ell}{\lambda} \tag{161}$$

It is seen that this condition does not depend on droplet size. Since λ_ℓ/λ is of the order of 10 to 100 (cf. Fig. 5), Eq. (161) is fulfilled in most situations.

In processes with sudden change of carrier environment (such as passage of a shock wave) or injection of droplets into a new environment (as in spraying), the *initial droplet size* may be corrected in the following way in order to account for the growth or evaporation taking place during the first time period of the order of Δt_{int}:

$$r_0' = r_0 \sqrt[3]{1 + (T_\ell - T_{\ell 0})\, c_\ell / L} \tag{162}$$

Here r_0 and $T_{\ell 0}$ are the initial values given by Eq. (1), r_0' is the corrected initial value to be used in connection with the quasi-steady droplet growth laws given in Sec. 5.3, and T_ℓ is the quasi-steady droplet temperature, for which the following estimate can be made:

$$T_\ell \approx T_r = T_\infty + (T_{s\infty} - T_\infty)\, Q \tag{163}$$

Here Q is to be calculated from the pertinent expression given in Sec. 5.2.5. For pure vapor environments ($Q \approx 1$), one finds that $T_\ell \approx T_r = T_{s\infty}$. In cold, lean gases, $T_\ell \approx T_\infty$ and in hot, lean gases, $T_\ell \approx T_{b\infty}$ may serve as tentative estimates until a better value of T_r is known.

3.7 Drift Velocity

A droplet carried along in a uniformly accelerated flow and/or subjected to a body force assumes a relative velocity $\Delta\vec{w}_{drf}$ that is different from zero. The amount of this "drift", "slip", or "sedimentation" velocity is a function of droplet size. Droplets injected into the carrier with relative velocities $\Delta\vec{w}_\infty \neq \Delta\vec{w}_{drf}$ change their velocity until equality is attained.

According to the equation of motion, Eq. (17),

$$m_\ell \frac{d\Delta\vec{w}_\infty}{dt} = -2\pi r\eta \cdot \mathrm{Nu}_F(\Delta w_\infty)\cdot\Delta\vec{w}_\infty + m_\ell\left(\vec{b} - \frac{1}{\rho_\ell}\,\mathrm{grad}\,p_\infty - \frac{d\vec{w}_{carr}}{dt}\right) \tag{164}$$

where $\vec{w}_\ell$ has been expressed by Eq. (18) and F by Eq. (70). The symbol $\mathrm{Nu}_F(\Delta w_\infty)$ refers to the dependence of Nu_F on the scalar of $\Delta\vec{w}_\infty$. By defining the drift velocity $\Delta\vec{w}_{drf}$ by the condition

$$\left(\frac{d\Delta\vec{w}_\infty}{dt}\right)_{\Delta\vec{w}_\infty = \Delta\vec{w}_{drf}} \triangleq 0 \tag{165}$$

we can write

$$0 = -2\pi r\eta\ \mathrm{Nu}_F(\Delta w_{drf})\cdot\Delta\vec{w}_{drf} + m_\ell\left(\vec{b} - \frac{1}{\rho_\ell}\,\mathrm{grad}\,p_\infty - \frac{d\vec{w}_{carr}}{dt}\right) \tag{166}$$

For droplets ($\rho_\ell \gg \rho_\infty$), in contrast to bubbles, the pressure gradient term is usually negligibly small. If m_ℓ is expressed by Eq. (2), the drift velocity is obtained as

$$\Delta\vec{w}_{drf} \approx \frac{2}{3}\,\frac{r^2\rho_\ell}{\eta\ \mathrm{Nu}_F(\Delta w_{drf})}\left(\vec{b} - \frac{1}{\rho_\ell}\,\mathrm{grad}\,p_\infty - \frac{d\vec{w}_{carr}}{dt}\right) \tag{167}$$

It is seen tnat the drift velocity has the direction of the vectorial sum of the carrier deceleration and of the pressure gradient and body force fields. Its scalar value is proportional to r^2 and to the reciprocal of Nu_F, this latter being given, depending on the type of conditions, by one of the equations (79), (82), (105), or more generally by Eq. (125). Since the value of Nu_F depends in general (by way of Re or Ma) on the relative velocity, the above is an implicit equation for $\Delta\vec{w}_{drf}$.

It is of interest to know how fast the adaptation $\Delta\vec{w}_\infty \to \Delta\vec{w}_{drf}$ occurs. During this adaptation any initial relative velocity components perpendicular to $\Delta\vec{w}_{drf}$ decay to zero while the parallel component assumes the scalar value Δw_{drf}. By subtracting Eq. (166) from Eq. (164), assuming that $d\Delta w_{drf}/dt = 0$ and looking at the components parallel to $\Delta\vec{w}_{drf}$ only, we have

$$m_\ell \frac{d(\Delta w_\infty - \Delta w_{drf})}{dt} = 2\pi r\eta \left[\Delta w_{drf}\ \mathrm{Nu}_F(\Delta w_{drf}) - \Delta w_\infty\ \mathrm{Nu}_F(\Delta w_\infty)\right] \quad (168)$$

The time constant of velocity adaptation will be defined as

$$\Delta t_{vel} \mathrel{\hat{=}} - \frac{\Delta w_\infty - \Delta w_{drf}}{d(\Delta w_\infty - \Delta w_{drf})/dt} \quad (169)$$

In order to estimate its value, we neglect the variation of Nu_F with Δw_∞. Thus we obtain, by inserting m_ℓ from Eq. (2),

$$\Delta t_{vel} \approx \frac{2}{3} \frac{r^2\ \rho_\ell}{\eta\ \mathrm{Nu}_F} \quad (170)$$

This velocity adaptation time will now be compared to the time constant of droplet radius change, Δt_G, as defined by Eq. (140). Using Eq. (146) to estimate Δt_G and setting $T_r \approx T_{b\infty}$, we obtain

$$\frac{\Delta t_{vel}}{\Delta t_G} \approx \frac{2\ \bar{c}_p/R_v}{3\ \mathrm{Pr\ Cl\ Nu}_F} \frac{|T_\infty - T_{b\infty}|}{T_\infty} \quad (171)$$

The first factor on the right side is typically of the order of $2/\mathrm{Cl} \approx 0.3$–$0.1$, and the temperature ratio is much smaller than 1. Therefore we find that *velocity adaptation is usually much faster than droplet growth*. This is especially pronounced under continuum and near-continuum conditions (Kn<1), where Nu_F is of the order of 2 or even larger. An exception is the case of very large Knudsen numbers (Kn>10), where Nu_F may become much smaller than unity. The rapidity of velocity adaptation allows us to analyze droplet growth by assuming that the relative velocity $\Delta\vec{w}_\infty$ is, at all instants, substantially equal to the drift velocity, that is, that

$$\Delta w_\infty \approx \Delta w_{drf} \approx \frac{2}{3} \frac{r2\ \rho_\ell}{\eta\ \mathrm{Nu}_F(\Delta w_{drf})} \left| \vec{b} - \frac{d\vec{w}_{carr}}{dt} - \frac{1}{\rho_\ell} \mathrm{grad}\ p_\infty \right| \quad (172)$$

By this assumption the motion and the growth of the droplet are mathematically decoupled, and the latter can be analyzed in the simplified manner discussed in Sec. 5.1.

The body force and pressure gradient fields may be neglected in many instances. In highly swirling flows, such as in turbomachines, $d\vec{w}_{carr}/dt$ is determined mainly by the centrifugal force field. In any type of flow the acceleration of the carrier can be conveniently characterized by the time constant

$$\Delta t_{acc} = \frac{|\vec{w}_{carr}|}{|d\vec{w}_{carr}/dt|} \tag{173}$$

expressing the time in which the velocity would change by its own amount provided the acceleration stayed constant. Here the absolute values mean scalars of positive sign. According to Eq. (57), in an expanding (accelerating) *rectilinear* polytropic flow,

$$\Delta t_{acc} = \frac{\gamma \; Ma^2_{carr}}{\dot{P}} \tag{174}$$

where

$$Ma_{carr} \triangleq \frac{|\vec{w}_{carr}|}{\sqrt{\gamma \; \bar{R} \; T_\infty}} \tag{175}$$

is the Mach number of the carrier flow with respect to its boundaries.

4 EQUILIBRIUM AND STABILITY OF DROPLETS

The expressions of Sec. 3 describe the heat and mass transfer between a sphere and the carrier for a given temperature T_r and vapor pressure ratio $\Pi_r = p_r/p_\infty$ existing at the sphere surface; cf. Eqs. (68) and (69). In the case of volatile substances, as liquid droplets or ice particles, T_r and p_r are mathematically coupled as a result of the thermodynamic properties of phase interfaces. In Sec. 4.1 the relationship $p_r(T_r)$ will be formulated and the conditions of phase equilibrium (zero growth) discussed. Section 4.2 will treat the entirely different topic of mechanical and thermal stability of droplets exposed to fast relative flow or subjected to a fast decrease of carrier pressure.

4.1 Thermodynamic Equilibrium

A *flat liquid surface* is in thermodynamic equilibrium with the adjacent gaseous phase if both have the same temperature T and the vapor has (partial) pressure p_v, which equals the saturation pressure p_s pertaining to T. The vapor pressure function $p_s(T)$ is a material property of the liquid substance and obeys with very good approximation the equation

$$\ell n \; \frac{p_s(T)}{p_s(T_{ref})} = Cl \left(1 - \frac{T_{ref}}{T}\right) \tag{176}$$

see App. 1, Eq. (A-13). The Clausius-Clapeyron number is defined by Eq. (34) and has to be evaluated at the reference temperature T_{ref}.

For a *droplet* of radius r, thermodynamic equilibrium requires (Kelvin, 1869; Helmholtz, 1886) that both droplet and carrier have the same temperature T and the vapor has (partial) pressure p_v, which equals

$$p_r = p_s(T) \exp \frac{2\sigma}{r \rho_\ell R_v T} \tag{177}$$

Here the surface tension σ is assumed to be independent of the degree of surface curvature, that is, of radius r (assumption Ie). The exponent being positive, the *vapor pressure of a droplet* is always greater than that of a flat liquid surface of the same temperture T_r, and is given by

$$p_r = p_r(T_r, r) = p_s(T_r) \exp \frac{2\sigma(T_r)}{r \rho_\ell R_v T_r} \approx p_s(T_r)\, e^{\mathrm{Ke}} \tag{178}$$

The exponential factor represents the "capillarity correction." It is seen to be governed by the dimensionless group termed the *Kelvin number* Ke and defined by Eq. (62). The last equality is an approximation implying that $\sigma(T_{\mathrm{ref}}) \approx T_\infty\, \sigma(T_r)/T_r$. If $T_{\mathrm{ref}} \approx T_r$ is set, this approximation is sufficiently good in all cases of interest and will be used throughout this chapter.[17]

The Ke number thus emerges as a measure of droplet size in terms of the importance of capillarity effects. It is inversely proportional to r and is related to Kn according to Eq. (63). Ke = constant lines are plotted in the r, p_∞ plane in Fig. 6. As seen from Eq. (178), the capillarity correction exceeds 1% if Ke>0.01. Figure 6 shows that this condition is fulfilled for water drops only if their radius is smaller than about 0.1 μm.

The *saturation ratio* of the carrier phase, defined as

$$S_\infty \triangleq \frac{p_{v\infty}}{p_s(T_\infty)} = \frac{p_{v\infty}}{p_{s\infty}} \tag{179}$$

has a decisive effect on droplet behavior. If $S_\infty > 1$, the carrier is said to be *supersaturated* or *subcooled*. For the relationship between the subcooling

$$\Delta T_\infty \triangleq T_{s\infty} - T_\infty \tag{180}$$

and S_∞, cf. App. 1, Eqs. (A-18) and (A-19).

It will be seen in Sec. 5.2 that droplet growth is governed by an "apparent saturation ratio" S_r, which may noticeably differ from S_∞. The definition of S_r will be given in Sec. 5.2.1.

No droplet of any size and no flat interface ($r \to \infty$) can be in equilibrium with a super*heated* carrier, because the condition $T_r = T_\infty$ in Eq. (178) would automatically lead to $p_r > p_{v\infty}$. In super*saturated* (or subcooled) carriers, where $S_\infty > 1$, $p_{v\infty} > p_{s\infty}$ and $T_\infty < T_{s\infty}$, Eq. (178) fulfills the condition of equilibrium

---

[17] The approximation is poor in highly superheated carriers ($T_\infty \gg T_r$). There, however, errors made in p_r are of little significance.

$(T_r = T_\infty,\ p_r = p_{v\infty})$ for one value of r only. This value is termed the *critical radius* and follows from Eq. (178) as

$$r_{\text{crit}} = \frac{2\,\sigma(T_\infty)}{\rho_\ell\, R_v\, T_\infty\, \ell n\, S_\infty} \approx \frac{2\,\sigma(T_{\text{ref}})}{\rho_\ell\, R_v\, T_\infty\, \ell n\, S_\infty} \tag{181}$$

The approximation $\sigma(T_\infty) \approx \sigma(T_{\text{ref}})$ is permissible in all cases where r_{crit} is of importance and T_{ref} is probably chosen.

In the superheated case $(S_\infty < 1)$, r_{crit} becomes negative and loses its physical significance as a critical droplet size. Negative values of r_{crit} must, however, be allowed in order to validate the subsequent equations for superheated carriers.

A given droplet of radius r and of temperature $T_r = T_\infty$ will tend to grow if S_∞ is sufficiently high to make $r_{\text{crit}} < r$. In this case $p_{v\infty} > p_r$ giving $\beta_{v\infty} > \beta_r$, and the impacting flux of vapor molecules will be greater than the emitted one. On the other hand, if S_∞ has a value corresponding to $r_{\text{crit}} > r$, it is $p_{v\infty} < p_r$, $\beta_{v\infty} < \beta_r$, and the droplet tends to evaporate. It is seen that the equilibrium of the droplet in case of $r = r_{\text{crit}}$ is a labile one. The ratio r/r_{crit} can be given, using Eqs. (181) and (62), as

$$\frac{r}{r_{\text{crit}}} = \frac{\ell n\, S_\infty}{\text{Ke}} \tag{182}$$

It is seen that the droplet is in equilibrium $(r/r_{\text{crit}} = 1)$ if the saturation ratio has a value S_∞ given by

$$\ell n\, S_\infty = \text{Ke} \quad \text{(for critical droplet)} \tag{183}$$

For sake of later reference, we now express p_r of a droplet having temperature T_r in terms of r/r_{crit}, by replacing Ke in Eq. (178):

$$p_r = p_r(T_r, r) = p_s(T_r)\, \exp\left(\frac{r_{\text{crit}}}{r}\, \ell n\, S_\infty\right) = p_s(T_r)\, S_\infty^{r_{\text{crit}}/r} \tag{184}$$

The highest numerical values of S_∞ occurring in pure vapor expansions are of the order of 10 to 20. Higher values may be reached in dilute gas/vapor mixtures (Wegener, 1969; Moses and Stein, 1978). At high S_∞ values, spontaneous nucleation (see Wegener) is induced, and the fluid rapidly returns to thermodynamic equilibrium $(S_\infty \to 1)$.

The Kelvin-Helmholtz equation given by Eq. (177) is based on the premises that the droplet is of a pure incompressible liquid (Ib, Ic) of high density, the surface tension makes physical sense and has the same value as in a plane surface (Id, Ie), the droplet is not electrically charged (Ii), and the carrier fluid is a pure vapor of perfect-gas type.[18] In many situations of practical

---

[18] Numbers refer to Table 1.

importance these premises are not valid, for example, for droplets formed on ions in atmospheric clouds; for water droplets formed on salt molecules carried in steam flow; for droplets in carriers of very high density. For the nucleation and growth of tiny droplets the exact value of r_{crit} is of great importance. For more details on nucleating flows the reader is referred to the book by Wegener (1969).

4.2 Mechanical Stability

The mechanical stability of a spherical droplet, and ultimately its integrity, is endangered by aerodynamic forces and by bubble formation (boiling) in its interior,[19] whereas the surface tension tends to maintain integrity and spherical shape.

According to Hinze (1949), the aberrations from spherical shape due to *aerodynamic forces* become of the order of 10% if the Weber number defined in Eq. (64) reaches a value of

$$\mathrm{We}_{limit} \approx 5 \tag{185}$$

At higher (so-called "critical") We values, of the order of We_{crit} 10–20, the droplet desintegrates into numerous smaller droplets (see Lenard, 1904; von Freudenreich, 1927; Hässler, 1972). In Fig. 6 this limit has been plotted for two values of relative velocity. A remarkable recent study on droplet deformation is that of Anisimova and Stekolshchikov (1977). According to them, at We = 5 the droplet is flattened by 28% while its equatorial diameter increases by 17%. Miura et al. (1977) have found that, up to at least We = 3, Nu_H and Nu_M are not affected by the flattening of the droplet. Equations for c_D (or Nu_F) of flattened drops at high Re ($>10^3$) are given by Moore (1976) and by Lebedev and Solonenko (1976).

Internal boiling of a droplet occurs if the inside pressure $p_\ell = (p_\infty + 2\sigma/r)$ becomes smaller than the saturation pressure belonging to the temperature T_{cent} existing in the hottest part (usually the center) of the droplet:

$$p_\ell = p_\infty + \frac{2\sigma}{r} < p_s(T_{cent}) - \Delta p_{bl} \tag{186}$$

The correction term Δp_{bl} characterizes the *boiling lag*. Internal boiling is also termed "flashing." It causes the instantaneous breakup of the droplet into several smaller ones. It is convenient to formulate the condition given by Eq. (186) in terms of temperatures. Expressing p_∞ by the boiling-point temperature $T_{b\infty} \triangleq T_s(p_\infty)$ and $p_s(T_{cent})$ by T_{cent} with the help of Eq. (176), the condition can be brought into the approximative form

$$\frac{T_{cent} - T_{b\infty}}{T_{b\infty}} > \frac{1}{\mathrm{Cl}}\left(1 + \mathrm{Ke}\,\frac{\rho_\ell}{\rho_\infty}\frac{R_v}{\bar{R}} + \frac{\Delta p_{bl}}{p_\infty}\right) \tag{187}$$

[19] Collision with other droplets or objects is not considered here.

The data on the right side are material properties, with the exception of Ke, which represents the droplet size. The boiling lag Δp_{bl} is a material property, too, and must be obtained by experiment. A conservative limit for the stability is obtained by setting $\Delta p_{bl} = 0$. In this case the right side is essentially determined by Ke. For example, for water droplets of $r = 1\ \mu m$ in saturated steam of $p_\infty = 1$ bar, one has $T_\infty = T_{b\infty} = T_{ref} = 373$ K, $Cl = 13$, $\rho_\ell/\rho_\infty \approx 1600$, and $Ke \approx 0.001$; see Figs. 4 to 6. Thus the condition requires

$$T_{cent} - T_{b\infty} = \frac{373}{13}(1 + 1.06 + 0) = 75 \text{ K}$$

This shows that droplets having Ke = 0.001 or more must be quite considerably overheated in order to flash, even if the boiling lag is assumed to be zero. Flashing commonly occurs when large droplets of a hot liquid are injected into a low-pressure environment, such as in the case of spray cooling, or if the carrier stream is subjected to very rapid expansion, as in steam nozzles and turbines. An estimate for the value of T_{cent} for this latter case was given in Sec. 3.6.

5 GROWTH AND VAPORIZATION

5.1 Simplication of the Basic Equations

5.1.1 Outline of Approach

The problem of droplet growth — as stated in Sec. 2.1 and as described by the set of basic balance equations formulated in Sec. 2.2 — can now be further pursued, making use of the expressions given in Sec. 3 for the fluxes of heat, mass, and momentum, and of the interdependence of droplet surface temperature and surface vapor pressure as formulated in Sec. 4, Eq. (178). Assumptions VIa and VIb (Table 4, see page 136) are now abandoned.

It should be borne in mind that droplet growth or vaporization is usually of interest in order to predict the time behavior of nonequilibrium two-phase flows in which the liquid phase is dispersed and consists of *many droplets of very different sizes*. The collective effect of these droplets is required to predict the overall flow behavior. Even if the many various droplets are grouped into a limited number of categories comprising droplets of nominally equal size, the computational task involved in the prediction of droplet behavior is enormous, because the basic system of equations has to be solved individually for each category. For instance, in spontaneously condensing (nucleating) flows, typically 50 or more categories have to be allowed for. The pertinent 50 systems of basic equations then have to be solved simultaneously with the fluid dynamic equations governing the overall flow. The solution of these latter equations for two- or three-dimensional flows of *single*-phase media is known to be a computationally difficult problem in itself. The forthcoming extension of such computational schemes to two-phase flows imperatively requires optimal simplicity of the mathematics involved in the description of the dispersed (second) phase. Keeping this in mind, the main aim of this section will be to simplify mathematically the description of droplet behavior without incurring undue loss of physical generality or accuracy.

Of course, optimally simple expressions for the droplet growth rate are of interest also for simpler situations, such as where only a single droplet is of concern. The expressions given in Sec. 3 for continuum, free-molecular, and multirange situations will permit the construction of the appropriate growth rate equations for each particular problem.

In this section, the basic equations (three simultaneous differential equations and one algebraic equation) that govern the droplet's behavior will *first* be restated for the general case of nonsteady processes within and around the droplet (Sec. 5.1.2). This general problem, however, requires the calculation of heat, mass, and momentum fluxes by *time-dependent solution* of the near-field problem around the droplet (relative aerodynamic flow with superimposed heat and mass transfer) and of the heat conduction equation within the droplet, with droplet radius itself being a variable. Obviously, this full nonsteady case poses a mathematically formidable problem even if only a single drop is to be considered.

Therefore, in a *second* step toward simpler equations (Sec. 5.1.3), the external fluxes H, M, and F will be specified by using the expressions given in Secs. 3.1 to 3.4. Thereby the fluxes become known functions of the conditions in the far field and at the surface of the drop. This step involves the assumption that the *radius* of the drop and the *near-field* distributions are quasi-*constant* (cf. Table 4). The applicability of these assumptions has been discussed in Sec. 3.5.

For small droplets, also the internal heat conduction may be described, as discussed in Sec. 3.6, by quasi-steady equations, and the heat flux H_{int} (which is involved in changing the sensible heat of the droplet) may be sufficiently well prespecified on the basis of an estimated rate of change of the surface temperature T_r of the droplet. In addition, for such small droplets, the relative velocity will be assumed to have the quasi-steady value Δw_{drf} (drift velocity) given in Sec. 3.7. These simplications will be introduced, as a *third* step, in Sec. 5.1.4. (The case of very small droplets having uniform internal temperature and negligible inertia is then included as a special case.) The advantage of this *fully quasi-steady* treatment of the drop consists not only in the simplification of the internal heat conduction problem but also in the drastic reduction of the number of differential equations from three to one.

The remaining differential equation gives the droplet growth rate dr/dt in terms of drop size, far-field conditions, relative velocity, and droplet surface temperature. Analytic solution for $r(t)$ is possible only in some very simple cases, such as that of a stationary droplet growing or shrinking in a constant-state environment.

Usually numerical, step-by-step solution of the growth equation must be carried out. Parallel to this solution, relative velocity Δw_∞, droplet surface temperature T_r and internal average temperature T_ℓ are calculated from algebraic equations. The influence of relative velocity is negligible in many situations of practical interest, but the surface temperature T_r always plays a crucial role in determining the growth rate. Its value is also of interest for understanding the relative importance of heat and mass transfer during droplet growth. The behavior of T_r will be scrutinized in Sec. 5.2. The insights obtained will lead to the formulation of straightforward quasi-steady growth laws in Sec. 5.3. Finally, in Sec. 5.4 the calculation procedure for predicting droplet growth rate on the basis of the quasi-steady theory will be summarized.

5.1.2 General Nonsteady Case

The basic system of equations governing the change of r, Δw_∞, T_ℓ, and T_r with time t is given by Eqs. (16), (19), (20), and (23). It consists of three simultaneous first-order common differential equations and one algebraic equation:

$$\frac{dr}{dt} = \frac{M}{4\pi r^2 \rho_\ell} \tag{188}$$

$$-\frac{d\,\Delta w_\infty}{dt} = \frac{F}{m_\ell} + b_F - \frac{dw_{\text{carr}F}}{dt} - \frac{1}{\rho_\ell}\,\text{grad}_F\, p_\infty \tag{189}$$

$$\frac{dT_\ell}{dt} = \frac{H_{\text{int}}}{m_\ell c_\ell} - \frac{M}{m_\ell}(T_\ell - T_r) + \frac{1}{\rho_\ell c_\ell}\frac{dp_\infty}{dt} \approx \frac{H_{\text{int}}}{m_\ell c_\ell} \tag{190}$$

$$ML_r = -H + H_{\text{int}} \tag{191}$$

The time derivatives of $w_{\text{carr}F}$ and p_∞ and the material properties are considered to be given. In Eq. (190) the last terms are usually very small in comparison to the first one. Therefore negligible error is introduced by the approximation made by the second equality. Only this simplified version of Eq. (190) will be used below.

The surface temperature T_r does not occur explicitly in any of the significant terms of the above equations, but it is of course implicitly involved in the fluxes H, H_{int}, and M.

For the general nonsteady case, the *fluxes* ought to be calculated by solving the general time-dependent partial differential equations (not presented here) governing the velocity, temperature, and concentration fields surrounding the droplet and the temperature field in the interior of the droplet. (Flow and diffusion within the drop are neglected; cf. Table 1.) As mentioned in Sec. 5.1.1, this problem does not lend itself to reasonably economic solution. Such general formulation would be required to describe the transient effects occurring in the first instances following a drastic change of the environmental conditions (e.g., passage of a shock wave, injection into a new environment). For small droplets these transients, if they are induced at all, disappear very quickly (cf. Sec. 5.4). Their accurate knowledge is usually void of practical interest and has received little attention in the literature (Fuks, 1959; Rüping, 1959; Hubbard, et al., 1975).

5.1.3 Partially Quasi-Steady Formulation

As a first simplication, the fluxes H, M, and F will be expressed by equations based on quasi-steady heat, mass, and momentum transfer; see Sec. 3.1. The Nusselt numbers Nu_H, Nu_M, and Nu_F occurring in these equations have been given for the different types of carrier environment (continuum, free-molecular and generalized) in Secs. 3.2 to 3.4. The above simplication is justified if the inequality (151) is fulfilled. Introduction of Eqs. (68), (69), and (70) into the basic system of equations leads to the system

$$\frac{dr}{dt} = \frac{\lambda\,\mathrm{Nu}_H}{2r\rho_\ell L_r}(T_r - T^*_\infty) + \frac{H_{\mathrm{int}}}{4\pi r^2 \rho_\ell L_r} \tag{192}$$

$$-\frac{d\,\Delta w_\infty}{dt} = \frac{3\eta\mathrm{Nu}_F}{2r^2\rho_\ell}\Delta w_\infty + b_F - \frac{dw_{\mathrm{carr}F}}{dt} - \frac{1}{\rho_\ell}\mathrm{grad}_F\, p_\infty \tag{193}$$

$$\frac{dT_\ell}{dt} = \frac{3H_{\mathrm{int}}}{4\pi r^3 \rho_\ell c_\ell} \tag{194}$$

$$\frac{\mathcal{D}L_r\mathrm{Nu}_M}{p_\infty}\left[p_{v\infty} - p_r(T_r, r)\right] = \lambda\,\mathrm{Nu}_H\,(T_r - T^*_\infty) + \frac{H_{\mathrm{int}}}{2\pi r} \tag{195}$$

Since the diffusion coefficient $\mathcal{D}$ has the inconvenience of becoming infinite in continuum-type pure vapor carriers, it has been eliminated from the equation of dr/dt, by using Eq. (191) to substitute H and H_{int} in the place of M in Eq. (188). Note that Eq. (194) is based on the simplified form of Eq. (190). In Eq. (195), $p_r(T_r, r)$ is the vapor pressure of the droplet as given by Eq. (178). L_r is the latent heat at temperature T_r corrected for capillarity according to Eq. (24).

In the above four equations the singular behavior of $\mathcal{D}$ affects only Eq. (195), reducing this equation for the continuum-type pure vapor carrier ($\mathcal{D} \to \infty$) to the simple form

$$p_{v\infty} - p_r(T_r, r) = 0 \quad (\text{for } \Pi_{g\infty} = 0,\ \mathrm{Kn} \ll 1) \tag{196}$$

The Nusselt numbers and T^*_∞ are functions of relative velocity; r occurs explicitly in all equations and implicitly in the Nusselt numbers. Thus the four equations are mutually coupled.

The internal heat flux H_{int} varies in time and is not known. It can in principle be determined by solving the time-dependent heat conduction problem in the interior of the drop (Konorski, 1968; Hubbard et al., 1975; Kulić et al., 1975). This approach is only necessary, however, if relatively large droplets are subjected to sudden changes of the environment, as is typical for spray cooling or for the initial stages of liquid fuel combustion. For most two-phase flow situations, simple estimation (or neglect) of H_{int} is adequate. The above nonsteady conduction problem will not be treated further in the present study.

5.1.4 Quasi-Steady Formulation

If the droplet is small enough and the rate of environment change is slow enough to meet the condition $\Delta t_{int} \ll \Delta t_G$ and $\Delta t_{int} \ll \Delta t_{exp}$, then H_{int} and T_ℓ may be expressed by the quasi-steady approximations, Eqs. (158) and (155), respectively. Simultaneously we may assume that $\Delta t_{vel} \ll \Delta t_G$ [cf. Fig. 20 or Eq. (171)], and set $\Delta w_\infty = \Delta w_{drf}$. The drift velocity is given by Eq. (172). Thus following set of equations is obtained for a droplet growing under quasi-steady conditions:

$$\frac{dr}{dt} = \frac{\lambda\,\mathrm{Nu}_H}{2r\,\rho_\ell\,L_r}(T_r - T_\infty^*) + \frac{rc_\ell}{3L_r}\frac{d\,T_{b\infty}}{dt} = \frac{\lambda\,\mathrm{Nu}_H}{2r\,\rho_\ell\,L_r}(T_r - T_\infty^{**}) \quad (197)$$

$$\Delta w_\infty = \Delta w_{drf} = \frac{2}{3}\frac{r^2\,\rho_\ell}{\eta\,\mathrm{Nu}_F}\left(b_F - \frac{dw_{carrF}}{dt} - \frac{1}{\rho_\ell}\,\mathrm{grad}_F\,p_\infty\right) \quad (198)$$

$$T_\ell = T_r - \frac{r^2}{15\,a_\ell}\frac{d\,T_{b\infty}}{dt} \quad (199)$$

$$\frac{L_r\,\mathrm{Nu}_M}{p_\infty}\left[p_{v\infty} - p_r(T_r, r)\right] = \lambda\,\mathrm{Nu}_H(T_r - T_\infty^*) + \frac{2r^2\,\rho_\ell\,c_\ell}{3}\frac{d\,T_{b\infty}}{dt} = \lambda\,\mathrm{Nu}_H(T_r - T_\infty^{**}) \quad (200)$$

Concerning the approximation $dT_{b\infty}/dt$ for the rate of change of surface temperature the reader is referred to the comments made with Eqs. (156) and (157). For the sake of formal simplification of Eqs. (197) and (200), the heat capacity terms have been amalgamated in the second equalities with T_∞^*, resulting in an "*apparent carrier temperature*" T_∞^{**}, defined as

$$T_\infty^{**} \triangleq T^* - \frac{2}{3}\frac{\lambda_\ell/\lambda}{\mathrm{Nu}_H}\frac{dT_{b\infty}}{dt}\frac{r^2}{a_\ell} \quad (201)$$

The physical meaning of T_∞^{**} is a carrier temperature that would be required for the same rate of droplet growth if aerodynamic recovery were absent and the thermal inertia of the droplet were zero. Using Eq. (71), we have

$$T_\infty^{**} = T_\infty k_T(\mathrm{Ma}) - \frac{2}{3}\frac{\lambda_\ell/\lambda}{\mathrm{Nu}_H}\frac{dT_{b\infty}}{dt}\frac{r^2}{a_\ell} \quad (202)$$

and with Eqs. (154) and (160),

$$T_\infty^{**} = T_\infty k_T(\mathrm{Ma}) + \frac{2}{0.9}\frac{\lambda_\ell/\lambda}{\mathrm{Nu}_H\,\mathrm{Cl}}\,T_{b\infty}\frac{\Delta t_{int}}{\Delta t_{exp}} \quad (203)$$

In all cases where quasi-steady treatment is permissible, the second term is but a small correction to the first. The correction is positive in expanding flows. Thus in expanding flows $T^{**}_\infty \geq T^*_\infty > T_\infty$ holds, whereas in compression flows $T^{**}_\infty < T_\infty$ may in principle occur. Equations for the recovery temperature T^*_∞, that is, for the function k_T(Ma), are given in Secs. 3.2 to 3.4. For a stationary droplet in a constant-state environment one has $T^{**}_\infty = T_\infty$.

5.1.5 Numerical Solution Procedures

In the quasi-steady formulation, Eqs. (197) to (200), only r is governed by a (first-order, common) differential equation, wherea Δw_∞, T_ℓ, and T_r are given by algebraic equations. If the carrier conditions and r are known, Δw_∞ can be obtained from Eq. (198). (An iterative calculation may be required for this if Nu_F is sensitive to Re or Ma and thus to Δw_∞.) If, then, Δw_∞ and thus Re and Ma are known, Nu_H, Nu_M, and T^{**}_∞ can be obtained from Sec. 3.2, 3.3, or 3.4 and from Eq. (202), respectively. With these parameters Eq. (200) can be solved iteratively to give the droplet surface temperature T_r. (Approximate explicit expressions for T_r will be given in Sec. 5.2.) Once T_r is known, Eq. (199) can be used to calculate the droplet mean temperature T_ℓ, although the calculation of T_ℓ is not compulsory. Finally, Eq. (197) will give dr/dt. Thus the loop is closed, and stepwise numerical calculation of droplet growth is possible in any (changing or constant) environmental conditions.

If Re and Ma are so low that Nu_H, Nu_M, and T^{**}_∞ are insensitive to them (which is the case for Re<1, Ma<0.1), the calculations of Δw_∞ can be omitted altogether. If, in addition, the environment stays constant, the values of T_r and T_ℓ become identical. Thus there remain only two unknown variables, r and T_r, and it is sufficient to solve the Eqs. (197) and (200). If the latter equation were explicit in T_r (or if it could be sufficiently well approximated by an explicit formula), the computation would become significantly faster, and even analytic integration of dr/dt could become possible in certain cases. The aim of the next section is to find such explicit expressions that are suited for replacing Eq. (200).

5.2 The Surface Temperature of the Drop

As pointed out above, the surface temperature T_r is determined under quasi-steady growth conditions by Eq. (200). In the following this equation will be scrutinized with the double aim of making general conclusions with respect to the behavior of T_r and of finding explicit expressions from which the value of T_r can be calculate with sufficiently good approximation.

The treatment of Eq. (200) will remain general in the sense that nonzero relative velocities (up to Re ≈ 200 under continuum conditions and up to Ma ≈ 1 under free-molecular conditions) will be allowed for, and the environmental conditions will be allowed to change at some known rate, as pointed out in Sec. 5.1.4.

5.2.1 Thermodynamic Parameters of the Droplet

First, however, some parameters will be introduced in order to characterize the thermodynamic situation of the droplet. It is convenient to use the Clausius-Clapeyron equation, Eq. (176), to express vapor pressure with temperature and vice versa. In treating small droplets, the capillarity correction has to be accounted for, see Eq. (178). Thus the vapor pressure p_r of a droplet having temperature T and radius r is given by the function

$$p_r = p_r(T,r) = p_s(T_{\text{ref}}) \exp \left[\text{Ke} + \text{Cl} - \frac{\text{Cl}\, T_{\text{ref}}}{T} \right] \tag{204}$$

The equality would be exact if $\sigma(T_r)/T_r = \sigma(T_\infty)/T_\infty$ were fulfilled. Since, according to this equation, a droplet (Ke>0) has a higher vapor pressure than a flat liquid surface (Ke = 0) of the same temperature, the saturation point and the boiling point of a droplet are always lower than $T_{s\infty}$ and $T_{b\infty}$, respectively. We define size-dependent saturation and boiling temperatures for the droplet, T_{sr} and T_{br}, by the relations

$$p_r(T_{sr}, r) \triangleq p_{v\infty} \tag{205}$$

$$p_r(T_{br}, r) \triangleq p_\infty \tag{206}$$

Use of the Eqs. (204), (A-16*a*), and (A-166) leads to the following expressions for T_{sr} and T_{br}:

$$T_{sr} = \frac{T_{s\infty}}{1 + (T_{s\infty}/\text{Cl}\, T_{\text{ref}})\text{Ke}} \tag{207}$$

$$T_{br} = \frac{T_{b\infty}}{1 + (T_{b\infty}/\text{Cl}\, T_{\text{ref}})\text{Ke}} \tag{208}$$

It is seen that in the case of substances having a Clausius-Clapeyron number Cl of the order of 10, the corrections become negligibly small if Ke<0.01 (which in the case of water requires that $r > 10^{-7}$ m, cf. Fig 6). In such cases one may set $T_{sr} = T_{s\infty}$ and $T_{br} = T_{b\infty}$. For critical-size droplets ($r = r_{\text{crit}}$), the Kelvin number has the value Ke = ln S_∞, cf. Eq. (182), and T_{sr} therefore assumes the value $T_{sr} = T_\infty$, as can be verified with Eqs. (207) and (A-18). The ratio of T_{sr} and T_{br} is found to be related to the composition of the carrier by

$$\frac{T_{br}}{T_{sr}} = 1 - \frac{T_{br}}{\text{Cl}\, T_{\text{ref}}} \ln \Pi_{v\infty} \tag{209}$$

which expression is formally analogous to Eq. (A-21).

Another consequence of the capillarity effect is that the carrier phase is always less supersaturated (or more superheated) with respect to droplets than to the flat liquid interface. Let an "apparent saturation ratio" be defined as

$$S_r \triangleq \frac{p_{v\infty}}{p_r(T_\infty^{**},r)} \tag{210}$$

S_r is seen to characterize the saturation of the carrier medium with respect to a droplet of size r having a temperature equal to the apparent carrier temperature T_∞^{**}. Comparison of Eqs. (210) and (179) yields the relationship

$$S_r = S_\infty \exp\left[-\text{Ke} + \frac{\text{Cl}\, T_{\text{ref}}}{T_\infty}\left(\frac{T_\infty}{T_\infty^{**}} - 1\right)\right] \tag{211}$$

It appears that S_r may significantly (by > 5%) differ from S_∞ for very small droplets (where Ke>0.05), as well as for very large ones (where nonzero relative velocity and/or thermal inertia in a changing environment may cause a significant difference between T_∞^{**} and T_∞). For droplet sizes of the order of 1 µm, usually $S_r \approx S_\infty$ holds. Using Eqs. (207), (211), and (A-18), S_r can be generally expressed by T_{sr} and T_∞^{**} as

$$\ell n\, S_r = \frac{\text{Cl}\, T_{\text{ref}}}{T_\infty^{**}} - \frac{\text{Cl}\, T_{\text{ref}}}{T_{sr}} \tag{212}$$

which is formally analogous to the flat-liquid expression, Eq. (A-18).

5.2.2 Parameter Domains of Practical Interest

It is now useful to review for a given substance and given carrier pressure p_∞ those domains of temperature and carrier composition where the analysis of droplet growth is of practical interest. Carrier temperature T_∞ will be referred to the boiling point $T_{b\infty}$ (which is fixed by p_∞), and composition will be expressed by the ratio of the saturation and boiling temperatures, $T_{s\infty}/T_{b\infty}$, making use of Eq. (A-21). Thus, in Fig. 16, horizontal lines mean constant carrier composition, with the top boundary ($T_{s\infty}/T_{b\infty} = 1$) corresponding to *pure vapor* and the bottom boundary ($T_{s\infty}/T_{b\infty} = 0$) to the limiting case of *pure gas*. Rich carriers (with vapor contents above 10%) fall near the top, whereas "lean" or "dry" carriers near or under mid-height. (It should be noted that 1 part vapor in 1 million parts gas ($g_\infty = 10^6$) typically corresponds to $T_{s\infty}/T_{b\infty} = 0.4$–$0.6$.) The horizontal direction expresses carrier temperature, *hot* carriers falling to the right and *cool* ones to the left; the boiling-point carrier ($T_\infty/T_{b\infty} = 1$) may be used as the division line. The 45° diagonal comprises the *saturated* conditions ($T_{s\infty} = T_\infty$). This line also separates the domains of droplet growth from the domains of evaporation. Growth occurs in the supersaturated (subcooled) domain, see the upper triangular half of the "cool" area, whereas evaporation takes place in the remaining domains, including the entire "hot" range, where the carrier is superheated (undersaturated). The lines of constant saturation ratio S_∞ flank the saturation diagonal as indicated by the dash-and-dot curves.

In Fig. 16 the shaded areas indicate following domains of particular

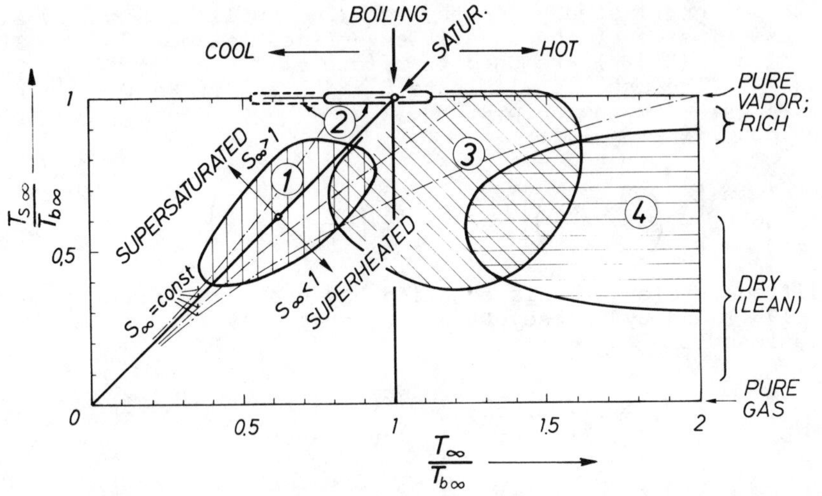

Fig. 16. Classification of carrier conditions in the composition versus carrier-temperature plane. [See Eq. (A-21) for the relationship between $T_{s\infty}/T_{b\infty}$ and composition.] The shaded areas indicate the domains of practical interest, as explained in the text.

interest:

1. Meteorological clouds, cooling tower plumes, and wind tunnel flows with moisture precipitation

2. Steam in steam turbines (or other condensible-vapor working fluids); the dashed extension indicates hypersonic wind tunnels with condensation of the main fluid

3. Spray cooling or spray drying

4. Vaporization in a very hot (e.g., combustion) environment

The conclusion may be drawn that approximate solutions to Eq. (200) are of interest for all conditions where $T_{s\infty}/T_{b\infty} \geqslant 0.4$ and $T_\infty/T_{b\infty} \geqslant 0.4$ are simultaneously fulfilled.

It should be pointed out that Fig. 16 has been drawn in terms of the thermodynamic parameters referred to a flat liquid surface. Use of the individual, droplet-specific parameters defined in Sec. 5.1.1 (i.e., of T_{br}, T_{sr}, T_∞^{**}, and S_r instead of $T_{b\infty}$, $T_{s\infty}$, T_∞ and S_∞) would lead to the same diagram, with the only essential difference that the boundaries of the shaded areas might be slightly shifted as a result of $T_\infty^{**} \neq T_\infty$ and $T_{br} \neq T_{b\infty}$.

5.2.3 Discussion of the Equation Giving T_r

In order to determine the droplet surface temperature T_r under conditions of quasi-steady growth, Eq. (200) will be formally simplified by use of some of the quantities defined in Sec. 5.2.1. If Eqs. (204) and (A-16*b*) are used to relate $p_r(T_r, r)$ to p_∞, if Eqs. (51) and (47) are used to eliminate $\mathcal{D}$ and λ and if Ke is eliminated from Eq. (204) with the help of Eq. (208), one can bring Eq. (200) into the nondimensional form

$$\exp\left[C\left(1 - \frac{T_{br}}{T_r}\right)\right] = \Pi_{v\infty} - \frac{A}{C}\left(\frac{T_r}{T_{br}} - \frac{T_\infty^{**}}{T_{br}}\right) \tag{213}$$

A simplified version of this equation has been analyzed by Todes et al. (1966) and by Fedoseyeva and Glushkov (1973). A and C are abbreviations for

$$A \triangleq \frac{\gamma}{\gamma-1}\,\frac{L}{L_r}\,\frac{\mathrm{Sc}}{\mathrm{Pr}}\,\Pi_{g\infty}\,\frac{\mathrm{Nu}_H}{\mathrm{Nu}_M} \tag{214}$$

$$C \triangleq \frac{\mathrm{Cl}\; T_{\mathrm{ref}}}{T_{br}} = \frac{\mathrm{Cl}\; T_{\mathrm{ref}}}{T_{b\infty}} + \mathrm{Ke} \tag{215}$$

It is seen that Eq. (213) contains the unknown T_r in the nondimensional form T_r/T_{br}. No explicit solution in possible. The value of the unknown is determined by the values of the four parameters occurring in the equation. These parameters and their approximate ranges of physical interest are

$$\begin{aligned} A &= 0 - 10 \\ C &= 5 - 20 \\ \Pi_{v\infty} &= 10^{-6} - 1 \\ T_\infty^{**}/T_{br} &= 0.5 - 3 \end{aligned} \tag{216}$$

All four parameters are inherently positive. $\Pi_{v\infty}$ indicates the composition of the carrier medium, which may range from nearly vaporless gas ($\Pi_{v\infty} \approx 0$) to pure vapor ($\Pi_{v\infty} = 1$). The range of interest of T_∞^{**}/T_{br} may be seen from Fig. 16 (it is $T_\infty^{**}/T_{br} \approx T_\infty/T_{b\infty}$). The value of C is substantially determined by the material property Cl, because the choice of T_{ref} is usually such that $T_{\mathrm{ref}} \approx T_{b\infty} \approx T_{br}$. The factor A is the really crucial parameter in the problem. It represents the influence of the heat and mass transfer processes on droplet temperature.

A is seen to be a function of material properties, carrier composition, and the Nusselt numbers for heat and mass transfer. As set forth in Sec. 3, these again are determined by droplet size and relative velocity, that is, by Kn, Re, and Ma. Thus A has individual values for each different droplet in a given carrier and varies during droplet growth. Numerical values of A will be discussed in Sec. 5.3.1. Since Nu_M may be a function of $\Pi_{g\infty}$, as seen from Eq. (81) or (104), A is not necessarily proportional to $\Pi_{g\infty}$, in contrast to the first impression given by its definition. Instead of using a complicated general (multirange) expression to calculate A in terms of Kn, Re, Ma, and so on, it is more convenient to use simpler, limited-range expressions wherever permissible. A review of the expressions giving A will be given in Sec. 5.3, Tables 5 and 6.

The mathematical structure of Eq. (213) is quite a bit more complicated in cases when A itself is a function of droplet temperature. Such may be the case in a free-molecular environment. (See the dependence of Nu_M on T_r and Π_r by way of Eqs. (104) and (107); a similar influence exists in the general multirange case, cf. Eq. (124) with B from Eq. (107). Fortunately, the dependence of A on T_r in these instances is sufficiently weak to make a simplified treatment permissible. This problem is considered in more detail in App. 4, Sec. A4.7 and A4.8.

The simple case $A = 0$ arises if there is no resistance to mass transfer, that is, if $Sc = 0$ or $\Pi_{g\infty}/Nu_M = 0$. In this case Eq. (213) simplifies to

$$C\left(1 - \frac{T_{br}}{T_r}\right) = \ln \Pi_{v\infty} \tag{217}$$

If $\ln \Pi_{v_\infty}$ is eliminated with Eq. (209), the following simple solution is found:

$$T_r = T_{sr} \quad \text{(valid for } Sc\ \Pi_{g\infty}/Nu_M = 0\text{)} \tag{218}$$

In words: *If the resistance to mass transfer is zero, the droplet is at its saturation temperature.* This situation is achieved only in pure vapor carriers under continuum conditions.

For critical-size droplets $T_{sr} = T_\infty$, cf. Sec. 5.2.1. Because of the smallness of r_{crit}, $T_\infty^{**} = T_\infty$ can be written. The solution of Eq. (213) for this case is trivial and gives $T_r = T_\infty$. In words: *Critical droplets assume the carrier temperature.*

5.2.4 Iterative calculation of T_r

Equation (213) can be solved iteratively with any desired degree of accuracy if the parameters have known values. Figure 17 gives the results of such calculations for the case $A = 2$, $C = 10$, with composition and temperature varied in the range $0 < \Pi_{v\infty} < 1$, $0 < T_\infty^{**}/T_{br} \leqslant 2$. In this figure T_r/T_{br} is plotted as a surface spanned over the composition/temperature plane. Profiles of T_r/T_{br} as functions of temperature T_∞^{**}/T_{br} are shown for seven compositions ranging from the limiting case of pure gas (foreground) to pure vapor (background). The diagonal line with open circles marks the saturation condition for each composition. The shape of the surface remains qualitatively the same for other combinations of A and C, except in the pure-vapor limit. Here the case $A \neq f(\Pi_{g\infty})$ leads to a (weak) dependence of T_r/T_{br} on carrier temperature, whereas the case $A \sim \Pi_{g\infty}$ leads to $T_r/T_{br} = 1$ for all carrier temperatures. This latter situation is pertinent to continuum-type carriers.

A number of general conclusions can be drawn from the plot shown in Fig. 17. Carrier pressure, droplet size, and relative

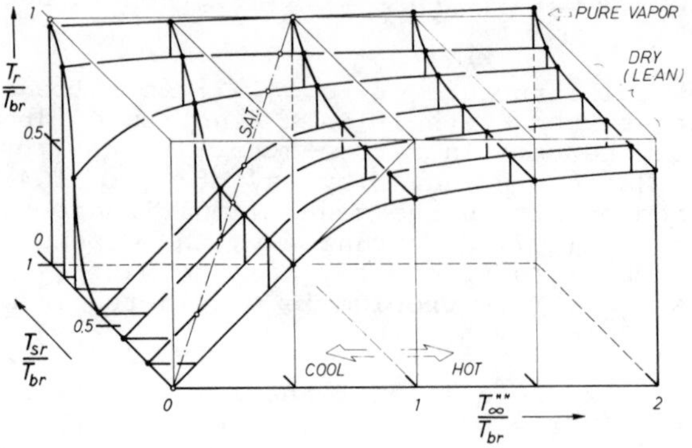

Fig. 17. Droplet surface temperature as a function of carrier composition and apparent carrier temperature, for $A = 2$, $C = 10$. All temperatures referred to T_{br}. For the relationship between T_{sr}/T_{br} and composition see Eq. (209).

velocity are considered to be given.

a. In pure vapor atmospheres the droplet surface assumes very nearly its boiling *and* saturation[20] temperature: $T_r \approx T_{br}$. The equality is exact under continuum conditions.

b. At a given carrier temperature T_∞^{**}, the drop is the colder the less vapor is contained in the carrier.

c. At a given carrier composition, T_r increases with T_∞^{**} but never exceeds its boiling point T_{br}.

d. In very cool and lean carrier mixtures (see the left foreground in Fig. 17), T_r closely equals T_∞^{**}.

Two last conclusions become evident from the mathematical discussion of Eq. (213) in App. 4:

e. In supersaturated ($T_{sr} > T_\infty^{**}$) carriers $T_r > T_\infty^{**}$, and in superheated ($T_{sr} < T_\infty^{**}$) carriers $T_r < T_\infty^{**}$.

f. The droplet always assumes a surface temperature lying *between* its saturation temperature and the apparent carrier temperature: $|T_r - T_\infty^{**}| < |T_{sr} - T_\infty^{**}|$.

Although most of these conclusions refer to well-known facts, – for example, the second merely expresses the wet-bulb temperature

---

[20] $T_{sr} = T_{br}$ in pure vapors, by definition.

effect – a plot like the one shown gives a useful overview about the various influences on T_r.

5.2.5 The Caloric Effectiveness Q

It follows from the conclusions (e) and (f) given above that the ratio

$$Q \triangleq \frac{T_r - T_\infty^{**}}{T_{sr} - T_\infty^{**}} \tag{219}$$

is always positive and smaller than 1. This ratio will be termed the *caloric effectiveness* of the carrier. Its numerator is the temperature difference that drives the growth process, that is, which generates the heat flow needed to carry away or bring in the latent heat involved in the condensation or evaporation process. The denominator of Q, on the other hand, is the hypothetical temperature difference in the case of zero mass transfer resistance, that is, the (positive or negative) maximum temperature difference possible at the supersaturation or superheat level in question. Their ratio expresses how effectively the subcooling or superheat existing with respect to the drop is able to induce the heat flow required for growth or evaporation. The denominator of Q is usually closely equal to the subcooling ΔT_∞. However, for very small or very fast-moving droplets and in changing carrier environments, corrections may be necessary. With Eqs. (A-17) and (207), one finds that

$$T_{sr} - T_\infty^{**} = \Delta T_\infty - \frac{T_{s\infty}\ \mathrm{Ke}}{(\mathrm{Cl}\ T_{\mathrm{ref}}/T_{s\infty}) + \mathrm{Ke}} - \left(T_\infty^{**} - T_\infty\right) \tag{220}$$

Usually the first term is predominant. The second is the capillarity correction. For critical droplets ($r = r_{\mathrm{crit}}$, $\mathrm{Ke} = \ln S_\infty$), it becomes *equal* to ΔT_∞. The last term corrects for relative velocity and thermal inertia and has to be determined from Eq. (202) or (203).

From the above temperature difference and from Q, the quasi-steady temperature of the droplet surface is obtained as

$$T_r = T_\infty^{**} + \left(T_{sr} - T_\infty^{**}\right) Q \tag{221}$$

This procedure to determine T_r is justified if Q is calculated from an explicit expression such as the ones given in the next section. If iterative solution of Eq. (213) is preferred, T_r is obtained directly, without recourse to Q.

5.2.6 Approximate Expressions for Q

With the aim of finding explicit approximations for the quasi-steady surface temperature T_r and for the caloric effectiveness Q, Eq. (213) has been analyzed mathematically in App. 4. The results of this analysis are briefly summarized below.

Four approximate expressions will be given. Each expression

is best suited for a different class of combinations of carrier composition and carrier temperature. Since all expressions are defined in terms of the droplet-specific parameters discussed in Sec. 5.2.1, they are valid for any Knudsen number and take into account the effects of capillarity, relative velocity, and thermal inertia.

Of course, many kinds of approximations can be constructed, and in limited ranges some may be better than the ones presented. The four presented below have been selected to fulfill at least two of the following criteria:

Wide applicability in terms of temperature and/or composition range

Simplicity

Domain of applicability complementing those of the other approximations (overlap is welcome)

The domain of applicability is defined in terms of the percentage error of Q. In general, less than 10% error in Q should be asked for. Above 20%, the explicit formula must be abandoned in favor of a better approximation or of the iterative solution of Eq. (213).

In these approximated expressions the main influence of carrier composition will be associated with a dimensionless group, the *gas influence factor*

$$G \triangleq \frac{A/\Pi_{v\infty}}{(C - \ln \Pi_{v\infty})^2} \tag{222}$$

introduced in Eq. (A-93). For vaporless gases ($\Pi_{v\infty} \approx 0$), G tends to infinity; for pure vapors ($\Pi_{v\infty} = 1$), G is below unity or even zero (if $A = 0$; case of continuum conditions). Some numerical values of A and $G = (G/A)\cdot A$ will be given in Table 7.

Vapor-Rich Carrier Media:

If $\Pi_{v\infty} > (0.1 \text{ to } 0.2)$, a good approximation is obtained by setting

$$Q \approx Q_{\text{vap}} = \frac{1}{1 + G} \tag{223}$$

cf. Sec. A.4.2 in App. 4. If the carrier is not far from saturation (meaning $0.5 < S_r < 2$), Q_{vap} may be used down to about $\Pi_{v\infty}=0.02$, provided that $A<10$. For pure vapors, the approximation is good regardless of the value of S_r.

Vaporless (Lean), Cool Mixtures:

If $\Pi_{v\infty} < 0.01$ and the carrier temperature is far below the boiling point of the droplet ($T_\infty^{**}/T_{br} \lesssim 0.6$), the growth process is controlled by diffusion. The very low Q values typical of such "nonvolatile droplet" situations can be approximated down to essentially pure-gas carriers ($\Pi_{v\infty} \to 0$) by setting

$$Q \approx Q_{gas} = \frac{1}{G} \frac{S_r - 1}{S_r \ \ell n \ S_r} \frac{C - \ell n \ \Pi_{v\infty} + \ell n \ S_r}{C - \ell n \ \Pi_{v\infty}} \qquad (224)$$

cf. Sec. A4.3.

Synthesis into an Approximation for Mixture Carriers:

By combination of the approximations Q_{vap} and Q_{gas}, an approximation can be obtained that covers all gas compositions of interest, provided that the degree of superheat is not excessive ($S_r > 0.01$). As shown in Sec. A4.4, one can set

$$Q \approx Q_{syn} = \frac{1 + (T_{sr}/T_{\infty}^{**})\ G}{Q_{vap}^{-1} + (1 + Q_{gas}^{-1})(T_{sr}/T_{\infty}^{**})\ G} \qquad (225)$$

Here T_{sr}/T_{∞}^{**} is related to S_r by Eq. (A-98). By this approximation almost all cases of practical interest can be covered, except hot lean carrier atmospheres as encountered in combustion problems.

Hot, Lean Carriers:

In order to cope with combustion-type situations, a high-temperature approximation has been devised in Sec. A4.5. This approximation, Q_{gen}, is general in the sense that it may be used from high temperatures down to the saturation temperature and below, unless the carrier is both cold and very lean. Above the boiling point ($T_{\infty}^{**} > T_{br}$), Q_{gen} is superior to the other approximations. Following Eq. (A-102)

$$Q \approx Q_{gen} = 1 - \frac{T_{br} - T_{sr}}{T_{sr} - T_{\infty}^{**}} \left[\exp\left(\frac{T_{sr} - T_{\infty}^{**}}{T_{br} - T_{sr}} \frac{G}{1 + G} \right) - 1 \right] \qquad (226)$$

Error Analysis:

The errors of the above approximations with respect to the exact value of Q resulting from Eq. (213) have been numerically controlled for a number of parameter combinations corresponding to the ranges indicated in Eq. (216). The percentage errors of Q_{vap}, Q_{gas}, Q_{syn}, and Q_{gen}, as defined by Eq. (A-103), are plotted in Fig. 18 for three combinations of the parameters C and A, the carrier composition and temperature covering in each case the entire range of interest down to $\Pi_{v\infty} = 10^{-6}$. The three examples concern three different physical situations, which will be briefly introduced.

Case (a): $C = 20$ or $C1 \approx 20$ means a liquid of large latent heat (e.g., water at room temperature). $A = 2\Pi_{g\infty}$ is fulfilled (as seen from Eq. (214) when inserting data from Tables A-1 and A-2), for instance, by a water droplet in air under continuum conditions (when $Nu_H \approx Nu_M$).

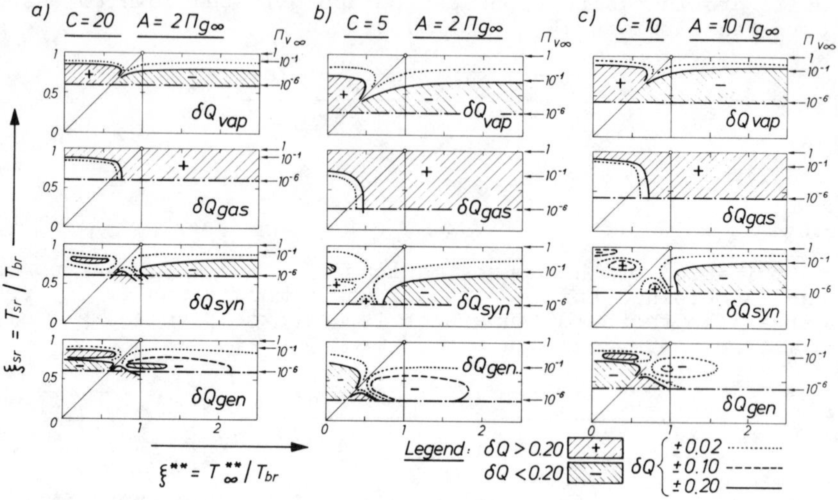

Fig. 18. Errors of the approximate expressions for Q, for three parameter value combinations in the composition versus apparent carrier temperature plane.

Case (b): $C \approx Cl \approx 5$ means very low latent heat, as may be the case near the critical temperature (cf. Fig. 3). A has not been altered.

Case (c): With $C = 10$, this case is intermediate in terms of latent heat. The high value $A = 10\ \Pi_{g\infty}$ may be caused by higher Sc numbers (as are typical of organic liquids in air, Table A-2) and/or lower γ of the mixture. As discussed in Sec. 5.3, high A values of the order of 10 may also result from free-molecular conditions[21], especially if the vapor has a higher molecular weight than the noncondensing gas.

For each of these cases, Fig. 18 shows the four composition versus carrier temperature fields with constant-error lines of one of the approximations in each field. In the shaded areas the approximations give unacceptable errors (>20%). Vapor contents below $\Pi_{v\infty} = 10^{-6}$ have not been considered. Constant composition corresponds to horizontal lines. The values of $\Pi_{v\infty}$ corresponding to the vertical scale are given at the right edge of each field. Pure vapor is at the upper limit of each field. Saturation conditions are on the diagonal line ending in the point representing saturated pure vapor (circle). "Hot" carriers are to the right of

[21] The proportionality of A to $\Pi_{g\infty}$ does not accurately represent such conditions.

this point, "cool" ones to the left. The fields are identical to the plane shown in Fig. 16, except that droplet-specific temperature criteria are now used.

The errors of Q_{vap} are seen to be continually increasing toward leaner mixtures. The error changes sign at the saturation diagonal. If $\Pi_{v\infty}$ is given, the error is in general smaller if C is large.

The errors of Q_{gas} are prohibitive everywhere except in the "cool, lean" corner, where this approximation is superior to all the others.

Q_{syn} is seen to be a compromise between Q_{vap} and Q_{gas}. Its irregularities in the "supersaturated" triangle (upper left corner) are of no physical concern, because such high supersaturations (e.g., $S_r>50$) cannot be generated. Q_{syn} is the best approximation for cool (below boiling point) carriers. It breaks down, however, in hot lean carriers.

For these latter cases Q_{gen} is seen to work satisfactorily. Problems may arise only if C or A (or both) are large, but the approximation for very hot carriers ($T_\infty^{**}/T_{br} > 2$) remains good in all events. The plots show that in the neighborhood or slightly under the boiling point ($T_\infty^{**}/T_{br} \leqslant 1$), Q_{syn} and Q_{gen} are of comparable accuracy

5.2.7 Conclusions and Critical Remarks

Stodola (1922), in the chapter of his book concerned with the thermodynamic behavior of very small droplets in steam under free-molecular conditions, was the first to point out that small droplets *assume very quickly a well-defined temperature* that they substantially maintain during subsequent growth or evaporation. It has been shown in the foregoing analysis that this *quasi-steady surface temperature*[22] can be defined for arbitrary carrier types and droplet sizes. The quasi-steady treatment of the transfer processes associated with droplet growth was shown in Secs. 3.5 and 3.6 to be applicable to all situations in which the size of the droplet undergoes *significant* change, that is, those that last for periods comparable to Δt_G. Thus the field of applicability of the above results is reasonably wide.

The abbreviations introduced in Sec. 5.2.1 provide a relatively simple, uniform formalism applicable to all droplet sizes, relative flow velocities, and carrier types, including changing carrier conditions. In cases where the various corrections (for surface tension effects, temperature recovery, and thermal inertia) may be neglected, the same formalism is maintained, except that the individual (and size-dependent) parameters T_{sr}, T_{br}, T_∞^{**}, and S_r

---

[22] Stodola has assumed that the surface and internal temperatures are equal. The present treatment accounts for their difference, which may be nonnegligible for larger droplet sizes.

of the droplet may be replaced by the parameters $T_{s\infty}$, $T_{b\infty}$, T_{∞}, and S_{∞} of the carrier medium.

Because of the explicit approximations given in Sec. 5.2.6, an iterative solution of Eq. (200) or (213) can be avoided, and a simple explicit growth rate equation can be formulated. Before treating this task in Sec. 5.3, two critical comments with respect to the validity of these approximations will be made.

The first remark concerns continuum-type situations. The rigorous treatment of mass transfer in vapor-*rich* carriers requires the Nusselt number Nu_M to depend on the partial pressures, cf. Eq. (83). This dependence causes a basic change in the mathematical structure of Eq. (213) from which the present results and approximations were derived. What errors are incurred by adopting the simplified (linear) diffusion law? This question is discussed in App. 4, Sec. A4.7. It is found that the simplification is permissible. The errors (being the highest for low values of Cl and moderate values of $\Pi_{v\infty}$) do not exceed 3% in the value of Q, and are mostly below 1%. The advantage of the linearized diffusion law is that the mathematical formalism remains uniform for all carrier types.

A second remark concerns free-molecular conditions. The direct influence of T_r on mass transfer can be accounted for by using an estimated value ($T_{r,\mathrm{est}}$) in Eq. (107). Then all results of this section remain applicable, except that the definition of T_{br} has to be replaced by T_{br}^{fm} as defined in Eq. (A-110). For further details on the free-molecular case, see Sec. A4.8.

An outline of the calculation procedure to obtain T_r and T_ℓ from the quasi-steady theory will be given in Sec. 5.4.

5.3 Laws of Quasi-Steady Droplet Growth

5.3.1 General Formulation

The term "*growth law*" usually refers to an expression that specifies dr/dt, the growth (or vaporization) rate of the droplet, as a function of its size, relative velocity, the carrier state, and material properties. The quasi-steady growth rate equation, Eq. (197), is seen to be an expression of this type, provided that Nu_H, T_r, and T_{∞}^{**} are known functions of the above parameters. The general form of quasi-steady growth laws will be treated first. In Sec. 5.3.2 simplified growth laws will be given for a number of special cases.

Expressions for Nu_H are given in Sec. 3, and T_{∞}^{**} can be calculated from Eq. (202) or (203). As discussed in Sec. 5.2, however, the value of T_r must either be iteratively calculated or approximated by one of the expressions given in Sec. 5.2.6. In the latter case it is convenient to express T_r by Q, the caloric efficiency of the carrier as defined in Eq. (219), because the latter is substantially determined by the carrier state alone and varies only slightly with droplet size.

If Eq. (197) is combined with Eq. (219), the *quasi-steady growth law* of a droplet can be written in the general form

$$\frac{dr}{dt} = \frac{\mathrm{Nu}_H}{2r} \frac{\lambda T_{\mathrm{ref}}}{\rho_\ell L_r} \frac{T_{sr} - T^{**}_\infty}{T_{\mathrm{ref}}} Q \tag{227}$$

The driving force for droplet growth is represented here by $(T_{sr} - T^{**}_\infty)Q$. It is convenient to nondimensionalize all temperatures by the constant reference value T_{ref} at which the material properties L, σ, ρ_ℓ, and so on, are specified. Maintaining the sub- or superscripts of T, we generally write $T/T_{\mathrm{ref}} \triangleq \boldsymbol{T}$. Analogously writing $\Delta T_\infty/T_{\mathrm{ref}} \triangleq \Delta\boldsymbol{T}_\infty$, one can deduce from Eqs. (220) and (A-18),

$$\boldsymbol{T}_{sr} - \boldsymbol{T}^{**}_\infty = \frac{\Delta\boldsymbol{T}_\infty - \mathrm{Ke}\,\boldsymbol{T}_\infty \boldsymbol{T}_{s\infty}/\mathrm{Cl}}{1 + \mathrm{Ke}\,\boldsymbol{T}_{s\infty}/\mathrm{Cl}} - (\boldsymbol{T}^{**}_\infty - \boldsymbol{T}_\infty)$$

$$= \frac{\boldsymbol{T}_\infty \boldsymbol{T}_{s\infty}}{\mathrm{Cl}} \frac{\ln S_\infty - \mathrm{Ke}}{1 + \mathrm{Ke}\,\boldsymbol{T}_{s\infty}/\mathrm{Cl}} - (\boldsymbol{T}^{**}_\infty - \boldsymbol{T}_\infty) \tag{228}$$

Except for droplets of far submicroscopic size ($r<10^{-9}$ m), $\mathrm{Ke}\,\boldsymbol{T}_{s\infty}/\mathrm{Cl} \ll 1$. Making use of Eq. (A-18) and introducing r_{crit} from Eq. (182), Eq. (228) is written as

$$\boldsymbol{T}_{sr} - \boldsymbol{T}^{**}_\infty = \Delta\boldsymbol{T}_\infty \frac{1 - r_{\mathrm{crit}}/r}{1 + (r_{\mathrm{crit}}/r)\,\Delta T_\infty/T_\infty} - (\boldsymbol{T}^{**}_\infty - \boldsymbol{T}_\infty)$$

$$= \ln S_\infty \frac{\boldsymbol{T}_\infty \boldsymbol{T}_{s\infty}}{\mathrm{Cl}} \frac{1 - r_{\mathrm{crit}}/r}{1 + (r_{\mathrm{crit}}/r)\boldsymbol{T}_{s\infty} \ln S_\infty/\mathrm{Cl}} - (\boldsymbol{T}^{**}_\infty - \boldsymbol{T}_\infty) \tag{229}$$

For a critical droplet this gives $\boldsymbol{T}_{sr} - \boldsymbol{T}^{**}_\infty = \boldsymbol{T}_\infty - \boldsymbol{T}^{**}_\infty$, that is, practically zero (because $\boldsymbol{T}^{**}_\infty = \boldsymbol{T}_\infty$ for such very small drops). In these equations r_{crit} has negative values if $S_\infty<1$, cf. Eq. (181). If the droplet is large enough to make capillarity effects negligible (Ke<0.01) but small enough to have negligible relative speed and thermal inertia (i.e., $\boldsymbol{T}^{**}_\infty = \boldsymbol{T}_\infty$), then Eq. (229) simplifies to

$$\boldsymbol{T}_{sr} - \boldsymbol{T}^{**}_\infty \approx \Delta\boldsymbol{T}_\infty = (\boldsymbol{T}_\infty \boldsymbol{T}_{s\infty}/\mathrm{Cl}) \ln S_\infty \tag{230}$$

For the calculation of time histories of growing or evaporating drops, it is convenient to nondimensionalize Eq. (227). We employ the *nondimensional droplet size* $\boldsymbol{r} = r/r_{\mathrm{ref}}$ introduced in Sec. 2.3 and define the *nondimensional time* τ as

$$\tau \triangleq \frac{\lambda T_{\mathrm{ref}}}{2 \rho_\ell L_r r^2_{\mathrm{ref}}} \mathrm{Nu}_{H0\mathrm{ref}}\, t \tag{231}$$

Thus the *nondimensional quasi-steady growth law* is obtained in the form

$$\frac{d\boldsymbol{r}}{d\tau} = \frac{\mathrm{Nu}_H}{\mathrm{Nu}_{H\mathrm{Oref}}} \frac{(\boldsymbol{T}_{sr} - \boldsymbol{T}^{**}_\infty)\ Q}{\boldsymbol{r}} \tag{232}$$

Here Nu_H, $\boldsymbol{T}_{sr}$, $\boldsymbol{T}^{**}_\infty$ and Q are functions of drop size $\boldsymbol{r}$ in general. The reference size r_{ref} may be chosen, for example, as the initial size (at time $t = t_0$) of the droplet considered. $\mathrm{Nu}_{H\mathrm{Oref}}$ is the value of Nu_H for a drop of size $r = r_{\mathrm{ref}}$ and of zero relative velocity at the carrier conditions at $t = t_0$. Note that the above definition of τ amounts to following choice of the time-scaling factor ϑ, which has been introduced in Eq. (47) but the meaning of which has been left open up to here:

$$\vartheta \triangleq \frac{2\ \rho_\ell\ L_r}{\eta\ \lambda\ T_{\mathrm{ref}}} \frac{p_\infty r^2_{\mathrm{ref}}}{\mathrm{Nu}_{H\mathrm{Oref}}} = \frac{\gamma-1}{\gamma} \frac{R_v}{\bar{R}} \frac{\rho_\ell}{\rho_\infty} \frac{L_r}{L} \frac{2\ \mathrm{Cl}\ \mathrm{Pr}}{\mathrm{Kn}^2_{\mathrm{ref}}\ \mathrm{Nu}_{H\mathrm{Oref}}} \tag{233}$$

The relationship between t and τ can be expressed with nondimensional quantities (except η/p_∞) as

$$t = \vartheta \frac{\eta}{p_\infty} \tau = \left(2\ \mathrm{Pr}\ \frac{\gamma-1}{\gamma} \frac{R_v}{\bar{R}} \frac{L_r}{L}\ \mathrm{Cl}\right) \frac{\eta\ \rho_\ell/\rho_\infty\ p_\infty}{\mathrm{Kn}^2_{\mathrm{ref}}\ \mathrm{Nu}_{H\mathrm{Oref}}}\ \tau \tag{234}$$

Above,

$$\mathrm{Kn}_{\mathrm{ref}} \triangleq \frac{\eta\ \sqrt{\bar{R}\ T_\infty}}{p_\infty\ r_{\mathrm{ref}}} \tag{235}$$

means the Knudsen number of the reference-size droplet. Note that the group multiplying τ in Eq. (234) has the meaning of a time unit. Its value is determined mainly by carrier pressure and density and by the reference values of Kn and Nu_H. The group in parantheses has a value of the order of 10 for most substances. Note that $L_r/L = 1$ may be set, the error being shown in App. 2 to be practically always negligible. Numerical values of the above time unit at near-atmospheric carrier conditions are of the order of

$$\vartheta \frac{\eta}{p_\infty} \approx \frac{(10^{-5}-10^{-6}\ \mathrm{s})}{\mathrm{Kn}^2_{\mathrm{ref}}} \quad (\text{at } p_\infty \approx 1 \text{ bar}) \tag{236}$$

Taking values of Kn at $p_\infty = 1$ bar from Fig. 6, it is seen that for large drops (Kn<0.001) the time unit is of the order of several seconds, whereas for very small droplets (Kn>1) its values are of the order of microseconds.

The growth law according to Eq. (227) or (232) is valid for any droplet size and carrier type. The complete calculation procedure to obtain dr/dt or $d\boldsymbol{r}/d\tau$ will be reviewed in Sec. 5.4. In many cases of practical significance the iterative calculation of Q can be avoided by using one of the approximative expression given in Sec. 5.2.6. For a number of such particular cases simplified growth laws will be given below. The approximation $L_r = L$ will be used. Note that the definition of τ will be adapted to each case.

5.3.2 Simplified Growth Laws for Pure Vapor or Vapor-Rich Carriers

If the carrier is a pure vapor or a vapor-rich mixture, Q may be substituted by Q_{vap}, as pointed out in Sec. 5.2.6. The domain of applicability of this approximation is seen in the error plots $\delta\ Q_{vap}$ shown in Fig. 18. Thus we set

$$Q = \frac{1}{1 + G} \quad \text{(pure-vapor or vapor-rich)} \tag{237}$$

For pure vapors, $G \approx 0$. The dependence of G on droplet size is weak and usually negligible.[23]

Large Drop in Pure-Vapor or Vapor-Rich Carrier:

For large drops, that is, for Kn<0.01 and Ke<0.01, continuum conditions apply and capillarity may be neglected. (As shown by Fig. 6, the Kn condition is usually the more stringent one.) Nu_H and $\mathrm{Nu}_{H0\mathrm{ref}}$ are to be obtained from Eq. (80). This gives

$$\frac{d\boldsymbol{r}}{d\tau} = \frac{\psi_1}{1 + G} \frac{\boldsymbol{T}_{s\infty} - \boldsymbol{T}_\infty^{**}}{\boldsymbol{r}} \tag{238}$$

where

$$\psi_1 \mathrel{\widehat{=}} 1 + 0.30\ \mathrm{Re}^{1/2}\ \mathrm{Pr}^{1/3} \tag{239}$$

and

$$\tau = \frac{\lambda\ T_{\mathrm{ref}}}{\rho_\ell\ L\ r_{\mathrm{ref}}^2}\ t \quad (\text{for } \mathrm{Kn}_{\mathrm{ref}} \ll 0.1 \text{ only}) \tag{240}$$

Note that Δw_∞, Re, ψ_1, and $\boldsymbol{T}_\infty^{**}$ are functions of time in general. If Re varies little, or Re<1, one may set ψ_1 = constant. If, in addition, $\boldsymbol{T}_\infty^{**} = \boldsymbol{T}_\infty$ and Eq. (230) may be applied, one obtains

$$\frac{d\boldsymbol{r}}{d\tau} = \frac{\psi_1}{1 + G} \frac{\Delta \boldsymbol{T}_\infty}{\boldsymbol{r}} = \frac{\psi_1}{1 + G} \frac{\boldsymbol{T}_{s\infty}\ \boldsymbol{T}_\infty\ \ell n\ S_\infty}{\mathrm{Cl}\ \boldsymbol{r}} \tag{241}$$

This is the familiar "r^2 law" of continuum-type media, making r^2 (or $\boldsymbol{r}^2$) a linear function of time for constant-state carriers.

Small Droplet in Pure-Vapor or Vapor-Rich Carrier:

For small droplets, that is, for Kn>5, free-molecular laws are

[23] The values of G will be discussed in Sec. 5.4.

applicable. Capillarity will not be neglected, since large Kn also means Ke>0.01, except if carrier pressure is very low, cf. Fig. 6. Often, however, Re≪ 1, Ma<0.1 will hold, and the thermal inertia of the droplet will be negligible. These leads to the simplifications $k_H(\mathrm{Ma}) = k_M(\mathrm{Ma}) = 1$ and $\boldsymbol{T}^{**}_\infty = \boldsymbol{T}_\infty$. Setting $Q = Q_{vap}$ and neglecting the small term Ke $\boldsymbol{T}_{s\infty}$/Cl in the denominator of Eq. (228), we obtain

$$\frac{dr}{d\tau} = \frac{\Delta \boldsymbol{T}_\infty - \mathrm{Ke}\, \boldsymbol{T}_\infty \boldsymbol{T}_{s\infty}/\mathrm{Cl}}{1 + G} \tag{242}$$

where

$$\tau = \frac{\lambda\, T_{ref}\, B_{HO}}{2\, \rho_\ell\, L\, r^2_{ref}\, \mathrm{Kn}_{ref}}\, t \quad \text{(for } \mathrm{Kn}_{ref} \gg 1 \text{ only)} \tag{243}$$

with B_{HO} given by Eq. (114). Since Ke is a function of droplet size, it is convenient to express it as $\mathrm{Ke} = \mathrm{Ke}_{ref}/\boldsymbol{r}$, where, in analogy to Kn_{ref}, the constant

$$\mathrm{Ke}_{ref} \triangleq \frac{2\, \sigma(T_{ref})}{r_{ref}\, \rho_\ell\, R_v\, T_\infty} \tag{244}$$

is the Kelvin number corresponding to r_{ref}. This gives

$$\frac{d\boldsymbol{r}}{d\tau} = \frac{\Delta \boldsymbol{T}_\infty - \mathrm{Ke}_{ref}\, \boldsymbol{T}_\infty \boldsymbol{T}_{s\infty}/\mathrm{Cl}\, \boldsymbol{r}}{1 + G} = \frac{\boldsymbol{T}_\infty \boldsymbol{T}_{s\infty}}{(1 + G)\, \mathrm{Cl}} \left(\ell n\, S_\infty - \frac{\mathrm{Ke}_{ref}}{\boldsymbol{r}} \right) \tag{245}$$

In the case of negligible capillarity ($\mathrm{Ke}_{ref} \ll 1$), $d\boldsymbol{r}/d\tau$ is seen to become independent of $\boldsymbol{r}$, resulting in linear change of $\boldsymbol{r}$ with time. For very small droplets the capillarity term is important. In supersaturated carriers ($\ln S_\infty > 0$), for critical-size droplets the growth rate is zero as a result of Eq. (182).

Multirange Expression for Pure-Vapor or Vapor-Rich Carriers:

Let relative speed be neglected (Re = Ma = 0) and the carrier state supposed to be constant (thus $\boldsymbol{T}^{**}_\infty = \boldsymbol{T}_\infty$). Capillarity will be considered, and no limitation on Kn will be imposed. Nu_H and Nu_{HOref} are to be obtained from Eq. (128). This leads to the following multirange expression for τ:

$$\tau = \frac{\lambda\, T_{ref}}{\rho_\ell\, L\, r^2_{ref}}\, \frac{1}{1 + (2/B_{HO})\, \mathrm{Kn}_{ref}}\, t \tag{246}$$

Note that this choice of τ includes the choices Eq. (240) and (243) as special cases. Setting $Q = Q_{vap}$ and neglecting Ke $\boldsymbol{T}_{s\infty}$/Cl beside 1 in Eq . (228), we get

$$\frac{d\boldsymbol{r}}{d\tau} = \frac{B_{HO}/2 + \mathrm{Kn}_{\mathrm{ref}}}{(B_{HO}/2)\boldsymbol{r} + \mathrm{Kn}_{\mathrm{ref}}} \, \frac{\Delta \boldsymbol{T}_\infty - \mathrm{Ke}_{\mathrm{ref}} \, \boldsymbol{T}_\infty \, \boldsymbol{T}_{s\infty}/\mathrm{Cl}\,\boldsymbol{r}}{1 + G}$$

$$= \frac{B_{HO}/2 + \mathrm{Kn}_{\mathrm{ref}}}{(B_{HO}/2)\boldsymbol{r} + \mathrm{Kn}_{\mathrm{ref}}} \, \frac{\boldsymbol{T}_\infty \, \boldsymbol{T}_{s\infty}}{(1 + G)\,\mathrm{Cl}} \left(\ln S_\infty - \frac{\mathrm{Ke}_{\mathrm{ref}}}{\boldsymbol{r}} \right) \tag{247}$$

The character of the dependence of $\boldsymbol{r}$ on τ is governed here primarily by the denominator of the first quotient.

5.3.3 Simplified Growth Laws for Lean, Cold Carriers

In lean carriers $\Pi_{g\infty} \approx 1$, $\Pi_{v\infty} \approx 0$, and $G \gg 1$. The term "cold" will mean that $T_\infty < T_{b\infty}$. The value of Q may be approximated by Q_{gas}, Eq. (224), provided that $T_\infty < 1.1\, T_{s\infty}$ and $\Pi_{v\infty} < 10^{-3}$, cf. the error plots δQ_{gas} in Fig. 18. If, however, T_∞ or $\Pi_{v\infty}$ is higher, it is preferable to use the more complicated expression Q_{syn} given by Eq. (225), which is good for vapor-rich and pure-vapor carriers as well. Note that Q_{gas} and Q_{syn} are functions of droplet size, because G, S_r, T_{sr}, and T_∞^{**} are influenced by Nu_H, Nu_M, and Ke, respectively. However, these dependences are normally weak, and in cases where the droplet size does not change by orders of magnitude, Q_{gas} and Q_{syn} may be treated as constants. The conditions for using Q_{gas} are fulfilled according to Eqs. (A-16*a*) and (A-16*b*) if simultaneously

$$\frac{T_{s\infty}}{T_{b\infty}} \lesssim \frac{1}{1 + 7/\mathrm{Cl}} \tag{248}$$

and

$$\frac{T_\infty}{T_{b\infty}} \lesssim \frac{1.1}{1 + 7/\mathrm{Cl}} \tag{249}$$

Such carrier conditions are encountered in the low-temperature part of domain 1 in Fig. 16.

Setting $Q = Q_{\mathrm{gas}}$, using Eq. (224), and eliminating T_{sr}, S_r, and G with the help of Eqs. (207), (211), and (222), Eq. (232) becomes, with τ as defined by Eq. (231),

$$\frac{d\boldsymbol{r}}{d\tau} = \frac{\mathrm{Nu}_H}{\mathrm{Nu}_{HO\mathrm{ref}}} \, \frac{p_{s\infty}/p_\infty}{A} \, \frac{\mathrm{Cl}}{\boldsymbol{r}} \left\{ S_\infty - \exp\left[\mathrm{Ke} - \frac{\mathrm{Cl}}{\boldsymbol{T}_\infty} \left(\frac{T_\infty}{T_\infty^{**}} - 1 \right) \right] \right\} \tag{250}$$

It is seen that the sign of $d\boldsymbol{r}/d\tau$ is determined by the difference of the saturation ratio S_∞ and of an exponential term. The latter is determined by Ke, that is, the capillarity, and by T_∞^{**}/T_∞, that is, the relative velocity, and, in changing carriers, the thermal inertia of the droplet; it has the value unity for large droplets (Ke = 0) in a stationary, constant environment ($T_\infty^{**} = T_\infty$). In terms of real time t, the growth rate can be shown to depend little on the Clausius-Clapeyron number Cl, that is, on the latent heat.

Carrier temperature T_∞ has a great effect on $d\boldsymbol{r}/d\tau$, however, because it determines $p_{s\infty}$. Note that Eqs. (A-16*b*) and (A-16*c*) give

$$\frac{p_{s\infty}}{p_\infty} = \exp\left(\frac{\mathrm{Cl}}{\boldsymbol{T}_{b\infty}} - \frac{\mathrm{Cl}}{\boldsymbol{T}_\infty}\right) \tag{251}$$

For lean carriers and with $L_r = L$, Eq. (214) gives

$$A \approx \frac{\gamma}{\gamma-1}\,\frac{\mathrm{Sc}}{\mathrm{Pr}}\,\frac{\mathrm{Nu}_H}{\mathrm{Nu}_M} \quad (\text{for } \Pi_{g\infty} \approx 1 \text{ only}) \tag{252}$$

By inserting this expression into Eq. (250), one finds that the growth rate is determined by the value of Nu_M whereas Nu_H plays no role. This is the well-known fact that growth or evaporation in cold, lean carriers is controlled by the mass transfer process.

Various expressions for Nu_M, as given in Sec. 3, lead to the following special growth laws.

Large Droplet in Lean, Cold Carriers:

Continuum conditions will be supposed, capillarity will be neglected (Ke = 0), as well as the effects of recovery temperature and thermal inertia ($\boldsymbol{T}_\infty^{**} = \boldsymbol{T}_\infty$).

Taking τ as given by Eq. (240), Eq. (250) is simplified to

$$\frac{d\boldsymbol{r}}{d\tau} = \psi_2\,\psi_3\,\frac{p_{s\infty}}{p_\infty}\,\frac{S_\infty - 1}{\boldsymbol{r}} = \psi_2\,\psi_3\,\frac{p_{v\infty} - p_{s\infty}}{p_\infty\,\boldsymbol{r}} \tag{253}$$

where ψ_2 is an abbreviation in analogy to ψ_1, defined as

$$\psi_2 \triangleq 1 + 0.30\mathrm{Re}^{2/3}\,\mathrm{Sc}^{1/3}\,\Pi_{g\infty}^{1/3} \tag{254}$$

and where the dimensionless constant

$$\psi_3 \triangleq \frac{\gamma-1}{\gamma}\,\frac{\mathrm{Pr}\ \mathrm{Cl}}{\mathrm{Sc}} \tag{255}$$

is seen to be a material property. Typically ψ_3 is of the order of 2 to 10.

Very Small Droplet in Lean, Cold Carriers:

Assuming free-molecular conditions, accounting for capillarity (Ke>0), but neglecting relative velocity and thermal inertia ($T_\infty^{**} = T_\infty$), taking Nu_M from Eq. (104) and using Eq. (243) for τ, one obtains from Eq. (250)

$$\frac{d\boldsymbol{r}}{d\tau} = \psi_4\,\frac{p_{s\infty}}{p_\infty}\,(S_\infty - e^{\mathrm{Ke}}) = \psi_4\,\frac{p_{v\infty} - p_{s\infty}e^{\mathrm{Ke}_{\mathrm{ref}}/\boldsymbol{r}}}{p_\infty} \tag{256}$$

where ψ_4 is a constant signifying

$$\psi_4 \triangleq \frac{\gamma-1}{\gamma} \frac{\text{Sc Cl}}{\text{Pr}} \frac{B_{HO}}{B_{MO}} \tag{257}$$

Here B_{HO} and B_{MO} are given by Eqs. (114) and (115). Typical numerical values of ψ_4 are around 3 to 5.

Multirange Expression for Lean, Cold Carriers:

Again, relative velocity and thermal inertia will be neglected. Taking the multirange expression for Nu_M as given by Eq. (129) and defining τ in accordance with Eq. (246), we obtain

$$\frac{d\boldsymbol{r}}{d\tau} = \psi_5 \frac{B_{MO}/2 + \text{Kn}_{\text{ref}}}{(B_{MO}/2)\boldsymbol{r} + \text{Kn}_{\text{ref}}} \frac{p_{s\infty}}{p_\infty} \left(S_\infty - e^{\text{Ke}_{\text{ref}}/\boldsymbol{r}}\right) \tag{258}$$

where the abbreviation

$$\psi_5 \triangleq \frac{\gamma-1}{\gamma} \frac{\text{Sc Cl}}{\text{Pr}} \frac{1 + (2/B_{MO})\ \text{Kn}_{\text{ref}}}{1 + (2/B_{HO})\ \text{Kn}_{\text{ref}}} \tag{259}$$

normally has some constant value between 2 and 10. The character of the dependence of $\boldsymbol{r}$ on time is decided by the value of Kn_{ref}. The influence of capillarity is strong if $\text{Ke}_{\text{ref}}/\boldsymbol{r} \approx \ln S_\infty$.

5.3.4 Simplified Growth Laws for Lean, Hot Carriers

The term "hot" will mean $T_\infty > T_{b\infty}$ so that Q can be approximated as Q_{gen}. The carrier being lean ($\Pi_{v\infty} < 10^{-3}$), Eq. (222) gives $G \gg 1$. We consider only the case of "large" droplets (Kn<0.01, Ke<0.01). Then τ is defined as in Eq. (240), and Eq. (232) can be written as

$$\frac{d\boldsymbol{r}}{d\tau} = -\frac{\psi_1\ \psi_6\ \boldsymbol{T}_{b\infty}}{\boldsymbol{r}} \tag{260}$$

where ψ_1 is a weak function of Re, and ψ_6 is a function of the temperatures T_∞^{**}, $T_{s\infty}$, and $T_{b\infty}$ according to

$$\psi_6 \triangleq \frac{T_\infty^{**}}{T_{b\infty}} - 1 + \left(1 - \frac{T_{s\infty}}{T_{b\infty}}\right) \exp\left(\frac{T_{s\infty}/T_{b\infty} - T_\infty^{**}/T_{b\infty}}{1 - T_{s\infty}/T_{b\infty}}\right) \tag{261}$$

Note that $T_\infty^{**} = T_\infty$ holds if Ma<0.1 and if the carrier state is constant. It is seen that the change of $\boldsymbol{r}^2$ with τ is nearly linear, and proportional to $\psi_1\ \psi_6$. If carrier composition (or $T_{s\infty}/T_{b\infty}$) is fixed, ψ_6 is an increasing function of the ratio $T_\infty^{**}/T_{b\infty}$; cf. Fig. 19. It is seen that $T_{s\infty}/T_{b\infty}$ has little effect on ψ_6 in the composition range considered. The nondimensional evaporation rate ($-\,d\boldsymbol{r}/d\tau$) is seen to increase with carrier temperature T_∞^{**}, if carrier pressure (and therefore $T_{b\infty}$) and composition are fixed.

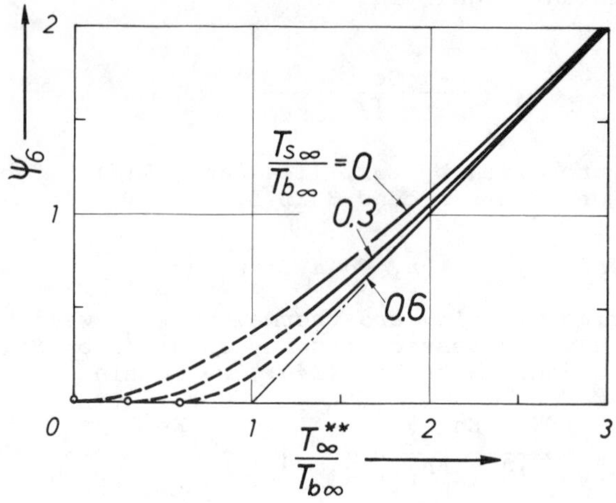

Fig. 19. Influence of $T_\infty^{**}/T_{b\infty}$ on the evaporation rate factor ψ_6 in hot, lean carriers.

Using Eqs. (26) and (240) to return to dimensional radius and time, Eq. (260) gives

$$- \frac{dr}{dt} = \frac{\lambda\ T_{b\infty}}{\rho_\ell\ L} \left(1 + 0.30\ \mathrm{Re}^{1/2}\ \mathrm{Pr}^{1/3}\right) \frac{\psi_6}{r} \qquad (262)$$

Here ρ_ℓ and L are specified at T_{ref} and λ at $T_{1/3}$. In hot carriers $T_{ref} \approx T_{b\infty} = T_s(p_\infty)$ is a reasonable choice. Since λ increases, ρ_ℓ and L decrease with increasing T_{ref}, the group $\lambda\ T_{b\infty}/\rho_\ell\ L$ will be a markedly increasing function of p_∞. If this increase is fast enough, the real evaporation rate may increase rather than decrease with increasing carrier pressure (Kent, 1973). It may be noted that in very hot carriers ($T_\infty > 2T_{b\infty}$), the one-third rule is likely to become inaccurate, and the variation of λ must be incorporated in heat conduction theory, making Nu_H a function of temperature and material properties. This problem has been analyzed in detail in connection with burning droplets by Kassoy and Williams (1968) and by Shyu et al. (1972). In combustion problems further complications arise as a result of the chemical reaction and radiative heat transfer. These aspects have been treated by, among others, Rüping (1959), Williams (1973), and Sangiovanni (1978). As a literature survey on droplet combustion, the paper of Williams is recommended.

5.4 Outline of Calculation Procedure

In view of the complexity of the equations given in Secs. 3 to 5 and of the simplifications that may be made in many different cases, it is useful to review the steps required to obtain the growth rate dr/dt or $d\mathbf{r}/d\tau$ and the surface and interior temperatures, T_r and T_ℓ, of a droplet, on the basis of the quasi-steady growth theory outlined in Secs. 5.1.4, 5.2, and 5.3. We assume that the use of the quasi-steady theory is justified (see Sec. 5.1.4 or 6.1).

As given, we suppose the momentary size r and relative velocity of the drop, the thermodynamic state and the composition of the carrier medium, including the saturation ratio S_∞ and the subcooling ΔT_∞. The property data will be specified at some reference temperature T_{ref} which is preferably chosen as $T_{\text{ref}} \approx T_{b\infty}$ or, in cold carriers ($T_\infty < T_{b\infty}$), somewhere in the range $T_\infty < T_{\text{ref}} < T_{b\infty}$. If the carrier state is subjected to continuous change, the expansion rate $\dot{P}$ or the Stodola number Sd should be known; however, their effect on T_r, T_ℓ, and dr/dt becomes noticeable only for large droplets in fast-changing, rapidly accelerating carriers where the drift velocity or $(T_\infty^{**} - T_\infty)$ become significant.

The property data may include the nondimensional groups Pr, Sc, Cl, and Eö. Except in continuum-type situations, the coefficients B_H and B_M and the accommodation coefficients α_{Hv}, α_{Hg}, and α_{Mv} must be specified, cf. Sec. 3.3. It should be noted that certain material properties occurring in the growth laws refer, in mixture carriers, to the mixture rather than to any of its components. The transport coefficients λ, η, and D (which determine Pr and Sc) are genuine mixture properties, whereas the data R, $\bar{c}_p$ (which is required for γ), γ, and $(\overline{R^{3/2}\, K\, \alpha_H})$ are to be determined by mass averaging as treated in App. 1, Sec. A1.2. The influence of mixture composition on the property values is especially important if the gas and vapor components have very different molecular masses. In the case of large temperature differences ($T_\infty/T_{b\infty} \gg 1$), the carrier data have to be specified according to the "one-third rule" (cf. Sec. 5.2) at a temperature equaling $(2T_{\text{ref}} + T_\infty)/3 = T_{1/3}$.

For the sake of nondimensional formulation, a reference size r_{ref} will be assumed. It is recommended to make r_{ref} be representative of the r values encountered in the problem; in dealing with the growth or evaporation of a single drop, the initial size $r(t_0) = r_{\text{ref}}$ is a convenient choice.

The calculation of Δw_∞ was discussed in Sec. 5.1, especially in Secs. 5.1.4 and 5.1.5.

Knowing r, r_{ref}, and Δw_∞, the following values are to be determined:

Re from Eq. (45)

Ma from Eq. (60)

Kn from Eq. (43)

Ke from Eq. (62)

Kn_{ref} from Eq. (235)

Ke_{ref} from Eq. (244)

The next step is the determination of the Nusselt numbers Nu_H, $\mathrm{Nu}_M \triangleq \mathrm{Sh}$, and $\mathrm{Nu}_{H0\mathrm{ref}}$. Since the general expressions for Nu_H and Nu_M, which are valid for any droplet size and for nonzero relative velocity, are algebraically complicated, it is recommended to utilize simpler, though limited-range, expressions wherever permissible. Tables 5 and 6 summarize the equations of Nu_H and Nu_M as given in Sec. 3, and show the corresponding expressions for A. The simplification $L_r = L$ has been used. Table 5 concerns the simpler case of stationary droplets (Re ≈ 0, Ma ≈ 0), whereas Table 6 gives the more general expressions, which are valid up to Re⩽200 and/or Ma⩽0.5–(1). In Table 5 and 6 no corrections have been made for simultaneous heat and mass transfer. The effect of T_r on the free-molecular and multirange expressions for Nu_H has been neglected altogether, by setting $T_\infty/T_r = 1$ in Eq. (107).

Next one calculates

T_∞^{**} from Eq. (202)

T_{sr} from Eq. (207)

T_{br} from Eq. (208)

S_r from Eq. (212)

A from Eq. (214)

C from Eq. (215)

Note that $T_\infty^{**} = T_\infty$ can be set if Ma<0.1, and $T_{sr} \approx T_{s\infty}$, $T_{br} \approx T_{b\infty}$ will result if Ke<0.01. In pure vapors the calculation of S_r is unnecessary. Since A involves the Nusselt numbers Nu_H and Nu_M, it is a function of Kn, Ma, and/or Re, in general. Expressions giving A are listed in Tables 5 and 6.

The caloric effectiveness Q has to be calculated only if T_r and T_ℓ are of direct interest, or if $d\mathbf{r}/d\tau$ cannot be calculated directly from any of the simplified expressions given in Secs. 5.3.2 to 5.3.4. In the latter case Q is obtained from Eq. (219) after having calculated

T_r from Eq. (213)

by an iteration process. This method is only justified, however, if the carrier composition is drastically changing (from $\Pi_{v\infty} \approx 1$ to $\Pi_{v\infty} \approx = 0$ or vice versa) during the process. Otherwise, one will prefer to avoid the iterative method by employing an explicit equation given in Sec. 5.2.6. In this case one will determine first

G from Eq. (222)

and then use one of the approximations

Table 5. *Equations Giving* Nu_H, Nu_M *and* A *for Stationary Droplets* (Applicable if $Re \ll 1$ and $Ma < 0.1$)

Knudsen Number Range	Nu_H and $Nu_M \widehat{=} Sh$	From Eq.	A
Continuum $Kn<0.01$	$Nu_H = 2$	(77)	$A = \frac{\gamma}{\gamma-1}\frac{Sc}{Pr}(1-\Pi_{v\infty})$
	$Nu_M = 2$	(78)	
Free-molecular, $Kn>5$	$Nu_H = \sqrt{\frac{2}{\pi}}\,\frac{\gamma-1}{\gamma}\,\frac{\overline{(R^{3/2}\,K\,\alpha_H)}}{\bar{R}^{3/2}}\,\frac{Pr}{Kn}$	(103) (114)	$A = \frac{\overline{(R^{3/2}\,K\,\alpha_H)}}{\bar{R}\,\sqrt{R_v}\,\alpha_{Mv}}$
	$Nu_M = \sqrt{\frac{2}{\pi}}\,\sqrt{\frac{R_v}{\bar{R}}}\,\alpha_{Mv}\,\frac{Sc}{Kn}\,(1-\Pi_{v\infty})$	(104) (115)	
Multirange, $0<Kn<\infty$	$Nu_H = \frac{2}{1+\frac{\sqrt{2\pi}\,\gamma}{\gamma-1}\,\frac{\bar{R}^{3/2}}{\overline{(R^{3/2}\,K\,\alpha_H)}}\,\frac{Kn}{Pr}}$	(127)	$A = \frac{Sc\,(1-\Pi_{v\infty})+\frac{\sqrt{2\pi}}{\alpha_{Mv}}\sqrt{\frac{\bar{R}}{R_v}}\,Kn}{\frac{\gamma-1}{\gamma}\,Pr+\frac{\sqrt{2\pi}\,\bar{R}^{3/2}}{\overline{(R^{3/2}\,K\,\alpha_H)}}\,Kn}$
	$Nu_M = \frac{2}{1+\sqrt{2\pi}\,\frac{\sqrt{\bar{R}/R_v}}{\alpha_{Mv}\,(1-\Pi_{v\infty})}\,\frac{Kn}{Sc}}$	(128)	

Table 6. *Equations giving Nu_H, Nu_M and A for Moving Droplets*
(Applicable if Re<200 and Ma<0.5)

Knudsen Number Range	Nu_H and $Nu_M \mathrel{\hat{=}} Sh$	From Eq.	A
Continuum Kn < 0.01	$Nu_H = 2\psi_1 \qquad \psi_1 \mathrel{\hat{=}} 1 + 0.30\,Re^{1/2}\,Pr^{1/3}$	(80)	$A = \frac{\gamma}{\gamma-1}\,\frac{Sc}{Pr}\,(1-\Pi_{v\infty})\,\frac{\psi_1}{\psi_2}$
	$Nu_M = 2\psi_2 \qquad \psi_2 \mathrel{\hat{=}} 1 + 0.30\,Re^{1/2}\,Pr^{1/3}\,\Pi_{g\infty}^{1/3}$	(81)	
Free-Molecular Kn > 5	$Nu_H = \sqrt{\frac{2}{\pi}}\,\frac{\gamma-1}{\gamma}\,\frac{\sqrt{R^{3/2}\,K\,\alpha_H}}{\bar{R}^{3/2}}\,k_H(Ma)\,\frac{Pr}{Kn}$	(103) (106) (109)	$A = \frac{\sqrt{R^{3/2}\,K\,\alpha_H}}{\bar{R}\,\sqrt{R_v}\,\alpha_{Mv}}\,\frac{k_H(Ma)}{k_M(Ma)}$
	$Nu_M = \sqrt{\frac{2}{\pi}}\,\sqrt{\frac{R_v}{\bar{R}}}\,\alpha_{Mv}\,k_M(Ma)\,\frac{Sc(1-\Pi_{v\infty})}{Kn}$	(104) (107) (110)	
Multirange, 0 < Kn < ∞	$Nu_H = \frac{2\psi_1}{1 + [\sqrt{2\pi}\,\gamma/(\gamma-1)][\bar{R}^{3/2}/\sqrt{(R^{3/2}\,K\,\alpha_H)}][\psi_1/k_H(Ma)](Kn/Pr)}$	(123) (80) (106) (109)	$A = \frac{[Sc\,(1-\Pi_{v\infty})/\psi_2] + (\sqrt{2\pi}/\alpha_{Mv})\,\sqrt{\bar{R}/R_v}\,[Kn/k_M(Ma)]}{[(\gamma-1)/\gamma](Pr/\psi_1) + [\sqrt{2\pi}\,\bar{R}^{3/2}/\sqrt{(R^{3/2}\,K\,\alpha_H)}][Kn/k_H(Ma)]}$
	$Nu_M = \frac{2\psi_2}{1 + \sqrt{2\pi}\,[\sqrt{\bar{R}/R_v}/\alpha_{Mv}\,(1-\Pi_{v\infty})][\psi_2/k_M(Ma)](Kn/Sc)}$	(124) (81) (107) (110)	

$$Q \approx \begin{cases} Q_{\text{vap}} & \text{from Eq. (223)} \\ Q_{\text{gas}} & \text{from Eq. (224)} \\ Q_{\text{syn}} & \text{from Eq. (225)} \\ Q_{\text{gen}} & \text{from Eq. (226)} \end{cases}$$

the choice to be based on Fig. 18, aiming at minimum errors δQ.

One should note that the same expression of Q should be employed during the entire growth or evaporation process.

In order to illustrate the numerical values of G, values of G/A and of A are presented in Table 7 as functions of Knudsen number and carrier composition for water drops in steam/air mixtures. Two overall pressure levels, corresponding to $T_{b\infty}$ = 300 K and 500 K, are shown. Stationary droplets are supposed, and the following assumptions were made: Ke = 0; $\alpha_{Hv} = \alpha_{Hg} = \alpha_{Mv} = 1$; $T_{\text{ref}} = T_{b\infty}$. Also, all gas-phase property data were evaluated at the temperature T_{ref}.

It is seen from Table 7 that G/A is a strong function of mixture composition. For vapor-rich carriers $G/A \ll 1$, whereas for gas-rich carriers $G/A \gg 1$. The effect of $T_{b\infty}$ on G/A is relatively slight.

On the other hand, the values of A lie around 2, except in pure or nearly pure steam, where A is zero for the continuum case (Kn = 0) and climbs to A = 3–4 in the free-molecular limit. The influence of the Knudsen number is slight, excepting the pure-vapor, near-continuum situation. Pressure and temperature have little effect on A.

As seen, the value of $G = (G/A) \cdot A$ may lie anywhere between zero and very large ($\to\infty$) values, but in pure vapors $G \ll 1$ holds.

Having determined Q, one obtains

T_r from Eq. (221)

T_ℓ from Eq. (199)

$d\boldsymbol{r}/d\tau$ from Eq. (232)

t/τ from Eq. (231) or Eq. (234)

Here t/τ is the time unit required to convert τ into real time t. Finally, if desired, $dr/dt = (r_{\text{ref}}\ \tau/t)\, d\boldsymbol{r}/d\tau$ can be calculated.

If the carrier conditions are such that one of the simplified explicit growth laws given in Secs. 5.3.2 to 5.3.4 may be used and if T_r and T_ℓ are of no interest, one will avoid the calculation of C, G, and Q and determine $d\boldsymbol{r}/d\tau$ directly from one of the equations

Table 7. *Numerical Values of the Gas Influence Factor* $G = (G/A) \cdot A$ *for Water Drop in Steam/Air Mixture Carriers of Various Vapor Partial-Pressure Ratios* $\Pi_{v\infty}$ *(Assumptions: See Text)*

$\Pi_{v\infty}$	$T_{b\infty} = 300$ K ($p_\infty = 0.0356$ bar, Cl = 17.6)						$T_{b\infty} = 500$ K ($p_\infty = 26.4$ bar, Cl = 9.3)					
	$\frac{G}{A}$	A					$\frac{G}{A}$	A				
		Kn→0	Kn=0.1	Kn=1	Kn=10	Kn→∞		Kn→0	Kn=0.1	Kn=1	Kn=10	Kn→∞
1 (steam)	3.228×10^{-3}	0	0.72	2.30	2.94	3.04	1.156×10^{-2}	0	1.07	3.13	3.89	3.99
0.9	3.544×10^{-3}	0.20	0.87	2.30	2.86	2.94	1.256×10^{-2}	0.26	1.22	3.01	3.65	3.74
0.5	5.976×10^{-3}	1.15	1.53	2.25	2.49	2.53	2.002×10^{-2}	1.30	1.76	2.49	2.72	2.75
0.1	2.524×10^{-2}	2.25	2.21	2.14	2.12	2.12	7.428×10^{-2}	2.33	2.11	1.84	1.77	1.76
0.01	2.028×10^{-1}	2.52	2.35	2.11	2.03	2.02	5.172×10^{-1}	2.55	2.12	1.66	1.55	1.54
10^{-4}	$1.392 \times 10^{+1}$	2.54	2.37	2.10	2.02	2.01	$2.919 \times 10^{+1}$	2.58	2.13	1.64	1.52	1.51
10^{-6}	$1.013 \times 10^{+3}$	2.54	2.37	2.10	2.02	2.01	$1.872 \times 10^{+3}$	2.58	2.13	1.64	1.52	1.51

given in these sections. One should note that t/τ is given by different expressions, depending on the growth law chosen.

6 APPLICATION

6.1 Time Constants

The time duration of droplet growth (or shrinkage) can be characterized by the time constant Δt_G defined in Eq. (140). At a constant growth rate the droplet would double its radius in this time – or would vaporize completely in the case of evaporation. If the droplet size and the carrier conditions are known Δt_G can be calculated as

$$\Delta t_G = \frac{\boldsymbol{r}}{|d\boldsymbol{r}/d\tau|} \; \frac{\rho_\ell \; L \; r_{\text{ref}}^2}{\lambda \; T_{\text{ref}}} \; \frac{2}{\text{Nu}_{HO\text{ref}}} \tag{263}$$

where $d\boldsymbol{r}/d\tau$ is the dimensionless growth rate as expressed by Eq. (234) or by one of the growth laws given in Sec. 5.3.2 to 5.3.4. Note that Eq. (231) has been used to convert τ into t, and $L_r = L$ has been set.

It was pointed out in Sec. 5.1.4 that the necessary condition for applying the quasi-steady growth theory presented in Secs. 5.2 and 5.3 is that $\Delta t_{\text{int}} \ll \Delta t_G$ and $\Delta t_{\text{int}} \ll \Delta t_{\text{exp}}$, where Δt_{int} is the time constant of internal heat conduction defined by Eq. (154) and Δt_{exp} is the time constant of carrier state change (for instance, expansion) defined by Eq. (50). Typical values of Δt_{exp} are indicated in Fig. 8. If the duration of the process is limited to some "residence time" Δt_{res} (for instance as a result of mechanical removal of the droplets from the flow), significant growth only occurs if Δt_{res} is comparable to or larger than Δt_G.

If furthermore $\Delta t_{\text{exp}} \gg \Delta t_G$, the carrier state may be regarded as constant during the growth process, provided the latter has no appreciable influence on the carrier (see below).

During its growth or vaporization, the droplet is in general moving with respect to the carrier, although the drag force tends to reduce its relative speed. The time constant of velocity adaptation, Δt_{vel}, is given by Eq. (170). If $\Delta t_{\text{vel}} \ll \Delta t_G$, the droplet will have a quasi-steady drift velocity Δw_{drf} during most of the growth process. The value of Δw_{drf}, given by Eq. (172), is determined by the body force (gravitational) field and by carrier acceleration, being zero in their absence. In accelerated flows (including curvilinear ones) the droplets closely follow the streamlines of the carrier (i.e., $\Delta w_{\text{drf}} \ll |\vec{w}_{\text{carr}}|$ is maintained) if $\Delta t_{\text{vel}} \ll \Delta t_{\text{acc}}$, where Δt_{acc} is the acceleration time constant defined by Eq. (173).

Whereas the values of Δt_{exp}, Δt_{res} and Δt_{acc} in general result from the macroscopic properties of the system in question, the values of Δt_{int}, Δt_{vel} and Δt_G are determined by droplet size and carrier state.

In order to give an overview about the values of Δt_{int}, Δt_{vel}, and Δt_G, Fig. 20 shows curves of constant values of these time constants for quiescent water droplets of sizes ranging between and 10^{-3} m. Various carrier pressures p_∞ and two carrier substances were considered: pure steam and dry air, the latter at

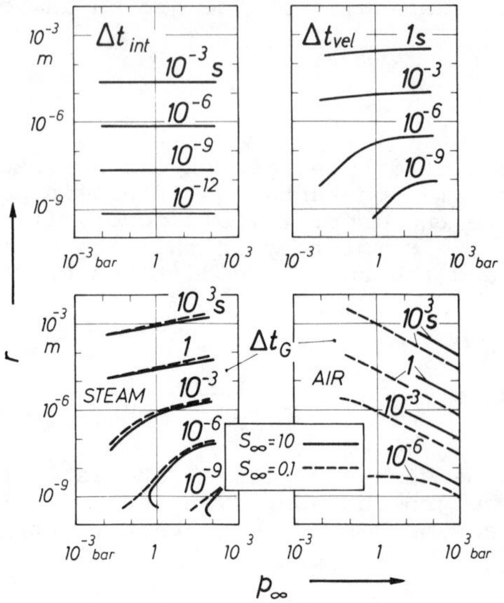

Fig. 20. Time constants of quiescent water drops in steam and in air carriers, as functions of carrier pressure and drop radius. The curves of Δt_{int} and Δt_{vel} are common for both carriers. The curves of Δt_G are plotted for a supersaturated and an undersaturated state.

constant temperature of $T_\infty = 300$ K.

The curves of Δt_{int} and Δt_{vel} are common for both carriers. (The former is independent of the carrier; the latter has very similar values in steam and air.) The bottom two fields show Δt_G, in steam and air, respectively. For both carriers two examples are used: a highly supersaturated ($S_\infty = 10$; solid lines) and an undersaturated condition ($S_\infty = 0.1$; dashed lines). The growth rate $d\delta/d\tau$ was calculated from the pertinent multirange expressions, that is, Eq. (247) for steam and Eq. (258) for air. Since Eq. (258) presupposes that the air is in a "dry, cold" state (meaning that $p_{v\infty} = S_\infty \cdot p_{s\infty} < 10^{-3}\, p_\infty$), the air data are good only for high pressures p_∞. In order to show the trends, the curves for air have been drawn slightly beyond their domain of accuracy.

At such great deviations from saturation, droplet growth or vaporization is very fast, making Δt_G relatively small. Still, it appears from Fig. 20 that Δt_G exceeds Δt_{int} and Δt_{vel} for any droplet size and carrier pressure, typically by factors of 1000 and of 100, respectively. Thus *sustained growth or evaporation are seen to be typically quasi-steady processes even in carriers of high super- or subsaturation.* If S_∞ is closer to unity, the values of Δt_G becomes still higher.

If the liquid mass fraction x_ℓ is high enough to exert a significant influence on the state of the carrier, the latter may of course not be regarded as staying constant during growth, regardless of whether $\Delta t_{exp} \gg \Delta t_G$ is fulfilled or not. Such droplet/carrier interaction has to be accounted for typically if $x_\ell > 10^{-3}$; cf. Eq. (15). The influence on the carrier state can be estimated from Eq. (59).

In cases where large amounts of liquid having an initial temperature very different from $T_{g\infty}$ are being injected into the carrier (as in spray cooling), the mere change of the liquid temperature is apt to influence the carrier markedly. Then the carrier state will change on a time scale of the order of Δt_{int}, during which time growth or evaporation are nonsteady. After this initial period, quasi-steady growth will occur, provided that the droplets have sufficient residence time.

The data of Fig. 20 refer to stationary *water* droplets only, but the orders of magnitudes are valid for other substances as well. As seen from Fig. 4 and Table A-1, chemically very different substances have in fact quite similar values of the nondimensional material properties Pr, Sc, Cl, Eö (also α_H, α_{Mv}, and α_F), and therefore comply reasonably well with the above conclusions. Nonzero relative velocities will tend to reduce the values of Δt_{vel} and Δt_G. Greatly different values of Δt_G may result if S_∞ and/or carrier temperature are different from the values indicated especially if the carrier is hot and highly undersaturated ($S_\infty \approx 0$).

6.2 Spray Droplet Vaporization

This first example deals with a situation that is typical of spray evaporator or fuel prevaporizer equipment. A relatively large droplet (of many micrometers diameter) is injected into a hot, dry air stream where it becomes vaporized. The time t_E and flight distance ξ_E needed for complete evaporation are to be determined. The problem is complicated by the great velocity and temperature differences existing initially between the droplet and the carrier.

We assume a rectilinear arrangement, as sketched in Fig. 21, and we consider a single droplet of radius r_0 and temperature $T_{\ell 0}$, which is injected at time $t = 0$ with an absolute velocity $\vec{w}_{\ell 0}$ that is parallel to the carrier velocity $\vec{w}_{carr}$. The droplet will be of methyl alcohol (methanol, CH_3OH). The following data are given:

Droplet: $r_0 = 10\ \mu m = 10^{-5}\ m$

$T_{\ell 0} = 313\ K$

$w_{\ell 0} = 12\ m/s$

Carrier: $p_\infty = 10\ bar = 10^5\ Pa$

$T_\infty = 600\ K$

$w_{carr} = 30\ m/s$

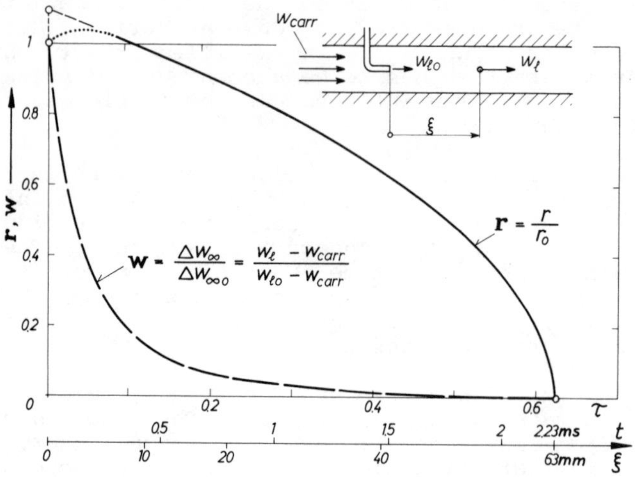

Fig. 21. Velocity and size history of the evaporating spray droplet.

The air being "dry" (which means that there is no methanol vapor in it), we have $p_{v\infty} = S_{\infty} = g_{\infty} = T_{s\infty} = 0$. The influence of the vaporization process on the carrier will be neglected, which amounts to setting for all times

$$x_{\ell} \equiv 0$$

We assume that the carrier pressure, temperature, and velocity stay constant during the process.

The first thing to decide is the kind of drag and growth laws to be applied: continuum, free-molecular, or multirange.[24] The next question concerns the comparative values of the time constants Δt_{int}, Δt_{vel}, and Δt_G in order to determine the degree of simplification that is permissible. The history of relative velocity and size will be obtained by solution of the differential equation giving $\Delta w_{\infty}(t)$ and $r(t)$. The coordinates t_E and ξ_E finally result from the condition $r(t_E) = 0$ and from the integration of the droplet velocity $w_{\ell}(t)$ from zero to t_E.

The following materials data must be provided. [Use of standard compilations such as Landolt-Börnstein (1960), Weast and Selby (1966), etc., are recommended.] Note that air is treated as a single gas in the present context. Data pertinent to free-molecule

[24] Of course, each process leading to complete vaporization of a droplet ends in the free-molecular range, since $r \to 0$ means that $Kn \to \infty$. However, continuum laws may be used right down to $r = 0$ provided that they are satisfactory in the range $r_0 \geq r > 0.1 r_0$.

processes (such as B_H, B_M, B_F, α_H, α_{Mv}, α_F) will not be specified unless the need for multirange or free-molecular equations will appear.

Note that the vapor pressure curve of methyl alcohol gives $T_{b\infty} = T_s(p_\infty) = 411$ K. Since one obtains $T_\infty/T_{b\infty} = 1.46 > 1$ and it is $Y_{v\infty}<10^{-3}$, the carrier is a "hot, lean gas." The liquid properties will be evaluated first at a reference temperature tentatively chosen as $T_{ref} = T_{b\infty} = 411$ K. We have

$$\begin{aligned} \rho_\ell &= 670 \text{ kg/m}^3 \\ c_\ell &= 2700 \text{ J/kg K} \\ a_\ell &= 1\text{x}10^{-7} \text{ m}^2\text{/s} \\ L &= 850\text{x}10^3 \text{ J/kg} \\ \sigma &= 0.0106 \text{ N/m} \end{aligned}$$

Since the carrier is "hot," the one-third rule will be applied and the carrier properties will be evaluated (instead of at T_{ref}) at

$$T_{1/3} = T_r + \frac{T_\infty - T_r}{3} \approx T_{ref} + \frac{T_\infty - T_{ref}}{3} = 474 \text{ K}$$

at which for pure air we find

$$\begin{aligned} \eta &= 2.6\text{x}10^{-5} \text{ kg/s m} \\ \lambda &= 3.9\text{x}10^{-2} \text{ J/s m K} \\ \text{Pr} &= 0.7 \\ \text{Sc} &= 1.2 \quad \text{(estimated from Table A-2)} \end{aligned}$$

Using $R_g = 287.2$ J/kg K and $\gamma = 1.4$, we calculate

$$\rho_\infty = \frac{p_\infty}{\sqrt{R_g \, T_{1/3}}} = 7.35 \text{ kg/m}^3$$

$$\sqrt{\gamma \, R_g \, T_{1/3}} = 437 \text{ m/s}$$

The reference radius is chosen as

$$r_{ref} = r_o = 1 \times 10^{-4} \text{ m}$$

Initially (time $t = 0$) the relative velocity is

$$\Delta w_{\infty o} = |w_{\ell o} - w_{carr}| = 18 \text{ m/s}$$

Note that $\Delta w_\infty(t)$ will decrease from this value toward zero during the evaporation process.[25]

The nondimensional parameters Re, Ma, We, Kn, and Ke will vary as r and Δw_∞ are diminishing. (We is of interest here in order to

---

[25] The settling velocity in the gravity field can be shown to be negligible for this droplet, since Eq. (167) yields, with $r = r_o$, $|\vec{b}| = 9.81$ m/s^2, $d\vec{w}_{carr}/dt = 0$ and $\text{Nu}_F \approx 3$, merely $\Delta w_{drf} = 0.008$ m/s.

check if the droplet remains spherical.) The following limits are found, the first values being obtained with $r = r_o = r_{ref}$ and $\Delta w_\infty = \Delta w_{\infty o}$, the second with $r = 0.1 r_o$ and $\Delta w_\infty = 0$:

$$\begin{aligned} \mathrm{Re} &= 102 && - 0 \\ \mathrm{Ma} &= 0.04 && - 0 \\ \mathrm{We} &= 4.5 && - 0 \\ \mathrm{Kn} &= 0.0010 && - 0.010 \\ \mathrm{Ke} &= 3 \times 10^{-5} && - 3 \times 10^{-4} \end{aligned}$$

We conclude that velocity effects are initially important (since Re>1); recovery temperature effects are always negligible (since Ma<0.1); the droplet is approximately spherical (since We<5).[26]

From the fact that Kn<0.01, the important conclusion is drawn that continuum-type conditions exist during most of the evaporation process, which means that $\mathrm{Kn}_{ref} = 0$ may be set. Since Ke<0.01, capillarity may be neglected all the way through (set $\mathrm{Ke}_{ref} = 0$). Since continuum laws are valid, the Nusselt numbers Nu_H, Nu_M, and Nu_F will be calculated from Eqs. (80) to (82), and the quasi-steady droplet growth (if applicable) from Eq. (260). We find that in the above range of Re, the following values apply:

$$\begin{aligned} \mathrm{Nu}_H &= 7.4 - 2.0 \\ \mathrm{Nu}_M &= 8.4 - 2.0 \\ \mathrm{Nu}_F &= 13.8 - 3.0 \end{aligned}$$

Note that Ma ≈ 0 and Ke ≈ 0 result in $T^{**}_\infty = T_\infty = 600$ K, $T_{sr} = T_{s\infty} = 0$, $T_{br} = T_{b\infty} = 411$ K. The values of A will range, according to Eq. (214), between $A = 5.3$ and 6.0 depending on the quotient $\mathrm{Nu}_H/\mathrm{Nu}_M$. The value of G is infinite.

Next we calculate the time constants Δt_{int}, Δt_{vel}, and Δt_G from Eqs. (154), (170), and (263). For the latter, dimensionless growth rate has to be determined from Eq. (260), with the value $\psi_6 = 0.69$ read from Fig. 19. The time constants will be calculated for the size $r = r_o = r_{ref}$. (Other r values would lead to the same ratio of the three time constants, even if their absolute values would change.) However, the variation of the Nusselt numbers affects the various time constants differently, and will be considered. We find that

$$\begin{aligned} \Delta t_{int} &= 0.30 \text{ ms} \\ \Delta t_{vel} &= 0.12 \text{ ms} && - 0.58 \text{ ms} \\ \Delta t_G &= 1.40 \text{ ms} && - 5.15 \text{ ms} \end{aligned}$$

[26] Distortions resulting from oscillations due to the injection process are not considered here.

It is seen that temperature and velocity adaptations need approximately equal times, but the time required for vaporization will be considerably longer. The result $\Delta t_{int} \ll \Delta t_G$ means that quasi-steady growth laws are applicable during the greatest part of the vaporization process and therefore the nonsteady heat conduction within the droplet need not be analyzed in detail.

The result $\Delta t_{vel} \ll \Delta t_G$, on the other hand, implies that r may be assumed to stay constant during the velocity adaptation process. Thus motion is decoupled from growth and can be analyzed separately.

In order to determine $\Delta w_\infty(t)$, one must integrate Eq. (193) over time. Since r = constant = r_{ref} may be assumed, analytic solution is possible. Gravity and carrier acceleration are zero. With Nu_F taken from Eq. (82) and introducing dimensionless relative velocity $\boldsymbol{w}$ and a reference Reynolds number,

$$\boldsymbol{w} \;\hat{=}\; \frac{\Delta w_\infty}{\Delta w_{\infty o}} \tag{264}$$

$$Re_{ref} \;\hat{=}\; \frac{2\, r_{ref}\, \rho_\infty\, \Delta w_{\infty o}}{\eta} \tag{265}$$

we have

$$\frac{d\boldsymbol{w}}{dt} = \frac{9\,\eta}{2\, r_{ref}^2\, \rho_\ell}\, \boldsymbol{w}\cdot\left(1 + 0.15\, Re_{ref}^{0.687}\, \boldsymbol{w}^{0.687}\right) \tag{266}$$

Separation and integration leads to following implicit equation for $\boldsymbol{w}(t)$:

$$\ln\left[\frac{1}{\boldsymbol{w}}\left(\frac{1+0.15\, Re_{ref}^{0.687}\, \boldsymbol{w}^{0.687}}{1 + 0.15\, Re_{ref}^{0.687}}\right)^{1/0.687}\right] = \frac{9\,\eta\, t}{2\, r_{ref}^2\, \rho_\ell} = \frac{t}{0.575\ \text{ms}} \tag{267}$$

The plot of this function is shown in Fig. 21 over the physical time scale t. Conversion to nondimensional time τ will be discussed later. The flight distance ξ has been determined from

$$\xi(t) = \int_0^t \left(w_{carr} - \Delta w_{\infty o}\, \boldsymbol{w}\right) dt \tag{268}$$

The resulting flight distances ξ are marked under the t scale.

In order to calculate the evolution of particle size, the simplified growth law given by Eq. (260) may be used in the present case. Using Eqs. (239) and (240), we get for the dimensionless size $\boldsymbol{r} = r/r_{ref} = r/r_o$ the differential equation

$$\frac{d\boldsymbol{r}}{d\tau} = -\,\psi_6\, \frac{1 + 0.30\, Re^{1/2}\, Pr^{1/3}}{\boldsymbol{r}} \tag{269}$$

where

$$\mathrm{Re} = \mathrm{Re}_{ref}\, \boldsymbol{w}\, \boldsymbol{r} \tag{270}$$

$$\tau = \frac{\lambda\, T_{ref}}{\rho_\ell\, L\, r_{ref}^2}\, t = \frac{t}{3.58 \text{ ms}} \tag{271}$$

It is seen that Re depends both on $\boldsymbol{r}$ and [due to $\boldsymbol{w}$, and via Eq. (267)] on τ. Numerical computation is required. An approximate solution can, however, be obtained by setting Re = constant = Re_{repr} where the latter is a representative mean value for the changing values occurring during the process. Based on the Δt_G values shown above, one may expect that evaporation will be completed in about 3 ms. During most of this time $\boldsymbol{w}(t)$ is seen to have very low values We shall choose the value $\mathrm{Re}_{repr} = 4$ for the sake of an approximate sclution.

The initial condition is not $\boldsymbol{r}(0) = 1$, because the quasi-steady solution implies that droplet temperature transients have disappeare These transients, however, cause an increase of droplet size from r_o to the value r_o' given by Eq. (162). Estimating T_ℓ as equal to T_{ref}, the initial condition is

$$\boldsymbol{r}(0) = r_o'/r_{ref} = r_o'/r_o = \sqrt[3]{1 + (T_{ref} - T_{\ell o})\, c_\ell / L} = 1.095 \tag{272}$$

The solution of Eq. (269) is now obtained by simple integration as

$$\boldsymbol{r}(t) = \sqrt{\boldsymbol{r}(0)^2 - 2\psi_6\, (1 + 0.30\, \mathrm{Re}_{repr}^{1/2}\, \mathrm{Pr}^{1/3})\, \tau}$$

$$= \sqrt{1.199 - 1.9228\, \tau} \tag{273}$$

Evaporation is complete if $\boldsymbol{r} = 0$, giving

$$\tau_E = 0.624 \quad \text{and} \quad t_E = 0.624 \cdot 3.58 \text{ ms} = 2.23 \text{ ms}$$

The function $\boldsymbol{r}(\tau)$ is plotted in Fig. 21. This curve is an approximation in several respects. First, the solution is not valid in the range $0 < t < \Delta t_{int}$, because the transient growth from r_o to r_o' does not occur instantaneously. The physically correct evolution is illustrated by the dotted line starting from $\boldsymbol{r} = 1$ and joining the $\boldsymbol{r}(\tau)$ curve at $\tau \approx \Delta t_{int}/(3.58 \text{ ms}) \approx 0.08$. Second, the assumption Re = constant = Re_{repr} falsifies the curve $\boldsymbol{r}(\tau)$, making it too flat at low values of τ, where Re > Re_{repr}, and too steep at higher τ, where Re < Re_{repr} would apply.

The evaporation distance is calculated from Eq. (268) as $\xi_E = 63$ mm.

The question posed at the outset is now answered. It may be of interest to calculate the quasi-steady droplet temperature, since this had to be estimated at the beginning to select the temperatures

T_{ref} and $T_{1/3}$ on which the material properties were based. According to Eq. (199), we have for nonexpanding flows ($dT_{b\infty}/dt = 0$) $T_\ell = T_r$. The surface temperature T_r has to be calculated from Eq. (221), setting $Q = Q_{gen}$ for hot, lean vapors. Thus we obtain

$$T_\ell = T_r = T_\infty^{**} + (T_{sr} - T_\infty^{**})\, Q_{gen} = 316 \text{ K} \tag{274}$$

independent of droplet size. The value of Q_{gen} was calculated from Eq. (226) as 0.474. It is seen than T_ℓ is unexpectedly low! Thus it turns out that T_{ref} and $T_{1/3}$ have been assigned much too high values. The materials data should now be reevaluated[27] and the calculation repeated. This second calculation would follow the path of the first one, and will not be discussed further here. It is to be noted that the correct value of T_ℓ happens to be very nearly equal to the injection temperature $T_{\ell o}$. Thus the droplets almost have the correct quasi-steady temperature from the beginning and $r_o' \approx r_o$ will be obtained from Eq. (272), where T_{ref} will now have to be replaced by the better estimate $T_{ref} = 316$ K.

The calculation procedure outlined in this section appears to be too lengthy for the simple problem posed. Indeed, if the similarity of the materials data for methyl alcohol and water (cf. Table A-1) is considered, Figs. 6 and 20 may be used to estimate the values of Kn, Ke, Δt_{int}, and Δt_{vel}, making much of the calculation unnecessary. The values of Δt_G found in Fig. 20 (for $S_\infty = 0.1$) are too large, however, because $S_\infty = 0$ in the present case.

6.3 The Influence of Droplet Size on Droplet Temperature

It is of interest to investigate the quasi-steady temperature assumed by droplets of various sizes in a given carrier environment. Very small droplets (of $r \approx 10^{-9}$ m) are of particular interest because of the strong effect of capillarity on their surface temperature, which was discussed in Sec. 5.2.

As an example, we take water droplets suspended in a nonmoving, constant-state, infinite carrier fluid consisting of water vapor and air, and we shall consider a superheated and a subcooled carrier state:

1. Superheated: $\Delta T_\infty = -50$ K

2. Subcooled: $\Delta T_\infty = 20$ K

The vapor content and the overall pressure of the carrier will be the same in both cases, corresponding to $\Pi_{v\infty} = 0.2$ and $p_\infty = 1$ bar. From the above data and from the vapor partial pressure $p_{v\infty} =$

---

[27] Note that no corrections due to compositional variations are necessary, because $\Pi_{v\infty} = 0$ and $\Pi_r \approx p_s(T_r)/p_\infty = 0.035$, making $\Pi_{v\infty} - \Pi_r \ll 1$ and $\Pi_{1/3} = 0.01 \approx 0$. Thus the use of pure-air data is justified.

$\Pi_{v\infty}\, p_\infty$ = 0.2 bar, we find the following temperature values and saturation ratios:

1. $T_{b\infty}$ = 372.8 K, $T_{s\infty}$ = 333.2 K, T_∞ = 383.2 K, S_∞ = 0.139, r_{crit} = - 35x10^{-9} m

2. $T_{b\infty}$ = 372.8 K, $T_{s\infty}$ = 333.2 K, T_∞ = 313.2 K, S_∞ = 2.623, r_{crit} = 0.88x10^{-9} m

The dilution ratio of the carrier is found to be g_∞ = 6.43. The materials data are taken from standard steam tables (Schmidt, 1969), assuming that T_{ref} = 372.8 K. For air we set R_g = 287.2 J/kg K, c_{pg} = 1020 J/kg K, η_g = 2.17x10^{-5} kg/s m, Pr_g = 0.70. For the mixture one obtains $\bar{R}$ = 310.7 J/kg K, $\bar{c}_p$ = 1156 J/kg K, γ = 1.37, Pr = 0.74, Sc = 0.49. The values $\alpha_{Hv} = \alpha_{Hg}$ = 1 and α_{Mv} = 0.9 for the accommodation coefficients lead to the free-molecule coefficients B_{HO} = 0.52 and B_{MO} = 0.35. The steam tables yield Cl = 13.5, Eö = 12.0, and $(2\,\sigma/\rho_\ell\, R_v\, T_{ref})$ = 7.12x10^{-10} m. As a result of the assumptions made above, we have Re = Ma = 0 and $T^{**}_\infty = T_\infty$. The caloric effectiveness Q of the carrier will be calculated from Eq. (223).

The main results of the calculation made for five droplet sizes ranging between r = 10 μm and r = 0.0005 μm are summarized in Table 8. The Kn data shown in the table indicate that continuum conditions (Kn<0.01) exist only for the largest size considered. Capillarity is seen from the Ke values to become noticeable (Ke>0.01) for the size 10^{-8} m, and strong (Ke≈1) for 10^{-9} m and below. The Nusselt numbers vary strongly, as dictated by Kn; in the noncontinuum range Nu_H is slightly higher than Nu_M, as a result of $B_{HO} > B_{MO}$ for the carrier in question. The characteristic temperatures T_{sr} and T_{br} deviate noticeably from the flat surface values $T_{s\infty}$ and $T_{b\infty}$, provided that Ke>0.01. It is interesting to note that relative undersaturation (S_r<1) may occur even in the supersaturated (subcooled) carrier, provided that the capillarity effect is strong enough. (r = 0.5x10^{-9} m is here a subcritical size!)

The factors A and C are fairly independent of drop size, and the gas influence factor G has low (≪ 1) values. As a result, the caloric effectiveness Q is high (≈1).

The values of T_r obtained in Table 8 are used in Fig. 22 to illustrate the temperature fields existing in the vicinity of drops of various sizes. (Note that the sizes are not shown in proper proportion, and the temperature gradients are not correct (are too steep) with respect to the diameters.) The upper part of the figure shows the case of the *superheated carrier*. All drops are colder than their environment. Their temperature is close to the flat-liquid saturation temperature $T_{s\infty}$, except if they are very small and the capillarity effect causes them to assume still lower temperatures. The lower part shows the *subcooled carrier* case. Here the droplets are warmer than the carrier, except if they are of subcritical size. Large droplets have temperatures close to $T_{s\infty}$.

It is seen that for large droplets (>10^{-8} m), neither size nor

Table 8. *Calculation of the Quasi-Steady Droplet Temperature* T_r *(Water Droplets; Carrier Data as Given in Sec. 6.3)*

Case	Superheated					Subcooled				
Drop size r (m)	10^{-4}	10^{-6}	10^{-8}	10^{-9}	0.5×10^{-9}	10^{-4}	10^{-6}	10^{-8}	10^{-9}	0.5×10^{-9}
Kn	0.0007	0.0704	7.04	70.4	141	0.0006	0.0636	6.36	63.6	127
Ke	0.0000	0.0007	0.069	0.693	1.39	0.0000	0.0008	0.085	0.847	1.69
Nu_H	1.995	1.574	0.071	0.0074	0.0037	1.995	1.606	0.078	0.0081	0.0041
Nu_M	1.992	1.429	0.049	0.0050	0.0025	1.993	1.469	0.054	0.0055	0.0028
T_{sr} (K)	333.2	333.2	331.7	318.6	305.2	333.2	333.2	331.3	315.5	299.6
T_{br} (K)	372.8	372.8	370.9	354.6	338.1	372.8	372.8	370.5	350.8	331.2
S	0.139	0.139	0.130	0.070	0.035	2.623	2.621	2.410	1.124	0.482
A	1.973	2.154	2.873	2.903	2.905	1.972	2.154	2.870	2.903	2.904
C	13.50	13.50	13.57	14.19	14.89	13.50	13.50	13.58	14.35	15.19
G	0.043	0.047	0.062	0.058	0.053	0.043	0.047	0.062	0.057	0.051
Q	0.959	0.956	0.941	0.945	0.949	0.959	0.955	0.942	0.946	0.951
T_r (K)	335.3	335.4	334.7	322.1	309.1	332.4	332.3	330.3	315.4	300.3

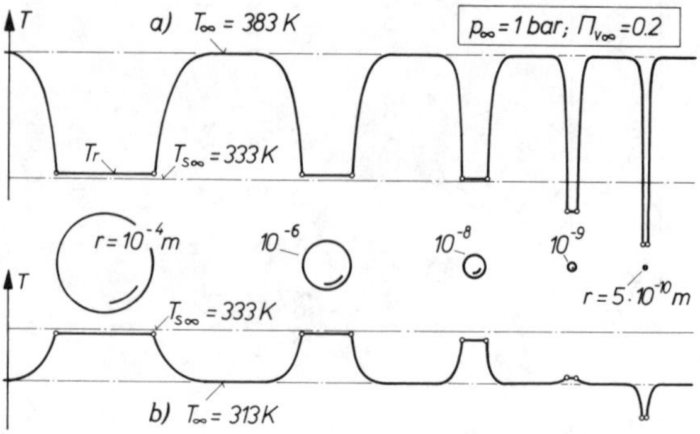

Fig. 22. Temperature fields around water droplets in steam/air mixtures carriers. (a) Superheated carrier. (b) Subcooled carrier.

carrier temperature has much influence in the temperature T_r assumed by the droplet. However, calculation of T_r values in carriers of different composition would show that $T_r \approx T_{s\infty}$ is typical only of vapor-rich carriers. (The composition $\Pi_{v\infty} = 0.2$ or $g_\infty = 6.43$ still counts as very rich!) For low vapor contents ($\Pi_{v\infty}<0.01$), T_r would gradually approach the carrier temperature.

A comment will be added regarding the effects of relative velocity on T_r. Because of the recovery effect, a moving droplet "sees" an enhanced carrier temperature T_∞^*. Assuming that Ma = 0.2 (which corresponds to high relative speeds of the order of 70 m/s), one finds from Eq. (126) for the continuum case $T_\infty^* = 1.010T_\infty$ and for the free-molecule case $T_\infty^* = 1.011T_\infty$. This change of T_∞ has negligible effect on T_r in vapor-rich carriers. Another question is what influence the changes in Nu_H and Nu_M have on T_r. Since both Nu_H and Nu_M change with Re or Ma in the same sense, as seen from the multirange equations given in Table 6, the changes in A and G are small. Thus the values of Q and T_r are left substantially unaffected by the changes in Nu. Summarizing, Ma = 0.2 causes practically no change in T_r if the carrier is rich in vapor or is hot, and about 1% increase in T_r if the carrier is a cool, lean gas.

6.4 Analytic Solutions for Growth in Carriers of Constant Conditions

As we pointed out in Sec. 5.1.5, the quasi-steady growth laws formulated in Sec. 5.3 can be integrated over time t (or τ) for certain simple cases. Thus growth or evaporation histories can be obtained analytically (rather than numerically).

Such integration will now be carried out for the case of a single droplet in a pure-vapor or vapor-rich carrier. Supersaturated ($S_\infty>1$) and undersaturated ($S_\infty<1$) conditions will both be considered. Relative velocity will be neglected (Re = Ma = 0, and the carrier state will be constant. It is assumed that at time t = 0 the droplet temperature T_r has the correct quasi-steady value, so that the quasi-steady growth law is applicable at any time. The initial size $r(0) = r_{ref}$ will be used as reference radius.

It is of interest to discuss the influence of rarefaction effects (i.e., of deviations from continuum-type conditions), and of the capillarity effect, on the evolution of droplet size in time. Both effects are accounted for in the multirange growth law given by Eq. (247). This expression will now be written, with $\boldsymbol{r}_{crit} \triangleq r_{crit}/r_{ref}$, as

$$\frac{d\boldsymbol{r}}{d\tau} = \frac{1 + 2\,\mathrm{Kn}_{ref}/B_{H0}}{\boldsymbol{r} + 2\,\mathrm{Kn}_{ref}/B_{H0}} \left(1 - \frac{\boldsymbol{r}_{crit}}{\boldsymbol{r}}\right) \frac{\Delta\boldsymbol{T}_\infty}{1 + G} \qquad (275)$$

where $\mathrm{Ke}_{ref}/\Delta T_\infty$ has been expressed in terms of r_{crit} by Eqs. (244), (181), and (A-18). The critical radius r_{crit} is a characteristic of the carrier state and is inversely proportional to $\ell n\, S_\infty$; cf. Eq. (181). τ is defined by Eq. (246). The initial condition at $\tau = 0$ is

$$\boldsymbol{r}(0) = 1 \qquad (276)$$

Before integrating, the parameters contained in Eq. (275) will be discussed. The gas influence factor G is substantially zero for pure vapor carriers and has some positive value for mixture carriers. Typical values of G can be constructed from Table 7. Note that the use of Eq. (275) excludes lean ($\Pi_{v\infty} < 10^{-3}$) carriers, and therefore G will not exceed the order of unity. It is seen from Table 7, or from Eqs. (214) and (222), that droplet size has an influence on the value of G, via the influence of the Knudsen number on the ratio $\mathrm{Nu}_H/\mathrm{Nu}_M$ occurring in A and via the dependence of C on the Kelvin number. However, both influences are very weak, as seen from the numerical values of A and G/A given in Table 7 and from the fact that $\mathrm{Cl} \gg \mathrm{Ke}$ holds in all cases of practical interest. (Note that A is sensitive to Kn only if $\Pi_{v\infty} \approx 1$, i.e., in the case of negligible values of G.) Therefore one may set

$$(1 + G) = \text{constant} = 1 + G_{ref} \qquad (277)$$

where G_{ref} is the value of G obtained by setting $\mathrm{Nu}_H/\mathrm{Nu}_M = \mathrm{Nu}_{H0ref}/\mathrm{Nu}_{M0ref}$. From Eq. (222) and the multirange expression of A given in Table 5, we find that

$$G_{ref} = \frac{\Pi_{v\infty}^{-1}}{[(\mathrm{Cl}\; T_{ref}/T_{b\infty}) + \mathrm{Ke}_{ref} - \ell n\, \Pi_{v\infty}]^2} \; \frac{(1-\Pi_{v\infty})\mathrm{Sc} + (\sqrt{2\pi}/\alpha_{Mv})\sqrt{\bar{R}/R_v}\;\mathrm{Kn}_{ref}}{[(\gamma-1)/\gamma]\mathrm{Pr} + [\sqrt{2\pi}\bar{R}^{3/2}/(R^{3/2}\;\mathrm{K}\alpha_H)]\mathrm{Kn}_{ref}} \qquad (278)$$

This expression allows us to calculate $1 + G$ for any carrier fluid.

Now Eq. (275) can be written in the form

$$\frac{\boldsymbol{r} + 2\,\mathrm{Kn}_{ref}/B_{H0}}{1 + 2\,\mathrm{Kn}_{ref}/B_{H0}} \frac{\boldsymbol{r}}{\boldsymbol{r} - \boldsymbol{r}_{crit}}\, d\boldsymbol{r} = \frac{\Delta\boldsymbol{T}_\infty}{1 + G_{ref}}\, d\tau \qquad (279)$$

It is seen that the solution $\boldsymbol{r}(\tau)$ will contain the following three constants as parameters:

$$C_1 \triangleq \frac{\Delta T_\infty / T_{\text{ref}}}{1 + G_{\text{ref}}} \tag{280a}$$

$$C_2 \triangleq \boldsymbol{r}_{\text{crit}} = \frac{\text{Ke}_{\text{ref}}}{\ell n \, S_\infty} \tag{280b}$$

$$C_3 \triangleq \frac{\text{Kn}_{\text{ref}}}{B_{HO}} \tag{280c}$$

of which the constant C_1 is a scaling factor for the nondimensional time τ.

Integration leads to the following implicit expression for $\boldsymbol{r}(\tau)$:

$$\frac{\boldsymbol{r}^2 - 1}{2 + 4\,C_3} + \frac{C_2 + 2\,C_3}{1 + 2\,C_3}\left(\boldsymbol{r} - 1 + C_2\,\ell n\,\frac{\boldsymbol{r} - C_2}{1 - C_2}\right) = C_1\,\tau \tag{281}$$

Note that C_1 characterizes the state and composition of the carrier phase, and C_2 expresses the capillarity effect; both are positive for supersaturated (subcooled) and negative for undersaturated (superheated) carrier fluids. C_3 expresses the rarefaction effect. Note that B_{HO} is a material constant (typically about 0.5). In the case of *continuum* carriers, $C_3 = 0$ and Eq. (281) simplifies to

$$\frac{\boldsymbol{r}^2 - 1}{2} + C_2\left(\boldsymbol{r} - 1 + C_2\,\ell n\,\frac{\boldsymbol{r} - C_2}{1 - C_2}\right) = C_1\,\tau \tag{282}$$

whereas in the *free-molecular* limit $C_3 \to \infty$, and we have

$$\boldsymbol{r} - 1 + C_2\,\ell n\,\frac{\boldsymbol{r} - C_2}{1 - C_2} = C_1\,\tau \tag{283}$$

The produce $C_1\,\tau$ has the direct physical meaning

$$C_1\,\tau = \frac{t}{\Delta t_{G,\text{ref}}} \tag{284}$$

where

$$\Delta t_{G,\text{ref}} = \left(1 + G_{\text{ref}}\right)\left(1 + \frac{2\,\text{Kn}_{\text{ref}}}{B_{HO}}\right)\frac{\rho_\ell\,L\,r_{\text{ref}}^2}{\lambda\,\Delta T_\infty}$$

$$= \frac{\eta}{p_\infty}\left(1 + G_{\text{ref}}\right)\frac{B_{HO} + 2\,\text{Kn}_{\text{ref}}}{\text{Kn}_{\text{ref}}^2}\,\frac{\rho_\ell}{\rho_\infty}\,\frac{\gamma - 1}{\gamma}\,\frac{R_v}{\bar{R}}\,\frac{\text{Pr Cl}\,T_{\text{ref}}}{B_{HO}\,\Delta T_\infty} \tag{285}$$

is the time constant of droplet growth for a droplet of size r_{ref}, if capillarity is neglected (r_{crit}=0). Equation (285) follows from Eqs. (235), (246), (263), and (275). Typical values of $\Delta t_{G,ref}$ can be read from Fig. 20.

The plots of the equations (281) to (283) are shown in Figs. 23 and 24 for the case of supersaturated and superheated carriers, respectively. Three values of Kn_{ref}/B_{HO} (∞, 2, 0) and several values of $\mathbf{r}_{crit} = r_{crit}/r_{ref}$ are shown. The thick lines refer to the case of negligible capillarity ($\mathbf{r}_{crit}$ = 0). These curves are typical for large droplets in carriers of considerable super- or undersaturation, that is, for cases where $Ke_{ref}/\ln S_\infty$<0.1.

For *subcooled carriers* it is evident from Fig. 23 that the increase of $\mathbf{r}(\tau)$ is linear if free-molecular conditions exist and parabolic if the carrier behaves as a continuum. This statement is exact if $\mathbf{r}_{crit}$ = 0. If capillarity is noticeable (say, $\mathbf{r}_{crit}$>0.1), droplet growth is slowed down in its early stages. If the droplet is critical initially ($\mathbf{r}_{crit}$ = 1), it is in equilibrium and its size stays constant. Subscritical droplets ($r_{ref} < r_{crit}$)

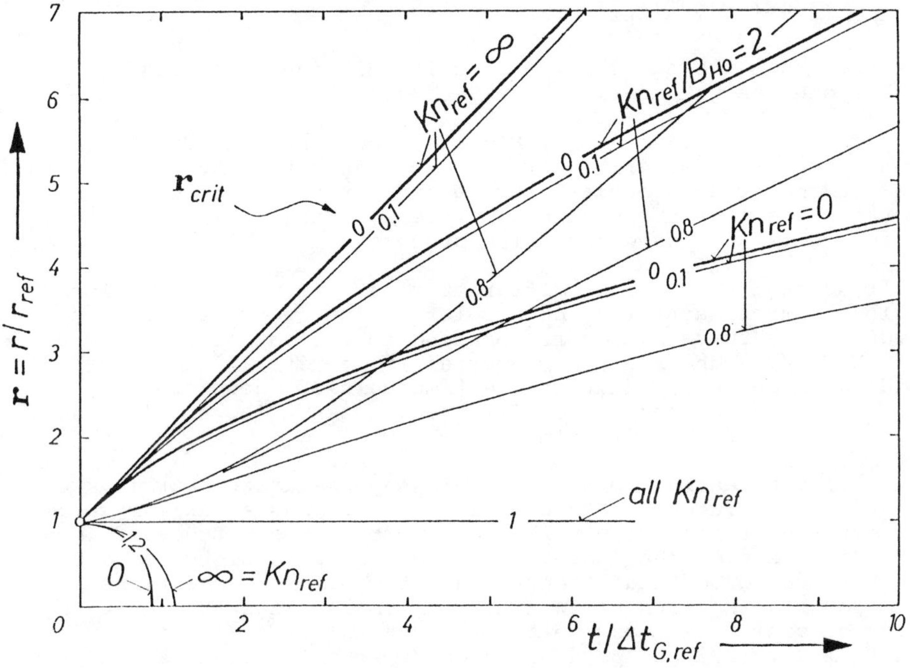

Fig. 23. Nondimensional growth histories of general validity for stationary droplets in *supersaturated* pure-vapor of vapor-rich carriers of constant state.

Thick curves: negligible capillarity (r_{crit}/r_{ref} = 0).

Thin curves: capillarity considered (r_{crit}/r_{ref} > 0).

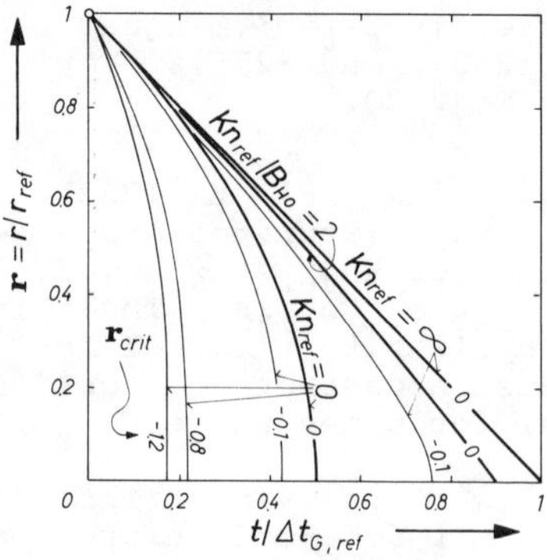

Fig. 24. Nondimensional evaporation histories of general validity for stationary droplets in *superheated* pure-vapor or vapor-rich carriers of constant state. Thick curves: negligible capillarity ($r_{crit}/r_{ref} = 0$). Thin curves: capillarity considered ($r_{crit}/r_{ref} < 0$).

vaporize. Thus the equilibrium of critical-sized droplets is an unstable one, as was mentioned in Sec. 4.1.

For *superheated carriers*, Fig. 24 shows that all droplets (whether super- or subcritical) tend to evaporate. Again, $\boldsymbol{r}(\tau)$ is linear in the free-molecular extreme and parabolic (in conformation with the r^2 law) in the continuum extreme. Capillarity effects tend to accelerate the evaporation process.

If several droplets of different sizes are present at time zero in a common carrier, each droplet will grow or evaporate according to a curve $\boldsymbol{r}(\tau)$ corresponding to the values of the parameters Kn_{ref}/B_{HO} and r_{crit}/r_{ref} pertaining to its initial size. It should be noted, however, that the time unit of these curves, $\Delta t_{G,ref}$, is very different for different sizes. The smaller the droplet, the faster is its time scale.

It is interesting to note that in supersaturated carriers the absolute growth rate in physical time, dr/dt, has a maximum value for a certain droplet size. The reason for the existence of this maximum is the fact that for large sizes dr/dt is slowed down by the deterioration of heat transfer under continuum conditions (decrease of Nu_H/r with r) and that for small sizes the retarding effects of capillarity are felt. The maximum of dr/dt can be shown from Eq. (275) to occur if droplet size is such that Kn_{ref} has the value

$$Kn^*_{ref} = \frac{B_{HO}}{2}\left(\sqrt{1 + \frac{2\,Kn_{crit}}{B_{HO}}} - 1\right) \qquad (286)$$

where

$$\mathrm{Kn}_{crit} \triangleq \frac{\eta \sqrt{\bar{R}\, T_\infty}}{p_\infty\, r_{crit}} \tag{287}$$

means the Knudsen number pertaining to the critical radius. For instance, let the carrier be air (B_{HO} = 0.48) of p_∞ = 1 bar at room temperature and have a supersaturation of S_∞ = 2.72. This S_∞ gives Ke_{crit} = 1, cf. Eq. (183). Figure 6 shows that this Ke value corresponds to a droplet radius of $r_{crit} \approx 1.2\times10^{-9}$ m. For this size at p_∞ = 1 bar, the Knudsen number curves roughly indicate $\mathrm{Kn}_{crit} \approx 20$. Inserting into Eq. (286), one finds that the fastest-growing droplets are those for which Kn = Kn^*_{ref} = 0.24 ($\sqrt{1 + 2\times20/0.48} - 1) \approx 2$. It is seen from Fig. 6 that at 1 bar this means that $r \approx 1.5\times10^{-8}$ m.

In superheated carriers ($-\,dr/dt$) has no maximum, and the absolute rate of evaporation will be the faster the smaller the droplet is.

6.5 Entropy Production in Mist-Type Flows

6.5.1 Polytropic Loss Coefficients

If a two-phase fluid is subjected to an expansion or compression process of finite rate, condensation or evaporation is taking place in thermodynamic disequilibrium. Similarly, if the fluid is being accelerated or decelerated, the velocity of the droplets becomes different from that of the carrier and internal friction arises. Both processes are irreversible and give rise to an increase of the specific entropy s of the fluid. The entropy production in two-phase mist flows of steam has been analyzed with respect to steam turbines by Gyarmathy (1962) and in a more general manner by Konorski (1969). In the following we shall apply the drag and growth laws formulated in Secs. 3 and 5 to the calculation of entropy production during steady-rate adiabatic processes. The case of fluctuating pressure (as is typical for acoustical phenomena, cf. e.g., Jaeschke et al., 1975) is a closely related problem, but will not be dealt with here.

We consider a mist-type two-phase fluid subjected to a continuous expansion or compression process. The liquid phase (of mass fraction x_ℓ) will consist of uniform-sized droplets[28] of radius r, and the carrier phase will be either pure vapor or a rich ($\Pi_{v\infty}$>0.01) vapor-gas mixture. No limitation on Knudsen number will be imposed.

We consider, as in Sec. 2.1, a representative sample of the fluid of mass m_s containing one droplet of mass m_ℓ, but assume that the sample is adiabatic (H_{ext} = 0) and closed (M_{ext} = 0). If ds/dt is the increase of the specific entropy s of the fluid in unit

[28] For the extension to several droplet categories, cf. the comment made in Sec. 2.3.5.

time, the dissipation rate in the sample is $T_\infty\, m_s\, ds/dt$. The dissipated power can be related to the isentropic work $v_s\, dp_\infty/dt$ produced or absorbed in unit time. This ratio is the *polytropic loss coefficient*:

$$\phi \mathrel{\hat{=}} \left| \frac{T_\infty\, m_s\, ds/dt}{V_s\, dp_\infty/dt} \right| = \frac{\pm\, ds/dt}{(1-x_\ell)\, \bar{R}\, \dot{P}} \tag{288}$$

Here and below the plus sign refers to expansion, the minus sign to compression processes. In the second equality the approximation $V_s/m_s \approx (1-x_\ell)/\rho_\infty$ was introduced, that is, the volume x_ℓ/ρ_ℓ of the droplets neglected, cf. Eq. (A-30) in App. 2. The *expansion rate* $\dot{P}$ is assumed to have a substantially *constant*, known value. The problem amounts to determining the entropy production rate ds/dt, which is in general caused both by thermodynamic and mechanical irreversibilities:

$$\frac{ds}{dt} = \left(\frac{ds}{dt}\right)_{\text{thd}} + \left(\frac{ds}{dt}\right)_{\text{mec}} \tag{289}$$

The thermodynamic term is due to heat and mass transfer between the phases having different temperatures and is given as

$$\left(\frac{ds}{dt}\right)_{\text{thd}} = H\left(\frac{1}{T_{s\infty}} - \frac{1}{T_\infty} + \frac{\text{Ke}}{\text{Cl}\; T_{\text{ref}}}\right) + H_{\text{int}}\left(\frac{1}{T_\ell} - \frac{1}{T_{s\infty}} - \frac{\text{Ke}}{\text{Cl}\; T_{\text{ref}}}\right) \tag{290}$$

For droplets with negligible capillarity (Ke = 0), a similar expression has been formulated by Konorski (1969). H and H_{int} are the heat fluxes introduced in Sec. 2.2. Setting $H_{\text{int}} = 0$, that is, neglecting the heat conduction within the droplets, and using Eq. (25), we have

$$\left(\frac{ds}{dt}\right)_{\text{thd}} = L\,\frac{dx_\ell}{dt}\left(\frac{\Delta T_\infty}{T_{s\infty}\; T_\infty} - \frac{\text{Ke}}{\text{Cl}\; T_{\text{ref}}}\right) = L\,\frac{dx_\ell}{dt}\,\frac{\ln S_\infty - \text{Ke}}{\text{Cl}\; T_{\text{ref}}} \tag{291}$$

A similar expression (with Ke = 0) has been used by Gyarmathy (1962) to express thermodynamic losses in wet-steam turbines. Equations (290) and (291) are valid for mixtures, too, where part of the irreversibility is due to a diffusion process.

The second term of Eq. (289) is the entropy production caused by mechanical disequilibrium (i.e., slip velocity) and is obtained from the frictional work as

$$\left(\frac{ds}{dt}\right)_{\text{mec}} = \frac{x_\ell}{m_\ell}\,\frac{F\,\Delta w_\infty}{T_\infty} \tag{292}$$

Here $F\,\Delta w_\infty$ is the power dissipated per droplet and per unit time as a result of friction, T_∞ is the temperature at which the dissipation occurs, and x_ℓ/m_ℓ is the number of droplets per unit mass of fluid. $F\,\Delta w_\infty$ can also be expressed by the momentum balance, if Eq. (17) is used, as

$$F\ \Delta w_\infty \equiv -\vec{F}\ \Delta\vec{w}_\infty = -\ m_\ell\ \Delta\vec{w}_\infty\ \frac{d\vec{w}_\ell}{dt} + m_\ell\ \Delta\vec{w}_\infty\ \vec{b} - \frac{m_\ell\ \Delta w_\infty}{\rho_\ell}\ \mathrm{grad}\ p_\infty \quad (293)$$

The last two terms are small, except in very strong body-force fields and very sudden pressure changes (shocks), respectively. Neglecting them and inserting Eq. (293) into Eq. (292), we have

$$\left(\frac{ds}{dt}\right)_{\mathrm{mec}} = -\ \frac{x_\ell}{T_\infty}\ \frac{d\vec{w}_\ell}{dt}\ \Delta\vec{w}_\infty \quad (294)$$

Inserting Eqs. (291) and (294) into (289) and the latter into (288), we can write

$$\phi = \phi_{\mathrm{thd}} + \phi_{\mathrm{mec}} \quad (295)$$

where the two terms represent the thermodynamic and mechanical contributions to the polytropic loss coefficient. They are given by

$$\phi_{\mathrm{thd}} = \frac{\pm 1}{1-x_\ell}\ \frac{R_v}{\bar{R}}\ \frac{\ell n\ S_\infty\ -\ \mathrm{Ke}}{\dot{P}}\ \frac{dx_\ell}{dt} \quad (296)$$

$$\phi_{\mathrm{kin}} = \frac{\pm x_\ell}{1-x_\ell}\ \frac{(-\ \Delta\vec{w}_\infty)}{\bar{R}\ T_\infty\ \dot{P}}\ \frac{d\vec{w}_\ell}{dt} \quad (297)$$

The plus sign pertains to expansion/acceleration processes, the minus sign to compression/deceleration processes. Both ϕ_{thd} and ϕ_{mec} are always positive.

In the following, ϕ_{thd} and ϕ_{mec} will be calculated on the basis of the data characterizing the two-phase dispersed-droplet fluid by assuming adiabatic flow.

6.5.2 Thermodynamic Disequilibrium

In a steady-rate expansion of a mist-type fluid, S_∞ and dx_ℓ/dt will assume nearly constant values. In order to determine ϕ_{thd}, these values must be expressed in terms of the expansion rate.

We consider a frictionless two-phase fluid in the absence of external heat or mass transfer. Writing $S_\infty = p_\infty\ \Pi_{v\infty}/p_{s\infty}$ and applying Eqs. (A-3), (49), and (58) and noting that $dx_v = -dx_\ell$, one finds for the rate of change of the saturation ratio S_∞,

$$\frac{1}{S_\infty}\ \frac{dS_\infty}{dt} = \dot{\bar{P}}\left(\frac{\gamma-1}{\gamma}\ \mathrm{Cl}^* - 1\right) - \frac{1}{1-x_\ell}\ \frac{dx_\ell}{dt}\left[\frac{\gamma-1}{\gamma}\ \frac{R_v}{\bar{R}}\ \mathrm{Cl}^{*2} + \frac{x_g}{x_v}\right] \quad (298)$$

This equation is valid for perfect-gas type carriers only. The group

$$\mathrm{Cl}^* \triangleq \frac{\mathrm{Cl}\, T_{\mathrm{ref}}}{T_\infty} = \frac{L}{R_v\, T_\infty} \qquad (299)$$

is practically a constant of the order of Cl. It is seen that a decrease of pressure ($\dot{P}>0$) causes an increase[29] of S_∞, whereas an increase of x (i.e., condensation) is seen to reduce S_∞. The steady-state level of S_∞, that is $dS_\infty/dt = 0$, is coupled according to Eq. (298) to following value of dx_ℓ/dt:

$$\left(\frac{dx_\ell}{dt}\right)_{S_\infty=\mathrm{const}} = (1-x_\ell)\, \dot{P}\, \frac{\bar{R}/R_v}{\mathrm{Cl}^*} \left(1 - \frac{\gamma}{\gamma-1}\frac{1}{\mathrm{Cl}^*}\right)\left(1 + \frac{\gamma}{\gamma-1}\frac{\bar{R}/R_v}{\mathrm{Cl}^{*2}}\frac{x_g}{x_v}\right)^{-1} \qquad (300)$$

Inserting this into (296), we obtain

$$\phi_{\mathrm{thd}} = \pm \frac{\ell n\, S_\infty - \mathrm{Ke}}{\mathrm{Cl}^*} \left(1 - \frac{\gamma}{\gamma-1}\frac{1}{\mathrm{Cl}^*}\right)\left(1 + \frac{\gamma}{\gamma-1}\frac{\bar{R}/R_v}{\mathrm{Cl}^{*2}}\frac{x_g}{x_v}\right)^{-1} \qquad (301)$$

The expression in the first set of parantheses has a value between 0.5 and 1 for most substances (see footnote 29). The second set of parantheses has the value 1 for pure-vapor carriers ($x_g = 0$), and is not very different from unity even in rather dilute vapors. Approximately we may write, neglecting also the capillarity effect:

$$\phi_{\mathrm{thd}} \approx \pm \frac{\ell n\, S_\infty}{\mathrm{Cl}^*} = \pm \frac{\Delta T_\infty}{T_{s\infty}} \qquad (302)$$

This is a remarkably simple result, stating that the *thermodynamic irreversibilities are simply proportional to the level of logarithmic supersaturation* (and of the *subcooling*) existing during the expansion process.

If the droplet size is given, this level is determined in the following way. The increase of x_ℓ being due to droplet growth, the growth law provides a link between dx_ℓ/dt and the supersaturation required to maintain condensation at this rate. Using the multirange growth law Eq. (247) and noting that $dx_\ell/dt = (3x_\ell/r)dr/dt$, we find that

$$\ell n\, S_\infty - \mathrm{Ke} = \dot{P}\, \frac{\bar{R}}{R_v}\, \frac{1-x_\ell}{x_\ell}\, \frac{\rho_\ell\, L\, r^2}{3\,\lambda\, T_{s\infty}} \left(1 + \frac{2\mathrm{Kn}}{B_{HO}}\right) \frac{\{1 - [\gamma/(\gamma-1)](1/\mathrm{Cl}^*)\}(1 + G)}{1+[\gamma/(\gamma-1)](\bar{R}/R_v\, \mathrm{Cl}^{*2})(x_g/x_v)} \qquad (303)$$

Expressing $\dot{P}$ by Sd, L by Cl, and r by Kn, this can be written as

[29] For some organic vapors one finds $(\gamma-1)\, L/\gamma\, R_v\, T_\infty<1$. These so-called *retrograde* vapors become superheated upon expansion.

$$\ell n\, S_\infty = \mathrm{Ke} \pm \frac{1-x_\ell}{x_\ell}\,\frac{\rho_\ell}{\rho_\infty}\,\frac{Z_s}{\mathrm{Sd\ Kn}^2}\left(1 + \frac{2\mathrm{Kn}}{B_{HO}}\right) \tag{304}$$

where the coefficient

$$Z_s = \frac{\gamma-1}{3\gamma}\,\frac{\mathrm{Pr\ Cl}\ T_{\mathrm{ref}}}{T_{s\infty}}\left(1 - \frac{\gamma}{\gamma-1}\,\frac{1}{\mathrm{Cl}^*}\right)(1 + G)\left(1 + \frac{\gamma}{\gamma-1}\,\frac{\bar{R}/R_v}{\mathrm{Cl}^{*2}}\,\frac{x_g}{x_v}\right)^{-1} \tag{305}$$

is substantially constant and independent of expansion rate, carrier pressure, and droplet size. Its value is determined by the material properties and carrier composition, and has the order of unity for most pure vapors. (For steam one calculates at $T_{\mathrm{ref}} = T_{s\infty}$ = 300 K and 373 K the values Z_s = 0.98 and 0.69, respectively; for nitrogen at 65 K it is Z_s = 0.59. The approximation $\mathrm{Cl}^* \approx \mathrm{Cl}$ was made in both cases.) Note that the above expression for Z_s is valid for perfect-gas carriers only and that Z_s is in no way related to the compressibility factor Z appearing in Eq. (41).

Equations (303) and (304) show that the level of supersaturation in expanding two-phase flow is the higher the faster the expansion takes place (i.e., high $\dot{P}$, low Sd). The level of $\ell n\, S_\infty$ is seen to be quite sensitive to the amount of liquid (x_ℓ) and to droplet size (r or Kn); it is the higher the *less* liquid is present and the larger the droplets are. Since the product $\rho_\infty\ \mathrm{Sd\ Kn}^2$ can be shown to be independent of p_∞, carrier pressure can have an influence on $\ell n\, S_\infty$ only if Kn is high (i.e., $2\mathrm{Kn}/B_{HO} \geqslant 0.5$).

A natural upper limit to the value of $\ell n\, S_\infty$ is set by spontaneous nucleation, which leads to the formation of new droplets and to the collapse of supersaturation. [See, e.g., Wegener (1969) or Gyarmathy (1976) for a detailed description of the process.] The maximum possible constant levels are about $\ell n\, S_\infty$ = 1 to 3, depending on the nature of the medium and the expansion rate. In compressed flow, $\ell n\, S_\infty$ - Ke is negative, and $(-\ \ell n\, S_\infty)$ is not subjected to any natural limits; however, the eventual vaporization of droplets will set an end to the steady-rate two-phase process being considered here.

The thermodynamic part of the polytropic loss coefficient is obtained, by inserting Eq. (304) into (301), as

$$\phi_{\mathrm{thd}} = \frac{1-x_\ell}{x_\ell}\,\frac{\rho_\ell}{\rho_\infty}\,\frac{1 + 2\mathrm{Kn}/B_{HO}}{\mathrm{Cl}^*\ \mathrm{Sd\ Kn}^2}\,\frac{\{1 - [\gamma/(\gamma-1)](1/\mathrm{Cl}^*)\}\ Z_s}{1 + [\gamma/(\gamma-1)](\bar{R}/R_v\ \mathrm{Cl}^{*2})(x_g/x_v)} \tag{306}$$

For pure vapors the last quotient has typically the value 0.5; also we may typically set B_{HO} = 0.5 and $\mathrm{Cl}^* \approx 12$. Therefore we have, for pure vapors such as steam, as a typical expression,

$$\phi_{\mathrm{thd}} \approx 0.04\,\frac{1-x_\ell}{x_\ell}\,\frac{\rho_\ell}{\rho_\infty}\,\frac{1 + 4\mathrm{Kn}}{\mathrm{Sd\ Kn}^2} \tag{307}$$

The above expressions give the thermodynamic loss for sustained, constant-rate adiabatic expansion or compression processes in terms of the known parameters characterizing the two-phase fluid (x_ℓ, Kn, and the material properties) and the rate of the process (Sd). It is seen that ϕ_{thd} is greater if the process is faster (low Sd), if the liquid content is lower (low x_ℓ), and if the droplets are larger (low Kn). Again, as above, p_∞ has a direct influence only if Kn is high. Capillarity is seen to have no effect on ϕ_{thd}. The above expressions are based on the growth law given by Eq. (247), and the validity of Eq. (306) is therefore limited to pure-vapor or vapor-rich carriers.

As a typical numerical example, let us consider wet steam at $p_\infty \approx 0.1$ bar containing $x_\ell = 0.05$ of water in the form of droplets of $r = 4\times10^{-8}$ m, and being expanded at a continuous rate of $\dot{P} = 1000\ s^{-1}$. (Expansion time constant: $\Delta t_{exp} = 1$ ms.) One calculates or reads from Figs. 5, 6, and 8 the values $\rho_\ell/\rho_\infty \approx 2\times10^{-4}$, Kn ≈ 5, Ke ≈ 0.02, Sd $\approx 10^{-6}$. With these, Eq. (307) gives $\phi_{thd} \approx 0.013$. The thermodynamic loss is about 1%. The level of supersaturation is obtained from Eq. (304), using $Z_s \approx 0.95$ and $B_{HO} \approx 0.5$, as $\ell n\ S_\infty = 0.30$. This is a moderate value showing that spontaneous nucleation of the vapor will not be induced under the fluid and expansion rate conditions assumed. As the expansion will proceed at constant $\dot{P}$ toward lower p_∞, x_ℓ and r will gradually increase, Kn will increase almost proportionally to $1/p_\infty$, and ρ_∞ Sd Kn^2 will stay practically constant. The resulting change of ϕ_{thd} will be a gradual increase.

6.5.3 Mechanical Disequilibrium

In order to calculate ϕ_{mec} we consider an adiabatic, steadily expanding and accelerating two-phase fluid comprising droplets of radius r, and assume that the slip velocity has stabilized itself at a substantially constant vectorial value

$$\Delta\vec{w}_\infty = \Delta\vec{w}_{drf} = -\Delta t_{vel}\frac{d\vec{w}_{carr}}{dt} \tag{308}$$

The last expression was obtained from Eqs. (167) and (170) by neglecting the body-force and pressure-gradient terms. The constancy of $\Delta\vec{w}_\infty$ implies that

$$\frac{d\vec{w}_\ell}{dt} = \frac{d\vec{w}_{carr}}{dt} \tag{309}$$

Now the scalar product appearing in Eq. (297) can be represented as

$$-\Delta\vec{w}_\infty\frac{d\vec{w}_\ell}{dt} = \Delta t_{vel}\left(\frac{dw_{carr}}{dt}\right)^2 \tag{310}$$

If we furthermore assume that $\Delta\vec{w}_\infty$ is small compared to the carrier velocity $\vec{w}_{carr}$, it follows that

$$w_\ell \approx \vec{w}_{carr} \tag{311}$$

If the body-force field is negligible, $d\vec{w}_{carr}/dt$ is proporational to the expansion rate. According to Eqs. (A-37) and (311),

$$\bar{R}\ T_\infty\ \dot{P} \approx \vec{w}_{carr}\ \frac{d\vec{w}_{carr}}{dt} = \cos\nu\ w_{carr}\ \frac{dw_{carr}}{dt} \tag{312}$$

where ν is the angle enclosed by the vectors $\vec{w}_{carr}$ and $d\vec{w}_{carr}/dt$.

In rectilinear accelerating flows $\cos\nu = 1$, in rectilinear decelerating flows $\cos\nu = -1$; intermediate values of $\cos\nu$ pertain to curvilinear flows. Inserting Eq. (310) into Eq. (297) and eliminating the scalar dw_{carr}/dt by use of Eq. (312), one obtains

$$\phi_{mec} = \frac{\pm\ x_\ell}{1-x_\ell}\ \frac{\Delta t_{vel}}{\cos^2\nu}\ \frac{\bar{R}\ T_\infty\ \dot{P}}{w^2_{carr}} = \frac{\pm x_\ell}{1-x_\ell}\ \frac{\Delta t_{vel}}{\cos^2\nu}\ \frac{\dot{P}}{\gamma\ \mathrm{Ma}^2_{carr}} \tag{313}$$

It is seen that ϕ_{mec} is the higher the *more* liquid is being carried, the greater the droplets are (higher Δt_{vel}, cf. Fig. 20), the faster the expansion and the *lower* the Mach number of the carrier flow.

Now Δt_{vel} will be expressed by Eq. (170) and by the multirange low velocity expression (129) for Nu_F:

$$\Delta t_{vel} = \frac{2\ \rho_\ell\ r^2}{\eta}\left(1 + \frac{3\mathrm{Kn}}{B_{FO}}\right) = \frac{2\ \rho_\ell}{\rho_\infty}\ \frac{1}{\mathrm{Kn}}\left(\frac{1}{\mathrm{Kn}} + \frac{3}{B_{FO}}\right)\frac{\eta}{p_\infty} \tag{314}$$

Note that according to Sec. 3.3, B_{FO} has a common value of about 1.5 for most substances. Inserting this into the above expression for ϕ_{mec}, we have, using Eq. (51),

$$\phi_{mec} = \frac{x_\ell}{1-x_\ell}\ \frac{\rho_\ell}{\rho_\infty}\ \frac{(2/\ \gamma\ \cos^2\nu)}{\mathrm{Ma}^2_{carr}}\ \frac{1}{\mathrm{Sd}\ \mathrm{Kn}}\left(\frac{1}{\mathrm{Kn}} + 2\right) \tag{315}$$

This expression relates ϕ_{mec} to the (nondimensional) data of the fluid and of the process. It is the mechanical counterpart of Eq. (306).

As a numerical example we again take water droplets of $r = 4\mathrm{x}10^{-8}$ m, $x_\ell = 0.05$, in steam of $p_\infty = 0.1$ bar expanding at $\dot{P} = 1000\ \mathrm{s}^{-1}$. We assume a momentary carrier Mach number of $\mathrm{Ma}_{carr} = 0.3$. The flow will be rectilinear ($\cos\nu = 1$). With $\rho_\ell/\rho_\infty \approx 2\mathrm{x}10^4$, $\mathrm{Kn} \approx 5$, $\mathrm{Sd} \approx 10^6$, and $\gamma \approx 1.3$, we have $\phi_{mec} \approx 0.0008$. Thus in this example the kinematic disequilibrium leads to a loss that is about 15 times smaller than the thermodynamic loss of $\phi_{thd} \approx 0.013$. Both together give $\phi \approx 0.014$, that is, a polytropic efficiency of 98.6%. In the above example Δw_{drf} is found to correspond to a relative-flow Mach number of Ma = 0.003, showing that the approximation made in Eq. (311) was justified.

6.5.4 Similarity Law of Entropy Production

For some applications, for example, the flow of steam in turbines, it is of interest to formulate the conditions under which the entropy production caused by interphase nonequilibrium is equal in two geometrically similar flow fields of different scale. The macroscopic thermodynamic parameters, including x_ℓ and Ma_{carr}, are assumed to have equal values in all corresponding points of the two flow fields considered. The expansion rate is inversely proportional to the geometric scale, and the Stodola number is directly proportional to it. The droplet size (i.e., the value of Kn) will be considered to be a free parameter.

The similarity of the entropy production field is maintained if the polytropic loss coefficient ϕ has equal values in corresponding points. If we express $\phi = \phi_{thd} + \phi_{mec}$ by the simplified expressions given by Eqs. (307) and (315), which are valid for low-pressure steam, we have

$$\phi \approx \frac{1-x_\ell}{x_\ell}\,\frac{\rho_\ell}{\rho_\infty}\left[0.04\,\frac{1+4\mathrm{Kn}}{\mathrm{Sd\ Kn}^2} + 0.17\left(\frac{x_\ell/(1-x_\ell)}{Ma_{carr}\cos\nu}\right)^2\frac{1+2\mathrm{Kn}}{\mathrm{Sd\ Kn}^2}\right] \qquad (316)$$

Here Ma_{carr}, x_ℓ, and ρ_ℓ/ρ_∞ have identical values in both fields and ν is the same as a result of geometric similarity. The relative magnitude of the two terms in the brackets is seen to be determined mainly by the value of the squared quotient; it was shown above that the first term is likely to predominate in rectilinear flows.

If the second term is indeed negligible, equality of ϕ exists if the values of $(1+4\mathrm{Kn})/\mathrm{Sd\ Kn}^2$ are equal in the two systems, that is if the product $(1+4\mathrm{Kn})\mathrm{Kn}^{-2}$ is proportional to geometric size. If, however, the second term is the predominant one, then the product $(1+2\mathrm{Kn})\mathrm{Kn}^{-2}$ is required to be proportional to the geometric scale. It is obvious that the two conditions lead to very similar ratios of Kn values required for similarity in ϕ.

As hypothesized by Moore (1976) and Kreitmeier et al. (1980), the efficiency losses due to steam wetness in turbines may be characterized by two parameters of the type of Damköhler numbers, one for thermodynamic relaxation and one for mechanical relaxation[30]. These Damköhler numbers are closely related to ϕ_{thd} and ϕ_{kin}, respectively. From the above it is apparent that the equality of both criteria, of the thermodynamic and of the mechanical similarity, are nearly simultaneously satisfied if the values of Kn have the proper relationship in the systems.

---

[30] The Damköhler number (see, e.g., Weast and Selby, 1966) expresses the importance of nonequilibrium phenomena; it is the ratio of overall process duration to the relaxation time constant.

7 CRITICAL DISCUSSION AND CONCLUSIONS

7.1 Classification of Droplet Growth Theories

It appears from Sec. 5 that the theory describing droplet growth is mathematically complicated even in the case of severe idealizations, if it is required to be applicable to any carrier composition (ranging from the pure vapor of the droplet to a substantially vaporless, noncondensing gas), to any carrier temperature (from far below the boiling point of the droplet to far above), to any carrier density (ranging from continuum-type to free-molecular behavior), and if it is supposed to account for various complicating effects such as relative flow velocity and interface curvature (i.e., capillarity). This complexity may be the reason why the growth theories published in the literature deal mostly with certain special situations in which appropriate simplications are possible.

In order to discuss the various existing droplet growth theories, it is convenient to adopt the following classification:

1. Theories accounting for nonsteady transfer processes outside of and within the drop

2. Theories accounting for nonsteady heat conduction within the drop, but treating the processes in the carrier phase as quasi-steady

3. Quasi-steady theories using a differential equation to determine the droplet surface temperature

4. Quasi-steady theories using an algebraic equation for the droplet surface temperature. If this equation is explicit, a growth law is obtained in terms of droplet size, relative velocity, and carrier state.

Within these four types, distinction may be made according to the restrictions relating to carrier type, droplet size, relative velocity, and so on.

7.1.1 Fully Nonsteady Treatment.

In the case of fully nonsteady treatment the transfer rates of heat and mass must be calculated from time-dependent temperature and concentration profiles (cf. Sec. 5.1.2). The problem is irrelevant for the free-molecular case. For the continuum case the differential equations pertinent to the heat conduction and diffusion in the carrier have been solved by Fuks (1959). The deviations from quasi-steady behavior become significant in high-temperature, high-pressure carriers (Torda, 1973). An analysis of nonsteady heat conduction within and around a stationary droplet was made by Konorski (1969) and by Cooper (1977), in the latter case in connection with thermal relaxation in a liquid/liquid dispersion. For the normal case of droplet growth in moderate-temperature, moderate-density gaseous carriers as defined by Eq. (147), the corrections due to the time dependence of r and of the carrier-side distributions are found to be negligible except in the short initial period (of the order of Δt_{int}) in which an injected droplet adjusts its temperature to the conditions of the carrier (Rüping, 1959; Hubbard et al., 1975).

7.1.2 Partially Quasi-Steady Treatment

In the partially quasi-steady treatment, time-dependent differential equations are applied only to the heat conduction within the droplet; cf. Sec. 5.1.3. This may be necessary, for example, if cold droplets are injected into a vapor-rich, warm environment, and their effect is caused mainly by the adjustment of their temperature to the new environment rather than by the change of their size. This approach has been applied to spray cooling by Nitsch (1971) and by Kulic and co-workers (1975, 1977, 1978), whereas Agosta and Hammer (1975) have analyzed the evaporation of oxygen drops in a combustion-gas environment. Even though $\Delta t_{int} \ll \Delta t_G$ may hold, the time-dependent treatment of the droplet interior becomes necessary in these cases, because the residence time is of the order of Δt_{int} only and the process is determined mainly by these transients. Such situations occur if the droplets leave the carrier again, such as by falling out. In the above applications the carrier was of continuum type. There are some indications that internal heat conduction may be considerably enhanced by convective liquid motion within the droplet (Dibelius and Voss, 1979).

7.1.3 Quasi-Steady Formulation Using a Differential Equation for T_r

If the temperature within the drop is taken to be uniform ($T_\ell = T_r$) or is represented by some quasi-constant distribution (as we proposed in Sec. 3.6), the heat flux H_{int} into the droplet is expressed simply as the product of the rate of surface temperature change and of the heat capacity of the drop; cf. Eq. (155c). By inserting this expression for H_{int} into Eq. (195), a first-order differential equation is obtained for the surface temperature T_r:

$$\frac{\mathcal{D}\, L_r\, \mathrm{Nu}_M}{p_\infty}\left[p_{v\infty} - p_r(T_r, r)\right] - \lambda\, \mathrm{Nu}_H (T_r - T_\infty^*) = \frac{2\, r^2\, \rho_\ell\, c_\ell}{3} \frac{dT_r}{dt} \qquad (317)$$

Such a simplifying assumption with regard to the temperature distribution within the drop may be made, of course, only if Δt_{int} is much smaller than the duration of the process considered. In such cases Eq. (317) inevitably produces the result that T_r very rapidly approaches a quasi-steady value and thereafter stays substantially constant (Stodola, 1922; Buhler and Nagamatsu, 1952; Hill et al., 1963). The quasi-steady value is obtained from Eq. (317) by simply setting $dT_r/dt = 0$ or, better, by estimating its value on the basis of the rate of change of the carrier conditions (Buhler and Nagamatsu,

1952). The latter approach was adopted in Eq. (200). Thus T_r is obtained from an algebraic equation. The use of a differential equation of the type of Eq. (317) to calculate T_r is seen to be an unnecessary complication: Namely, if the process considered is of short duration (of the order of Δt_{int}), Eq. (317) is rather inaccurate in describing the heat conduction within the droplet, and is better abandoned in favor of the approach of type 2; if the process lasts much longer than Δt_{int}, Eq. (317) is valid, but an algebraic equation leads to substantially identical results. The range of process duration where Eq. (317) is an advantageous compromise is very narrow. One must note, however, that growth theories using equations of the type of (317) have considerably contributed to the understanding of droplet behavior.

Stodola, in his book on steam turbines (1922), has analyzed the growth and temperature of very small water droplets in steam (free-molecular case, no relative velocity). This problem has been reformulated by Buhler and Nagamatsu (1952), Hill (1966), and Puzyrewski (1969, 1976) on the basis of a more rigorous kinetic theory.

The equations thus obtained correspond to Eqs. (194), (196), and (197) if there the Nusselt numbers Nu_H and Nu_M are expressed by the free-molecular relationships given for Ma = 0 in Sec. 3.3. Puzyrewski's treatment differs from the other theories in that the accommodation coefficients α are replaced by $\alpha/(2-\alpha)$[31] and in some particular approximations made with respect to temperatures. Hill's formulation formally agrees with Sec. 3.3, but his use of mass accommodation (or "condensation") coefficient values as low as $\alpha_{Mv} = 0.04$ is based on erraneous experimental data.

7.1.4 Quasi-Steady Theories Using an Algebraic Equation for T_r

By setting dT_r/dt in Eq. (317) equal to zero or to an estimated finite value, the theories of class 3 change over into this class 4. Here distinction has to be made between the iterative solution of an implicit equation for T_r (Stodola, 1922; Buhler and Nagamatsu, 1952; Hill et al., 1963, 1966; Campbell and Bakhtar 1971) and the approximation of T_r by an explicit equation. Such approximations were made for a great variety of physical conditions.

In the *free-molecular* extreme, Wu (1972) and Wegener et al. (1972) have considered the case of vaporless carriers at low temperature as typical for condensation in wind tunnels. By setting $T_r = T_\infty$ they obtained the simple growth equation

[31] This formulation, originally conceived by Knudsen (1911, p. 597) for molecular processes between *two* imperfectly accommodating surfaces, is not applicable to a single drop in an infinite free-molecule environment.

$$\frac{dr}{dt} = \frac{\alpha_{Mv}\ p_{v\infty}}{\rho_\ell \sqrt{2\ \pi\ R_v\ T_\infty}} \left(1 - S_\infty^{(r_{crit}/r)\ -\ 1}\right) \tag{318}$$

which is identical to the dimensional form of Eq. (256), with Ke being expressed according to Eq. (182).

The *continuum* case has been the subject of numerous analyses. Evaporation or growth in low-temperature gaseous carriers, as given by Eq. (253), was considered by, for example, Mason (1951), Fuks (1959), Rüping (1959), and Nitsch (1971). In the case of *high* carrier temperatures, such as occur in combustion environments, the growth rate is usually represented as

$$\frac{dr}{dt} = \frac{\rho_\infty\ D}{r\ \rho_\ell}\ \ell n\left(1 + \frac{\bar{c}_p\,(T_\infty - 0.9\ T_{b\infty})}{L} + \Phi\right) \tag{319}$$

where the estimate $T_r = 0.9 T_{b\infty}$ is made for the droplet surface temperature (cf., e.g., the review by Williams, 1973). Here Φ is a term that is zero when no chemical reactions occur. This equation does not correspond to any of the equations obtained in Sec. 5.3, because it rigorously accounts for the convective heat transfer caused by the radial outflow of vaporized liquid. (In the Nu_H expressions given in Tables 5 and 6 this convection effect has been neglected; its influence on the Nusselt number is given by Eq. (85).) Bringing Eq. (260) into dimensional form, setting $T_\infty^{**} = T_\infty$ and making the approximation $\psi_6 = (T_\infty - 0.9 T_{b\infty})/T_{b\infty}$, which is justified as evident from the curves shown in Fig. 19, and setting $\psi_1 = 1$ for stationary drops, we obtain

$$\frac{dr}{dt} = \frac{\lambda}{r\ \rho_\ell\ L}\ (T_\infty\ - 0.9\ T_{b\infty}) \tag{320}$$

An identical equation is obtained from Eq. (319) if $\Phi = 0$ is set (no chemical reaction), the mathematical approximation $\ell n\ (1 + x) \approx x$ is made, and $Pr/Sc = \bar{c}_p\ \rho_\infty\ D/\lambda = 1$ is assumed.

Several authors have been concerned with *multirange* expressions covering continuum, free-molecular and intermediate conditions. Gyarmathy (1962) established an expression for stationary water droplets in pure steam, setting $T_r = T_{s\infty}$, using a Cunningham-type dependence on the Knudsen number and taking capillarity into account. In an extension to gas-vapor mixture environments (Gyarmathy, 1963), T_r was calculated by accounting for the combined effect of heat and mass transfer, but the Kn-dependent factor was assumed to be the same for both transfer processes. The growth law thus obtained is substantially identical to Eq. (247). Note that the two definitions of Kn are different, cf. Eq. (A-12) in App. 1.

Kang (1967) used a Kn dependence of the kind of Eq. (121), assigned different values to the constant c in the expressions of the two Nusselt numbers Nu_H and Nu_M, neglected capillarity and relative flow, and used a simplified diffusion law. His results are therefore valid only for relatively large (Ke<0.01) droplets in vaporless (dilute) carriers. Fukuta and Walter (1970) followed a similar approach. Their treatment accounts for the capillarity effect and for other factors such as the presence of solutes in the droplet, neglects relative velocity, and disregards Stephan flow. The latter simplification limits the validity of their growth law to dilute, gaseous carriers. Due to a misunderstanding in the definition of the condensation coefficient, they advocate the use of very low values (of 0.0265 to 0.2), although their equation is based on an α_{Mv} of which the correct empirical values are close to unity. Apart from these differences, the treatment of Fukuta and Walter corresponds to Eq. (224).

7.2 Conclusions

Determinant factors for the thermodynamic and kinematic behavior of a droplet is gaseous carrier flow are the *heat, mass, and momentum transfer processes* resulting from the superheat or subcooling of the carrier and its relative velocity, and the abnormal thermodynamic properties of the curved liquid surface (capillarity or Kelvin effect). It has been shown that a close analogy of heat, mass, and momentum transfer is maintained at all Knudsen numbers, that is, at any carrier density and almost any droplet size.

The motion of the droplet is substantially independent of concomitant heat and mass transfer processes. Compared to the time required for a substantial change in droplet size r, the droplet velocity is quickly adapted to the carrier flow. Thus the relative velocity may usually be assumed in growth calculations to equal the drift velocity Δw_{drf}, the latter being a known function of droplet size, carrier properties, carrier flow acceleration, and/or body-force field. In the absence of the latter factors, the drift velocity is zero.

On the other hand, heat and mass transfer (and growth) are markedly influenced by relative flow, provided that Δw_∞ is large enough to make Re>1 and/or Ma>0.1. Therefore the calculation of droplet growth rates may require knowledge of the momentary value of Δw_∞, or at least of Δw_{drf}.

The question of whether a droplet grows, stays unaltered, or evaporates in a carrier stream is substantially determined by the saturation ratio S_∞. If $S_\infty > e^{Ke}$, the droplet grows, if $S_\infty < e^{Ke}$, it shrinks.[32] Here Ke is the Kelvin number, which characterizes the importance of the capillarity effect. For droplets having $r > 10^{-8}$ m, Ke is for most substances a very small number, and $e^{Ke} = 1$ may be set.

At given value of $(S_\infty - e^{Ke})$, the rate of growth or evaporation

---

[32] For more rigorous statements see Sec. 5.3, Eqs. (227) and (228).

is determined by the efficiency of heat and mass transfer, that is, by the Nusselt numbers Nu_H and Nu_M. The transfer rates are strongly influenced by the ratio of molecular mean free path in the carrier fluid and the droplet diameter, that is, the Knudsen number Kn. At low Kn (continuum-type carriers) the rates are inversely proportional to r, whereas at large Kn (free-molecular carriers) the rates are independent of r.

The relative importance of heat and mass transfer depends mainly on the composition of the carrier. In pure vapors the heat transfer process is determinant. In cold, dilute carriers the mass transfer is the controlling factor. At intermediate mixture compositions and in hot, lean carriers both processes are significant. For most substances, for example, water, it may be stated quite generally[33] that the time constant Δt_G for doubling the radius of the droplet (or reducing it to zero) is much longer than the time Δt_{int} characterizing the thermal inertia of the droplet interior. Thus processes involving substantial changes of droplet size can be described satisfactorily by the quasi-steady growth theory, using a simple model of the droplet interior.

The surface temperature T_r of droplets always lies between the carrier temperature T_∞ existing far from the drop and the saturation temperature $T_{s\infty}$ pertaining to the partial pressure of the vapor in the carrier. (Capillarity and other effects may shift these limits slightly.) In pure vapors $T_r \approx T_{s\infty}$, whereas in dilute carriers $T_r \approx T_\infty$. Approximate expressions can be specified in order to calculate the value of T_r in terms of these limits and the composition of the carrier. In extreme cases T_r is also influenced by relative velocity and droplet size.

The effect of thermal inertia of droplets in a changing environment can be formally accounted for in a convenient way by defining an apparent carrier temperature $T_\infty^{**} \neq T_\infty$. In the value of T_∞^{**} the recovery heating of a moving droplet can also be included.

All of the above effects can be included with reasonable accuracy in *droplet growth laws* that give dr/dt in terms of droplet size, carrier composition, and state and relative velocity. The growth laws given in the present chapter cover following parameter ranges: Kn = 0 – ∞, Re = 0 – 200, Ma = 0 – 0.5 – (1), arbitrary carrier compositions ($g_\infty = 0 - \infty$), and arbitrary ratios of carrier temperature T_∞ and liquid boiling point $T_{b\infty}$.

In order to complete the set of nondimensional parameters required to describe droplet growth and evaporation processes, a Clausius-Clapeyron group Cl and an Eötvös group Eö have been defined, which relate the latent heat and the surface tension to other materials data. The Kelvin number mentioned above is a convenient measure of capillarity effects. The rate of environ-

[33] An exception is liquids of very low heat of vaporization (Cl<5) in carriers of very high supersaturation or superheat.

mental change (for example, the expansion rate) can be made nondimensional with the collision frequency of a carrier-phase molecule with the other carrier molecules. We propose calling the resulting nondimensional expansion rate the Stodola group Sd.

By applying the laws governing droplet growth or evaporation and momentum exchange to *two-phase fluids* with droplets distributed uniformly over the carrier phase (mist-type fluids), the entropy production in such fluids can be described. The nondimensional parameters governing the *entropy production* rate have been formulated for two-phase fluids with pure-vapor or mixture-type carrier phases undergoing an adiabatic expansion or compression process.

7.3 Suggestions for Further Work

The correct empirical values of the accommodation coefficients entering free-molecular transfer-rate expressions are still subject to some controversy. Further data, theoretical work, and clarification of any interdependence among α_H, α_M, and α_F would be useful.

The influence of Knudsen number on transfer rate is expressed in the multirange expressions in a very crude way. More rigorous theoretical treatments than the ones leading to expressions of the type of Eq. (121) or (122), and experimental determination of growth rates of very small drops, could lead to improvements in the growth laws, especially for Knudsen numbers of the order of one.

The boiling lag required to predict the thermal bursting limits for droplets is not known for most liquids. Relevant data would be useful, among other domains also for steam turbine engineering.

Mass transfer and growth rate measurement in vapor-rich carriers are scarce, even for continuum-type conditions. The use of the product (Sc $\Pi_{g\infty}$) in the mass transfer correlation expressed by Eq. (81) should be verified by experiments.

The influence of finite droplet environment on the growth rate can be only crudely estimated (proximity effect). More general solutions for constant and varying carrier conditions would be of interest for the analysis of high-pressure, high-density condensing flow and of very dense sprays, cf. Sec. 3.5.4.

The thermodynamics of very small droplets (containing less than 100 molecules) is still not completely resolved. Uncertainties with respect to surface tension and latent heat values hamper the evolution of nucleation theories. The effects of very small amounts of solute molecules (e.g., one salt molecule in a droplet containing 50 molecules of the liquid) can only be estimated by very crude approximations; the same is true for the effect of an ion in the drop. No attempt has been made to cope with these problems in the present study.

The analysis of vaporization in very hot carriers with due account taken of the temperature dependence of ρ, c_p, λ, and so on today requires numerical computation. It is desirable to correlate such solutions by a simple growth law of the kind given by Eqs. (260) or (262), but in which additional correction is made for the temperature dependence effects and their nature (as described, for example, by the the exponents of the temperature variances). In certain problems also the influence of composition variations may be significant. Such corrections may remove the poor performance of Eq. (262) at high carrier pressure.

In the case of an individually burning droplet, the radius r_∞ of the sample domain is given by the position of the flame front, and the mass and heat fluxes M_{ext} and H_{ext} crossing the sample boundary are significant. For such cases the Nusselt numbers obtained from the simple diffusion and conduction theory applied to the case of $r_\infty \to \infty$ and $M_{ext} = H_{ext} = 0$, cf. Eqs. (77) and (78), are not valid. Again, the influence of the modified boundary conditions should be expressed as corrections to the classical case.

Experimental data concerning the diffusivity of various vapors and gases in wide ranges of temperature, pressure, and composition are scarce.

It is to be hoped that the nondimensional characterization of two-phase media, as advocated in the present study, will become widely accepted. In order to promote this process, it would be helpful to make available values of the Clausius-Clapeyron number and of the Eötvös number for a greater variety of substances as those shown in Fig. 4. Also it is desirable to include Cl and Eö into standard property data tables, as, for example, the Prandtl number is included today in the internal formulation of steam properties (Schmidt, 1969).

NOMENCLATURE

Symbols occurring on a single place only are defined in the text. Equations in the text are designated with running numbers. Designations "(A-1)" etc. indicate Equations in the Appendix. The symbol "$\hat{=}$" means "equal by definition". Vectors are designated by $\to$, their positive scalars by $|\to|$. Otherwise $|...|$ means absolute value. Boldface type (e.g., $\mathbf{r}$) indicates referred (nondimensional) quantities.

a	thermal diffusivity (Eqs. 56a, 56b), m^2/s
A	abbreviation defined in Eq. (214)
b	body-force field strength, N/kg
B_F, B_H, B_M,	free-molecular transfer-law coefficients (Eqs. 106 to 108)
Bi	Biot number (Eq. 66)
c	constant introduced in Eq. (121)
c_D	drag coefficient of sphere (Eq. 76)
c_ℓ	specific heat of liquid phase, J/kg K
c_p	isobaric specific heat, J/kg K

C	abbreviation defined in Eq. (215)
$C_1 - C_3$	constant coefficients in Eq. (280)
Cl	Clausius-Clapeyron number (Eq. 34 or Eq. 37), see Fig. 4
D	diffusivity of vapor and gas components, m^2/s
$\mathcal{D}$	modified diffusion coefficient (Eq. 53), kg/s m
E	excess thermal energy of field (Eq. 142), J
Eö	Eötvös number (Eq. 38 or Eq. 40), see Fig. 4
f_L, f_R	functions defined by Eqs. (A-87) and (A-88)
F	momentum transfer rate to the drop (i.e., drag force) N
F_{ext}	resultant external force acting on boundary of sample, N
Fo	Fourier number (Eq. 67)
Fr	Froude number (Eq. 46)
g_∞	dilution ratio (Eq. 9)
G	gas influence factor (Eq. 222)
h	specific enthalpy, J/kg
H	heat transfer rate to the drop, J/s
H_{ext}	net heat transfer rate to the sample from the exterior, J/s
H_{int}	heat conduction rate from drop surface toward the interior, J/s
k_T	nondimensional recovery temperature function (Eq. 71)
k_F, k_H, k_M	kinetic theory functions representing Mach number influence, cf. Sec. 3.3.3 and Eqs. (109) to (111)
K_g, K_v, K	kinetic theory coefficients (Eqs. 95 and A-65)
Ke	Kelvin number (Eq. 62); values in Fig. 6
Kn	Knudsen number (Eq. 43); conversion to other definitions, see Eq. (A-12); values in Fig. 6
ℓ	molecular mean free path (Eq. 44); for conversion to other definitions, see Eq. (A-11)
L	latent heat (i.e., enthalpy) of condensation $L = L(T_{ref})$, J/kg
L_r	latent heat corrected for capillarity (Eq. 24), J/kg
m_g, m_ℓ, m_v	mass of gas, liquid, vapor components in the sample, kg
m_s	total mass of fluid sample, kg
m_{mol}	mass of one molecule (vapor or gas), kg
m_{Mol}	molar mass of liquid, kg
M	mass transfer rate to the drop, kg/s
M_{ext}	net vapor mass transfer rate to the sample from the exterior, kg/s

Ma	Mach number (Eq. 61 and Eq. 94)
Ma_{carr}	Mach number of absolute carrier flow (Eq. 175)
n	polytropic exponent
Nu_F	Nusselt number for momentum transfer (Eq. 54), equaling c_DRe/8; has no special name in literature
Nu_H	Nusselt number for heat transfer (Eq. 55), commonly called Nusselt number
Nu_M	Nusselt number for mass transfer (Eq. 52), commonly called Sherwood number Sh
p	pressure, Pa
p_∞	overall pressure of carrier phase, Pa
$p_{g\infty}$, $p_{v\infty}$	partial pressures in the carrier phase (Eq. 6), Pa
p_ℓ	pressure in droplet interior (Eq. 186), Pa
p_r	vapor pressure of droplet surface (Eq. 177), Pa
p_s	vapor pressure of flat surface (Eq. (A-15), Pa
$p_{s\infty}$	saturation vapor pressure $p_{s\infty} = p_s(T_\infty)$, Pa
Δp_{bl}	boiling lag (Eq. 186), Pa
$\dot{P}$	expansion rate (Eq. 49), s^{-1}
Pr	Prandtl number (Eq. 32)
q	correction factors for mixture carriers (App. 3)
δQ	relative error in Q
Q	caloric effectiveness of the carrier (Eq. 219)
r	radius of droplet, m
r_∞	radius of spherical sample (Fig. 1b), m
r_{crit}	critical radius (Eq. 181), m
r_{mol}	radius of a liquid molecule (for water, $r_{mol} \approx 2x10^{-10}$ m) m
$\boldsymbol{r}$	nondimensional radius ($\boldsymbol{r} \triangleq r/r_{ref}$); subscripts as for r
R_g, R_v	specific gas constant of gas and vapor component, J/kg K
$\bar{R}$	specific gas constant of carrier mixture, J/kg K
$\hat{R}$	R_v and R_g averaged according to Eq. (A-52), J/kg K
Re	Reynolds number (Eq. 45)
s	specific entropy, J/kg K
S_∞	saturation ratio of the carrier (Eq. 179)
S_r	apparent saturation ratio (Eq. 210)
Sc	Schmidt number (Eq. 33)
Sd	Stodola number (Eq. 51)

Sh	Sherwood number, identical to Nu_M
t	time, s
Δt	characteristic time constants; its subscripts have following meanings:
	coll = time interval between molecular collisions (Eq. 48)
	exp = of carrier expansion; $\Delta t_{exp} \triangleq \lvert \dot{P}^{-1} \rvert$, see Eq. (50)
	acc = of carrier-flow acceleration (Eq. 173)
	G = of droplet growth (Eq. 140)
	int = of internal temperature distribution in drop (Eq. 154)
	vel = of velocity adaptation (Eq. 169)
T	temperature, K
$\bar{T}$	mean carrier temperature, K ($\bar{T} = T_\infty$ is set, cf. Sec. A2.5)
$T_{1/3}$	reference temperature according to the one-third rule (Sec. 3.2.4), K
T_∞	far-field carrier temperature, K
T_∞^*	adiabatic body temperature (Eq. 71, 75, 112, or 126), K
T_∞^{**}	apparent carrier temperature (Eq. 201), K
$T_{b\infty}$	boiling-point temperature $T_{b\infty} = T_s(p_\infty)$, K
T_{br}	boiling temperature of droplet, capillarity corrected (Eq. 206 or 208), K
T_c	critical temperature of liquid, K
T_{cent}	temperature at center of droplet, K
T_ℓ	interior mean temperature of droplet, K
T_r	surface temperature of droplet, K
T_{ref}	reference temperature, K
$T_{s\infty}$	saturation temperature $T_{s\infty} = T_s(p_{v\infty})$, K
T_{sr}	saturation temperature of droplet, capillarity corrected (Eq. 205 or 207), K
ΔT_∞	carrier-phase subcooling (Eq. 180), K
ΔT_ℓ	excess internal temperature of droplet, (Eq. 21), K
$\boldsymbol{T}$	nondimensional temperature ($\boldsymbol{T} \triangleq T/T_{ref}$); subscripts and superscripts as for T
$\Delta\boldsymbol{T}_\infty$	nondimensional subcooling ($\Delta\boldsymbol{T}_\infty \triangleq \Delta T_\infty/T_{ref}$)
U	energy, J
v	specific volume m^3/kg
V	volume, m^3

w_ℓ	absolute velocity of droplet, m/s
w_{carr}	absolute velocity of carrier stream, m/s
Δw_{ref}	reference value of Δw_∞, m/s
Δw_∞	relative velocity (Eq. 18), m/s
$\boldsymbol{w}$	nondimensional relative velocity ($\boldsymbol{w} \triangleq \Delta w_\infty / \Delta w_{ref}$)
Δw_{drf}	drift velocity (Eq. 167), m/s
W_{diss}	work dissipated in the sample per unit time, J/s
We	Weber number (Eq. 64)
x	mass fraction (static value)
Z	compressibility factor (Eq. 41)
Z_s	coefficient defined by Eq. (305)

Greek Letters

α_F	coefficient of diffuse reflection (Eq. 92)
α_H	coefficient of thermal accommodation (Eq. 88)
α_M	coefficient of mass accommodation (Eq. 91)
β_∞, $\beta_{g\infty}$, $\beta_{v\infty}$	molecular mass impingement rate (Eq. 87), kg/m^2 s
β_r	value of β_v corresponding to equilibrium with the droplet (Eq. 90), kg/m^2 s
γ	specific heat ratio (Eq. A-8)
Δ	relative deviation of Nu_F expressions (Sec. 3.4.6)
ζ	radial space coordinate, m
η	dynamic viscosity (without index: of carrier mixture), kg/s m
ϑ	time scaling factor (Eqs. 47 and 233)
λ	thermal conductivity (without index: of carrier mixture kg m/s^3 K
ν	angle characterizing streamline curvature (Sec. 6.5), deg.
$\Pi_{g\infty}$, $\Pi_{v\infty}$	partial pressure ratios (Eq. A-1)
Π_r	nondimensional vapor pressure of drop (Eq. 30)
ξ	nondimensional temperature coordinates in App. 4; flight distance [m] in Sec. 6.2, m
ρ	density, kg/m^3
ρ_∞	density of carrier phase, kg/m^3
$\rho_{g\infty}$, $\rho_{v\infty}$	partial densities (Eq. 8), kg/m^3
ρ_ℓ	density of liquid phase $\rho_\ell = \rho_\ell(T_{ref})$, kg/m^3
σ	surface tension of liquid $\sigma = \sigma(T_{ref})$, kg/s^2
τ	nondimensional time (Eq. 57)
ϕ	polytropic loss coefficient
Φ	combustion term in Eq. (319)

$\psi_1-\psi_6$	various abbreviations defined in Sec. 5.3

Subscripts

E	when droplet completely evaporated
est	estimated value
F	pertinent to momentum transfer (except in b_F, $w_{\mathrm{carr}F}$, and grad_F, where it means projections on the direction of $\vec{F}$, cf. Sec. 2.2)
g	of noncondensing gas component(s)
gas	for almost pure gas, cool carriers (Eq. 224)
gen	for saturated or hot carriers (Eq. 226)
H	pertinent to heat transfer
l	of liquid phase (droplet)
mec	due to mechanical nonequilibrium
M	pertinent to mass transfer; as second index: in simultaneous presence of mass transfer
nl	based on nonlinear law (Eq. 104)
r	of droplet surface (exception: L_r)
ref	reference value
s	at saturation (phase equilibrium with *plane* interface)
S	of sample
syn	synthesis of Q_{vap} and Q_{sat} (Eq. 225)
thd	due to thermodynamic nonequilibrium
v	of vapor component
vap	for vapor-rich carriers (Eq. 223)
o	at time t_o
0	for zero relative velocity and $T_r = T_\infty$
∞	of the carrier phase far from the drop (exception: r_∞)

Superscripts and Other Symbols

ct	continuum conditions
fm	free-molecule conditions
nl	based on nonlinear law, cf. Eq. (83)
prx	corrected for proximity effects (Eq. 153)
' "	of saturated liquid and vapor, respectively (exception: α'_M and the derivatives f' in App. 4)
‾	mass-averaged mean value for the carrier phase (Eq. A-7); exception: $\bar{T}$
⌒	special mean value according to App. 3
→	vector quantity

REFERENCES

Agosta, V. D., and S. S. Hammer 1975, Vaporization Response of Evaporating Drops with Finite Thermal Conductivity. NASA CR-2510, Jan. 1975.

Ashley, H. 1949, Application of the Theory of Free Molecule Flow to Aeronautics. *J. Aero. Sci.* vol. 16, pp. 95–104.

Anisimova, M. P., and E. V. Stekolshchikov 1977, Deformational Break-up of Drops in a Gas Stream (in Russian). *Izv. Akad. Nauk USSR, Energ. i Transport* vol. 15, no. 3, pp. 141–148. English translation: *Power Eng.* vol. 15, no. 3, pp. 126–133 (Allerton Press, 1977).

Barschdorff, D. 1975, Carrier Gas Effects on Homogeneous Nucleation of Water Vapor in a Shock Tube. *Phys. Fluids* vol. 18, no. 5, pp. 529–535.

Brock, J. R. 1964a, Evaporation and Condensation of Spherical Bodies in Noncontinuum Regimes. *J. Phys. Chem.* vol. 68, no. 10, pp. 2857–2862.

Brock, J. R. 1964b, Free-Molecule Drag on Evaporating or Condensing Spheres. *J. Phys. Chem.* vol. 68, no. 10, pp. 2862–2864.

Bryson, C. E., V. Cazcarra, and L. L. Levenson 1974, Condensation Coefficient Measurements of H_2O, N_2O and CO_2. *J. Vac. Sci. Technol.* vol. 11, no. 1, pp. 411–415.

Buhler, R. D., and H. T. Nagamatsu 1952, Condensation of Air Components in Hypersonic Wind Tunnels—Theoretical Calculations and Comparison with Experiment. GAL-CIT memo. no. 13, Pasadena, Calif., Dec. 1, 1952.

Burrows, G. 1975, Evaporation at Low Pressures. *J. Appl. Chem.* vol. 7, no. 7, pp. 375–384.

Campbell, B. A., and F. Bakhtar 1971, Condensation Phenomena in High Speed Flow of Steam. *Proc. Inst. Mech. Eng. (London)* vol. 185, no. 25/71, pp. 185–405.

Carlson, D. J., and R. F. Hoglund 1964, Particle Drag and Heat Transfer in Rocket Nozzles. *AIAA J.* vol. 2, no. 11, pp. 1980–1984.

Carlslaw, H. S., and J. C. Jaeger 1959, *Conduction of Heat in Solids*, 2nd ed. Oxford: Clarendon Press.

Chodes, N., J. Warner, and A. Gagin 1974, A Determination of the Condensation Coefficient of Water from the Growth Rate of Small Cloud Droplets. *J. Atmos. Sci.* vol. 31, p. 1351.

Cooper, F. 1977, Heat Transfer from a Sphere to an Infinite Medium. *Int. J. Heat Mass Transfer* vol. 20, pp. 991–993.

Crowe, C. T., W. R. Babcock, P. G. Willoughby 1971, Drag Coefficient for Particles in Rarefied, Low-Mach-Number Flows. International Symposium on Two-Phase Systems, Haifa, Aug.–Sept. 1971, Paper No. 3–3.

Cunningham, E. 1910, On the Velocity of Steady Fall of Spherical Particles Through Fluid Medium. *Proc. Roy. Soc., Series A* vol. 83, p. 357.

Daum, F. L., and G. Gyarmathy 1968, Condensation of Air and Nitrogen in Hypersonic Wind Tunnels. *AIAA J* vol. 6 (1968), no. 3, pp. 458–465.

Dibelius, G., and H. Voss 1979, Mischkondensation. Final Report on Research Contract DI 13/21 (15,17), Technical University of Aachen.

Dzung, L. S. 1944, Beiträge zur Thermodynamik der realen Gase. *Schweiz. Archiv* vol. 10, pp. 305–313.

Dzung, L. S. 1953, The Medium in Fluid Mechanics. *J. Aeronaut. Sci.* vol. 20, no. 9, pp. 650–651.

Dzung, L. S. 1955, Thermostatische Zustandsänderungen des trockenen und nassen Dampfes. *Z. Angew. Math. Mech.* vol. 6, pp. 207–223.

Einstein, A. 1911, Bemerkung zu dem Gesetz von Eötvös. *Ann. Phys. 4. F.* vol. 34, no. 4, pp. 165–169.

Eötvös, L. 1866, Ueber den Zusammenhang der Oberflächenspannung der Flüssigkeiten mit ihrem Molekularvorkommen. *Ann. Phys. Chem., Neue Folge* vol. 27, pp. 448–459.

Fedoseyeva, N. V., and V. E. Glushkov 1973, Computation of the Evaporation Rate of Liquid Droplets. *Adv. Aerosol Phys. (USSR)* vol. 6, pp. 36–40.

von Freudenreich, J. 1927, Der schädliche Einfluss der Dampfnässe in Dampfturbinen. *Brown Boveri Rev.* vol. 14, p. 119.

Frössling, N. 1938, Ueber die Verdunstung fallender Tropfen. *Gerlands Beiträge Geophy.* vol. 52, pp. 170–215.

Fuks, N. A. 1959, *Evaporation and Droplet Growth in Gaseous Media.* New York: Pergamon.

Fukuta, N., and J. A. Armstrong 1974, A New Method for Precision Measurements of the Deposition Coefficient of Water Vapor onto Ice. *Proceedings of the 9th International Symposium on Rarified Gas Dynamics, Göttingen* vol. 2, Paper C5, DFVLR Press.

Fukuta, N., and L. A. Walter 1970, Kinetics of Hydrometeor Growth from a Vapor-Spherical Model. *J. Atmos. Sci.* vol. 27, no. 8, pp. 1160–1172.

Gardner, G. C. 1964, Evaporation of Drops: Some Factors Limiting the Rate of Evaporation. CEGB (London), Laboratory Memorandum No. RD/L/M70.

Gibbs, J. W. 1878, On the Equilibrium of Heterogeneous Substances. *The Collected Works of J. Willard Gibbs, Vol. 1.* New Haven, Conn.: Yale University Press.

Gollub, J. P., I. Chabay, and W. H. Flygare 1974, Laser Heterodyne Study of Water Droplet Growth. *J. Chem. Phys.* vol. 61, no. 5, p. 2139.

Golovin, M. N., and A. A. Putnam 1962, Inertial Impaction on Single Elements. *Ind. Eng. Chem. Fundamentals* vol. 1, no. 1, pp. 264–273.

Grassmann, P. 1961, *Physikalische Grundlagen der Chemie-Ingenieur-Technik.* Aarau: Sauerländer.

Grassmann, P., F. Widmer, and K. Ritsonis 1965, Günstigste Tropfengrösse bei der Luftbefeuchtung. *Schweiz. Blätter für Heizung und Lüftung* no. 4/1965, pp. 1–8.

Gyarmathy, G. 1962, Grundlagen einer Theorie der Nassdampfturbine. Swiss Federal Institute of Technology, Doctorial Thesis No. 3221, Juris-Verlag, Zürich. English Translation: CEGB (London) T-781, and USAF-FTT (Dayton, Ohio) TT-63-785.

Gyarmathy, G. 1963, Zur Wachstunsgeschwindigkeit kleiner Flüssigkeitstropfen in einer übersättigen Atmosphäre, *ZAMP* vol. 14, no. 3, pp. 280–293.

Gyarmathy, G. 1976, Condensation in Flowing Steam. *Two-Phase Steam Flow in Turbines and Separators*, M. J. Moore and C. H. Sieverding, Eds. Washington, D.C.: Hemisphere, Chap. 3, pp. 127–189.

Gyarmathy, G., and H. Meyer 1965, Spontane Kondensation. *Ver. Dtsch. Ing.*, Forschungsheft no. 508, Düsseldorf: VDI-Verlag.

Hamielec, A. E., and T. W. Hoffmann 1967, Numerical Solution of the Navier-Stokes Equation for Flow Past Spheres, Part I: Viscous Flow Around Spheres with and without Radial Mass Efflux. *AIChE J.* vol. 13, no. 2, pp. 212–219.

Hässler, G. 1972, Untersuchung zur Zerstörung von Wassertropfen durch aerodynamische Kräfte. *Forsch. Ing.-Wes.* vol. 38, no. 6, pp. 183–192.

Hayward, G. L., and D. C. T. Pei, 1978, Local Heat Transfer from a Single Sphere to a Turbulent Air Stream. *Int. J. Heat Mass Transfer* vol. 21, no. 1, pp. 35–41.

Heinemann, M. 1948, Theory of Drag in Highly Rarefied Gases. *Commun. Pure Appl. Math.* vol. 1, pp. 259–273.

von Helmholtz, R. 1886, Untersuchungen über Dämpfe und Nebel, besonders über solche von Lösungen. *Wied. Ann.* vol. 27, pp. 508–543.

Henderson, C. B. 1976, Drag Coefficients of Spheres in Continuum and Rarefied Flows. *AIAA J.* vol. 14, no. 6, pp. 707–708. For comments by M. J. Walsh (1977), see *AIAA J.* vol. 15, no. 6, pp. 893–895.

Hertz, H. 1882, Ueber die Verdunstung der Flüssigkeiten, insbesondere des Quecksilbers, im luftleeren Raume. *Ann. Phys. Chem.* vol. 17, no. 10, pp. 177–193.

Hill, P. G. 1966, Condensation of Water Vapor During Supersonic Expansion in Nozzles. *J. Fluid Mech.* vol. 25, part 3, pp. 593–620.

Hill, P. G., H. Witting, and E. P. Demetri 1963, Condensation of Metal Vapors During Rapid Expansion. *Trans. ASME, J. Heat Transfer* vol. 85, pp. 303–317.

Hinze, J. O. 1949, *Turbulence.* New York: McGraw-Hill.

Hirschfelder, J. O., C. F. Curtiss, and R. B. Bird, 1954, *Molecular Theory of Gases and Liquids.* New York: Wiley.

Hobson, J. P. 1973, Cryopumping. *J. Vac. Sci. Technol.* vol. 10, no. 1, pp. 73–79.

Hubbard, G. L., V. E. Denny, and A. F. Mills 1975, Droplet Evaporation: Effects of Transients and Variable Properties. *Int. J. Heat Mass Transfer* vol. 18, pp. 1003–1008.

Ingebo, R. D. 1951, Vaporization Rates and Heat-Transfer Coefficients for Pure-Liquid Drops. NASA TN 2368, Washington, July 1951.

Jaeschke, M., W. J. Hiller, and G. E. A. Meier 1975, Acoustic Damping in a Gas Mixture with Suspended Submicroscopic Droplets. *J. Sound Vibration* vol. 43, no. 3, pp. 467–481.

Kang, S. -W. 1967, Analysis of Condensation Droplet Growth in Rarefield and Continuum Environments. *AIAA J.* vol. 5, no. 7, pp. 1288–1295.

Kassoy, D. R., T. C. Adamson, and A. F. Messiter 1966, Compressible Low Reynolds Number Flow Around a Sphere. *Phys. Fluids* vol. 9, no. 4, pp. 671–681.

Kassoy, D. R., and F. A. Williams 1968, Variable-Property Effects on Liquid Droplet Combustion. *AIAA J.* vol. 6, no. 10, pp. 1961–1965.

Kavanau, L. L. 1955, Heat Transfer from Spheres to a Rarefied Gas in Subsonic Flow. *Trans. ASME, J. Heat Transfer* vol. 77, pp. 617–623.

Kelvin, Lord: See Thomson, Sir W.

Kemp, N. H. 1979, Comment on Hypersonic Molecular Heating of Micron Size Particulate. *AIAA J.* vol. 17, no. 4, pp. 445–446.

Kent, J. C. 1973, Quasi-steady Diffusion-Controlled Droplet Evaporation and Condensation. *Appl. Sci. Res.* vol. 28, pp. 315–360.

Knudsen, M. 1911, Die Molekulare Wärmeleitung der Gase und der Akkommodationskoeffizient. *Ann. Phys., 4. Folge* vol. 34, no. 4, pp. 593–656.

Knudsen, M. 1915, Die maximale Verdampfungsgeschwindigkeit des Quecksilbers. *Ann. Phys.* vol. 47, pp. 697–708.

Konorski, A. 1968, The Calculation of Internal Mass and Energy Transport in a Two-Phase Medium Containing Large Drops. (in Polish). *Proc. Inst. Fluid-Flow Machinery, Polish Acad. Sci.* no. 39, pp. 27–60.

Konorski, A. 1969, Dynamik der Zustandsänderungen im Nassdampf. *Proc. Inst. Fluid-Flow Machinery, Polish Acad. Sci.* no. 42–44, pp. 479–501.

Konorski, A. 1976, Thermodynamics of Rapid Non-Equilibrium Expansion of Two-Phase Media. *Proc. Inst. Fluid-Flow Machinery, Polish Acad. Sci.* no. 70–72, pp. 387–418.

Konorski, A. 1977, Non-Equilibrium Phase Change and Heat Transfer Processes in Two-Phase Media Flows. *Proc. Inst. Mech. Eng. London,* pp. 245–250.

Kramers, H. 1946, Heat Transfer from Spheres to Flowing Media. *Physica* vol. 12, no. 2–3, pp. 61–80.

Kreitmeier, F., W. Schlachter, and J. Smutný 1980, Strömungsuntersuchungen in einer Niederdruck-Modellturbine zur Bestimmung der Nässeverluste. *Ver. Dtsch. Ing. Bericht* no. 361, pp.

Kulić, E., and E. Rhodes 1977, Direct Contact Condensation from Air-steam Mixtures on a Single Droplet. *Can. J. Chem. Eng.* vol. 55, pp. 131–137.

Kulić, E., and E. Rhodes 1978, Heat Transfer Rates to Moving Droplet in Air/Steam Mixtures. *Proc. International Heat Transfer Conference Toronto, Aug. 1978.*

Kulić, E., E. Rhodes, and G. Sullivan 1975, Heat Transfer Rate Predictions in Condensation on Droplets from Air-Steam Mixtures. *Can. J. Chem. Eng.* vol. 53, pp. 252–258.

Landolt-Börnstein, 1960, *Zahlenwerte und Funktionen aus Physik, Chemie, Astronomie, Geophysik und Technik.* 6th ed. Berlin: Springer

Langmuir, I. 1913, The Vapor Pressure of Metallic Tungsten. *Phys. Rev., 2nd Ser.* vol. 2, no. 5, pp. 329–342.

Langmuir, I. 1916, The Evaporation, Condensation and Reflection of Molecules and the Mechanism of Adsorption. *Phys. Rev.* vol. 8, no. 4, pp. 149–176.

Langmuir, I. 1932, Vapor Pressures, Evaporation, Condensation and Adsorption. *J. Am. Chem. Soc.* vol. 54, pp. 2798–2832.

Lebedev, O. N. and O. P. Solonenko 1976, Numerical Investigation of Some Parameters in Homogeneous Two-Phase Flows (in Russian). *Izvest. Sibir. Otdel. Akad. Nauk, Ser. Tekh. Nauk,* vol. 13, no. 3, pp. 66–75.

Lee, K., and H. Barrow 1968, Transport Processes in Flow Around a Sphere with Particular Reference to the Transfer of Mass. *Int. J. Heat Mass Transfer,* vol. 11, pp. 1013–1026.

Lee, K., and D. J. Ryley 1968, Evaporation of Water Droplets in Superheated Steam. *ASME Paper 68-HT-11.*

Lenard, P. 1904, Ueber Regen. *Meteorolog.* June 1904, pp. 249–262.

Mason, B. J. 1951, Spontaneous Condensation of Water Vapor in Expansion Chamber Experiments. *Proc. Roy. Soc. London, Ser. B* vol. 64, pp. 773–779.

Maxwell, J. C. 1877, *The Scientific Papers of James Clark Maxwell, Vol. 2* Cambridge: Cambridge University Press, 1980.

May, S. 1974, Mass, Momentum and Energy Exchange Between a Spherical Particle and Gaseous Medium (in Polish). *Mechan. Teoretyczna i Stosowana (Warsaw)* vol. 3, no. 12, pp. 313–326.

Millikan, R. A. 1911, The Isolation of an Ion, a Precision Measurement of Its Charge, and the Correction of Stokes's Law. *Phys. Rev.* vol. 32, pp. 349–397.

Millikan, R. A. 1923a, Coefficients of Slip in Gases and the Law of Reflection of Molecules from the Surfaces of Solids and Liquids. *Phys. Rev., 2nd Ser.* vol. 21, no. 3, pp. 217–238.

Millikan, R. A. 1923b, The General Law of Fall of a Small Spherical Body Through a Gas, and Its Bearing upon the Nature of Molecular Reflection from Surfaces. *Phys. Rev.* vol. 22, no. 1, pp. 1–23.

Mills, A. F., and R. A. Seban 1967, The Condensation Coefficient of Water. *Int. J. Heat Mass Transfer* vol. 10, pp. 1815–1827.

Miura, K., T. Miura, and S. Ohtani 1977, Heat and Mass Transfer to and from Droplets. *AIChE Symp. Ser.* vol. 73, no. 163, pp. 95–102.

Montlucon, J. 1975, Heat and Mass Transfer in the Vicinity of an Evaporating Droplet. *Int. J. Multiphase Flow* vol. 2, no. 2, pp. 171–182.

Moore, M. J. 1976, Gas Dynamics of Wet Steam and Energy Losses in Wet-Steam Turbines. *Two-Phase Steam Flow in Turbines and Separators,* M. J. Moore and C. H. Sieverding, Eds. Washington, D.C.: Hemisphere, Chap. 2, pp. 59–126.

Moore, M. J., and C. H. Sieverding 1976, *Two-Phase Steam Flow in Turbines and Separators.* Washington, D.C.: Hemisphere.

Moses, C. A., and G. D. Stein 1978, On the Growth of Steam Droplets Formed in a Laval Nozzle Using Both Static Pressure and Light Scattering Measurements. *Trans. ASME, J. Fluids Eng.* vol. 100, no. 9, pp. 311–322.

Narusawa, U., and G. S. Springer 1975, Measurement of the Condensation Coefficient of Mercury by a Molecular Beam Method. *Trans. ASME, J. Heat Transfer* vol. 97, no. 2, pp. 83–87.

Nitsch, J. 1971, Der Wärme- und Stoffaustausch an Tropfen bei hohen Dampfpartialdrücken und Dampfpartialdruckdifferenzen am Beispiel eines Gleichstromeinspritzkühlers. DLR Forschungsbericht 71-100, DFVLR Lampoldshausen.

Ohta, Y., K. Shimoyama and S. Ohigashi 1975, Vaporization and Combustion of Single Liquid Fuel Droplets in a Turbulent Environment. *Bull. JSME* vol. 18, no. 115, pp. 47–56.

Oppenheim, A. K. 1953, Generalized Theory of Convective Heat Transfer in a Free-Molecule Flow. *J. Aeronaut. Sci.* vol. 20, no. 1, pp. 49–58.

Oswatitsch, K. 1942, Kondensationserscheinungen in Ueberschalldüsen. *ZAMM* vol. 22, no. 1, pp. 1–14.

Pattantyús-H., E. 1977, The Characteristic Functions of Multiphase Fluid Systems and the Thermodynamic Interpretation of Vapor Bubble Collapse. *Int. J. Heat Mass Transfer* vol. 20, pp. 1307–1314.

Paul, B. 1962, Compilation of Evaporation Coefficients. *Am. Rocket Soc. J.* vol. 32, no. 9, pp. 1321–1328.

Puzyrewski, R. 1969, *Condensation of Water Vapor in a Laval Nozzle* (in Polish). Warszawa-Poznan: Inst. Masz. Przepl. PAN, 1969.

Puzyrewski, R., and T. Król 1976, Numerical Analysis of the Hertz-Knudsen Model of Condensation upon Small Droplets in Water Vapor. *Proc. Inst. Fluid-Flow Machinery, Polish Acad. Sci.* no. 70–72, pp. 285–307.

Ranz, W. E., and W. R. Marshall 1952, Evaporation from Drops. *Chem. Eng. Prog.* vol. 48, no. 3, pp. 141–146, and no. 4, pp. 173–180.

Rose, M. H. 1964, Drag on an Object in Nearly-Free Molecular Flow. *Phys. Fluids* vol. 7, no. 8, pp. 1262–1269.

Rüdinger, G. 1965, Some Effects of Finite Particle Volume on the Dynamics of Gas-Particle Mixtures. *AIAA J.* vol. 3, No. 7, pp. 1217–1222.

Rüping, G. 1959, Die Theorie der Verdampfung und Verbrennung des einzelnen Kraftstofftropfens. Dtsch. Versuchsanstalt f. Luftfahrt, Report No. 93, Westdeutscher Verlag, Köln, Nov. 1959.

Sangiovanni, J. J. 1978, A Model for the Nonsteady Ignition and Combustion of a Fuel Droplet. *Evaporation-Combustion of Fuels*, J. T. Zung, Ed. Advances in Chemistry Series no. 166, Washington, D.C.: American Chemical Society, papers no. 2, pp. 27–53.

Schaaf, S. A. and P. L. Chambré 1958, Flow of Rarefied Gases. *Fundamentals of Gas Dynamics*, H. W. Emmons, Ed. Princeton, N. J.: Princeton University Press, Sec H.

Schlichting, H. 1965, *Grenzschichttheorie*, 5th ed. Karlsruhe: Braun-Verlag.

Schmidt, E. 1929, Verdunstung und Wärmeübergang. *Gesundheits-Ingenieur* vol. 52, pp. 525–536. Reprinted in *Int. J. Heat Mass Transfer* vol. 19, pp. 3–8.

Schmidt, E. 1969, *Properties of Water and Steam in SI-Units*. Berlin: Springer.

Schrage, R. W. 1953, *A Theoretical Study of Interphase Mass Transfer*. New York: Columbia University Press.

Sherman, F. S. 1963, A Survey of Experimental Results and Methods for the Transition Regime of Rarefied Gas Dynamics. *Rarefied Gas Dynamics* vol. 2, pp. 228–260.

Shyu, R. R., C. S. Chen, et al. 1972, Multi-Component Heavy-Fuel Drop Histories in a High Temperature Flow Field. *Fuel* vol. 51, no. 2, pp. 135–145.

Stodola, A. 1922, *Dampf- und Gasturbinen* 5th ed. Berlin: Springer English edition: "Steam and Gas Turbines", McGraw Hill, New York, 1927.

Stokes, G. G. 1851, On the Effect of the Internal Friction of Fluids on the Motion of Pendulums. *Trans. Cambridge Phil. Soc.* vol. IX, pt. II, pp. 8–106.

Takao, K. 1963, Heat Transfer from a Sphere in a Rarefied Gas. *Rarefied Gas Dynamics* vol. 2, pp. 102–111.

Thomson, J. J. 1888, Evaporation. *The Application of Dynamics to Physics and Chemistry*. Cambridge: Cambridge University Press, chap. 11.

Thomson, Sir W. 1969, On the Equilibrium of Vapor at a Curved Surface of Liquid. *Proc. Roy. Soc. Edinburgh* session 1869–70, vol. 7, pp. 63–68. Also *Phil. Mag.* vol. 42, no. 282, pp. 448–452 (1871).

Todes, O. M., V. A. Fedoseyev, and V. I. Zubkov 1966, Computation of the Rate of Evaporation and Growth of a Droplet (Sphere), Temperature Changes Being Taken into Account (in Russian). *Kolloid. Zh. (USSR)* vol. 28, no. 4, pp. 573–579.

Toei, R., et al. 1966, Evaporation from a Water Drop in the Stream of Steam-Air Mixtures. *Chem. Eng. (Japan)* vol. 4, no. 2, pp. 220–223.

Tolman, R. C. 1949, The Effect of Droplet Size on Surface Tension. *J. Chem. Phys.* vol. 17, no. 3, pp. 333–337.

Torda, T. P. 1973, Evaporation of Drops and Break-Up of Sprays. *Astronaut. Acta.* vol. 18, pp. 383–393.

Torobin, L. B., and W. H. Gauvin 1959, Fundamental Aspects of Solid-Gas Flow, Part I: Introductory Concepts and Idealized Sphere Motion in Viscous Regime. *Canad. J. Chemical Engrg.* vol. 37, pp. 129–141.

Traupel, W. 1952, Zur Dynamik realer Gase. *Forsch. Ing.-Wes.* vol. 18, no. 1, pp. 3–9.

Traupel, W. 1958, *Thermische Turbomaschinen Vol. 1,* 1st ed. Berlin: Springer.

Traupel, W. 1971, *Die Grundlagen der Thermodynamik.* Karlsruhe: Braun.

Walsh, M. J. 1975, Drag Coefficient Equations for Small Particles in High-Speed Flows. *AIAA J.* vol. 13, no. 11, pp. 1526–1528.

Weast, R. C., and S. M. Selby 1966, *Handbook of Chemistry and Physics* 47th ed. Cleveland, Ohio: The Chemical Rubber Co.

Wegener, P. P. 1969, Gasdynamics of Expansion Flows with Condensation, and Homogeneous Nucleation of Water Vapor. *Nonequilibrium Flows, Vol. 1,* P. P. Wegener, Ed., New York: Marcel Dekker, chap. 4.

Wegener, P. P., J. A. Clumpner, and B. J.-C. Wu 1972, Homogeneous Nucleation and Growth of Ethanol Drops in Supersonic Flow. *Phys. Fluids* vol. 15, no. 11, pp. 1869–1867.

Wegener, P. P. and J.-Y. Parlange 1967, Non-Equilibrium Nozzle Flow with Condensation. AGARD Conference Proceedings No. 12: *Recent Advances in Aerothermochemistry,* Vol. 2, I. Glassmann, Ed., pp. 607–634.

Wegener, P. P., and G. D. Stein 1969, Light-Scattering Experiments and Theory of Homogeneous Nucleation in Condensing Supersonic Flow. *12th Symposium (International) on Combustion.* Pittsburgh, Pa.: The Combustion Institute, pp. 1183–1191

Williams, A. 1973, Combustion of Droplets of Liquid Fuels: A Review. Vol. 21, no. 1, pp. 1–31.

Williams, A. L., J. C. Carstens, and J. T. Zung 1978, Drop Interaction in a Spray. *Evaporation-Combustion of Fuels,* J. T. Zung, Ed. Advances in Chemistry Series no. 166, Washington, D.C.: American Chemical Society, paper no. 3, pp. 54–62.

Willis, D. R. 1966, Sphere Drag at High Knudsen Number and Low Mach Number. *Phys. Fluids* vol. 9, pp. 2522–2524.

Wu, B. J.-C. 1972, A Study of Vapor Condensation by Homogeneous Nucleation in Nozzles. Ph.D. Thesis, Yale University, New Haven, Conn.

Yuen, U. C., and L. W. Chen 1976, On Drag of Evaporating Liquid Droplets. *Combustion Sci.* vol. 14, pp. 147–154.

Zierep, J. 1972, *Aehnlichkeitsgesetze und Modellregeln der Strömungslehre*. Karlsruhe: Braun.

Zung, J. T., Ed. 1978, *Evaporation-Combustion of Fuels*. Advances in Chemistry Series no. 166, Washington, D.C.: American Chemical Society.

APPENDIX 1 Summary of Some Elementary Relationships and of Representative Materials Data

A1.1 Carrier-Phase Composition

As is apparent from Sec. 2.1, the composition of the carrier phase can be alternatively characterized

By the component masses m_v, m_g, and m_ℓ in the sample mass m_s

By the mass fractions x_v, x_g, and x_ℓ

By the partial densities $\rho_{v\infty}$ and $\rho_{g\infty}$

By the partial pressures $p_{v\infty}$ and $p_{g\infty}$ (or $\Pi_{v\infty}$ and $\Pi_{g\infty}$)

By the gas/vapor mass ratio (or dilution ratio) g_∞

Here $\Pi_{v\infty}$ and $\Pi_{g\infty}$ are the partial pressure ratios

$$\Pi_{v\infty} \hat{=} \frac{p_{v\infty}}{p_\infty} \tag{A-1a}$$

$$\Pi_{g\infty} \hat{=} \frac{p_{g\infty}}{p_\infty} \tag{A-1b}$$

Using Eqs. (3) to (11), following relationships can be established:

$$g_\infty = \frac{m_g}{m_v} = \frac{x_g}{x_v} = \frac{\rho_{g\infty}}{\rho_{v\infty}} = \frac{R_v}{R_g}\,\frac{\rho_{g\infty}}{\rho_{v\infty}} = \frac{R_v}{R_g}\,\frac{\Pi_{g\infty}}{\Pi_{v\infty}} \tag{A-2}$$

$$\frac{\rho_{v\infty}}{\rho_\infty} = \frac{m_v}{m_v + m_g} = \frac{x_v}{1-x_\ell} = \frac{1}{1 + g_\infty} = \frac{\bar{R}}{R_v}\,\Pi_{v\infty} \tag{A-3}$$

$$\frac{\rho_{g\infty}}{\rho_\infty} = \frac{m_g}{m_v + m_g} = \frac{x_g}{1-x_\ell} = \frac{g_\infty}{1 + g_\infty} = \frac{\bar{R}}{R_g}\,\Pi_{\ \infty} \tag{A-4}$$

$$\Pi_{v\infty} = \frac{R_v\ \rho_{v\infty}}{\bar{R}\ \rho_\infty} = \frac{R_v\ \rho_{v\infty}}{R_v\ \rho_{v\infty} + R_g\ \rho_{g\infty}} = \frac{R_v}{R_v + g_\infty R_g} = \frac{R_v}{\bar{R}}\,\frac{1}{1 + g_\infty} \tag{A-5}$$

$$\Pi_{g\infty} = \frac{R_g\ \rho_{g\infty}}{\bar{R}\ \rho_\infty} = \frac{R_g\ \rho_{g\infty}}{R\ \rho_{v\infty} + R_g\ \rho_{g\infty}} = \frac{g_\infty R_g}{R_v + g_\infty R_g} = \frac{R_g}{\bar{R}}\,\frac{g_\infty}{1 + g_\infty} \tag{A-6}$$

Here $\bar{R}$ is the mixture-average gas constant defined according to Eq. (10).

A1.2 Carrier-Phase Mixture Properties

The single overbar $^{-}$ denotes *averaged* carrier properties $\bar{X}$ obtained by mass averaging the corresponding component properties X_v and X_g [where X may stand for a material property or a group of properties] according to the expression

$$\bar{X} \triangleq \frac{m_v X_v + m_g X_g}{m_v + m_g} = \frac{\rho_{v\infty} X_v + \rho_{g\infty} X_g}{\rho_\infty} = \frac{X_v + g_\infty X_g}{1 + g_\infty} \qquad \text{(A-7)}$$

Such averages include, for example, $\bar{R}$ and $\bar{c}_p$.

The transport properties η, λ, and D of the mixture cannot be obtained accurately by mass averaging. These values should be taken from empirical data or from more elaborate theories (cf. Hirschfelder, et al., 1954).

The specific heat ratio γ of the mixture and of its components is defined as

$$\gamma \triangleq \frac{\bar{c}_p}{\bar{c}_p - \bar{R}} \quad \text{(A-8}a\text{)} \qquad \gamma_v \triangleq \frac{c_{pv}}{c_{pv} - R_v} \quad \text{(A-8}b\text{)} \qquad \gamma_g \triangleq \frac{c_{pg}}{c_{pg} - R_g} \quad \text{(A-8}c\text{)}$$

giving

$$\frac{\bar{R}}{\bar{c}_p} = \frac{\gamma - 1}{\gamma} \qquad \text{(A-9)}$$

By this definition of γ, the speed of sound c_s in the carrier becomes

$$c_s = \sqrt{\gamma \bar{R} T_\infty} \qquad \text{(A-10)}$$

as used in the definition of the Mach number, Eq. (62).

The mean free path ℓ and the Knudsen number Kn = $\ell/2r$ are given various definitions in the literature. According to the kinetic theory of rigid sphere molecule (rsm) gases (Hirschfelder et al., 1954),

$$\ell_{\text{rsm}} \triangleq \frac{16}{5\sqrt{2\pi}} \frac{\eta\sqrt{\bar{R} T_\infty}}{p_\infty} = 0.64\ell \qquad \text{(A-11}a\text{)}$$

where ℓ is the value given by Eq. (38). Other values occurring in the literature are (Millikan, 1923*a* and *b*; Schaaf and Chambré, 1958; Gyarmathy, 1962, 1963) as listed:

$$\ell_{\text{Mill}} \mathrel{\hat{=}} \frac{\sqrt{\pi/8}}{0.3502} \frac{\eta\sqrt{\bar{R}\,T_\infty}}{p_\infty} = 0.90\ell \tag{A-11b}$$

$$\ell_{\text{Sch}} \mathrel{\hat{=}} 2r\,\frac{\text{Ma}}{\text{Re}} = \frac{\ell}{2\sqrt{\gamma}} \tag{A-11c}$$

$$\ell_{\text{Gy}} \mathrel{\hat{=}} \frac{1.5\,\eta\sqrt{\bar{R}\,T_\infty}}{p_\infty} = 0.75\ell \tag{A-11d}$$

Knudsen numbers defined on the basis of these expressions are related to Kn given by Eq. (37) as

$$\text{Kn} = 1.56\text{Kn}_{\text{rsm}} = 1.11\text{Kn}_{\text{Mill}} = \sqrt{4\gamma}\text{Kn}_{\text{Sch}} = 1.33\text{Kn}_{\text{Gy}} \tag{A-12}$$

A1.3 Vapor Pressure, Saturation Ratio, and Subcooling

According to the Clausius-Clapeyron equation, the derivative of the vapor pressure function $p_s(T)$ is related to L, v'' and v' as

$$\frac{dp_s}{dT} = \frac{L}{(v'' - v')T} \mathrel{\hat{=}} \frac{p_s}{T}\,\text{Cl} \tag{A-13}$$

The property group Cl is defined by the above. For most liquids the product $T\cdot\text{Cl}$ is constant over wide ranges of temperature. Setting $T\cdot\text{Cl} = T_{\text{ref}}\cdot\text{Cl}(T_{\text{ref}})$ = constant (and choosing T_{ref} to lie in the domain of the temperatures T_∞, $T_{s\infty}$, and $T_{b\infty}$), Eq. (176) gives

$$\frac{T^2}{p_s}\frac{dp_s}{dT} = T\cdot\text{Cl} = \text{constant} \tag{A-14}$$

By integration between T and T_{ref} (with $T<T_c$ and $T_{\text{ref}}<T_c$), the vapor pressure function $p_s(T)$ is obtained as

$$\ell n\,\frac{p_s(T)}{p_s(T_{\text{ref}})} = \text{Cl}\left(1 - \frac{T_{\text{ref}}}{T}\right) \tag{A-15}$$

where Cl has to be evaluated at T_{ref}. Because of this equation, characteristic vapor pressure and temperature values of the two-phase fluid are related to each other in the following way:

The vapor partial pressure $p_{v\infty}$ and the saturation temperature $T_{s\infty}$ as

$$\ell n\,\frac{p_{v\infty}}{p_s(T_{\text{ref}})} = \text{Cl}\left(1 - \frac{T_{\text{ref}}}{T_{s\infty}}\right) \tag{A-16a}$$

The overall pressure p_∞ and the boiling-point temperature $T_{b\infty}$ as

$$\ell n \frac{p_\infty}{p_s(T_{ref})} = \mathrm{Cl}\left(1 - \frac{T_{ref}}{T_{b\infty}}\right) \tag{A-16b}$$

The saturation pressure $p_{s\infty}$ and the carrier temperature T_∞ as

$$\ell n \frac{p_{s\infty}}{p_s(T_{ref})} = \mathrm{Cl}\left(1 - \frac{T_{ref}}{T_\infty}\right) \tag{A-16c}$$

In pure vapor carriers ($p_\infty = p_{v\infty}$), $T_{s\infty}$, and $T_{b\infty}$ are obviously identical. For mixture carriers see Eq. (A-21).

The saturation ratio S_∞ is linked to the subcooling

$$\Delta T_\infty \triangleq T_{s\infty} - T_\infty \tag{A-17}$$

according to Eqs. (179), (A-16a), and (A-16c) as

$$\ell n\, S_\infty \triangleq \ell n \frac{p_{v\infty}}{p_{s\infty}} = \mathrm{Cl}\left(\frac{T_{ref}}{T_\infty} - \frac{T_{ref}}{T_{s\infty}}\right) = \frac{\mathrm{Cl}\; T_{ref}}{T_\infty\, T_{s\infty}}\, \Delta T_\infty \tag{A-18}$$

If $T_{ref} = T_\infty$ is set, this gives

$$\ell n\, S_\infty = \mathrm{Cl}\, \frac{\Delta T_\infty}{T_{s\infty}} \tag{A-19}$$

The subcooling is seen to be roughly proportional to $\ell n\, S_\infty$.

The time gradient of $T_{b\infty}$ is linked by Eq. (A16b) to the expansion rate $\dot{P}$ as

$$\frac{1}{T_{b\infty}}\frac{dT_{b\infty}}{dt} = \frac{\dot{P}}{\mathrm{Cl}} = \frac{\pm 1}{\mathrm{Cl}\;\Delta t_{exp}} \qquad \begin{pmatrix} + \text{ compression} \\ - \text{ expansion} \end{pmatrix} \tag{A-20}$$

The ratio of the boiling and saturation points can be expressed using Eqs. (A-16a), (A-16b), and (A-5), by mixture composition (e.g., $\Pi_{v\infty}$ or g_∞) as

$$\frac{T_{b\infty}}{T_{s\infty}} = 1 - \frac{T_{b\infty}}{\mathrm{Cl}\; T_{ref}}\, \ell n\, \Pi_{v\infty} = 1 - \frac{T_{b\infty}}{\mathrm{Cl}\; T_{ref}}\, \ell n \left(\frac{R_v/\bar{R}}{1 + g_\infty}\right) \tag{A-21}$$

A1.4 Representative Materials Data

In order to illustrate the order of magnitude of the dimensional and nondimensional material property data occurring in the equations used, some representative data are shown in Table A-1. Diffusion properties of some vapors in atmospheric pressure air are shown in Table A-2. The data were taken from standard reference books, for example, Landolt-Börnstein (1960), Weast and Selby (1966), Schmidt (1969).

Table A-1. *Representative Material Property Data*

Quantity	Unit	Air		Water/Steam			Methyl Alcohol		Heptane	Nitrogen
R	J/kg K	287		462			260		83	297
T_c	K	—		647.3			512.6		540.0	126.0
State: T	K	300	1000	300	373	523	313	393	373	65
	(°C)	(27)	(727)	(27)	(100)	(250)	(40)	(120)	(100)	(-208)
p_∞	Pa	1×10^{5}	1×10^{5}	0.0356×10^{5}	1×10^{5}	40×10^{5}	0.336×10^{5}	6.129×10^{5}	1.06×10^{5}	0.175×10^{5}
ρ_∞	kg/m^3	1.16	0.348	0.0258	0.598	20.0	0.445	7.14	3.6	0.89
c_p	J/kg K	1.00×10^{3}	1.14×10^{3}	1.87×10^{3}	2.03×10^{3}	3.77×10^{3}	1.42×10^{3}	1.59×10^{3}	2.03×10^{3}	1.04×10^{3}
γ	—	1.40	1.34	1.33	1.30	1.14	1.22	1.20	1.043	1.40
K	—	3.00	3.44	3.53	3.83	7.64	5.05	5.50	23.76	3.00
η	kg/s m	18.3×10^{-6}	41.8×10^{-6}	9.1×10^{-6}	12.0×10^{-6}	17.5×10^{-6}	10.3×10^{-6}	12.9×10^{-6}	7.2×10^{-6}	4.6×10^{-6}
λ[a]	J/K s m	0.0262	0.0675	0.0197	0.0248	0.0484	0.0183	0.0289	0.0176	0.0059
D	m^2/s	b	b	—	—	—	—	—	—	—
a	m^2/s	2.25×10^{-5}	1.70×10^{-4}	4.09×10^{-4}	2.05×10^{-5}	6.4×10^{-7}	2.9×10^{-5}	2.55×10^{-6}	2.4×10^{-6}	6.4×10^{-5}
ℓ	m	1.07×10^{-7}	4.5×10^{-7}	1.9×10^{-6}	0.99×10^{-7}	4.30×10^{-9}	1.77×10^{-7}	1.34×10^{-8}	2.4×10^{-8}	7.3×10^{-8}
p_∞/η	s-1	5.5×10^{9}	2.4×10^{9}	3.9×10^{8}	8.3×10^{9}	2.3×10^{11}	3.3×10^{9}	4.8×10^{10}	1.5×10^{10}	3.8×10^{9}
c_s	m/s	347	619	429	472	524	320	349	180	164
ρ_ℓ	kg/m^3			997	958	799	775	691	612	862
c_ℓ	J/kg K			4.2×10^{3}	4.2×10^{3}	4.9×10^{3}	2.7×10^{3}	?	2.7×10^{3}	2.0×10^{3}
L	J/kg			2.4×10^{6}	2.3×10^{6}	1.7×10^{6}	1.1×10^{6}	0.97×10^{6}	0.31×10^{6}	0.217×10^{6}
$p_s(v''-v')$	m^2/s^2			1.4×10^{5}	1.7×10^{5}	2.0×10^{5}	0.76×10^{5}	0.85×10^{5}	0.29×10^{5}	0.195×10^{5}
$\lambda_v/\rho_\ell R_v$	m^2/s			0.43×10^{-7}	0.56×10^{-7}	1.31×10^{-7}	0.91×10^{-7}	1.61×10^{-7}	3.46×10^{-7}	0.23×10^{-7}
σ	N/m			0.072	0.059	0.0262	0.021	0.0125	0.0125	0.012
λ_ℓ	J/K s m			0.61	0.68	0.62	0.21	0.20	0.16	0.19
a_ℓ	m^2/s			1.47×10^{-7}	1.69×10^{-7}	1.59×10^{-7}	1.01×10^{-7}	?	0.99×10^{-7}	1.1×10^{-7}
$\eta''\sqrt{p_s(v''-v')}$	N/m			3.4×10^{-3}	4.9×10^{-3}	7.7×10^{-3}	2.8×10^{-3}	3.8×10^{-3}	1.23×10^{-3}	0.64×10^{-3}
$\sigma/\rho_\ell R_v T$	m			5.2×10^{-10}	3.56×10^{-10}	1.36×10^{-10}	3.2×10^{-10}	1.77×10^{-10}	6.6×10^{-10}	7.2×10^{-10}
$Cl = L/[p_s(v''-v')]$				17.6	13.5	8.8	15.2	11.4	10.6	11.1
$E\ddot{o} = \sigma/\eta''\sqrt{p_s(v''-v')}$				21.2	12.0	3.40	7.4	3.29	10.2	18.6
$Pr = c_p\,\eta/\lambda$		0.70	0.71	0.87	0.98	1.36	0.80	0.71	0.83	0.81
$Sc = \eta/\rho_\infty D$		b	b	—	—	—	—	—	—	—

[a]In the gaseous (vapor) state.

[b]see Table A-2.

Table A-2. *Diffusion Properties of Dilute Vapors in Air at Atmospheric Pressure* ($p_\infty = 10^5$ Pa)

Vapor Substance	Air Temp. T (K)	Mass Diffusivity		Schmidt No. Sc
		Common D (m^2/s)	Modified $\mathcal{D}$ (kg/s m)	
Water	300	3.1×10^{-5}	2.24×10^{-5}	0.51
	373	4.6×10^{-5}	2.66×10^{-5}	0.49
	1000	27.4×10^{-5}	5.94×10^{-5}	0.32
Methyl alcohol	313	1.37×10^{-5}	1.68×10^{-5}	1.27
Heptane	373	1.06×10^{-5}	3.42×10^{-5}	2.16

One gas (room temperature or hot air) and four condensible substances (water, methyl alcohol, heptane, and liquid nitrogen) are shown. For the first two vapors, several temperature values T are considered. The vapors are always assumed to be saturated, that is, to be at a pressure $p_\infty = p_{v\infty} = p_s(T)$.

The lowest four lines of Table A-1 show the dimensionless material property groups Cl, Eö, Pr, and Sc. The essential conclusion from these data is that despite the great differences in the substances and pressure and temperature ranges in question, the nondimensional properties are, for all cases considered, of very similar magnitude.

APPENDIX 2 **Balance Equations for the Two-Phase Fluid Sample**

A2.1 Definitions

The definition of momentum and heat transfer rates between the gaseous (carrier) phase and the droplet is not trivial if mass transfer is present, because the mass exchanged at the interface has, in the general case of noncontinuum carriers, a temperature and velocity that are different from those of each phase. Starting out from the mass, momentum, and energy balance equations of a fluid sample comprising the drop and the amount of carrier fluid that is momentarily contained in the concentric sample domain (of radius r_∞), the aim of the following analysis is to find convenient definitions for the three exchange rates M, F, and H, and to formulate the balance equations of each phase in a form that is compatible with these definitions. The processes will be observed from a coordinate frame fixed to the droplet. The assumptions listed in Table 1 are made.

Figure A-1 shows the sample domain. Forces are shown by black arrows, whereas white arrows indicate the (net) fluxes affecting the two phases. The outer boundary $\mathcal{s}$ is traversed by a carrier flow of velocity $-\Delta\vec{w}_\infty$. The radii r and r_∞ may vary with time. The force components parallel to $-\Delta\vec{w}_\infty$ are designated by the subcript F.

The surface $\mathcal{s}$ encloses by definition a constant amount of total mass. However, $\mathcal{s}$ may be traversed by a vapor mass flux M_{ext} and an equal flux of opposing sense (dotted arrow in Fig. A-1) of the noncondensing gas. These fluxes are a result of diffusion and/or convection in the carrier. $\vec{F}_{ext}$ is the resultant of all external (viscous and convective) forces acting on the outer boundary in addition to pressure. The body force $\vec{b}$ is the resultant of gravitational and inertial force fields. The surface $\mathcal{s}$ is assumed to be diabatic, with H_{ext} being the net heat flow rate entering it from the outside by conduction and/or convection.

At this point the assumptions of far-field uniformity, cylindrical symmetry and uniform pressure (IIb, IIc, and IId in Table 1) have not yet been introduced, because IIb and IIc would imply that $M_{ext} = H_{ext} = 0$ and $\vec{F}_{ext} = 0$, and IId would imply that grad $p_\infty = 0$. These simplications would unnecessarily restrict the validity of the balance equations. (These assumptions are required, however, to

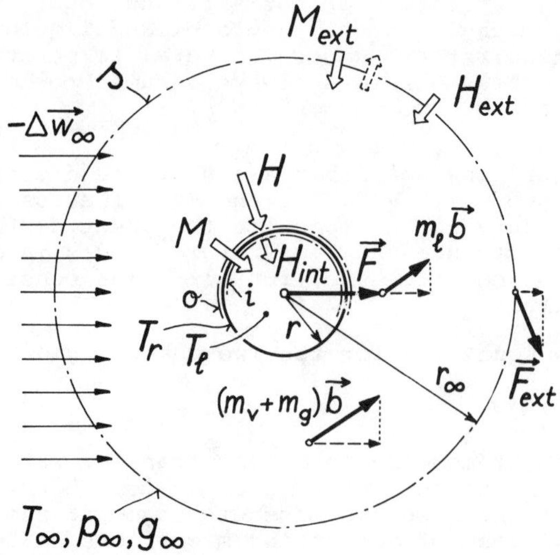

Fig. A-1. Definition of forces and fluxes in the fluid sample.

derive the expressions presented in Sec. 3 for the fluxes M, H, and F between the carrier and the drop.) The values of M_{ext}, H_{ext} and $\vec{F}_{ext}$ are supposed to be known from the equations determining the macroscopic flow field in which the sample is entrained. Their determination is beyond the scope of the present study. Except in shear layers and near flow boundaries, they may usually be set equal to zero.

The interface of the phases, that is, the surface of the drop, will be bounded by an outer and an inner surface (o and i in Fig. A-1), both lying infinitesimally close to it. Between these surfaces the mass flow M condensing on (or evaporating from) the drop is considered to change from vapor to liquid (or vice versa). No mass or momentum is stored or liberated between these surfaces, however. The momentum flux traversing o (and i) is identical to the drag force $\vec{F}$.

The heat flux entering o from the carrier side will be denoted by H, and the heat flux entering i (i.e., the interior of the drop) by H_{int}. Because of the liberation of latent heat between o and i, H and H_{int} will have drastically different values. The pressure has the value p_∞ at the surface o, and $(p_\infty + 2\sigma/r)$ at the surface i. The conservation laws will be formulated first for the overall sample. Each of the resulting three balance equations can then be separated into a balance equation of the droplet interior, of the interface, and of the carrier. Two of the nine balance equations (those for the mass and momentum at the interface) are trivial because, in contrast to energy, no mass or momentum is being stored or liberated during phase change. Therefore the exchange of mass and momentum will be treated as occurring directly between the carrier and the droplet interior.

A2.2 Mass Balance

The sample contains the amount $m_\ell = x_\ell\, m_s$ of liquid, $m_v = x_v m_s$ of vapor, and $m_g = x_g m_s$ of gas. The total mass m_s being constant by definition,

$$\frac{dm_s}{dt} = \frac{d}{dt}(m_\ell + m_v + m_g) = 0 \tag{A-22}$$

Introducing the net mass fluxes,

$$M \triangleq \frac{dm_\ell}{dt} \tag{A-23}$$

and

$$M_{\text{ext}} \triangleq \frac{d}{dt}(m_\ell + m_v) \tag{A-24}$$

where the latter accounts for net diffusion/convection of vapor across s, we may write

$$\frac{dm_v}{dt} = M_{\text{ext}} - M \tag{A-25}$$

$$\frac{dm_g}{dt} = -\,M_{\text{ext}} \tag{A-26}$$

$$\frac{d(m_v + m_g)}{dt} = -\,M \tag{A-27}$$

Equation (A-23) is the mass balance of the droplet, cf. Eq. (16), whereas Eq. (A-27) is the mass balance of the carrier. This latter can be written in terms of the liquid mass fraction x_ℓ as

$$\frac{1}{x_\ell}\frac{dx_\ell}{dt} = \frac{M}{m_\ell} \tag{A-28}$$

A2.3 Momentum Balance

The momentum balance of the sample will be formulated in vectorial form. The change of momentum is caused by the resultant of pressure forces, by the body-force field, and by the resultant external force:

$$\frac{d}{dt}\left[m_\ell\,\vec{w}_\ell + (m_v + m_g)\vec{w}_{\text{carr}}\right] = -\,V_s \text{ grad } p_\infty + m_s\vec{b} + \vec{F}_{\text{ext}} \tag{A-29}$$

Here

$$V_s \triangleq \frac{m_\ell}{\rho_\ell} + \frac{m_v + m_g}{\rho_\infty} \tag{A-30}$$

is the volume of the sample. It has been assumed that V_s is small enough to treat grad p_∞ as constant over the entire sample boundary. $\vec{w}_{carr}$ is the mass-averaged velocity of the carrier phase. $\vec{b}$ is assumed constant over V_s. Differentiation yields, if Eqs. (A-23) to (A-27) and (A-30) are taken into account.

$$m_\ell\left(\frac{d\vec{w}_\ell}{dt} + \frac{\text{grad } p_\infty}{\rho_\ell} - \vec{b}\right) + (m_v+m_g)\left(\frac{d\vec{w}_{carr}}{dt} + \frac{\text{grad } p_\infty}{\rho_\infty} - \vec{b}\right) - \vec{F}_{ext} + (\vec{w}_\ell - \vec{w}_{carr})M = 0 \quad \text{(A-31)}$$

In order to define the drag force $\vec{F}$ acting between the phases, one must decide which terms of the above overall momentum balance will be associated with the drop and which with the carrier phase. Obviously, the first term concerns the drop only, whereas the second and third terms refer to the carrier. The last term, however, involves the mass exchange rate M and the velocity of both phases; this term could be assigned to the drop or to the carrier or subdivided between the two. In cases when the adaptation of the velocity of the exchanged mass occurs outside the drop (as in all cases of vaporization and in condensation in viscous, continuum-type flow), the exchange term affects the carrier phase alone. If, however, the adaptation occurs within or on the drop (as in case of condensation in free-molecular or idealized (nonviscous) continuum flow), the term should figure in the momentum balance of the drop.

In the present study the interaction terms will be assigned wholly to the carrier phase, regardless of whether condensation or evaporation occurs. This convention is adopted because it is realistic in a greater variety of physical situations and does not violate the assumption of zero flow within the drop (assumption If). It implies that the mass exchanged at the interface is considered to have a momentum corresponding to $\vec{w}_\ell$. In cases where this is not true, the excess momentum of the mass exchanged will have to be accounted for in the expression of the drag force $\vec{F}$. (Such corrections are discussed briefly in Sec. 3.2.)

As a result of this convention, the drag force $\vec{F}$ is defined as

$$\vec{F} \triangleq m_\ell \left(\frac{d\vec{w}_\ell}{dt} + \frac{\text{grad } p_\infty}{\rho_\infty} - \vec{b}\right) \quad \text{(A-32)}$$

and the *momentum balance of the drop* follows as

$$m_\ell \frac{d\vec{w}_\ell}{dt} = \vec{F} + m_\ell \vec{b} - \frac{m_\ell}{\rho_\ell} \text{grad } p_\infty \quad \text{(A-33)}$$

giving Eq. (17). This equation is discussed further in Sec. 2.2.

The *momentum balance of the carrier phase* follows from Eqs. (A-31) and (A-32) as

$$(m_v + m_g) \frac{d\vec{w}_{carr}}{dt} = (m_v + m_g)\left(- \frac{\text{grad } p_\infty}{\rho_\infty} + \vec{b}\right) + \vec{F}_{ext} - \vec{F} - M\,\Delta\vec{w}_\infty \tag{A-34}$$

or, using the liquid mass fraction x_ℓ, as

$$\frac{d\vec{w}_{carr}}{dt} = - \frac{1}{\rho_\infty} \text{ grad } p_\infty + \vec{b} + \frac{x_\ell}{1-x_\ell}\left(\frac{\vec{F}_{ext}}{m_\ell} - \frac{\vec{F}}{m_\ell} - \frac{M}{m_\ell}\,\Delta\vec{w}_\infty\right) \tag{A-35}$$

In many technical situations $\vec{b}$ and x_ℓ are small enough to make the last two terms negligible. Then the acceleration of the carrier is substantially dictated by the pressure gradient alone:

$$\frac{d\vec{w}_{carr}}{dt} \approx - \frac{1}{\rho_\infty} \text{ grad } p_\infty \tag{A-36}$$

Since the product $(\vec{w}_\ell \cdot \text{grad } p_\infty)$ is obviously the rate of pressure change in the sample, we may write

$$\vec{w}_\ell \frac{d\vec{w}_{carr}}{dt} = - \frac{1}{\rho_\infty} \frac{dp_\infty}{dt} = \bar{R}\, T_\infty\, \dot{P} \tag{A-37}$$

The last equality follows from Eqs. (11) and (49). If the flow is rectilinear and $\vec{w}_\ell \approx \vec{w}_{carr}$, this equation yields the last member of Eq. (57).

A2.4 Energy Balance

According to the first law of thermodynamics, the change of the energy U_s of the fluid sample is given as

$$\frac{dU_s}{dt} = - p_\infty \frac{dV_s}{dt} + H_{ext} + W_{diss} \tag{A-38}$$

where H_{ext} is the net heat flux entering the sample from the exterior (and is supposed to be known) and W_{diss} is the work dissipated within the sample.

The energy U_s of the sample can be expressed by enthalpies, pressures, and volumes as

$$U_s \triangleq m_\ell h_\ell + m_v h_{v\infty} + m_g h_{g\infty} - \left(p_\infty + \frac{2\sigma}{r}\right) \frac{m_\ell}{\rho_\ell} - p_\infty \frac{m_v + m_g}{\rho_\infty} \tag{A-39}$$

where $h_{v\infty}$ and $h_{g\infty}$ are the specific enthalpies of the vapor and gas, respectively. In h_ℓ, the specific enthalpy of the "bulk," compressible liquid, surface effects are not included. Note that $(p_\infty + 2\sigma/r)$ is the pressure in the interior of the droplet. Equation

(A-39) involves the approximation that the carrier phase has everywhere the specific enthalpy corresponding to the far-field conditions. We have for incompressible liquid and perfect gases of constant specific heat,

$$
\begin{aligned}
dh_\ell &= c_\ell \, dT_\ell \\
dh_{v\infty} &= c_{pv} \, d\bar{T} \approx c_{pv} \, dT_\infty \\
dh_{g\infty} &= c_{pg} \, d\bar{T} \approx c_{pg} \, dT_\infty
\end{aligned}
\tag{A-40}
$$

Here $\bar{T}$ is the caloric mean carrier temperature, cf. Eq. (152). From Eq. (A-39) we obtain, if Eqs. (2), (A-23), (A-25), (A-26), and (A-30) are used, if the surface tension σ is assumed to be independent of the droplet radius r (assumption Ie), and if ρ_ℓ is set constant (assumption Ic), after differentiation,

$$
\begin{aligned}
\frac{dU_s}{dt} = {} & m_\ell c_\ell \frac{dT_\ell}{dt} + (m_v c_{pv} + m_g c_{pg}) \frac{dT_\infty}{dt} - (h_{v\infty} - h_\ell)\, M + (h_{v\infty} - h_{g\infty})\, M_{\text{ext}} \\
& - \frac{dp_\infty}{dt} V_s - p_\infty \frac{dV_s}{dt} + \frac{2\sigma}{3\rho_\ell r} M - \frac{2\sigma}{\rho_\ell r} M
\end{aligned}
\tag{A-41}
$$

Here the penultimate term is the energy associated with an increase of the droplet surface and the last term stems from the pressurization of the mass condensed.

If T_r is the temperature of the interface and $L = L(T_r)$ is the enthalpy of vaporization at this temperature, we may write, using the first equation of (A-40),

$$
h_{v\infty} - h_\ell = c_{pv}(T_\infty - T_r) + L + c_\ell (T_r - T_\ell) \tag{A-42}
$$

The dissipated work W_{diss} consists of the drag work $F\Delta w_\infty$, the dissipation of kinetic energy resulting from phase change, $|M|\Delta w_\infty^2/2$, and the work W_{shear} of shear stresses existing within the carrier flow independently of the viscous flow field of the droplet (as in boundary layers):

$$
W_{\text{diss}} = F\Delta w_\infty + \frac{1}{2} \Delta w_\infty^2 \, |M| + W_{\text{shear}} \tag{A-43}
$$

Inserting Eqs. (A-41) and (A-42) into Eq. (A-38), the thermal energy balance of the sample is obtained as

$$
\begin{aligned}
m_\ell c_\ell \frac{dT_\ell}{dt} + (m_v c_{pv} + m_g c_{pg}) \frac{dT_\infty}{dt} = {} & \left(\frac{m_\ell}{\rho_\ell} + \frac{m_v + m_g}{\rho_\infty}\right) \frac{dp_\infty}{dt} \\
& + M \left[c_{pv}(T_\infty - T_r) + c_\ell (T_r - T_\ell) + L - \frac{2\sigma}{3\,\rho_\ell\, r} + \frac{2\sigma}{\rho_\ell\, r} \right] \\
& + W_{\text{diss}} - M_{\text{ext}} (h_{v\infty} - h_{g\infty}) + H_{\text{ext}}
\end{aligned}
\tag{A-44}
$$

On the left side stands the enthalpy change of the droplet and of the carrier. On the right side the first term is the compression work. The second is due to phase change and mass transfer from the far field of the carrier to the interior of the drop. Its brackets comprise the contributions of the temperature drop on the carrier side and within the droplet, of liberation of latent heat, plus the terms associated with surface increase and condensate pressurization. The next term is the dissipative work of the drag force, and the last two terms represent the energy transport due to the mass and heat flux across the outer boundary.

Now this overall thermal energy balance will be broken down into three balance equations: one for the droplet interior, one for the interface, and one for the carrier medium.

The *energy balance of the droplet interior* will be written as

$$m_\ell c_\ell \frac{dT_\ell}{dt} - \frac{m_\ell}{\rho_\ell}\frac{dp_\infty}{dt} - M\, c_\ell (T_r - T_\ell) \mathrel{\hat{=}} H_{\text{int}} \tag{A-45}$$

This equation *defines* the heat flux H_{int} across the inner boundary i. It is seen that the mass M entering the droplet interior is considered to carry an amount of thermal energy corresponding to the temperature difference $T_r - T_\ell$.

We now define the heat transfer rate H across o to the drop surface by writing the *energy balance of the phase interface* as

$$H_{\text{int}} - M\, L_r \mathrel{\hat{=}} H \tag{A-46}$$

where

$$L_r \mathrel{\hat{=}} L\ (T_r) + \frac{4\,\sigma}{3\,\rho_\ell\, r} \tag{A-47}$$

comprises the latent heat of vaporization and the energy terms due to surface tension. L_r may be interpreted as the enthalpy of vaporization at a curved liquid surface of radius r. Using the nondimensional groups Cl and Ke as given by Eqs. (43) and (64), one may write, provided that $T_r \approx T_{\text{ref}}$ may be set with sufficient accuracy,

$$L_r = L\ (T_r)\ \left(1 + \frac{2}{3}\,\frac{\text{Ke}}{\text{Cl}}\right) \tag{A-48}$$

Since Cl is of the order of 10 for most substances and Ke $\ll$ 1 for most droplet sizes of interest (cf. Fig. 6), the capillarity correction is usually small. For example, in the case of small droplets of $r = 10^{-8}$ m and $T_r = 373$ K, one has Ke = 0.07 and Cl = 13.5, and the correction factor is only 1.0035, that is, negligible. The correction is noticeable only for the smallest droplet sizes that occur at all, say (for water) $r = 0.5 \times 10^{-9}$ m, in which cases Ke $\approx$ 1, and the correction amount to 5%. Therefore generally little error is incurred by neglecting the capillarity correction and setting $L_r = L(T_r)$.

With Eqs. (A-45) and (A-46), the *energy balance of the carrier phase* is obtained from Eq. (A-44) as

$$(m_v c_{pv} + m_g c_{pg}) \frac{dT_\infty}{dt} = \frac{m_v + m_g}{\rho_\infty} \frac{dp_\infty}{dt} - H$$

$$+ M c_{pv}(T_\infty - T_r) + W_{\text{diss}} - M_{\text{ext}} (h_{v\infty} - h_{g\infty}) + H_{\text{ext}} \qquad \text{(A-49)}$$

Using the liquid mass fraction x_ℓ and the specific heat $\bar{c}_p$ averaged according to Eq. (A-7), this can be written as

$$\bar{c}_p \frac{dT_\infty}{dt} = \frac{1}{\rho_\infty} \frac{dp_\infty}{dt} - \frac{x_\ell}{1-x_\ell} \frac{H}{m_\ell}$$

$$+ \frac{x_\ell}{m_\ell (1-x_\ell)} \left[M c_{pv}(T_\infty - T_r) + W_{\text{diss}} - M_{\text{ext}} (h_{v\infty} - h_{g\infty}) + H_{\text{ext}} \right] \qquad \text{(A-50)}$$

The first term is the enthalpy drop caused by compression or expansion. It is the predominant term in monophase flow. The second term expresses the liberation or absorption of energy during phase change. The bracketed last term, which is usually small in comparison to the two preceding terms, accounts for the energetic side effects of mass exchange, for dissipation (caused by drag or by the main flow), and for heat conduction/convection within the carrier.

If this last term may be neglected and $H = -ML$ may be set, Eq. (A-50) simplifies to

$$\bar{c}_p \frac{dT_\infty}{dt} = -\bar{R}\, T_\infty \dot{P} + \frac{L}{1-x_\ell} \frac{dx_\ell}{dt} \qquad \text{(A-51)}$$

This is equivalent to Eq. (58) and leads with Eq. (11) to Eq. (59).

A2.5 Comments on Temperature Definitions

It should be noted that *the definition of the heat fluxes* H_{int} and H by Eqs. (A-45) and (A-46) *implies the convention that any mass* M *exchanged between the phases carries an energy corresponding to the liquid or vapor state, respectively, at the temperature* T_r. This convention is analogous to the one introduced in Sec. A2.3. If in reality the mass exchanged has a different energy level, the excess energy has to be accounted for in the expression of H.

Another comment concerns the *meaning of the carrier temperature* T_∞. All temperatures T_∞ occurring in the above balance equations and in the equations given in Sec. 2 are to be interpreted, strictly speaking, as the caloric mean temperature $\bar{T}$ defined by Eq. (152). By the approximation made in Eq. (A-40), however, $\bar{T}$ has been replaced by T_∞, which designates the carrier temperature at the sample boundary and determines the heat transfer and growth rates dealt with in Secs. 3 and 5. The distinction between T_∞ and $\bar{T}$ is necessary only if the spatial density of droplets is high ($r_\infty/r<20$) and has been made only in the proximity correction treated in Sec. 3.5.

APPENDIX 3 **Free Molecular Transfer Processes in Mixture Carriers**

The free-molecular theories cited in Sec. 3.3 present the rates of heat and momentum transfer as the sum of two terms representing the contribution of the vapor and gas molecule streams, respectively, cf. Eqs. (96) and (101). Although this formulation is quite convenient for the free-molecular case itself, a single-term formalism is required for the construction of multirange expressions that link the free-molecular and the continuum extremes. In the following treatment the vapor and gas terms will be formally combined into a single "mixture" term. The aim of this appendix is to define the appropriate mixture-average data and to outline the derivation of the Nusselt numbers Nu_H^{fm}, Nu_M^{fm}, and Nu_F^{fm} as given by Eqs. (103) to (111).

A3.1 Mass Impingement Rate of Molecules

The molecular mass impingement rate β_∞ on unit surface area in contact with a gaseous mixture is given by Eq. (87*c*). The gas constant $\hat{R}$ introduced with the last expression of this equation is a mean obtained as

$$\hat{R} \mathrel{\hat{=}} \frac{\bar{R}^2}{(\overline{R^{1/2}})^2} = \frac{(R_v + g_\infty R_g)^2}{(\sqrt{R_v} + g_\infty \sqrt{R_g})^2} \tag{A-52}$$

as can be easily verified using Eqs. (87*a*), (87*b*), (A-5), and (A-6). It also follows from the latter equations that the β-weighted average of any quantities X_v and X_g can be expressed as

$$X_v \frac{\beta_{v\infty}}{\beta_\infty} + X_g \frac{\beta_{g\infty}}{\beta_\infty} = \frac{\overline{(R^{1/2} X)}}{\overline{(R^{1/2})}} \tag{A-53}$$

where the overbar means mass averaging according to Eq. (A-7).

A3.2 Heat Transfer

In formal analogy to the terms of Eq. (96) and to Eqs. (97*a*) and (97*b*), we write for the carrier mixture

$$H = 4\pi r^2 (T_\infty^* - T_r)\, \beta_\infty \bar{R} \hat{K} \hat{\alpha}_H k_H(\mathrm{Ma}) \tag{A-54}$$

$$T_\infty^* = T_\infty k_T^{fm}(\mathrm{Ma}) \tag{A-55}$$

$$k_T^{fm}(\mathrm{Ma}) = 1 + \frac{2\gamma}{3\hat{K}} \mathrm{Ma}^2 q_T = 1 + \frac{2\gamma}{3\hat{K}} \mathrm{Ma}^2 \tag{A-56}$$

$$k_H^{fm}(\mathrm{Ma}) = 1 + \frac{\gamma}{6}\,\mathrm{Ma}^2\, q_H \tag{A-57}$$

The last equality of Eq. (A-56) anticipates Eq. (A-63). The mean values $\hat{K}$ and $\hat{\alpha}_H$ as well as the coefficients q_T and q_H have to be determined such that Eq. (A-54) is mathematically identical to Eq. (96).

The equation determining these coefficients is found by inserting Eqs. (A-55) to (A-57) into Eq. (A-54), inserting Eqs. (99*a*), (97*b*), (98*a*), and (98*b*) into Eq. (96), and setting equal both expressions thus obtained for H.

The resulting equation, if arranged according to the dependence of the terms on the Mach number and on temperature, has the form

$$\begin{aligned}
&(T_\infty - T_r)\left[\beta_\infty \bar{R}\,\hat{K}\,\hat{\alpha}_H - \beta_{v\infty} R_v K_v \alpha_{Hv} - \beta_{g\infty} R_g K_g \alpha_{Hg}\right] \\
&+ T_\infty \frac{2}{3}\gamma \bar{R}\,\mathrm{Ma}^2\left[\beta_\infty \hat{K}\,\hat{\alpha}_H \frac{q_T}{\hat{K}} - \beta_{v\infty}\alpha_{Hv} - \beta_{g\infty}\alpha_{Hg}\right] \\
&+ (T_\infty - T_r)\frac{\gamma}{6}\bar{R}\,\mathrm{Ma}^2\left[\beta_\infty \hat{K}\,\hat{\alpha}_H q_H - \beta_{v\infty} K_v \alpha_{Hv} - \beta_{g\infty} K_g \alpha_{Hg}\right] \\
&+ T_\infty \frac{\gamma^2}{9}\bar{R}^2\,\mathrm{Ma}^4\left[\beta_\infty \hat{K}\,\hat{\alpha}_H \frac{q_T\, q_H}{\hat{K}\,\bar{R}} - \beta_{v\infty}\frac{\alpha_{Hv}}{R_v} - \beta_{g\infty}\frac{\alpha_{Hg}}{R_g}\right] = 0
\end{aligned} \tag{A-58}$$

Here use has been made of the relationship

$$\gamma\,\bar{R}\,\mathrm{Ma}^2 = \gamma_v R_v \mathrm{Ma}_v^2 = \gamma_g R_g \mathrm{Ma}_g^2 \tag{A-59}$$

which follows from Eqs. (60) and (94*a*, *b*).

Equation (A-58) is identically fulfilled if the four brackets are made zero by appropriate choice of $\hat{K}$, $\hat{\alpha}_H$, q_T and q_H. Bearing in mind that the Ma^4 term is small, especially if $\mathrm{Ma} \ll 1$, and that the Ma dependences are in any case approximations only, by reasons of formal simplicity we renounce making the last term identically zero. Thus one requires only that

$$\beta_\infty \bar{R}\,\hat{K}\,\hat{\alpha}_H = \beta_{v\infty} R_v K_v \alpha_{Hv} + \beta_{g\infty} R_g K_g \alpha_{Hg} \tag{A-60}$$

$$\beta_\infty \hat{\alpha}_H q_T = \beta_{v\infty}\alpha_{Hv} + \beta_{g\infty}\alpha_{Hg} \tag{A-61}$$

$$\beta_\infty \hat{K}\,\hat{\alpha}_H q_H = \beta_{v\infty} K_v \alpha_{Hv} + \beta_{g\infty} K_g \alpha_{Hg} \tag{A-62}$$

Now one of the unknowns can be freely chosen. It is convenient to set

$$q_T \mathrel{\hat{=}} 1 \tag{A-63}$$

One obtains from Eqs. (A-61) and (A-53)

$$\hat{\alpha}_H = \frac{\overline{(R^{1/2} \alpha_H)}}{\overline{(R^{1/2})}} \tag{A-64}$$

With this, Eq. (A-60) yields

$$\hat{K} = \frac{\overline{(R^{3/2} K \alpha_H)}}{\overline{(R^{1/2})}\ \bar{R}\ \hat{\alpha}_H} = \frac{\overline{(R^{3/2} K \alpha_H)}}{\bar{R}\ \overline{(R^{1/2} \alpha_H)}} \tag{A-65}$$

and Eq. (A-62)

$$q_H = \frac{\overline{(R^{1/2} K \alpha_H)}}{\overline{(R^{1/2})}\ \hat{K}\ \hat{\alpha}_H} = \frac{\bar{R}\ \overline{(R^{1/2} K \alpha_H)}}{\overline{(R^{3/2} K \alpha_H)}} \tag{A-66}$$

In order to derive the free-molecular Nusselt number for heat transfer, Eq. (A-54) is inserted into the general definition of Nu_H, Eq. (55), and Kn is introduced from Eq. (43), giving

$$Nu_H^{fm} = \sqrt{\frac{2}{\pi}}\ \frac{\gamma-1}{\gamma}\ \frac{\overline{(R^{3/2} K \alpha_H)}}{\bar{R}^{3/2}}\ k_H(\text{Ma})\ \frac{\text{Pr}}{\text{Kn}} \triangleq \frac{B_H}{\text{Kn}} \tag{A-67}$$

where the abbreviation B_H means

$$B_H \triangleq \sqrt{\frac{2}{\pi}}\ \frac{\gamma-1}{\gamma}\ \frac{\overline{(R^{3/2} K \alpha_H}}{\bar{R}^{3/2}}\ \text{Pr}\ k_H(\text{Ma}) \tag{A-68}$$

and where, according to Eqs. (A-57) and (A-66),

$$k_M(\text{Ma}) = 1 + \frac{\gamma}{6}\ \text{Ma}^2\ \frac{\bar{R}\ \overline{(R^{1/2} K \alpha_H)}}{\overline{(R^{3/2} K \alpha_H)}} \tag{A-69}$$

At low Mach numbers ($k_H = 1$), B_H depends only on the carrier-phase data and on the accommodation coefficients. With the typical values $\gamma = \gamma_v = \gamma_g = 1.4$, $\hat{\alpha}_H = \alpha_{Hv} = \alpha_{Hg} = 0.95$, $R_v \approx R_g \approx \bar{R}$, $\hat{K} = 3$, and Pr = 0.7, we find that $B_H = 0.45$.

A3.3 Mass Transfer

As seen from Eq. (99), the free-molecular mass transfer rate M does not depend directly on the gas component. Expressing k_{Mv}

by Eq. (100), introducing Ma from Eq. (A-59) and using Eqs. (87) and (90), one obtains

$$M = 4 \pi r^2 \alpha_{Mv} \frac{p_\infty}{\sqrt{2 \pi R_v T_\infty}} \left[k_M(\text{Ma})(\Pi_{v\infty}-\Pi_r) + \left(1 - \sqrt{\frac{T_\infty}{T_r}}\right) \Pi_r \right] \quad \text{(A-70)}$$

with

$$k_M(\text{Ma}) = 1 + \frac{\gamma}{6} \text{Ma}^2 \frac{\bar{R}/R_v}{\Pi_{v\infty}-\Pi_r} \Pi_{v\infty} \quad \text{(A-71)}$$

It is seen that M is not proportional to $(\Pi_{v\infty}-\Pi_r)$ as hypothesized in Eq. (69) and as fulfilled in the case of a continuum. Instead, for $\Pi_{v\infty}-\Pi_r = 0$, the value of M in general remains different from zero:

$$M(\Pi_{v\infty} = \Pi_r) = 4 \pi r^2 \alpha_{Mv} \frac{p_\infty}{\sqrt{2 \pi R_v T_\infty}} \left[\frac{\gamma}{6} \text{Ma}^2 \frac{\bar{R}}{R_v} \Pi_{v\infty} + \left(1 - \sqrt{\frac{T_\infty}{T_r}}\right) \Pi_r \right] \quad \text{(A-72)}$$

For the frequently occurring situation of Ma $\ll 1$ and $(1 - \sqrt{T_\infty/T_r}) \ll 1$, Eq. (A-70), however, yields

$$M_0 = 4 \pi r^2 \alpha_{Mv} \frac{p_\infty}{\sqrt{2 \pi R_v T_\infty}} (\Pi_{v\infty}-\Pi_r) \quad \text{(A-73)}$$

which is *proportional* to $(\Pi_{v\infty}-\Pi_r)$. The index 0 alludes to the special case of Ma $= T_r - T_\infty = 0$ (isothermal, stationary sphere).

Combining Eq. (A-70) with Eq. (52), we find the free-molecular Nusselt number for mass transfer to have in general the complicated form

$$\text{Nu}_M^{fm} = \frac{2r}{\mathcal{D}} \alpha_{Mv} \frac{p_\infty}{\sqrt{2 \pi R_v T_\infty}} \left[k_M(\text{Ma}) + \left(1 - \sqrt{\frac{T_\infty}{T_r}}\right) \frac{\Pi_r}{\Pi_{v\infty}-\Pi_r} \right] \quad \text{(A-74)}$$

Expressing $\mathcal{D}$ by Eq. (53) and using Eq. (43) this takes the form

$$\text{Nu}_M^{fm} = \sqrt{\frac{2}{\pi}} \sqrt{\frac{R_v}{\bar{R}}} \alpha_{Mv} \Pi_{g\infty} \frac{\text{Sc}}{\text{Kn}} \left[k_M(\text{Ma}) + \left(1 - \sqrt{\frac{T_\infty}{T_r}}\right) \frac{\Pi_r}{\Pi_{v\infty}-\Pi_r} \right] \mathrel{\hat{=}} \frac{B_M \Pi_{g\infty}}{\text{Kn}} \quad \text{(A-75)}$$

where

$$B_M \mathrel{\hat{=}} \sqrt{\frac{2}{\pi}} \sqrt{\frac{R_v}{\bar{R}}} \alpha_{Mv} \text{Sc} \left[k_M(\text{Ma}) + \left(1 - \sqrt{\frac{T_\infty}{T_r}}\right) \frac{\Pi_r}{\Pi_{v\infty}-\Pi_r} \right] \quad \text{(A-76)}$$

Here B_M is an abbreviation analogous to B_H in Eq. (A-67). In contrast to B_H, however, B_M also depends on the surface conditions of the droplet. In the special case of $\mathrm{Ma} = T_r - T_\infty = 0$, which is typical of small droplets in saturated or very diluted carriers, Eq. (A-73) yields

$$B_{MO} = \sqrt{\frac{2}{\pi}}\,\sqrt{\frac{R_v}{\bar{R}}}\,\alpha_{Mv}\,\mathrm{Sc} \tag{A-77}$$

Typical numerical values of B_{MO} in carriers having $\bar{R} \approx R_v$ are of the order of 0.5.

For pure-vapor carriers, $\Pi_{g\infty} = 0$ entails $\mathrm{Nu}_M^{fm} = 0$. Simultaneously, however, $\mathcal{D} \to \infty$, and the product $\mathrm{Nu}_M^{fm}\,\mathcal{D}$ remains finite [except if $\Pi_{v\infty} = \Pi_r$, see Eq. (A-72)], thus making Eq. (69) to give a finite value of $\dot{M}$ in compliance with Eq. (A-70).

A3.4 Momentum Transfer (Drag)

The procedure is similar to that exposed in Sec. A3.2. For the mixture carrier we write, in formal analogy to Eq. (101),

$$F = 4\,\pi\,r^2\,\Delta w_\infty\,\beta_\infty\left(\frac{4}{3}\,k_F + \frac{\pi}{6}\,\hat{\alpha}_F\,\sqrt{\frac{T_r}{T_\infty}}\right) \tag{A-78}$$

where $\hat{\alpha}_F$ is a mean diffuse reflection coefficient and

$$k_F = 1 + \frac{1}{10}\,\gamma\,\mathrm{Ma}^2\,q_F \tag{A-79}$$

Identity with the system (101), (102*a*, *b*) has to be ensured by proper choice of α_F and q_F. The results, obtained the same way as in Sec. A3.2, are

$$\hat{\alpha}_F = \frac{\overline{(R^{1/2}\,\alpha_F)}}{\overline{(R^{1/2})}} \tag{A-80}$$

$$q_F = \bar{R}\,\frac{\overline{(R^{-1/2})}}{\overline{(R^{1/2})}} \tag{A-81}$$

The latter gives, with Eq. (A-78),

$$k_F = 1 + \frac{1}{10}\,\gamma\,\mathrm{Ma}^2\,\bar{R}\,\frac{\overline{(R^{-1/2})}}{\overline{(R^{1/2})}} \tag{A-82}$$

Inserting Eq. (A-76) into Eq. (60) and making use of Eqs. (43) and (87*c*), one obtains

$$\mathrm{Nu}_F^{fm} = \sqrt{\frac{2}{\pi}} \left[\frac{4}{3} \frac{\overline{(R^{1/2})}}{\bar{R}^{1/2}} k_F + \frac{\pi}{6} \frac{\overline{(R^{1/2} \alpha_F)}}{\bar{R}^{1/2}} \sqrt{\frac{T_r}{T_\infty}} \right] \frac{1}{\mathrm{Kn}} \mathrel{\hat{=}} \frac{B_F}{\mathrm{Kn}} \tag{A-83}$$

where

$$B_F \mathrel{\hat{=}} \sqrt{\frac{2}{\pi}} \left[\frac{4}{3} \frac{\overline{(R^{1/2})}}{\bar{R}^{1/2}} k_F + \frac{\pi}{6} \frac{\overline{(R^{1/2} \alpha_F)}}{\bar{R}^{1/2}} \sqrt{\frac{T_r}{T_\infty}} \right] \tag{A-84}$$

The numerical value of B_F for a slow-moving ($k_F \approx 1$), diffuse reflecting ($\alpha_{Fv} = \alpha_{Fg} = 1$), nearly isothermal ($T_r \approx T_\infty$) droplet, in a carrier for which $R_v \approx R_g$, is of the order of $B_F = 1.5$.

APPENDIX 4 **The Analysis of Droplet Surface Temperature**

A4.1 Discussion of Eq. (213)

Introducing the abbreviations

$$\begin{aligned} \xi &\mathrel{\hat{=}} T_r/T_{br} \quad (\text{range: } \xi > 0) \\ \xi^{**} &\mathrel{\hat{=}} T_\infty^{**}/T_{br} \quad (\text{range: } 0 < \xi^{**} < \infty) \\ \xi_{sr} &\mathrel{\hat{=}} T_{sr}/T_{br} \quad (\text{range: } 0 < \xi_{sr} < 1) \end{aligned} \tag{A-85}$$

Eq. (213), from which the unknown ξ has to be determined, is written as

$$\exp\left(C - \frac{C}{\xi}\right) = \Pi_{v\infty} - \frac{A}{C}(\xi - \xi^{**}) \tag{A-86}$$

The left side is an indirect exponential function of ξ:

$$f_L(\xi) = e^{C - C/\xi} = \Pi_{v\infty}\, e^{(C/\xi_{sr} - C/\xi)} \tag{A-87}$$

whereas the right side is a simple linear function

$$f_R(\xi) = \Pi_{v\infty} - \frac{A}{C}(\xi - \xi^{**}) \tag{A-88}$$

In the second equality of Eq. (A-87), advantage has been taken of Eq. (209), which is now written as

$$\frac{1}{\xi_{sr}} = 1 - \frac{\ln \Pi_{v\infty}}{C} \tag{A-89}$$

Figure A-2 shows the graphs of these functions for three typical cases:

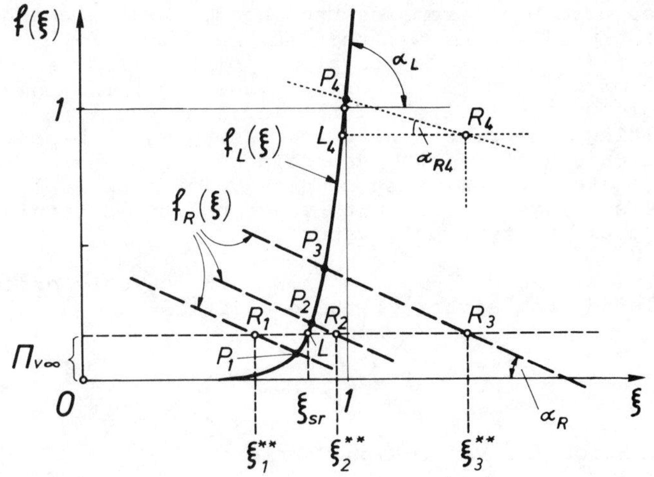

Fig. A-2. Graphical representation of the left and right sides of Eq. (A-86).

Case 1. Supersaturated carrier ($\xi^{**} = \xi_1^{**}$, with $\xi_1^{**} < \xi_{sr}$)

Case 2. Superheated, cool carrier ($\xi^{**} = \xi_2^{**}$, with $\xi_{sr} < \xi_2^{**} < 1$)

Case 3. Superheated, hot carrier ($\xi^{**} = \xi_3^{**}$, with $\xi_3^{**} > 1$)

The left side is independent of ξ^{**}. It is always positive, with $f_L(1) = 1$ and $f_L'(1) = t_g\ \alpha_L = C$. At the origin, $f_L(0) = 0$.

At low values of ξ, f_L remains very close to zero. Its slope $f_L'(\xi) = C\xi^{-2} f_L(\xi)$ is always positive. Two noteworthy features of the left side are that $f_L(\xi_{sr}) = \Pi_{v\infty}$, which is point L, and $f_L(\xi^{**}) = \Pi_{v\infty}/S_r$, as follows from Eq. (212).

The right side $f_R(\xi)$ is represented by a straight line of negative slope that is determined by A and C to $t_g\ \alpha_R = A/C$. At ξ^{**} it is $f_R(\xi^{**}) = \Pi_{v\infty}$; see point R_1, R_2, or R_3. By changing ξ^{**}, $f_R(\xi)$ is shifted without changing its inclination.

The solution ξ is defined geometrically by the intersection of $f_L(\xi)$ and $f_R(\xi)$; see point P_1, P_2, or P_3. It follows from $f_L(\xi_{sr}) = \Pi_{v\infty} = f_R(\xi^{**})$ and from the slopes of opposite sign that ξ always has a value between those of ξ_{sr} and ξ^{**}. In the case of supersaturation, $\xi_{sr} > \xi^{**}$ and therefore $\xi_{sr} > \xi > \xi^{**}$; conversely, $\xi_{sr} < \xi^{**}$ and $\xi_{sr} < \xi < \xi^{**}$ are true for superheated conditions. These properties of the solution lead to conclusions 1–5 in Sec. 5.2.4.

Next, let the composition of the medium (i.e., $\Pi_{v\infty}$) be changed while ξ^{**} is kept constant. An example of such a shift (with ξ^{**} = constant = ξ_3^{**}) is indicated in Fig. A-2) by the dotted lines going through the point R_4. The level of L_4 and R_4 corresponds to the new $\Pi_{v\infty}$. The curve $f_L(\xi)$ remains unchanged, but ξ_{sr} changes as required by the position of L_4, that is, by Eq. (A-89). The straight line $f_R(\xi)$ has R_4 as its pivot point. If A is influenced by the change of $\Pi_{v\infty}$ (which is normally the case, the only exception being free-molecular situations), the inclination of f_R is changed, this change being a decrease[1] with increasing $\Pi_{v\infty}$. Depending on this inclination α_{R4} at R_4, the new solution P_4 may lie at a value $\xi<1$ or $\xi>1$. In the extreme case of pure vapors ($\Pi_{v\infty} = 1$), the intersection is always at $\xi \geqslant 1$, with the equality pertaining to the case $A = 0$ (continuum-type pure vapor).

If the solution $\xi = T_r/T_{br}$ is known, the caloric effectiveness can be calculated from either of the equalities

$$Q = \frac{\xi-\xi^{**}}{\xi_{sr}-\xi^{**}} = 1 - \frac{\xi_{sr}-\xi}{\xi_{sr}-\xi^{**}} \tag{A-90}$$

A4.2 Approximation for Vapor-Rich Carriers

If $\xi^{**} \approx \xi_{sr}$, the solution ξ lies close to ξ_{sr}, and $f_L(\xi)$ may be linearized around $\xi = \xi_{sr}$, giving

$$f_L(\xi) \approx f_L(\xi_{sr}) + f_L'(\xi_{sr})(\xi-\xi_{sr}) = \Pi_{v\infty} + \frac{C\ \Pi_{v\infty}}{\xi_{sr}^2}(\xi-\xi_{sr}) \tag{A-91}$$

Introducing this approximation into Eq. (A-86), the equation becomes linear, and its solution yields as approximation for ξ:

$$\xi \approx \frac{\xi_{sr} + G\ \xi^{**}}{1 + G} \triangleq \xi_{\text{vap}} \tag{A-92}$$

where

$$G \triangleq \frac{\xi_{sr}^2}{C^2}\frac{A}{\Pi_{v\infty}} = (C - \ln \Pi_{v\infty})^{-2}\ \frac{A}{\Pi_{v\infty}} \tag{A-93}$$

It is seen that ξ_{vap} is a weighted mean of ξ_{sr} and ξ^{**}, with the dimensionless group G acting as weighting factor. G becomes infinite for the pure-gas extreme, causing $\xi_{\text{vap}} \to \xi^{**}$ for this case. Setting $\xi \approx \xi_{\text{vap}}$ in Eq. (A-90), an approximate expression for Q is obtained:

$$Q \approx \frac{1}{1 + G} \triangleq Q_{\text{vap}} \tag{A-94}$$

The first-order approximation, Eq. (A-91), is quite good at high

[1]This becomes evident from the expressions given for A in Sec. 5.3.

values where the curve of f_L is slightly curved. Therefore Eqs. (A-92) and (A-94) are satisfactory for vapor-rich carriers, as indicated by the index "vap." At nearly saturated conditions ($\xi_{sr} \approx \xi^{**}$), the approximation retains its validity over the entire composition range.

A4.3 Approximation for Cool, Lean Mixture Carriers

Since $\xi^{**} \ll 1$ and $\Pi_{v\infty} \ll 1$, the intersection of f_L and f_R lies in the nearly horizontal part of f_L and (f_R being relatively steep) closer to ξ^{**} than to ξ_{sr}. We therefore set

$$f_L(\xi) \approx f_L(\xi^{**}) = \frac{\Pi_{v\infty}}{S_r} \tag{A-95}$$

Replacing the left side of Eq. (A-85) by this leads to the approximation

$$\xi \approx \xi^{**} + \frac{C\,\Pi_{v\infty}}{A}\left(1 - \frac{1}{S_r}\right) \mathrel{\hat{=}} \xi_{\text{gas}} \tag{A-96}$$

The index "gas" notes the lean mixture compositions for which this approximation is valid. Insertion of ξ_{gas} for ξ in Eq. (A-90) yields the following approximation for Q:

$$Q \approx \frac{1}{G}\,\frac{\xi_{sr}}{\xi^{**}}\,\frac{S_r - 1}{S_r\,\ell n\,S_r} \mathrel{\hat{=}} Q_{\text{gas}} \tag{A-97}$$

The ratio ξ_{sr}/ξ^{**} can be expressed by Eqs. (A-89) and (212) as

$$\frac{\xi_{sr}}{\xi^{**}} = \frac{T_{sr}}{T^{**}_{\infty}} = \frac{C - \ell n\,\Pi_{v\infty} + \ell n\,S_r}{C - \ell n\,\Pi_{v\infty}} \tag{A-98}$$

A4.4 Synthesis of the Rich and Lean Approximations

If the carrier is saturated ($\xi_{sr} = \xi^{**}$), both approximations Q_{vap} and Q_{gas} yield the exact value of Q. At large deviations from saturation Q_{vap} retains its validity if $\Pi_{v\infty}$ is high (typically, $\Pi_{v\infty} > 0.2$) but gives great errors in gas-rich carriers. The behavior of Q_{gas} is just the opposite. It is therefore logical to combine the two approximations into a single expression that behaves as Q_{vap} if the carrier is rich (i.e., if G is small) and as Q_{gas} if the carrier is lean (i.e., if G is large). Such a "synthetic" expression is, for instance,

$$Q_{\text{syn}} \mathrel{\hat{=}} \frac{1 + (\xi_{sr}/\xi^{**})\,G}{(1/Q_{\text{vap}}) + (\xi_{sr}/\xi^{**})\,G + (\xi_{sr}/\xi^{**})\,G/Q_{\text{gas}}} \tag{A-99}$$

where ξ_{sr}/ξ^{**} is given by Eq. (A-98). With the latter, one obtains

$$Q_{\text{syn}} = \frac{(1+G)(C-\ell n\,\Pi_{v\infty}) + G\,\ell n\,S_r}{(1+2G)(C-\ell n\,\Pi_{v\infty}) + G\,\ell n\,S_r + G^2\,(C-\ell n\,\Pi_{v\infty})(S_r\,\ell n\,S_r)/(S_r-1)} \tag{A-100}$$

A corresponding approximation ξ_{syn} to ξ can be obtained by inserting Q_{syn} into Eq. (A-90). As it appears in Sec. A4.6, Q_{syn} is a satisfactory approximation to Q for all carrier conditions of interest, except for lean mixtures at high temperature. As a rough rule, Q_{syn} may be used if the superheat does not exceed the limit corresponding to $S_r = 0.01$.

A4.5 Approximation for Hot, Lean Carriers

As evident from Fig. 17, for lean carriers $\xi = T_r/T_{br} \to 1$ if $\xi^{**} = T^{**}/T_{br} \to \infty$. An approximation displaying this property can be constructed, for instance, according to $\xi \approx 1 - a\exp(-b\xi^{**})$, where a and b are appropriate constants. If we furthermore require that the approximation be exact to the first order in $(\xi^{**}_\infty - \xi_{sr})$ at the saturation point ξ_{sr}, a and b can be determined and we have

$$\xi \approx 1 - (1 - \xi_{sr})\exp\left(\frac{\xi_{sr} - \xi^{**}}{1 - \xi_{sr}}\,\frac{G}{1 + G}\right) \triangleq \xi_{\text{gen}} \tag{A-101}$$

With Eq. (A-90) we obtain

$$Q \approx 1 - \frac{1-\xi_{sr}}{\xi_{sr}-\xi^{**}}\left[\exp\left(\frac{\xi_{sr}-\xi^{**}}{1 - \xi_{sr}}\,\frac{G}{1 + G}\right) - 1\right] \tag{A-102}$$

The index "gen" is meant to refer to the generally satisfactory approximation resulting over a wide temperature range; cf. Sec. A4.6.

A4.6 Error Analysis of the Approximations

The exact value $Q = Q(\xi^{**}, \xi_{sr}, A, C)$ of the thermal effectiveness has been calculated for a large number of parameter combinations covering the domain of interest defined in Eq. (216) and compared to the approximate values Q_{vap}, Q_{gas}, Q_{syn}, and Q_{gen}. The errors of the four approximate expressions are characterized by

$$\delta Q_{\text{vap}} \triangleq \frac{Q_{\text{vap}} - Q}{Q} \qquad \delta Q_{\text{gas}} \triangleq \frac{Q_{\text{gas}} - Q}{Q}$$
$$Q_{\text{syn}} \triangleq \frac{Q_{\text{syn}} - Q}{Q} \qquad \delta Q_{\text{gen}} \triangleq \frac{Q_{\text{gen}} - Q}{Q} \tag{A-103}$$

The computation was made straightforward by using Eq. (A-86) to determine the value of ξ^{**} for specified values of ξ, ξ_{sr}, C, and A or $A/\Pi_{g\infty}$. Then Q was obtained from Eq. (A-90), G from Eq. (A-93), and the approximations to Q from Eqs. (A-94), (A-98), (A-99), and (A-102).

Typical results of these comparisons are shown in Fig. 18 and discussed in Sec. 5.2.5.

A4.7 The Case of Continuum Nonlinear Mass Transfer

If the simple expression for Nu_M^{ct}, the mass transfer Nusselt number in continuum carriers, as given by Eq. (78) or (81), is replaced in Eq. (213) by the more rigorous expression (83), according to which Nu_M^{ct} is a function of $\Pi_{v\infty}$ and Π_r, the mathematical structure of Eqs. (213) and (A-86) is altered. The value of ξ is then determined, instead of Eq. (A-86), by

$$\exp\left(C - \frac{C}{\xi}\right) = 1 - \Pi_{g\infty} \exp\left(\frac{A}{C} \frac{\xi - \xi^{**}}{\Pi_{g\infty}}\right) \qquad \text{(A-104)}$$

This equation is implicite in ξ, too. The difference to Eq. (A-86) consists in having a nonlinear function of ξ on the right side. It is easy to show that the linear function $f_R(\xi)$ is a first-order Taylor approximation to the above nonlinear function at $\xi = \xi^{**}$. Thus the solutions ξ resulting from Eqs. (A-86) and (A-104) may differ significantly only if ξ_{sr} and ξ^{**} are very different.

Equation (A-104) has been solved numerically to give $\xi_{n\ell}$ and $Q_{n\ell}$ for various values of the parameters ξ^{**}, ξ_{sr}, A, and C. (The index $n\ell$ refers to the nonlinear diffusion law used). The results were compared to the values obtained from Eq. (A-86). It was found that, for the entire parameter range of interest, the deviation

$$\delta Q_{n\ell} = \frac{Q_{n\ell} - Q}{Q} \qquad \text{(A-105)}$$

remains practically negligible. The largest deviations occur if the vapor has low heat of vaporization (e.g., $C \approx \mathrm{Cl} \approx 5$) and the carrier is a hot ($\xi^{**}>1.5$) mixture of comparable amounts of vapor and gas components ($\Pi_{v\infty} = 0.5$–0.1). Under such conditions, $\delta Q_{n\ell}$ reaches values between +2% and +3%. Under all other conditions of physical interest the deviations remain smaller. This leads to the conclusion that Eq. (A-86) is sufficiently accurate in predicting Q (and T_r) even when the nonlinear diffusion law, Eq. (83) has to be used. In such cases Nu_M in Eq. (213) has to be replaced by $Nu_M^{ct,nl}$. This refinement is rarely significant and will not be further considered here.

A4.8 Correction for Free Molecular Conditions

According to Eqs. (104) and (107) the free-molecular mass transfer Nusselt number Nu_M is a function of T_r and Π_r. For Nu_H it is appropriate to use the expression given by Eqs. (121) and (122), which takes into account the simultaneous presence of mass transfer. In this case Eq. (214) yields the following expression for the nondimensional factor A (the index fm denoting free-molecular conditions):

$$A^{fm} = \frac{\overline{(R^{3/2}\ K\ \alpha_H)}}{\bar{R}\ \sqrt{R_v}\ \alpha_{Mv}} \frac{k_H}{1 + K_v\ T_r/\mathrm{Cl}\ T_{\mathrm{ref}}} \frac{\Pi_{v\infty} - \Pi_r}{k_{Mv}\ \Pi_{v\infty} - \Pi_r\ \sqrt{T_\infty/T_r}} \qquad \text{(A-106)}$$

A^{fm} is seen to be a function of T_r, both directly and indirectly (though Π_r). The influence of relative velocity is represented by k_H and k_{Mv}. Insertion of this expression into Eq. (213) and use of the abbreviations introduced in this appendix leads to

$$\frac{\sqrt{T_\infty/T_{br}}}{k_{Mv}\sqrt{\xi}} \exp(C - \frac{C}{\xi}) = \Pi_{v\infty} - \frac{\overline{(R^{3/2}\; K\; \alpha_H)}}{\bar{R}\;\sqrt{R_v}\;\alpha_{Mv}} \frac{k_H}{k_{Mv}} \frac{\xi - \xi^{**}}{C + K_v\;\xi} \qquad \text{(A-107)}$$

instead of equation (A-86). As in Sec. A4.7, again Eq. (A-86) is replaced by an equation of different algebraic structure. It can be shown that in this case also the solution ξ will lie in the range $0.5<\xi\leqslant 1$ in all situations of physical relevance shown in Fig. 17. Furthermore, for many substances, $C > K_v$.[2] Therefore the influence of ξ by way of the terms $1/\sqrt{\xi}$ and $C + K_v\;\xi$ is weak. As an approximation one may set in these terms ξ = constant = ξ_{est} = $T_{r,est}/T_{br}$, where $T_{r,est}$ is an estimated value of T_r. [In pure vapors and in hot mixtures $T_{r,est} = T_{br}$ (i.e., $\xi_{est} = 1$) is a good estimate. In cold mixtures $T_{r,est} = T^{**}$ is more appropriate.] Thus Eq. (A-107) obtains the form

$$\frac{\sqrt{T_\infty/T_{r,est}}}{k_{Mv}} \exp(C - \frac{C}{\xi}) = \Pi_{v\infty} - \frac{A_{est}}{C} (\xi - \xi^{**}) \qquad \text{(A-108)}$$

with

$$A_{est} \triangleq \frac{\overline{(R^{3/2}\; K\; \alpha_H)}}{\bar{R}\;\sqrt{R_v}\;\alpha_{Mv}} \frac{k_H/k_{Mv}}{1 + K_v\; T_{r,est}/\text{Cl}\; T_{ref}} \qquad \text{(A-109)}$$

The formal analogy to Eq. (A-86) can be made complete by redefining C, ξ, and ξ^{**} by replacing T_{br} in their definitions by a modified boiling temperature of the droplet, T_{br}^{fm}, which is defined as

$$T_{br}^{fm} = \frac{T_{b\infty}}{1 + (T_{b\infty}/\text{Cl}\; T_{ref})(\text{Ke} + \ln\sqrt{T_\infty/T_{r,est}} - \ln k_{Mv})} \qquad \text{(A-110)}$$

If these redefinitions are carried through, all conclusions and approximate expressions obtained in Sec. A4.1 to A4.5 are applicable to the modified variable.

The difference between T_{br}^{fm} and T_{br} may become significant if the relative velocity is high (corresponding to Ma>0.1) or if $T_\infty/T_{r,est}$ is markedly different from unity. The latter situation always occurs except when the carrier is nearly saturated and/or is a cold, lean mixture.

---

[2]Vapors of high molecular weight (such as heptane) are an exception; cf. Table A-1.

APPENDIX 5 Biographic Notes Concerning the Non-Dimensional Named Groups Proposed

Rudolf Immanuel Clausius (1822-1888) was professor of physics at Zürich, Würzburg and Bonn. His contributions to the edifice of thermodynamics (mechanical theory of heat, Second Law, definition of entropy, kinetic theory of gases) were of epochal importance.

Benoit-Pierre-Emile Clapeyron (1799-1864), French mathematician and designer of steam locomotives, established the equation of state of perfect gases and discovered a fundamental law concerning vapor pressure that has been later completed by Clausius; cf. Eq. (A-13) or Eq. (35).

József, Baron Eötvös (1848-1919) was professor of experimental physics in Budapest, Hungary. His research activities centered on the surface tension of liquids and on measurements concerning the gravitational and geomagnetic fields of the Earth.

Aurel Stodola (1859-1942) pioneered the introduction of scientific methods into mechanical engineering. As professor of machinery design at the Swiss Federal Institute of Technology, Zürich, he made fundamental contributions to the vibration analysis and the thermodynamics of steam turbines operated with single-phase or two-phase fluids.

Sir William Thomson, Lord Kelvin (1824-1907) physicist, mathematician and inventor, professor at Glasgow for over 50 years, was one of the main architects of classical thermodynamics. His contributions include the Joule-Thomson effect, the mathematical formulation of the First Law, the concept of absolute temperature and the theory of capillarity.

Chapter 3

Boiling in Multicomponent Fluids

R. A. W. Shock
Engineering Sciences Division, A.E.R.E.
Harwell, Didcot, Oxon, England

1 INTRODUCTION

In the last few decades much effort has been expended, and much progress achieved, toward an understanding of the boiling process. The ultimate aim of such research is the ability to predict *a priori* the boiling characteristics of any combination of solid heater and fluid coolant; this goal is still some way in the future. Nevertheless, many aspects of the underlying phenomena of two-phase heat transfer are reasonably well understood and provide a sound basis for correlations used by designers of a wide variety of boilers when the fluids concerned are pure substances. However, the addition of further components to a pure substance increases the number of phenomena, and parameters, that govern the boiling process. The study of mixture boiling is becoming increasingly important in a number of fields. Some of the most important are (1) the chemical and petroleum industries, where many, if not most, boiler feeds are mixtures of two or more (and often many more) substances; (2) the cooling of high power-density electronic components by boiling liquids, where the use of mixtures offers a possible advantage in an increase of maximum heat load; (3) the evaporation of fuel for engines; and (4) spills of liquefied gases.

Since the boiling process is more complex in mixtures than in pure substances, mathematical modeling is also more complex and requires more initial information. The lack of general correlating equations for various features of the boiling process in mixtures, especially in flow boiling, is now one of the limitations in the design of efficient evaporation equipment. Considerable progress has been made, however, in understanding many of the basic phenomena, and this forms the subject of this review.

The mixture boiling literature shows that a number of effects

---

The author's grateful thanks are due to D. Butterworth, J. G. Collier, P. Hawtin, and D. B. R. Kenning for their helpful comments on the manuscript of this chapter. The facilities provided by the United Kingdom Atomic Energy Authority are acknowledged.

can be distinguished. Perhaps the most significant is a consequence of the differing volatilities of the components. During evaporation the liquid at the liquid-vapor interface is denuded of the lighter, more volatile components, causing a local increase in saturation temperature. The ultimate effect of this change is to reduce bubble growth rates, bubble departure diameters, and heat transfer coefficients. The heat flux at the crisis (departure from nucleate boiling) is often increased greatly for small heaters although it may be decreased for large ones. Since this feature of bubble growth involves the stripping of the lighter components at the bubble boundary and the consequent establishment of concentration gradients, it introduces mass diffusion as well as heat diffusion to the study of bubble growth.

Physical property gradients along, and toward, the bubble interface can also be more important for mixtures than for pure substances. In particular, surface tension gradients along bubble boundaries can often be larger than those due solely to temperature gradients (considerably larger for aqueous organic solutions). The resultant surface forces (the Marangoni effect) can influence the behavior of individual bubbles and the mutual interaction of neighboring bubbles. Similarly, in some mixtures, viscosity changes normal to the interface can affect bubble coalescence.

When comparing the boiling behavior of mixtures with that of pure substances, it is necessary to define a basis of comparison. The feature of mixtures that, directly or indirectly, renders their boiling different from that of pure substances is that the liquid and vapor compositions are different. Thus, a significant comparison of mixture boiling data is with the expected behavior of a pure substance that resembles the mixture in all its physical properties save that the liquid and vapor compositions are identical. Such a pure substance is referred to here as the equivalent pure fluid (EPF). Since the thermodynamic and transport properties of mixtures can be complex functions of the composition, it may be difficult to evaluate the predicted behavior of the EPF, hence the convenient supposition is often adopted that a parameter such as surface superheat or heat transfer coefficient would be linearly related to the composition in the absence of mixture effects (see Sec. 5).

The boiling process is inherently a departure from equilibrium. However, the discussion of boiling involves the use of the concepts and terminology of mixture equilibrium; Sec. 2 gives a brief review.

Up to the time of writing, the majority of published work has been on pool boiling of binary mixtures, which is therefore almost the sole concern of this review. The subject is treated in Secs. 3 to 7, which consider in turn the various features encountered along the boiling curve as the heater surface temperature is increased.

The superheats necessary to initiate or to maintain boiling in mixtures are discussed in Sec. 3. Once a bubble nucleus achieves a superheat in excess of this minimum, we are interested in the dynamics of the growth process and the departure of bubbles; these topics are discussed in Sec. 4. Nucleate boiling data for a wide variety of mixtures have been reported. These are summarized in Sec. 5 together with correlations for boiling superheats, some of which are based on the work presented in Sec. 4. The limit of the nucleate boiling region, the critical heat flux, is also a complex phenomenon in mixtures and is discussed in Sec. 6.

With suitable heater control, once the critical heat flux has been exceeded, the heat transfer mode enters the transition, or partial film boiling region. With a further increase in heater temperature the heat flux passes through a minimum and then enters the true film boiling region. Section 7 discusses the heat and mass transfer processes that occur at and around the minimum.

Although design correlations are not yet available for flow boiling, some underlying work that may indicate trends is described in Sec. 8.

Finally, in Sec. 9, conclusions from this review are presented together with recommendations for future work.

2 PHASE EQUILIBRIUM

Phase equilibrium is incidental to the study of the boiling of mixtures, but this short section describes some of its important features and defines the terms that are used in the following presentation.

Figure 1 shows a typical vapor/liquid equilibrium diagram for a binary mixture of species A and B at constant pressure, A being the lighter-boiling component, that is, that with the lower boiling point. Imagine an experiment in which a liquid mixture of A and B, mole fraction $\tilde{x}_A = 0.25$, is charged into a vessel sealed with a piston, so that the pressure can be kept constant. The liquid is heated from its initial temperature T_1 until the temperature reaches the bubble point T_{bub} of the liquid $\tilde{x}_A$. This is the temperature at which vapor is in equilibrium with liquid, the interface being flat. Although curvature has an effect on the equilibrium locus, the effect is negligible at all but the smallest radii and certainly within the range of interest of this review. At the bubble point T_{bub} the composition of the vapor in equilibrium with the liquid is given by the horizontal line from the point (0.25, T_{bub}) to the vapor dew point line, that is, vapor mole fraction $\tilde{y}_A = 0.7$. Hence the vapor is richer than the liquid in the light component. Further heating to T_2, which generates further vapor and requires withdrawal of the piston to maintain constant pressure, causes the liquid composition to fall to $\tilde{x}_A = 0.14$ and the vapor composition to fall to $\tilde{y}_A = 0.52$. T_2 is the bubble point of a liquid $\tilde{x}_A = 0.14$ and the dew point of vapor $\tilde{y}_A = 0.52$. The overall composition is maintained at 0.25. With further heating a point is reached where the liquid is completely vaporized, the temperature is T_{dew}, the dew point of a vapor of composition $\tilde{y}_A = 0.25$, and the final liquid is of composition $\tilde{x}_A = 0.05$. Further heating merely results in an increase of temperature by an amount dependent on the specific heat capacity of the vapor.

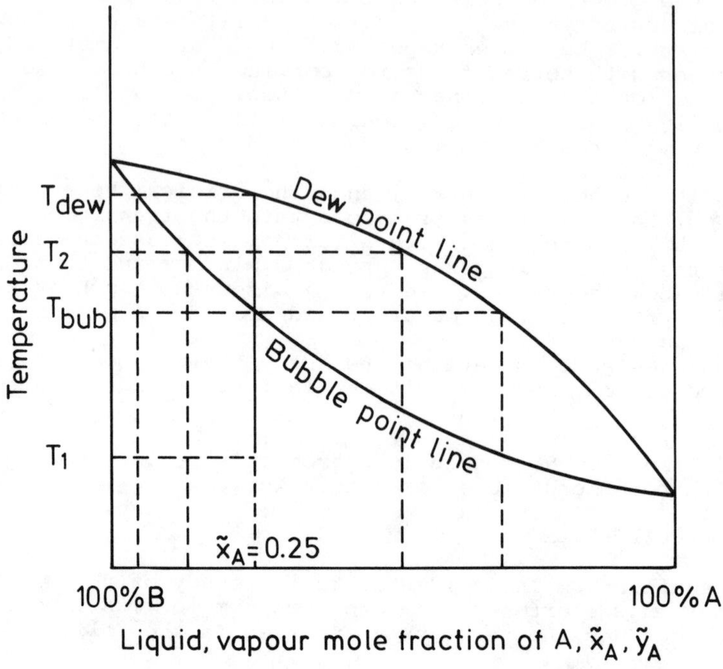

Fig. 1. Typical phase diagram for an ideal mixture.

This type of equilibrium diagram is exhibited mainly by ideal mixtures; those in which the A—B intermolecular forces are similar to the A—A or B—B interactions. Substances forming such mixtures are usually chemically similar. When significantly different interactive forces occur, the curves are distorted and in the limit can show the more complex form exhibited in Fig. 2. Mixtures that exhibit this extreme form of behavior are known as azeotropes or constant-boiling mixtures, which refer strictly to the particular composition at which the liquid and vapor composition are identical. At the azeotrope the dew and bubble point lines meet, and they do so at a minimum or, less commonly, a maximum in the equilibrium temperature. Consider a mixture of species C and D, C being the lighter. When $\tilde{x}_C < \tilde{x}_{C,\text{azeotrope}}$, that is, to the left of the azeotrope, the situation is identical to that in Fig. 1, and the vapor is richer than the liquid in the lighter-boiling component, $\tilde{y}_C > \tilde{x}_C$. When $\tilde{x}_C > \tilde{x}_{C,\text{azeotrope}}$, however, the situation is reversed and $\tilde{y}_C < \tilde{x}_C$. In effect the azeotrope has become the light-boiling component. Konovalov's rule summarizes this by stating that the vapor is always richer than the liquid with which it is in equilibrium in that component which on addition raises the vapor pressure (i.e., decreases the equilibrium temperature). Stated in mathematical terms, $\tilde{y} - \tilde{x}$ and $\delta x/\delta T_{\text{bub}}$ are always of opposite sign; the consequences of this will be seen in Secs. 4 and 5. The behavior illustrated in Fig. 2 is shown by many aqueous organic solutions, for example, of ethanol and i-propanol, and also by organic binary

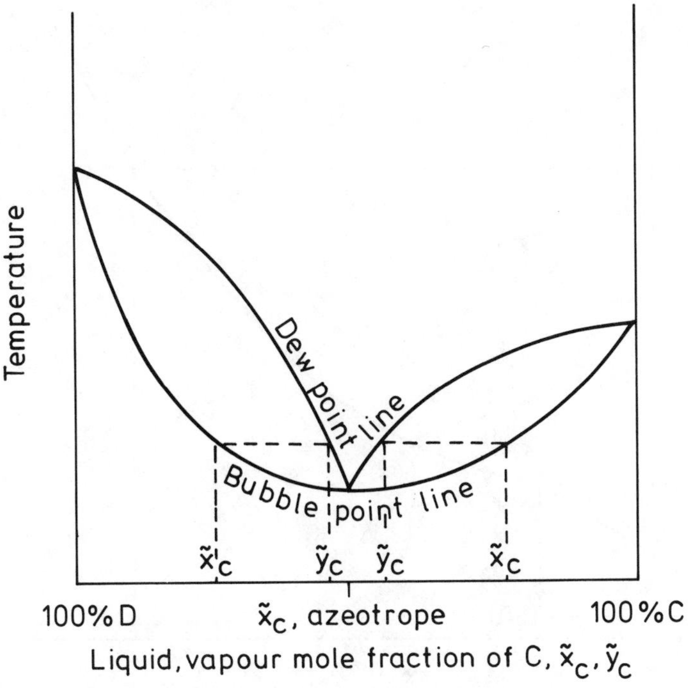

Fig. 2. Typical phase diagram for a mixture forming a minimum boiling azeotrope.

mixtures, for example, ethanol/benzene.

A further complication is exhibited by mixtures with regions of immiscibility where the liquid splits into two layers, each of which may contain both components. Such as mixture is n-butanol/water, which are miscible in the region $0 < \tilde{x}_{n\text{-butanol}} < 0.02$ and $0.49 > \tilde{x}_{n\text{-butanol}} > 1.0$, but in $0.02 < \tilde{x}_{n\text{-butanol}} < 0.49$ the liquid splits into one layer rich in water and a second rich in n-butanol.

So far we have examined the phase equilibrium diagram by considering the isobaric heating of a liquid mixture in a static reservoir, the heating being conducted so that each phase is of uniform composition (fully mixed). We can also consider the idealized case of isobaric evaporation of a liquid flowing in a tube, or heat exchanger, similarly conducted to eliminate concentration gradients within the phases. By reference to Fig. 1, we see that, for a liquid feed of composition $\tilde{x}_A = 0.25$, the initial bulk saturation temperature will be T_{bub}. The temperature will increase along the heat exchanger until it reaches T_{dew} at the point of complete evaporation. The curve of equilibrium temperature against distance, z, quality, $\dot{x}$, or specific enthalpy, h, Fig. 3 (known to condenser designers as the "condensation curve"), is required before the design of a mixture-fed boiler can be carried out.

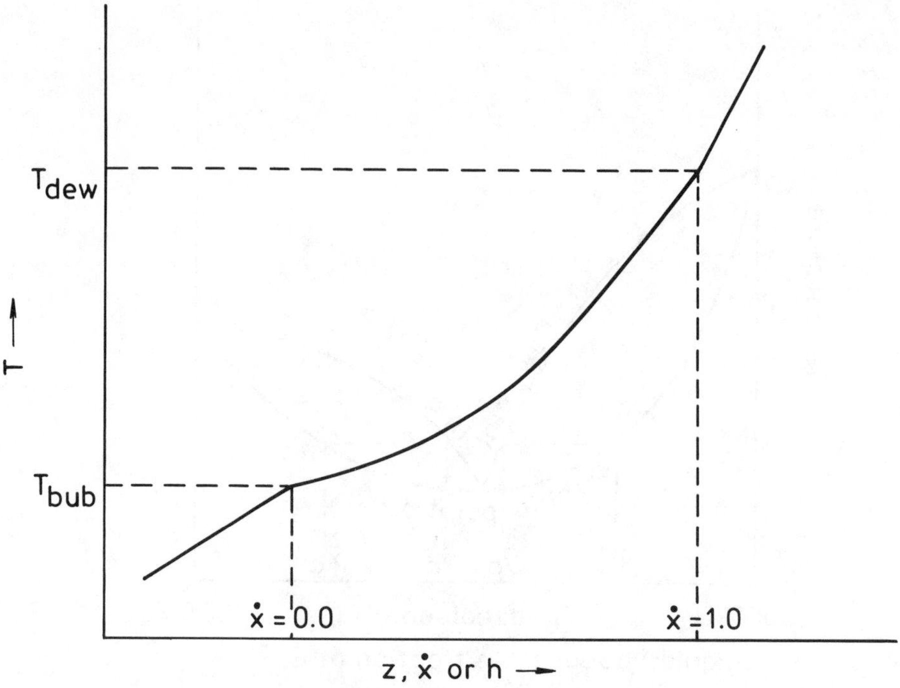

Fig. 3. Change of bulk equilibrium temperature along a heat exchanger.

The prediction of phase equilibrium conditions is a topic beyond the scope of this study, but it is fully documented in many texts such as that by Prausnitz (1969).

3 ONSET OF BOILING

3.1 Homogeneous Nucleation

Homogeneous nucleation is the process by which embryonic vapor nuclei of few molecules can survive and grow within the bulk of a superheated liquid. Because of the high superheats involved, the growth process is explosively violent and should usually be avoided. Blander and Katz (1975) show how the probability of nucleus formation is exponentially related to the superheat. In the vicinity of the nucleation point a small increase in temperature causes this probability to increase from a very small value to almost unity, thus narrowly defining the superheat limit. Pure component data agreed well with the nucleus formation theory where the rate of nucleation J is given by

$$J = N \left(\frac{2\sigma}{\pi \tilde{M}}\right)^{1/2} \exp\left[\frac{-16\pi\sigma^3}{3kT\delta^2(p_v - p_l)^2}\right] \qquad (1)$$

where N is the number density of molecules in the liquid, σ is the surface tension and $\tilde{M}$ is the molar mass, k is Boltzmann's content, T is the temperature, p_v and p_l are, respectively, the equilibrium vapor pressure and the liquid pressure. The term δ, which is unity except at high pressures, accounts for the effect of pressure on the vapor pressure. Equation (1) shows a slight tendency to overestimate the necessary superheat, due perhaps to premature nucleation on foreign particles, always present in liquids, which can act as micronuclei.

Holden and Katz (1978) later extended the methodology to mixtures of two or more components. By assuming chemical, but not necessarily mechanical, equilibrium in all assemblies, of whatever size, of vaporized molecules they obtained for a binary mixture

$$J = N\left(\frac{\tilde{y}_A}{\sqrt{\tilde{M}_A}} + \frac{\tilde{y}_B}{\sqrt{\tilde{M}_B}}\right)\left(\frac{2\sigma}{\pi}\right)^{1/2} \exp\left[\frac{-16\pi\sigma^3}{3kT\delta^2(p_v-p_l)^2}\right] \qquad (2)$$

For mixtures of more than two components, further ratios of the type $\tilde{y}/\sqrt{\tilde{M}}$ are inserted in the first term. In addition to the information required for Eq. (1), solution of this second equation requires phase equilibrium data and the function $\sigma(\tilde{x},T)$; the latter is not always readily available or easily calculable. Notwithstanding the uncertainties in the physical properties and the assumptions inherent in the derivation of Eq. (2), the trends with changing composition given by this method are shown by Holden and Katz to agree reasonably well with experimental data. This is true both for mixtures that are ideal, in the sense that they obey Raoult's law, as shown in Fig. 4, and for those that are not, Fig. 5. Although the data are classified into mixtures that obey Raoult's law and those that do not, it is more likely that the important factor is the form of the p_v-$\tilde{x}$-T relationship [as well as $\sigma(\tilde{x},T)$], irrespective of the physicochemical phenomena that govern its particular form.

All the data for nonideal mixtures reported by Holden and Katz are for those with positive deviations from Raoult's law (the more common type), and they all show a homogeneous nucleation superheat roughly equal to, or less than, that given by a linear relationship with $\tilde{x}$. The foregoing suggests, however, that in a mixture with negative deviation, such as acetone/chloroform, the different form of p_v-$\tilde{x}$-T relationship may tend to reverse this behavior, although data are needed to confirm this suggestion.

Data given by Porteous and Blander (1975) for ternary hydrocarbon mixtures show a linear superheat-mole fraction relationship.

3.2 Heterogeneous Nucleation

The more common form of boiling is initiated by growth of bubble nuclei from trapped gas pockets in the many cavities on a heater surface. When initial bubble nucleation is considered, the system is in equilibrium hence the onset of nucleate boiling (ONB) is a topic concerning only the physical properties of the

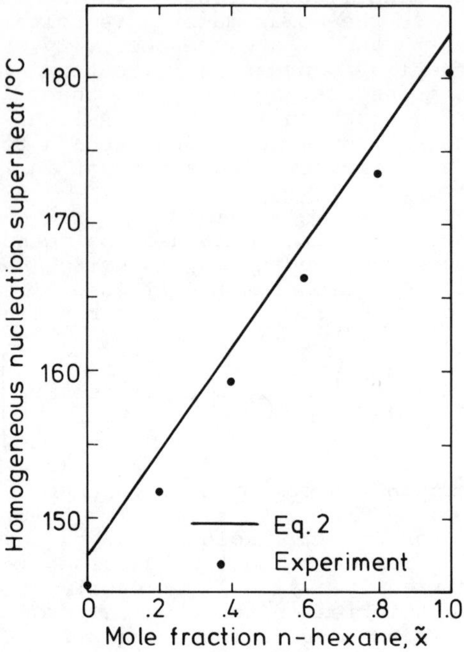

Fig. 4. Homogeneous nucleation superheat for *n*-pentane/*n*-hexane mixtures. Holden and Katz (1978).

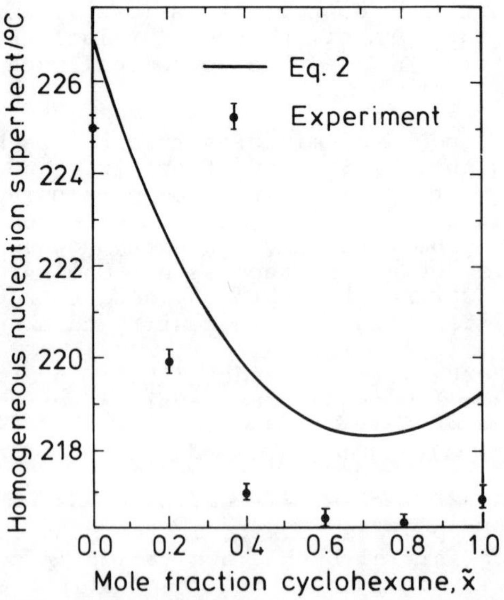

Fig. 5. Homogeneous nucleation superheat for benzene/cyclohexane mixtures. Holden and Katz (1978).

fluid and its interaction with the solid. The fact that the fluid is a mixture does not bring into play any new phenomena but can affect the necessary calculations by increasing the complexity of the physical property relationships. As will be seen, the contact angle is an important factor in ONB, and for surfactant solutions this can change rapidly for small changes in compositions; an important and much studied example is aqueous organic solutions.

In contrast to the initial growth, the successive initiation of bubbles from a nucleus takes place in a changing temperature field as liquid that fills the space left by a previous bubble is heated before the next bubble starts to grow. For mixtures, the fresh liquid at the heater surface may have been partially stripped of light component by the growth of the previous bubble, and thus during the delay time its concentration may be changing through mass diffusion from the bulk liquid. A new bubble is initiated after a suitable waiting time, which is dictated by requirements of mechanical and thermodynamic equilibrium, based on the nucleus geometry and local instantaneous liquid-vapor interface properties, x_i and T_i. Hence the maintenance of bubble nucleation in mixtures with decreasing heat flux is a more complex matter than the initial bubble growth. The delay time in bubble growth is discussed further in Sec. 4.3, and the onset of nucleate boiling in mixtures is examined here in detail.

Equilibrium of a spherical bubble requires an excess of pressure inside it, given by

$$\Delta p = \frac{2\sigma}{R} \tag{3}$$

where R is the bubble radius.

A bubble nucleus in a surface cavity takes the form of a spherical segment of radius R_e, hence

$$\Delta p = \frac{2\sigma}{R_e} \tag{4}$$

Equation (4) can be transformed, assuming dp_v/dT to be independent of p over the range Δp, to give the superheat requirement

$$\Delta T_{\text{sat,ONB}} = \frac{2\sigma}{R_e(dp_v/dT)} \tag{5}$$

In this equation R_e is taken to be the critical value beyond which the bubble will grow and depart, hence the corresponding superheat applies to ONB. The parameters σ, R_e, and dp_v/dT may be complex functions of the physical properties and of the composition of a mixture, and as we shall see, each can be important in the analysis of bubble nucleation.

There has been some confusion in the literature regarding dp_v/dT and its relationship to composition. It has been suggested from thermodynamic arguments that dp_v/dT should be lower for mixtures than for pure substances of otherwise identical physical properties. However, as shown by Denbigh (1966), the Clausius-Clapeyron equation,

$$\frac{dp_v}{dT} = \frac{\Delta\tilde{h}}{T(\tilde{V}_v - \tilde{V}_l)} \tag{6}$$

where $\Delta\tilde{h}$ is the molar latent heat and $\tilde{V}_v$ and $\tilde{V}_l$ are the vapor and liquid molar volumes, is equally valid for mixtures as for pure substances (to be strictly accurate, for mixtures $\Delta\tilde{h}$ should be taken as the differential latent heat). Thus, the particular form of the dp_v/dT - $\tilde{x}$ relationship will depend on the $\Delta\tilde{h}$ - $\tilde{x}$, T_{bub} - $\tilde{x}$, and $\tilde{V}$ - $\tilde{x}$ relationships. For some mixtures small changes in $\tilde{x}$ cause profound changes to $\Delta\tilde{h}$ and T_{bub} and hence to dp_v/dT; Eq. (5) shows that this could influence greatly the superheat required for ONB and thus the active nucleation site density. If there were to be a reduction in dp_v/dT for all mixtures, this would reduce the site density through an increase in $\Delta T_{sat,ONB}$ (for given heat flux), and this is sometimes held to explain the observed reduction in mixture boiling heat transfer coefficients.

An analysis of the variation of dp_v/dT with composition was carried out by Shock (1977). He considered the particular cases of ethanol/water and ethanol/benzene mixtures, for which boiling data are presented in Figs. 6 and 7 in the form of $\Delta T_{sat,w}$ - $\tilde{x}$ curves at constant heat fluxes. It is seen that addition of small amounts of ethanol to water and, to a lesser extent, of water to ethanol causes the wall superheat to increase, that is, the heat transfer coefficient to decrease. Ethanol/benzene mixtures show similar effects, more marked at the pure ethanol end of the composition range and more so also for higher pressures.

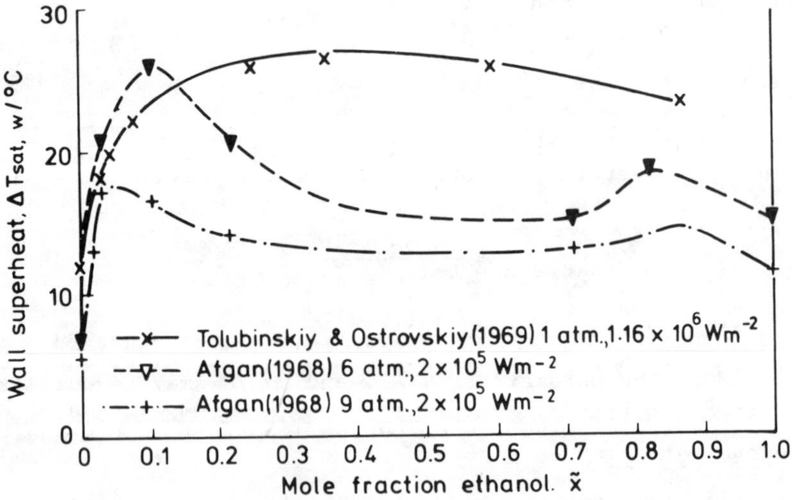

Fig. 6. Wall superheat requirement for various heat fluxes for ethanol/water mixtures.

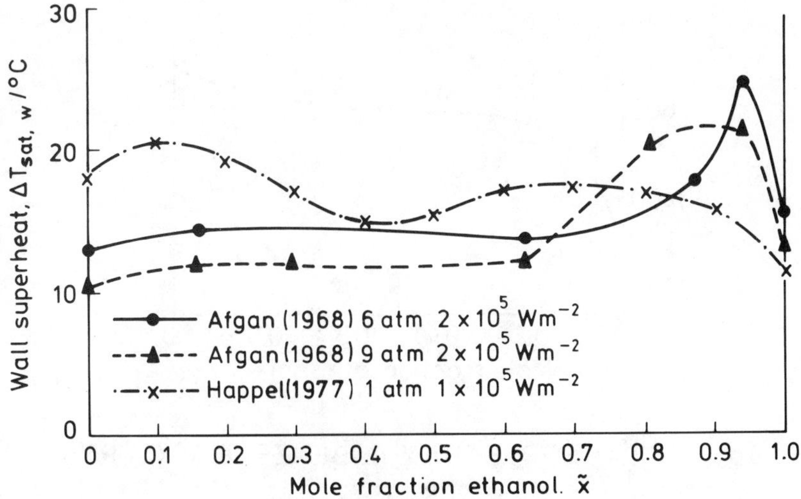

Fig. 7. Wall superheat requirement for various heat fluxes for ethanol/benzene mixtures.

Figure 8 shows dp_v/dT for ethanol/water and ethanol/benzene. In each case addition of ethanol in small quantities causes a steep increase in dp_v/dT, which according to Eq. (5) should lead to an increase in heat transfer coefficient through an increase in nucleation site density. That this does not occur shows that this effect is overwhelmed by other changes. For addition of small amounts of benzene to ethanol there is a slight decrease in dp_v/dT, and the consequent decreased nucleation site density could partly explain the observed sharp decrease of heat transfer coefficient for low $\tilde{x}_{benzene}$ in ethanol.

For ethanol/benzene, and for most organic systems, the explanation of the reduction in heat transfer coefficient lies predominantly in the mass transfer resistance effects that are described in the following three sections; these effects tend to cause a decrease in coefficient at all compositions. The general shape of Fig. 7 can thus be explained as follows. At low $\tilde{x}$ a decrease in coefficient, through mass transfer resistances, is opposed by an increase in nucleation site density, through the mechanism described above; these effects combine to cause a net small decrease in coefficient. Conversely, at high $\tilde{x}$, the two effects tend to act in concert, at least for higher pressures, and give a sharp decrease in coefficient for small additions of benzene to ethanol. These arguments suggest an experiment whereby the relationship between heat transfer coefficient and composition is measured for a binary organic pair over the whole composition range, simultaneously with the nucleation site density. As will be seen later, the general effect of pressure also must be considered.

In addition to the above arguments, it should be noted that,

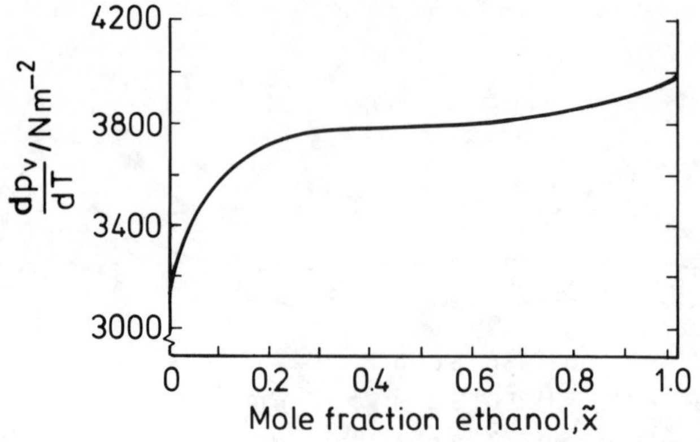

FIG. 8a. $\frac{dp_v}{dT}$ FOR ETHANOL/BENZENE MIXTURES SHOCK (1977)

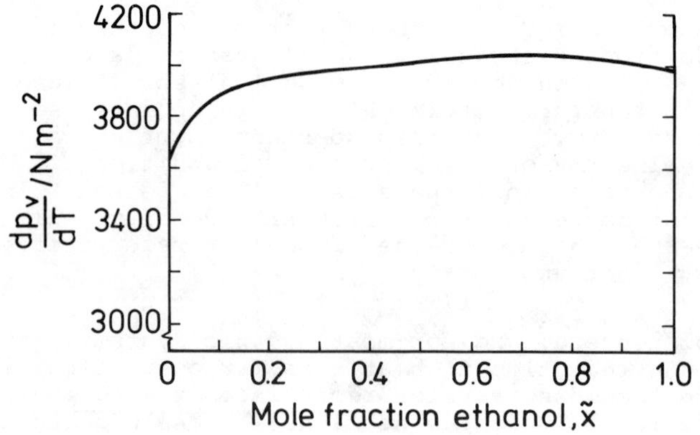

FIG. 8b. $\frac{dp_v}{dT}$ FOR ETHANOL/WATER MIXTURES SHOCK (1977)

Fig. 8. Variation of $\frac{dp_v}{dT}$ with composition.

as well as dp_v/dT, other physical properties, in particular viscosity and thermal conductivity, can change profoundly with composition, and this can complicate still further the interpretation of data such as those given in Fig. 7.

For aqueous solutions the above arguments in relation to dp_v/dT are further complicated by the significant effects of small changes in composition on the remaining terms in Eq. (5), that is, σ and R_e.

The addition of ethanol to water considerably decreases the surface tension. Addition of 10% (molar) ethanol to water at atmospheric pressure decreases the surface tension at the bubble point by some 40%; Eq. (5) indicates that this should tend to decrease $\Delta T_{sat,ONB}$ and thus to decrease the wall superheat required for given heat flux, the opposite behavior to that seen in Fig. 6.

In using Eq. (5) it is common to consider the simple situation where bubbles that are on the point of nucleating are hemispheres at the mouths of cavities with circular entrances of radius R_c, in which case $R_e = R_c$. This is usually satisfactory for high-contact-angle liquids such as water, but the situation is complicated by the use of low-contact-angle, well-wetting fluids. The contact angle of water on a metal surface is greatly reduced by addition of surface-active agents. Thus Eddington and Kenning (1978) have shown that the contact angle of an ethanol/water solution ($\tilde{x}_{ethanol} = 0.09$) on a brass or stainless steel surface is about 35% less than for pure water. This change has two consequences in relation to cavities on a boiling surface and their propensity to act as nucleating sites. First, the number of potential sites that are completely filled with liquid and are thus "snuffed out" will increase (Bankoff, 1958). Second, in those cavities where some gas has been trapped, the nucleus volume is greatly reduced. Robb and Cole (1970) and Singh et al. (1976) have discussed this situation, in the context of pure substances of low contact angle, where the critical nucleus for ONB may no longer be of hemispherical shape at the cavity mouth, $R_e = R_c$, but one deep within the cavity, $R_e \ll R_c$, requiring a greater superheat for stability. Robb and Cole calculated equivalent nucleus sizes from observed ONB wall superheats in artificially drilled conical cavities and found that they could be as low as one-tenth of the cavity mouth radius. The decrease of R_e upon addition of ethanol to water has been demonstrated by data of Shock (1973), shown in Table 1 for flow boiling on nickel-plated copper.

Table 1. *Data for Onset of Nucleate Boiling in Ethanol/Water Mixtures*

$\tilde{x}_{ethanol}$	$T_{w,ONB}$ (°C)	$\Delta T_{sat,w,ONB}$ (°C)	R_e (μm)	p (bar)
0	138.2	9.9	1.05	2.61
0.058	143.5	26.6	0.23	2.59
0.197	145.0	36.7	0.095	2.47

Evidence of the decreased number of active sites in aqueous organic solutions for given wall superheat can be drawn from visual observations carried out by Van Stralen (1959a). Table 2 shows

Table 2. *Nucleate Pool Boiling Data of Van Stralen (1959a)*

Number of Active Sites	Water		4.1% wt. MEK		1.3% wt. *n*-Butanol	
	$\Delta T_{sat,w}$ (°C)	$\dot{q} \times 10^{-5}$ (Wm^{-2})	$\Delta T_{sat,w}$ (°C)	$\dot{q} \times 10^{-5}$ (Wm^{-2})	$\Delta T_{sat,w}$ (°C)	$\dot{q} \times 10^{-5}$ (Wm^{-2})
1	10.5	1.67	18.0	2.93	18.0	2.59
10	17	2.93	23.0	3.97	21.5	3.98
20	20	4.81	24.0	4.68	22.0	5.27
30	21	6.06	26.0	5.44	22.5	6.27

active site populations drawn from his work on boiling of pure water, water/methyl ethyl ketone, and water/*n*-butanol on thin wires. These data confirm that the superheat required to cause activation of the first few nucleation sites is much greater in the aqueous organic solutions than in pure water. The difference diminishes as a greater proportion of the potential nucleation sites become activated.

From an analysis of the data in the above table, Shock (1977) has demonstrated that the large reduction in active site density cannot explain the entire diminution in heat transfer coefficient for aqueous ethanol solutions; the reduction of bubble growth rate (see Sec. 4) is also significant.

The onset of nucleate boiling can be affected by the presence of dissolved gases in an otherwise pure system. For this case Eq. (3) can be written

$$p_v + p_i - p_l = \frac{2\sigma}{R} \tag{7}$$

where p_i is the partial pressure of the gas.

Hence a significant contribution from the inert gas to the total pressure in the bubble can reduce the necessary p_v and thus the superheat requirement for nucleus stability. The superheat can be negative with respect to the saturation temperature of the pure fluid, as demonstrated by the gas bubbles in a standing glass of water.

McAdams et al. (1949) compare boiling curves for pure and air-saturated water (in flow boiling). The reduced $\Delta T_{sat,ONB}$ in the latter case is illustrated in Fig. 9. The nucleate boiling portion of the boiling curve for gassy fluid lies above that for degassed fluid. The curves approach each other with increasing heat flux and merge when the superheat is sufficient to activate most of the potential sites on the surface. Similar behavior is reported by Behar et al. (1966) for nitrogen dissolved in metaterphenyl and by Murphy and Bergles (1972) for flow boiling in fluorinated hydrocarbons.

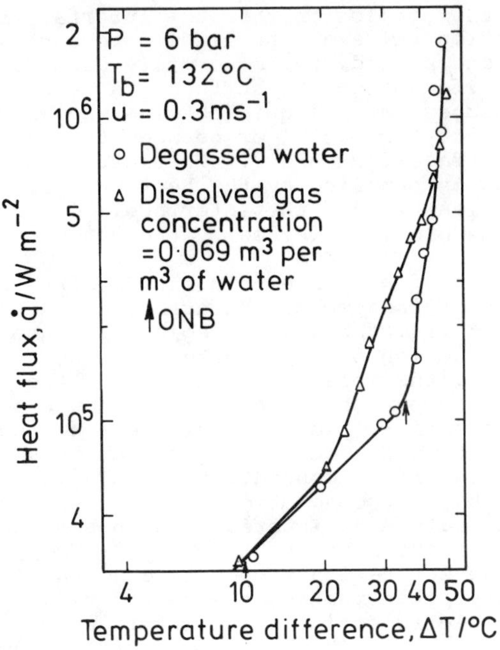

Fig. 9. Boiling curve data for water with and without dissolved gas. Reprinted with permission from McAdams et al. 1949, *Ind. Eng. Chem.* vol. 41, pp. 1945-1953. Copyright (1949) American Chemical Society.

Singer (1970) examined the onset of nucleate boiling in sodium and showed that the effect of dissolved gases depended on the history. If, after freezing and melting, the fluid was left for a sufficiently long period before approaching ONB, the gas diffused out of the cavities and had no effect on the ONB superheat. It did, however, reduce the superheat if the experiments were conducted so that enough remained in the cavities to exert a significant partial pressure.

4 BUBBLE GROWTH DYNAMICS

4.1 Introduction

In considering bubble growth, it is common to examine two somewhat idealized cases: (1) a spherical bubble growing in an infinite volume of uniformly superheated liquid or (2) an isolated bubble at a heated surface. In either case bubble growth in a pure fluid can be divided into two periods (Forster and Zuber, 1955). The first period is dominated by hydrodynamic and surface tension forces. It is usually short except for low-pressure, high-heat-flux conditions (Stewart and Cole, 1972). The inertia stresses initially rise rapidly after nucleation and then decline as $1/R^2$ as the bubble enters the second, "asymptotic," period. The radius increases rapidly during the first period and hence the excess pressure requirement, $2\sigma/R$, rapidly decreases. Making the common assumptions of uniform conditions within the bubble and equilibrium at the interface leads to the conclusion that the interface temperature falls to the saturation value at the liquid pressure. The growth then enters the second period, which normally occupies the greater part of the bubble lifetime on the surface. A temperature gradient exists between the superheated liquid outside the thermal boundary layer and the interface for case 1 or between the superheated solid surface and the interface for case 2. The latent heat requirement for vapor production is supplied by the heat that is conducted down the temperature gradient toward the interface. Consequently the growth rate of the bubble is governed by the conduction of heat.

During the growth of a bubble in a binary mixture the difference in volatilities of the two components causes a stripping of the lighter component from the liquid region close to the bubble. Thus, $\tilde{x}_i$, the mole fraction of the lighter component at the interface, is less than $\tilde{x}_b$, the mole fraction outside the mass diffusion boundary layer. As for pure substances, the growth period in mixtures can be divided into two portions. During the asymptotic period the interface temperature is the saturation value at the liquid pressure and interface concentration. The bubble growth is still controlled by heat conduction from the superheated bulk fluid or wall to the saturated interface and hence the driving force for supply of latent heat is T_b (or T_w) - T_{bub} $(\tilde{x}_i,p)$. As shown in Fig. 10, the saturation temperature at the interface, T_{bub} $(\tilde{x}_i,p)$ is greater than that of the bulk fluid, T_{bub} $(\tilde{x}_b,p)$. Although it is possible to evaluate the interface conditions for

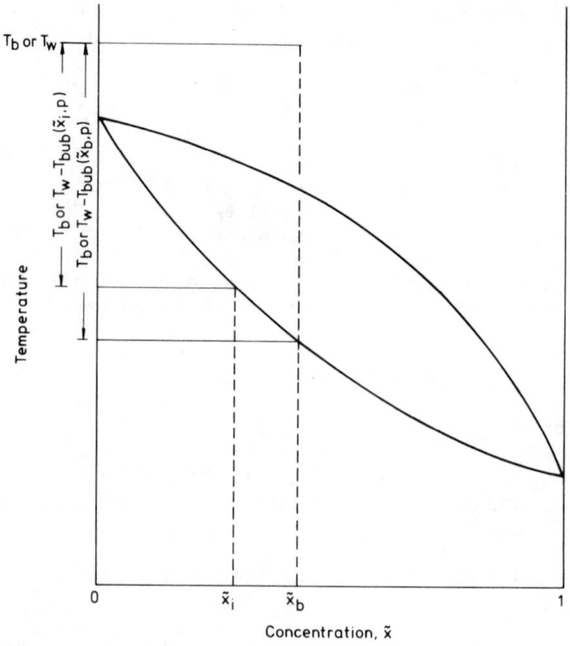

Fig. 10. Concentration differences around a growing bubble.

certain idealized cases, bubble growth rates are normally compared for a given bulk (or wall) superheat relative to the bulk fluid condition. Thus the effect of the increased interface (saturation) temperature is to make the mixture bubble growth rate less than that in the EPF by reducing the conduction driving force. The rule of Konovalov, Sec. 2, shows that in a mixture of two substances that can form an azeotrope (minimum or maximum boiling point), this effect happens at any composition except at the azeotropic point itself, where there is no preferential stripping and no bubble growth rate change.

4.2 Free Bubbles

The majority of previous work has been concerned with the asymptotic bubble growth period. Bruijn (1960) deduced the theoretical growth rate for a bubble in a binary mixture with a spherically symmetrical, initially uniform superheat. He extended the study to the case where neighboring bubbles show mutual interference and showed that this could still further reduce the bubble growth rate. Scriven (1959) performed a similar study for the same initial conditions and accounted for the changes in temperature and concentration fields around the growing bubble that are caused by the radial liquid motion due to the differing phase densities. Skinner and Bankoff (1964a) analyzed the growth of bubbles in pure fluids in spherically symmetric but nonuniform temperature fields, and they extended this work to the case of binary mixtures considering a nonuniform concentration field of arbitrary form (1964b). Since all these studies assume a spherically symmetrical system, they are more truly relevant to bubble growth far from a heating surface.

The following development of the bubble growth rate equation is basically that given by Scriven. Although it is less general than that of Skinner and Bankoff, it gives a deeper insight into the physical picture and has been found to be satisfactory in describing the behavior of real bubbles in several experimental studies.

Before examining the development of the model, it is helpful to note in full the assumptions that are necessary to produce a tractable system of equations:

1. The fluid is of not more than two, nonreacting, components. The heat of mixing of the components is negligible, and their specific heat capacities are equal.

2. The vapor-liquid equilibrium relationship is linear, and equilibrium is assumed at the interface. Dalton's law for gases holds.

3. Within the vapor the compressibility, inertia, and viscosity can be neglected; because of the high thermal and mass diffusivity, there are no gradients of pressure, temperature, and concentration. The presence of noncondensable gases is discounted.

4. The liquid is taken to be an incompressible Newtonian fluid in which viscous dissipation can be ignored and constant physical properties assumed. Heat and mass are transferred by ordinary conduction and diffusion only. There is no heat generation in the liquid.

5. There are no external body forces; the kinetic energy of the bubble is negligible.

6. The effects of liquid viscosity and inertia and of surface tension can be ignored.

The last set of assumptions limits the validity of the method to the asymptotic growth period.

It is perhaps worthwhile to note that in spite of this long list of assumptions and approximations, the method is often successful in describing bubble growth away from heating surfaces.

Continuity considerations around the bubble give

$$ur^2 = \varepsilon \dot{R} R^2 \tag{8}$$

where u is the velocity at radial distance r in the liquid, R is the bubble radius, and $\dot{R}$ is its rate of increase. The parameter $\varepsilon(=1-\rho_v/\rho_l)$ characterizes the convective motion of the liquid surrounding the bubble caused by the difference in densities of the phases. The effect diminishes as the pressure increases.

The full equation of motion can be written

$$\frac{(p_v - p_\infty - 2\sigma/R)}{\varepsilon\ \rho_l} = R\ddot{R} + \frac{3}{2}\dot{R}^2 + 4\frac{\mu_l}{\rho_l}\frac{\dot{R}}{R} \tag{9}$$

where μ_l is the liquid viscosity. This is the Rayleigh equation extended by the inclusion of the surface tension term.

The energy equation in the liquid gives

$$\frac{\delta T}{\delta t} = \kappa\left(\frac{\delta^2 T}{\delta r^2} + \frac{2}{r}\frac{\delta T}{\delta r}\right) - \frac{\varepsilon R^2 \dot{R}}{r^2}\frac{\delta T}{\delta r} \tag{10}$$

where t is time and κ is the thermal diffusivity.

A component mass balance in the liquid gives

$$\frac{\delta c}{\delta t} = \delta\left(\frac{\delta^2 c}{\delta r^2} + \frac{2}{r}\frac{\delta c}{\delta r}\right) - \frac{\varepsilon R^2 \dot{R}}{r^2}\frac{\delta c}{\delta r} \tag{11}$$

where c is the concentration in the liquid, the subscript l being omitted for clarity, and δ is the diffusivity.

The equations are solved subject to the following initial conditions:

$$\dot{R}(0) = 0 \tag{12}$$

$$T(r,0) = T_\infty \tag{13}$$

$$c(r,0) = c_\infty \tag{14}$$

and to the following boundary conditions:

1. At infinite distance

$$T(\infty,t) = T_\infty \tag{15}$$

$$c(\infty,t) = c_\infty \tag{16}$$

2. At the liquid-vapor interface

Component mass balance:

$$y\rho_v\dot{R} = c(R,t)(1-\varepsilon)\dot{R} + \delta\left(\frac{\delta c}{\delta r}\right)_{r=R} \tag{17}$$

Energy balance:

$$\rho_v\dot{R}\left\{\Delta h + c_v\left[T(R,t) - T_\infty\right]\right\} = \rho_l c_l\left[T(R,t) - T_\infty\right](1-\varepsilon)\dot{R} + \lambda\left(\frac{\delta T}{\delta r}\right)_{r=R} \tag{18}$$

The specific latent heat Δh is a mass fraction average, and λ is the thermal conductivity.

Since the total enthalpy of the system is constant, as is the amount of the light component in it, we can write

$$\frac{4}{3}\pi R^3\rho_v\left\{\Delta h + c_v\left[T(R,t) - T_\infty\right]\right\} + \rho_l c_l\int_R^\infty 4\pi r^2\left[T(r,t) - T_\infty\right]dr = 0 \tag{19}$$

$$4\pi R^3\rho_v\;(\rho_l - c_\infty) + \rho_l\int_R^\infty 4\pi r^2\left[c(r,t) - c_\infty\right]dr = 0 \tag{20}$$

Equation (9) can be simplified by noting the assumptions that the viscous, inertia, and surface tension terms can be ignored, which simply gives that $p_v = p_\infty$ (throughout the asymptotic growth period). From the assumption of equilibrium at the interface and using the first terms of a Taylor expansion, Scriven expresses p_v as a function of concentration and temperature.

$$p_v = p_v(T_{sat}, c_\infty) + \left[T(R,t) - T_{sat}\right]\frac{\delta p_v}{\delta T} + \left[c(R,t) - c_\infty\right]\frac{\delta p_v}{\delta c} \tag{21}$$

Since $p_v(T_{sat}, c_\infty) = p_v = p_\infty$, Eq. (21) gives

$$c(R,t) = c_\infty + \left[T(R,t) - T_{\text{sat}}\right]\left(\frac{\delta c}{\delta T}\right)_p \tag{22}$$

In solving the equations Scriven assumed a solution of the form

$$R = 2\beta\ (\kappa t)^{1/2} \tag{23}$$

An analytical solution is not possible, and in order to find the composition of the vapor in the bubble y and the growth constant β, the following two equations must be solved simultaneously:

$$y^2\omega\zeta\ (1-\alpha_i)\ \psi - y\left[\zeta\ (\alpha_i + \omega\ \psi) - \omega(1-\alpha_i)(1 + \omega\ \psi)\right] + \omega(1 + \omega\ \psi) = 0 \tag{24}$$

$$\tau = \frac{\xi\phi(\varepsilon,\beta)}{1-\omega\nu\phi(\varepsilon,\beta)} + \frac{(\omega-y\zeta)\ \phi(\varepsilon,\gamma\beta)}{\mu[1-\omega\phi(\varepsilon,\gamma\beta)]} \tag{25}$$

In Eqs. (24) and (25), α_i is the relative volatility at the interface; ω, ζ, and ξ are physical property functions given, respectively, by ρ_v/ρ_l, ρ_v/c_∞, and $\rho_v\Delta h/\rho_l c_l\ T_\infty$. τ is the dimensionless superheat

$$\frac{T_\infty - T_{\text{bub}}\ (x_\infty)}{T_\infty}$$

and ν is the concentration difference $(c_l-c_v)/c_l$.

The functions μ and ϕ are calculated from

$$\mu = \rho_l\left\{\frac{\delta[(x-x_\infty)/x_\infty]}{\delta[(T-T_\infty)/T_\infty]}\right\}_p \tag{26}$$

$$\phi(\varepsilon,\beta) = 2\beta^3\ \exp(\beta^2 + 2\varepsilon\beta^2)\int_\beta^\infty a^{-2}\ \exp(-a^2 - 2\varepsilon\beta^3 a^{-1})\ da \tag{27}$$

where a is a dummy integration variable and $\phi(\varepsilon,\Gamma\beta)$ is calculated as in Eq. (27) with β replaced by $\Gamma\beta$; Γ is the ratio of diffusivities κ/δ. Finally, the function $\psi(\varepsilon,\Gamma\beta)$ in Eq. (24) is calculated from

$$\psi(\varepsilon,\Gamma\beta) = \frac{\phi(\varepsilon,\Gamma\beta)}{1-\omega\phi(\varepsilon,\Gamma\beta)} \tag{28}$$

The interface concentration $c(R,t)$ is given by

$$c(R,t) = -(y\zeta - \omega)\phi(\varepsilon,\Gamma\beta) \tag{29}$$

To calculate the bubble growth constant β from this system of equations, the terms $\varepsilon, \omega, \zeta, \alpha, \xi, \nu$, and μ must be calculated from the physical properties of the system and τ must be calculated from the known superheat. With all the other terms known, simultaneous solution of Eqs. (24) and (25) gives β and y. Figure 11 shows a graph of the function $\phi(\varepsilon, \beta)$; Scriven gives analytical expressions for certain limiting cases.

The complete solution of this set of equations is possible only with a computer. However, simplified $R(t)$ relationships are available for some limiting cases.

For small superheats,

$$R \approx \left(\frac{2\, \Delta T_{\text{sat}}\, \lambda t}{\rho_v \left\{ \Delta \bar{h} - \lambda (y-x)\, \tilde{R}\, T^2_{\text{bub}}\, (1-\alpha_i)/\rho\, \delta x_\infty \Delta h_1 \left[\tilde{M}_2 x + (1-x_\infty)\tilde{M}_l\, (1+\alpha l) \right] \right\}} \right)^{1/2} \tag{30}$$

where l is defined by

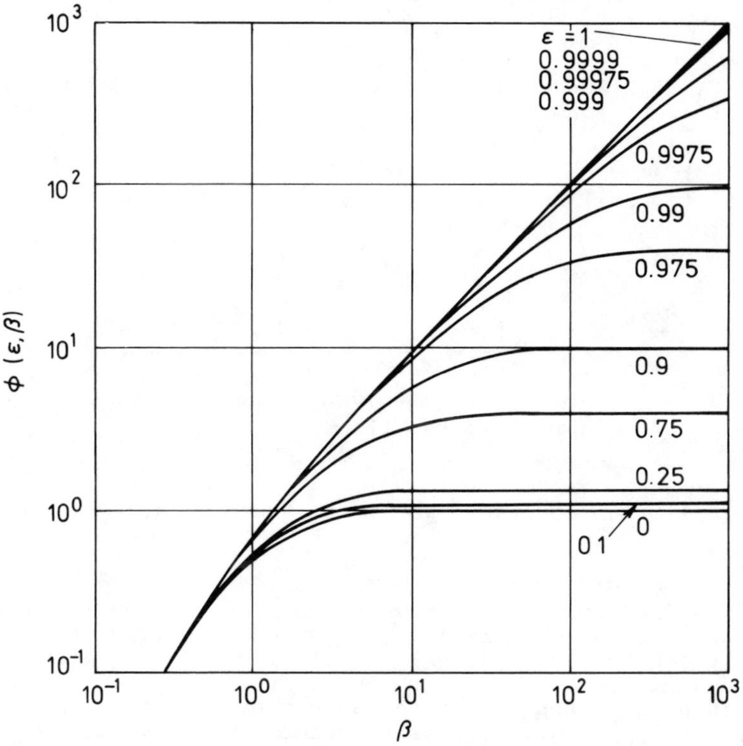

Fig. 11. The function $\phi(\varepsilon, \beta)$. Scriven (1959).

$$l = \frac{\Delta h_2}{\Delta h_1} \frac{(1-x_\infty)}{x_\infty} \tag{31}$$

and ΔT_{sat} is the superheat relative to the bulk liquid bubble point. The interface concentration, and hence temperature, can be evaluated from Eq. (29) using ψ, which is given by

$$\psi \approx \left(\frac{\kappa}{\delta}\right) \frac{\Delta T_{sat}}{(\rho_v/\rho_l)\{(\Delta h/c_l)[(y-x_\infty)/\delta x/\delta T)_p] \ (\kappa/\delta)\}} \tag{32}$$

For $\rho_v \ll \rho_l$ and for large β, but not too large because of the assumed linear equilibrium relationship,

$$R \approx \sqrt{\frac{12}{\pi}} \left(\frac{\Delta T_{sat} \ \sqrt{\kappa t}}{(\rho_v/\rho_l)(\Delta h/c_l) \ \{ \ 1 - (c_l/\Delta h) \ \sqrt{\kappa/\delta} \ (y-x_\infty)(dT/dx)\}} \right) \tag{33}$$

and the interface condition is given by Eq. (29) and

$$\psi \approx \sqrt{\frac{\kappa}{\delta}} \left\{ \frac{\Delta T_{sat}}{(\rho_v/\rho_l)(\Delta h/C_l)\{1 - (c_l/\Delta h) \ \sqrt{\kappa/\delta} \ (y-x_\infty)(dT/dx)\}} \right\} \tag{34}$$

Van Stralen (1966a) deduced a result similar to Eq. (33) but in a slightly different form:

$$R = \sqrt{\frac{12}{\pi}} \left\{ \frac{\Delta T_{sat} \ \sqrt{\kappa t}}{(\rho_v/\rho_l)(\Delta h/c_l) \ [1 + (c_l/\Delta h) \ \sqrt{\kappa/\delta} \ (\Delta T_{av}/G_d)]} \right\} \tag{35}$$

where

$$\frac{\Delta T_{av}}{G_d} = - \ x_\infty \ [K(x_i) - 1] \left(\frac{dT}{dx}\right)_{x=x_\infty} \tag{36}$$

$$\approx -x_\infty \ [K(x_\infty) - 1] \left(\frac{dT}{dx}\right)_{x=x_\infty} \tag{37}$$

K is the equilibrium constant, y^*/x; ΔT_{av} and G_d are defined below.

Substitution of Eq. (37) into Eq. (35) gives a result identical to Eq. (33). The formulation presented by Van Stralen is somewhat easier to use through the simple graphical device illustrated in Fig. 12. It can be seen that ΔT_{av} is the increase in saturation temperature at the bubble boundary due to the depletion of the light component. It represents the amount by wich the effective superheat, $T - T_{bub}(x_i)$, is less than the (known) value appropriate to the bulk fluid, $T - T_{bub}(x_\infty)$, and has been termed by Van Stralen the "loss of available superheat." The term G_d is the mass fraction

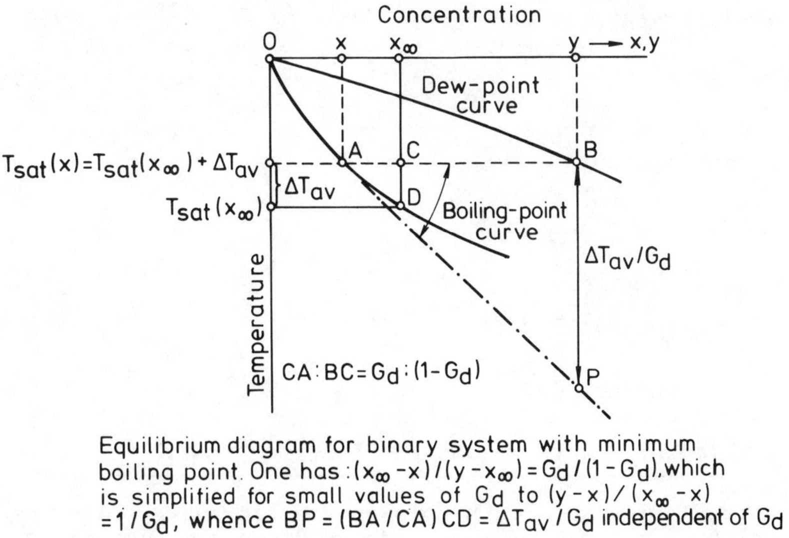

Fig. 12. Graphical calculation of $\Delta T_{av}/G_d$. Van Stralen (1966a).

of the liquid that is vaporized; Van Wijk et al. (1956) showed it to be in the range 0.001–0.01 for bubbles growing on a heating wire. The approximate solution for $\Delta T_{av}/G_d$ shown in Fig. 12 is independent of G_d (for small G_d), pressure (at low pressure), and ΔT_{sat}.

Conceptually, G_d could be in the range 0 to 1. As it tends to zero the loss of available superheat also tends to zero; as G_d tends to 1 (total vaporization), the loss of available superheat tends to the potential maximum $T_{dew}(x_\infty) - T_{bub}(x_\infty)$.

Examination of Eq. (33) shows that the bubble growth rate is reduced by a factor F, where

$$F = 1 - \frac{c_l}{\Delta h} \sqrt{\frac{\kappa}{\delta}} \, (y^*-x) \, \frac{dT}{dx} \tag{38}$$

that is, to obtain the bubble growth rate in a mixture, multiply that for the EPF by $1/F$. Since $y-x$ and dT/dx are always of opposite sign, this is in agreement with the reduction in bubble growth (through the volatility mechanism) for all mixtures. Since the mass diffusivity in liquids is often two orders of magnitude lower than the thermal diffusivity, the term $\sqrt{\kappa/\delta}$ can be large. For example, in dilute aqueous solutions of n-butanol it is 13.

Van Stralen has calculated F for several aqueous organic solutions, and Table 3 shows these results together with the growth constant for a superheat of 10°C. [Note that β is directly proportional to ΔT_{sat}, as shown by Eqs. (23) and (35).] Note the

Table 3. *Bubble Growth Constant for Water and Dilute Aqueous Solutions*

Fluid	F	β
Water	1	29.2
$x_{n\text{-butanol}} = 0.015$	1.33	21.9
$x_{n\text{-butanol}} = 0.06$	1.15	25.5
$x_{\text{methyl ethyl ketone}} = 0.041$	4.0	7.3

considerable slowing down of bubble growth for small additions of solute, which is typical of many aqueous organic solutions.

It can be seen from Eq. (35) that a maximum reduction in bubble growth rate occurs at a maximum in $\Delta T_{av}/G_d$, the value of which depends on the composition. As we shall see, the bubble departure diameter has a minimum and the critical heat flux a maximum at the same concentration as the minimum in the bubble growth rate. Van Stralen (1970) has termed this the "boiling paradox."

Van Stralen (1968a) has suggested that the effect of the volatility difference can be envisaged as a change in the effective Jakob number, Ja. The bubble growth rate is given as

$$R = \sqrt{\frac{12}{\pi}}\ \text{Ja}\ \sqrt{\kappa t} \tag{39}$$

which is equivalent to Eq. (35). The Jakob number for a mixture is then given by

$$\text{Ja} = \frac{\rho_l}{\rho_v}\frac{c_l}{\Delta h}\left(\Delta T_{\text{sat}} - \Delta T_{av}\right) \tag{40}$$

Zijl et al. (1977) extended these studies of free bubble growth to the more general case of inertia control (as in the early stages of growth). They wrote the full heat, mass, and momentum transfer equations, including the liquid inertia terms, and solved them, by the use of fractional derivatives, to obtain the following equation valid for all stages of growth of a spherical bubble:

$$R = \text{Ja}\ \sqrt{\kappa\ \gamma}\ e^{t/\gamma}\text{erfc}\left(\sqrt{\frac{t}{\gamma}}\right) + \frac{2}{\sqrt{\pi}}\ \text{Ja}\ \sqrt{\kappa t} - \text{Ja}\ \sqrt{\kappa\ \gamma} \tag{41}$$

where the Jakob number is given by Eq. (40) and γ, given by

$$\gamma = \frac{3}{2} \frac{\rho_l}{\rho_v} \frac{T_{bub}(x_\infty)}{\Delta T_{sat}} \frac{\kappa}{\Delta h} Ja^2 \tag{42}$$

is the characteristic time for transition from the initial to the asymptotic growth period. Equation (41) should be compared with Eq. (39), valid only for the asymptotic period. At large times they give identical results except for the constant multiplier which differs because Zijl et al. did not allow for the curvature of the bubble.

Van Stralen (1968a) has discussed bubble growth in aqueous salt solutions where addition of salt to water increases the boiling point. Here too the lighter component (water) is preferentially stripped from the interface, and there is a loss of available superheat and a reduction in bubble growth rate.

So far we have considered only theoretical analyses of the bubble growth process. An excellent experimental test of the models proposed was conducted by Florschuetz and Rashid-Khan (1970), who measured bubble growth rates in a system that closely resembled free bubble growth. A movie camera was focused at a point some way above a heating surface from which a stream of bubbles was rising. The pressure was suddenly reduced, causing an appreciable uniform bulk liquid superheat. The subsequent growth of the nuclei provided by the bubble stream was measured from the movie films.

Since they used conditions such that $\varepsilon \approx 1$ and $\beta \approx 6$, Eq. (33) is valid. For a pure substance, an azeotrope, or an EPF of a mixture, this reduces to

$$R = \sqrt{\frac{12}{\pi}} \frac{\rho_l c_l \Delta T_{sat} \sqrt{\kappa t}}{\rho_v \Delta h} \tag{43}$$

from which the theoretical growth constant

$$\beta = \sqrt{\frac{3}{\pi}} \frac{\rho_l c_l \Delta T_{sat}}{\rho_v \Delta h}$$

was calculated. Figure 13a shows results calculated on this basis, that is, for heat diffusion only. If heat diffusion alone could account for bubble growth, the results would lie on the straight line, which represents Eq. (43). Two groups of results can be distinguished. For $\tilde{x}_{i\text{-propanol}} = 0.40$ and 0.46, where the boiling range is fairly small (7 and 5.4°C), the loss of available superheat is small and the simplified heat diffusion model is adequate. For lower concentrations of *i*-propanol the boiling range is larger, up to 15.4°C for $\tilde{x}_{i\text{-propanol}} = 0.088$. The loss of available superheat is then significant and the simplified model, Eq. (43), overpredicts the bubble growth rates. When the results are presented in the form of Fig. 13b, with theoretical β calculated from the general formulation, Eq. (33), all the results are seen

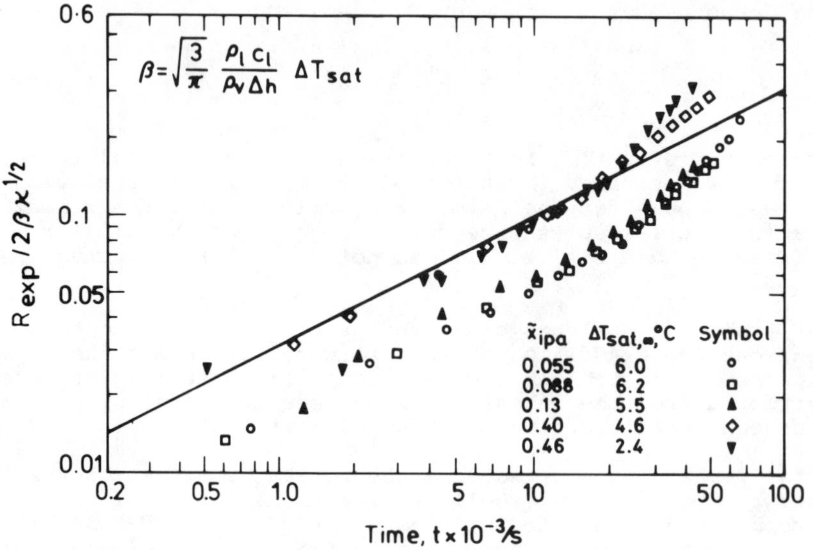

FIG. 13a. COMPARISON WITH HEAT DIFFUSION THEORY

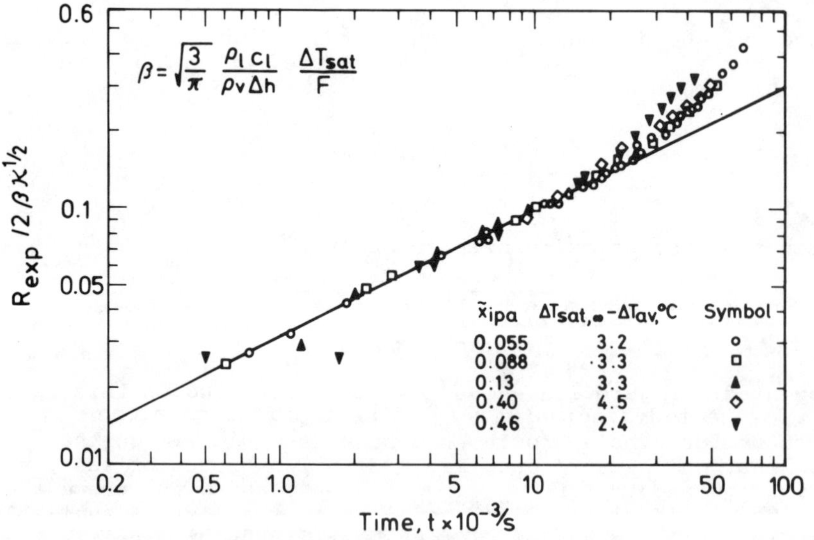

FIG. 13b. COMPARISON WITH HEAT AND MASS DIFFUSION THEORY

Fig. 13. Results of Florschuetz and Khan (1970) for growth rates of free bubbles in isopropanol/water.

to be well predicted by the model in all but the later stages of bubble growth. Florschuetz and Rashid-Khan suggest that this discrepancy is due to the effects of bubble translation, which are not accounted for in the model.

Shah and Sha (1978) analyzed the growth of bubbles rising in superheated liquid. They showed how the motion of the bubble decreases the thermal boundary layer thickness and hence increases the growth rate. An opposing tendency arises through the increased drag force arising from the increased size. Their analysis models the growth of the bubble by considering the interaction between net bubble motion through the liquid and the radial motion of the bubble interface. The model agrees well with data of Florschuetz et al. (1969) for pure ethanol and *i*-propanol. Tokuda (1970) presented a similar model for pure fluids and for binary mixtures. Although he does not evaluate the combined effects of mass transfer and translation, we may perhaps expect that translation would decrease the mass transfer boundary layer thickness and so enhance the bubble growth rate still further than for pure substances by reducing the loss of available superheat at the interface. Both these theoretical studies implicitly assume that there is no hindrance to circulation within the bubble and to the liquid motion at the interface. However, the velocity distribution around the interface can give rise to concentration differences. The resultant surface tension gradients can affect the flow patterns within and around the rising bubble and thus can, in turn, affect the rise velocity and the growth rate. Ruckenstein (1964) has considered this situation for spherical bubbles with laminar flow. He shows that, for small bubbles, the mass transfer coefficient in the liquid can be significantly affected by the Marangoni effect. An increase in coefficient results if the direction of mass transfer is such as to lead to a decrease of surface tension with distance measured in the direction of bubble motion, such as in the case of a steam bubble rising in liquid ethanol; the opposite effect occurs if an increase of surface tension arises through bubble motion. The work of Ruckenstein ignores the possibility that a compressed monolayer of solute molecules in, for example, aqueous surfactant solutions, can build up from the trailing edge of the bubble and eventually surround the entire bubble. This can severely restrict the flow patterns in and around the bubble and hence the mass transfer through the interface.

Further support for the bubble growth rate given by Eq. (33) is provided by the work of Van Stralen (1968b). He measured bubble growth rates in *n*-butanol/water on a heated wire, $\Delta T_{sat,\infty} = 0.31^0C$. The growth rate for free bubbles is seen in Fig. 14 to agree well with the theoretical rate calculated using the bulk superheat. The displacement in radius is due to the high superheat experienced by the bubble before it leaves the wire.

Figure 14 shows that a bubble can oscillate in shape and volume but that the equivalent spherical radius is in accord with Eq. (35). Zijl et al. (1977) discuss the behavior of bubbles with respect to small perturbations in radius and deduced that bubble growth was stable. Oscillations induced in the rapid initial rise tend to be damped out. The radius-time curves shown in Fig. 14 appear to support this conclusion. However, it should be noted

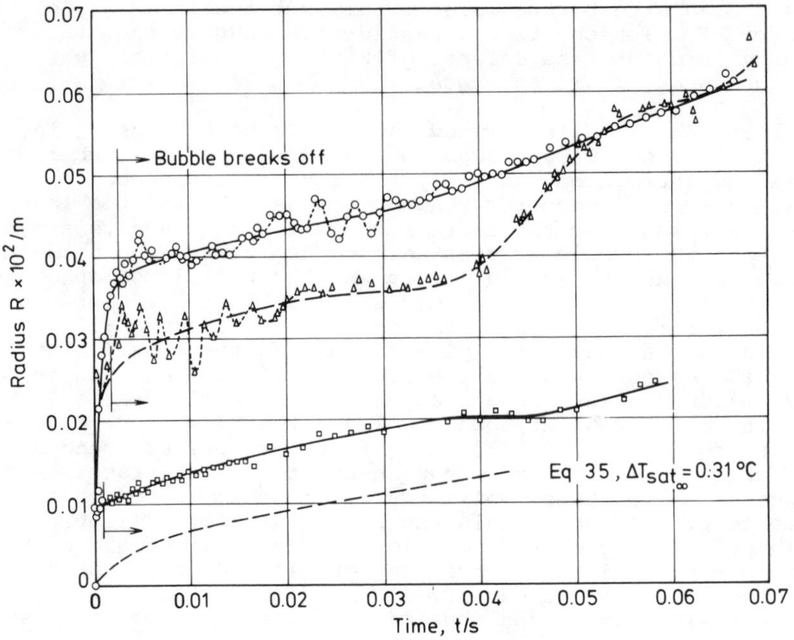

Fig. 14. Growth rate of bubbles on a heating wire and after departure water/methyl ethyl ketone, $x_{\text{methyl ethyl ketone}} = 0.06$ $\Delta T_{sat,w} = 22^0C$ $\Delta T_{sat,\infty} = 0.31^0C$ Van Stralen (1968b).

that Zijl et al. examined stability only with respect to the generalized momentum, energy, and mass balance equations. It is possible that in some systems of surface active agents the perturbations could set up local variations in mass transfer efficiency around the bubble. The consequent local variations in surface tension could produce forces that would maintain the oscillation.

Van Stralen's results (1968b) also show the mutual influence of neighboring bubbles. A large bubble can consume so much of the light component from the vicinity of a smaller neighbor that the growth of the latter may temporarily cease if the loss of available superheat equals $T_b - T_{bub}(x_b)$. In such a case the only means by which the bubble can grow is mass diffusion—a much slower process than heat diffusion. This effect could theoretically also occur for a single bubble if it rose from the heater surface with $T_{bub}(x_i)$ equal to the bulk temperature (quite possible, as bulk superheats in pool boiling are usually fairly small), no heat transfer could occur into the bubble until mass diffusion had increased x_i and consequently decreased the interface temperature.

4.3 Attached Bubbles

The picture presented in Sec. 4.2 becomes complex when the

distorting effect of a heating surface acts on the velocity, temperature, and concentration fields around a growing bubble. Experimental data have been reported for mixture bubble growth on surfaces providing a range of cavity sizes (Afgan, 1968; Tolubinskiy and Ostrovskiy, 1966, 1969; Tolubinskiy et al., 1970), and at prepared sites (Benjamin and Westwater, 1961; Yatabe and Westwater, 1966).

Yatabe and Westwater and Benjamin and Westwater tested the validity of the relationships assumed in Eqs. (23) and (33), by fitting their results to the relationship

$$R = \beta(\kappa t)^{n} \tag{44}$$

Yatabe and Westwater found values of n for ethanol/water and ethanol/i-propanol ranging from 0.25 to 0.45 and changing smoothly with composition. Benjamin and Westwater found values ranging from 0.2 to 0.8 for glycol/water mixtures with a sharp minimum at $x_{water} = 0.1$. They found that the trends in β with changing composition agreed with those predicted by Eq. (33), whereas Yatabe and Westwater did not even find qualitative agreement. Thus it is clear that the simple model proposed by Scriven and by Van Stralen requires modification for boiling on surfaces.

The literature shows many models for nucleate boiling of pure substances at solid surfaces. The following mechanisms have been held to contribute to the total heat transfer:

1. Evaporation of the superheated evaporation microlayer underneath a growing bubble

2. Evaporation of the superheated relaxation microlayer around a growing bubble

3. Transient conduction as liquid rushes in after bubble departure

4. Natural convection in the regions away from active sites

5. Enhanced convection in the liquid around an active site due to bubble growth and detachment

6. Vapor-phase convection above a dry spot due to microlayer evaporation

Since many of these mechanisms are interrelated, the changes due to the fact that the coolant is a mixture may be complex. The volatility effect has mainly been examined in relation to case 1–4.

The concept of an evaporation microlayer under a growing bubble was proposed by Moore and Mesler (1961) after the observation of temperature fluctuations at the heating surface. As the bubble grows outwards in an approximately hemispherical shape, it leaves a wedge-shaped layer of liquid beneath it. Heat is conducted, with little resistance, through this layer and causes evaporation at the interface. Complete evaporation at the center results in a circular dry spot of increasing radius. Fath and Judd (1978) showed, using water, that the contribution of the microlayer to the total heat load decreased as the heat flux decreased and as the pressure increased (30% at 51.5 kN m^{-2} down to 5% at 101.3 kN m^{-2}). Judd and Hwang (1976) showed the contribution of the microlayer to be independent of the subcooling.

Van Ouwerkerk (1971, 1972) analyzed growth of bubbles at a wall assuming nearly-hemispherical shape; that is neglecting effects of surface tension. He showed that a very good approximation is then provided by a self-similar solution in which the thermal properties of the wall do not appear. That arises because most of the evaporation from the microlayer occurs in its outer portion, near the growing perimeter of the base, where the liquid-vapor interface has been freshly created. The temperature at a fresh interface in pure liquid falls almost immediately to the saturation value, causing rapid heat flow and evaporation. After subsequent time t, there is a thermal diffusion layer of thickness $\sqrt{\kappa t}$ within the liquid, constituting a region of decreasing temperature, a temperature "wave", which travels through the microlayer. When that "wave" reaches the wall, it is affected by the thermal properties of the wall (if the relevant group $\lambda \rho c$ differs in liquid and wall) and a reflected "wave" then starts from the wall. Evaporation from the interface is unaffected by the wall until that reflected "wave" reaches it. By that time, however, the bubble has usually grown much more and the significant contribution to bubble growth comes from regions near its new perimeter. Thus the physical properties of the wall, which affect the reflecting back of the wave, have little effect on the overall bubble growth rate.

Van Ouwerkerk extended his analysis of nearly-hemispherical bubbles to binary mixtures by showing that, in such mixtures, the temperature at a freshly created interface will again fall to a steady value. Again, the argument holds that, until the reflected "wave" reaches the interface, the initial value can be determined from the conditions of boiling and the properties of the mixture and lies above the bubble point temperature for the bulk fluid. There will also be a concentration "wave" moving through the microlayer at a much slower speed than the temperature "wave". Provided that initial interface temperature is substituted for saturation temperature, the argument above for pure liquids can be applied for binary mixtures. This implies that the growth rate of a hemispherical bubble in a binary mixture is slowed down by the same amount as for a spherical bubble, Sec. 4.2. This conclusion is in agreement with the earlier experimental results of Benjamin and Westwater (1961).

It has also received support and extension to other bubble shapes from recent work by Stone (1980). He grew individual bubbles into initially stagnant isothermal (superheated) hexane, octane and their mixtures, on a glass surface on which there were thin film resistance thermometers. Single bubbles were induced by pulsing current through an unused surface thermometer, involving energy input much less than the latent heat used in the bubble. He found that bubble behavior in mixtures, including growth and also change of shape and departure, were all very close to those in pure liquids provided the initial interface temperature was substituted for saturation temperature as described above. In particular, he thus confirmed that, for bubbles of any shape in initially stagnant isothermal liquid, the growth rate was reduced for mixtures to the same extent as for spherical bubbles. Behavior of the thermometers on the wall was also consistent with the discussion above, with due allowance for thermal "waves" and also mass diffusion and depletion at longer times. He obtained further confirmation of the model from fine thermocouples within the growing bubbles. For pure liquids the temperatures fell rapidly from the initial superheat to the saturation value and stayed constant until bubble departure. For mixtures they fell rapidly, but only to about the predicted initial interface temperature. Thereafter they rose, and more detailed analysis showed that the rise could be attributed to effects of the thermal "waves" and the longer-term diffusion and depletion of volatile component in the microlayer, causing changes in composition and temperature at its interface. Successive layers of vapor were thus produced at higher and higher temperature, which eventually reached the thermocouple. Assuming vapor temperatures did not vary on the way, there was general agreement between observation and theory, though mixing and diffusion in the vapor presumably could affect details. He also noted that, as expected, a binary microlayer would not dry out completely if the initial temperature of the system was less than the boiling point of the less volatile component at the prevailing pressure.

A rather different analysis was given earlier by Zeugin et al. (1975) who investigated the effect of coolant composition on microlayer behavior. They measured temperature-time traces of a thermocouple under a growth site on a pyrex surface for water, water/methanol and water/butanol mixtures, see Fig. 15. Heating was carried out by shining a powerful light at the region of the nucleation site; the heat flux was unknown. For water the initial sharp drop in temperature is attributed to the growth of bubble over the thermometer and formation of the microlayer. The temperature then approaches the saturation level and rises sharply when the increasing dry patch, marking complete exhaustion of the microlayer, reaches the thermometer. In the case of the mixtures there appears to be no drying out of the microlayer and a thin liquid layer exists above the thermometer, its temperature increasing gradually with the loss of butanol (note the change of scale in the graph for the mixture). The thermocouple traces showed good agreement, as shown in the figure, with a mathematical model based on heat, mass and momentum considerations. The model also requires a value for microlayer thickness chosen to match the thermometer traces. This model did not account for the heat capacity of the liquid microlayer, so the thermal "waves" considered by Van Ouwerkerk were effectively assumed to travel at infinite velocity.

It has been suggested that there can be significant effects of surface pressure gradients along the liquid-vapor interface of the microlayer (the Marangoni effect). Such questions are discussed in Sec. 5.

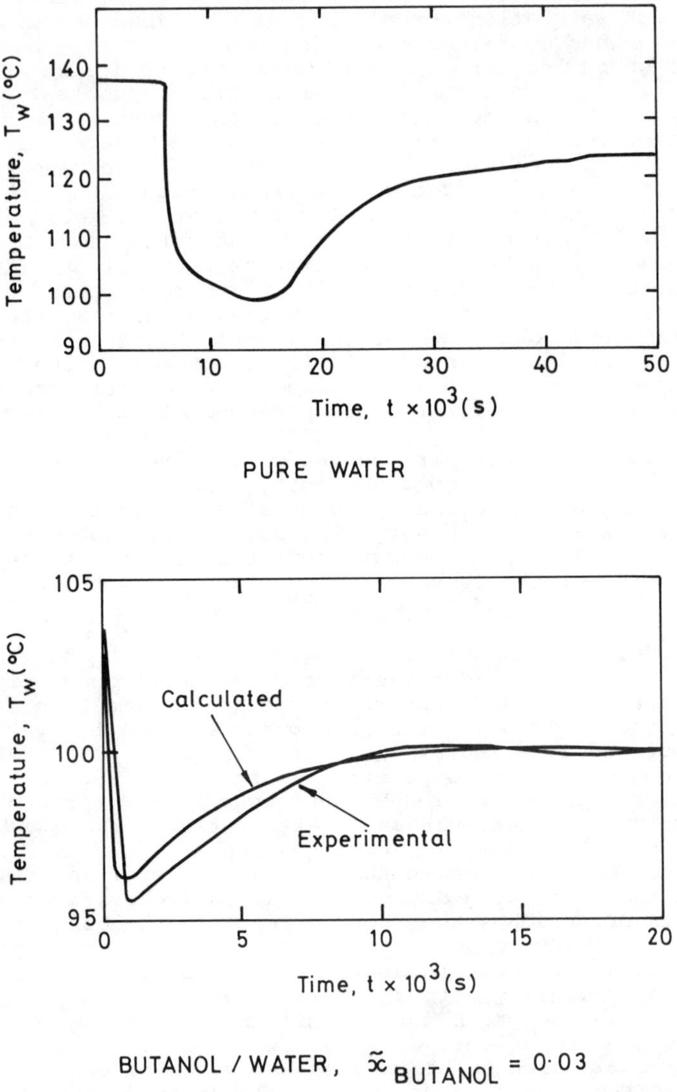

Fig. 15. Temperature recorded by a fast-response thermocouple underneath a growing bubble: effect of composition, Zeugin et al. (1975).

An alternative approach concerned mainly with the relaxation microlayer contribution has been extensively developed by Van Stralen using experimental evidence from wires of diameter 0.02 cm. For such small wires the contribution of the thin evaporation microlayer between bubbles and surface might be expected to be small, since the bubbles quickly grow to a size considerably greater than the wire diameter.

The relaxation microlayer model presumes that after the departure of a bubble the site is covered by fresh liquid that is saturated, or slightly superheated. During a "delay time," t_d, the superheat of the liquid layer increases. When the bubble starts to grow, the superheated liquid layer is pushed out by the bubble and surrounds it, releasing its superheat to the saturated bubble. When the bubble departs, the cycle recommences. Considerable refinements have been made to the model (Van Stralen, 1966a, 1966b), such as allowing the bubble to protrude through the relaxation microlayer. However, whatever may be the growth rate, it is predicted to be reduced in mixtures in the same proportion as for free bubbles, that is, by the factor F defined by Eq. (38).

By considering the unsteady-state heat transfer in the various stages of the bubble cycle, Van Stralen (1966b) deduced that the length of the growth period, t_g, should be identical in mixtures to that in the equivalent pure fluids (for given heat flux) but that the delay time in the mixtures should be reduced. Thus, in pure substances,

$$t_d = 3t_g \tag{45}$$

whereas in mixtures,

$$t_d = \frac{3}{\left[1 + (c_l/\Delta h)(\kappa/\delta)^{1/2}\,(y-x_\infty)(dT/dx)\right]}\, t_g \tag{46}$$

Hence the frequency of bubble growth $[1/(t_g + t_g)]$ should be increased in mixtures and evidence taken from films and summarized in Table 4 confirms this. These data are a summary of those given

Table 4. *Selection of Results Given by Van Stralen (1966b) for Boiling on Platinum wires, Diameter = 0.2 mm, Heat Flux = 44.8 x 10^4 W m^{-2}*

Fluid	F	R_d (mm)	$t_g \times 10^4$ (s)	$t_d \times 10^4$ (s)	ν (s)	$q_{bi} \times 10^4$ (Wm^{-2})
Water $\Delta T_{sat,w} = 20^0C$	1	1.19	71.7	128.3	50.0	15.1
$x_{methyl\ ethyl\ ketone} = 0.041$, $\Delta T_{sat,w} = 24^0C$	4.0	0.26	43.3	25.0	146.3	8.8
$x_{n\text{-butanol}} = 0.15$, $\Delta T_{sat,w} = 21^0C$	1.33	0.7	48.3	131.7	55.5	13.4

by Van Stralen for several bubbles in each of the mixtures; for each mixture the various bubbles display a fairly wide range of measured parameters. In general, even for pure water, Eq. (45) does not appear to be well adhered to. A reduced delay time in methyl ethyl ketone/water, for which $F = 4.0$ (see Table 3), is clearly seen, although t_d in butanol/water, $F = 1.33$, is little affected compared with pure water. The growth time is predicted by the model to be unaffected by composition, and this is not entirely borne out by the results. It is perhaps not surprising that these discrepancies between measured and predicted t_g and t_d arise since bubble growth and departure, and delay before regrowth, are not completely susceptible to treatment on a single-bubble basis—the effect of previous bubbles from the same, or neighboring, sites can be profound.

Since the bubble growth rate is reduced in mixtures, supposedly with no change in t_g, it follows that for mixtures with given initial superheating, a smaller proportion of the heat that is used in bubble growth is released while they are at the surface; see Table 4. Van Stralen (1966b) suggested that for pure fluids the entire excess enthalpy of the superheated layer (referred to saturation conditions) diffuses into the bubble while it is at the surface; hence the entire increase in heat flow above that due to natural convection results from direct vaporization at the surface. For mixtures the proportion of the excess enthalpy released at the surface is reduced by the factor F; the remainder of the excess in the relaxation boundary layer passes to the bulk liquid and hence to the bubbles as they rise. The consequent increase in the convection contribution to the total heat flow influences the critical heat flux (see Sec. 6.2).

Another parameter examined by Van Stralen (1966a) is the bubble departure size. He pointed out that from a purely thermal model it is directly proportional to the bubble growth rate for constant growth time. A reduction in growth rate in mixtures should cause a reduction in bubble departure size, a hypothesis confirmed by experiment, as shown in Table 4. Further support for the reduction of bubble departure diameter is given by Tolubinskiy and Ostrovskiy (1969). Their data for ethanol/water and ethanol/benzene (which can form azeotropes) boiling on a vertical steel tube are shown in Fig. 16. It is encouraging to note that in this respect the thermal model due to Van Stralen is in agreement with the work of Beer et al. (1977), who examined bubble departure from a dynamic viewpoint. The departure event is governed by static forces (surface tension, buoyancy, capillary, and excess internal pressure) and dynamic forces (liquid drag and inertia). At departure we can express an equality between those forces tending to keep the bubble on the surface and those tending to remove it.

$$\frac{1}{2} C_d \, \rho_l \, \dot{R}^2 \pi R_s^2 + 2\pi R_s \sigma \sin\phi$$

$$= -V_D \rho_l \ddot{R} + \left(\frac{2\sigma}{R_D} + \Delta p\right) \pi \, R_s^2 + V_D (\rho_l - \rho_v) g \qquad (47)$$

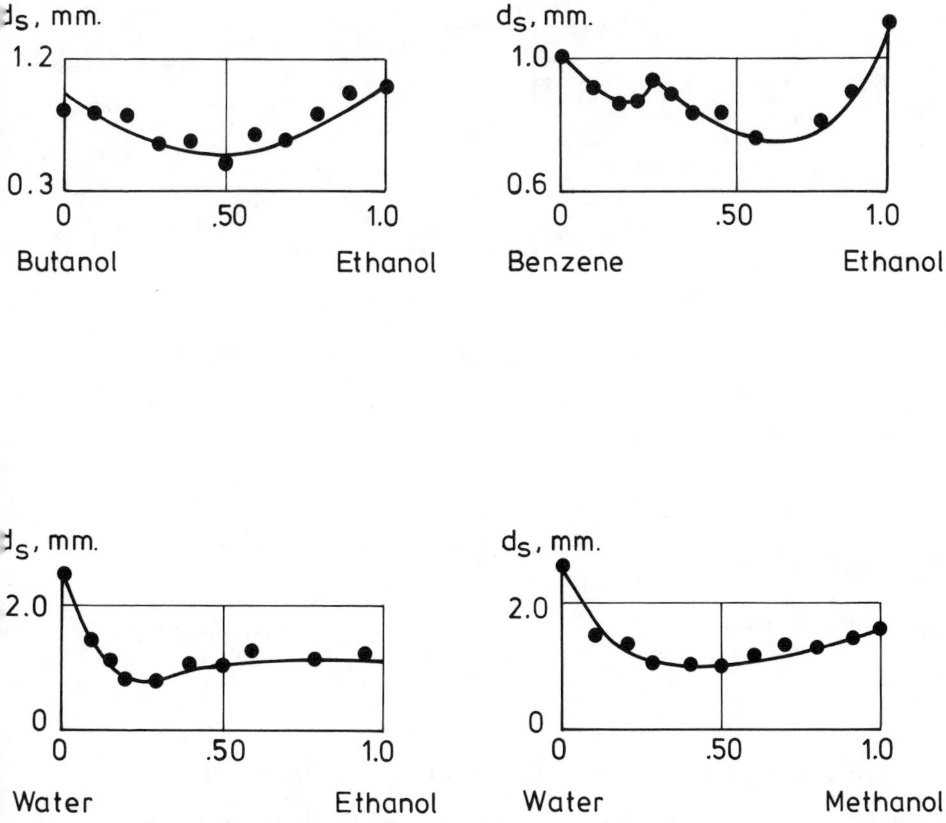

Fig. 16. Departure size of bubbles in binary mixtures. Tolubinskiy and Ostrovskiy (1969).

where C_d is the drag coefficient on the bubble, R_s is the radius of the dry patch under the bubble, ϕ is the contact angle of the fluid, V_D and R_D are the volume and radius of the bubble at departure, and g is the acceleration due to gravity.

The geometry is illustrated in Fig. 17. It should be noted that although certain arguments can be leveled against the use of this equation in this form [for example, for a spherical bubble inertia terms, as used here, may play no part in determining departure (Cooper and Lloyd, 1969); furthermore Cooper et al. (1978) showed that surface tension can assist departure by promoting sphericity], it does nevertheless give reasonable agreement for pure bubble data and can be used to suggest trends for mixtures. The Appendix to this chapter shows calculated values of the forces involved and illustrates the effect of the changing growth rate. The most important effects are on the drag term (tending to retain

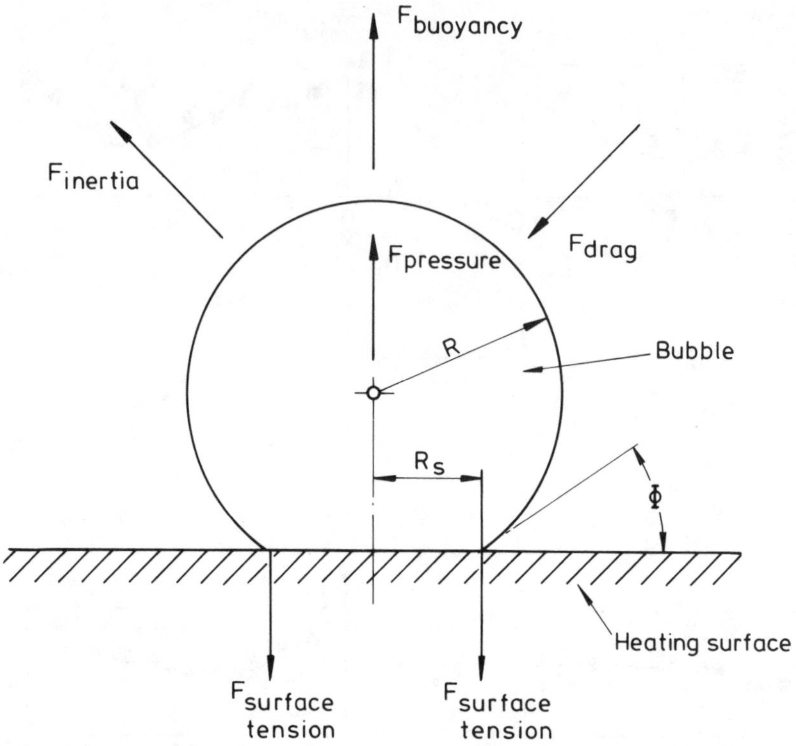

Fig. 17. Forces on a growing bubble at a heating surface.

the bubble), which is reduced, assisting early bubble departure, and on the liquid inertia term. For a bubble obeying Eq. (43), the inertia acts to remove the bubble. For mixtures, the deceleration in interface velocity is reduced and hence departure is delayed. However, this only partially offsets the effect of reduced drag, and Eq. (47) confirms the reduced bubble departure diameter in mixtures for given superheat, as illustrated in the Appendix.

Finally, the relaxation microlayer concept has been extended by Moalem et al. (1977) to boiling in immiscible liquid mixtures. They examined the growth of a bubble at a liquid-liquid interface due to heat transfer from both liquid layers. The asymptotic growth period of a stationary bubble is described by

$$R = \frac{2}{\sqrt{\pi}} \frac{\rho_l}{\rho_v} \frac{\Delta h}{c_l} \Delta T_{sat,\infty} \sqrt{\kappa t} \tag{48}$$

where the properties refer to the light, upper phase. This equation is identical to the asymptotic form of Eq. 41.

If the liquids are totally immiscible, there is no change of interface concentration during the growth. Fig. 18 shows experimental data for n-pentane/water that support the model.

Recently attention has been paid to the combined effect of the evaporation microlayer and the evaporation round the dome of the hemispherical bubble. Since the initial microlayer thickness is governed by the bubble growth rate, Van Stralen et al. (1975) considered the interaction between the two growth modes. They argued from a heat conduction model that if the evaporation microlayer is the dominant heat transfer mode, the growth rate is reduced by a factor F^2; that is, the growth rate is even lower than if the relaxation microlayer is dominant. This is due to the fact that not only is the available superheat reduced, but the microlayer thickness is increased by the lower growth rate; this conclusion contrasts that of Van Ouwerkerk discussed earlier. The growth rate for an adhering bubble in the asymptotic period, taking into account the contributions from both microlayers, is predicted by Van Stralen et al. to be given by

$$R = \left[(1.954b^* + 0.373\mathrm{Pr}^{-1/6}/F)\right] \exp\left(\frac{t}{t_d}\right) + 1.954\,\frac{\Delta T_{\mathrm{sat},\infty}}{\Delta T_{\mathrm{sat},w}}\,\mathrm{Ja}(\kappa t)^{1/2}/F \qquad (49)$$

where b^* is the height of the bubble to which the relaxation microlayer extends divided by the diameter. It allows for the fact that the bubble can erupt through the microlayer. The three terms inside the bracket refer respectively to the contributions of the

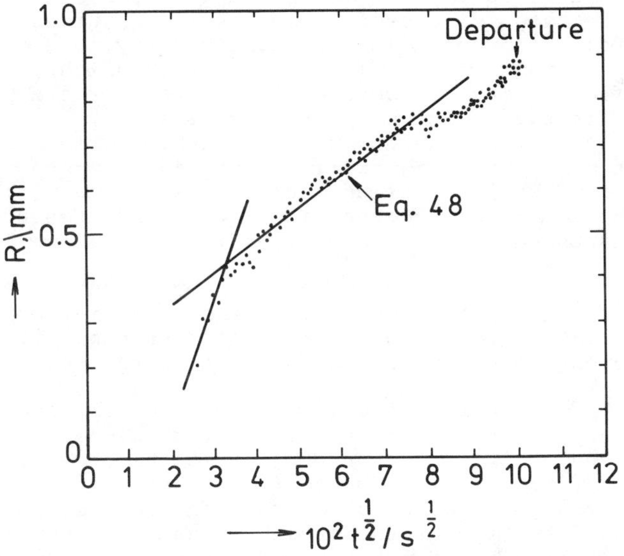

Fig. 18. Bubble radius against $\sqrt{\text{time}}$ for nucleate boiling of n-pentane at water interface. Moalem et al. (1977).

relaxation microlayer, the evaporation microlayer, and the bulk superheating. Van Stralen et al. (1977) found good agreement with experimental data in the pressure range 2.04 to 26.7 kN m^{-2}, although the geometry of the heating surface was not stated.

A different model is offered by Van Ouwerkerk (1971, 1972). He suggested that there was no interaction of the two growth contributions (dome and microlayer) and assumed that the rate of increase of the volume was given by simple addition of the volume increases from the two mechanisms. This is one area where further study appears to be required.

5 NUCLEATE BOILING HEAT TRANSFER

5.1 Introduction

Since the complexity of mixture boiling was recognized some 30 years ago, a wealth of data giving nucleate boiling superheats and heat transfer coefficients have been published. Many of the data concern miscible organic mixtures or aqueous organic solutions. Other systems studied include aqueous inorganic solutions and cryogenic mixtures. Table 5 summarizes the systems examined in all these studies. The dominant feature of these systems is the relative volatility effect, but we also consider in this section systems where Marangoni effects and physical property changes are dominant.

Typical nucleate boiling results for volatility-dominated situations are shown in Fig. 19 for an organic binary pair and in Fig. 20 for an aqueous organic solution; see also Fig. 6 and 7. The volatility difference of the components causing the reduction in bubble growth rate, discussed in Sec. 4, leads to an increase in superheat for given heat flux and consequently to a decrease in heat transfer coefficient. If the effect is slight, the heat transfer coefficient may lie between the values for the pure components or, if the relative volatility is large, it may lie outside these bounds and the α-x plot will show a minimum, or the $\Delta T_{\mathrm{sat},w}$-$x$ plot a maximum (Fig. 19). Figure 19 also shows that an increase in heat flux intensifies the decrease in heat transfer coefficient—the same trend occurs with increasing pressure (these effects are discussed further in Sec. 5.2.1).

One exception to this general conclusion is suggested by the data of Calus and Leonidopoulos (1974), who found the expected decrease in heat transfer upon addition of *n*-propanol to water but an improvement upon addition of water to pure *n*-propanol.

Most of the data summarized in Table 5 are for saturated pool boiling of binary mixtures with single horizontal heaters—either a tube in the center of a pool or a plate at the base of a vessel. Leppert et al. (1958) studied forced flow of dilute aqueous solutions of methanol or propanol over a stainless steel tube. The maximum concentration of the organics was 2.65% by weight. They too found a result against the general trend, with an increase by as much as 30% in heat transfer coefficient compared with pure water.

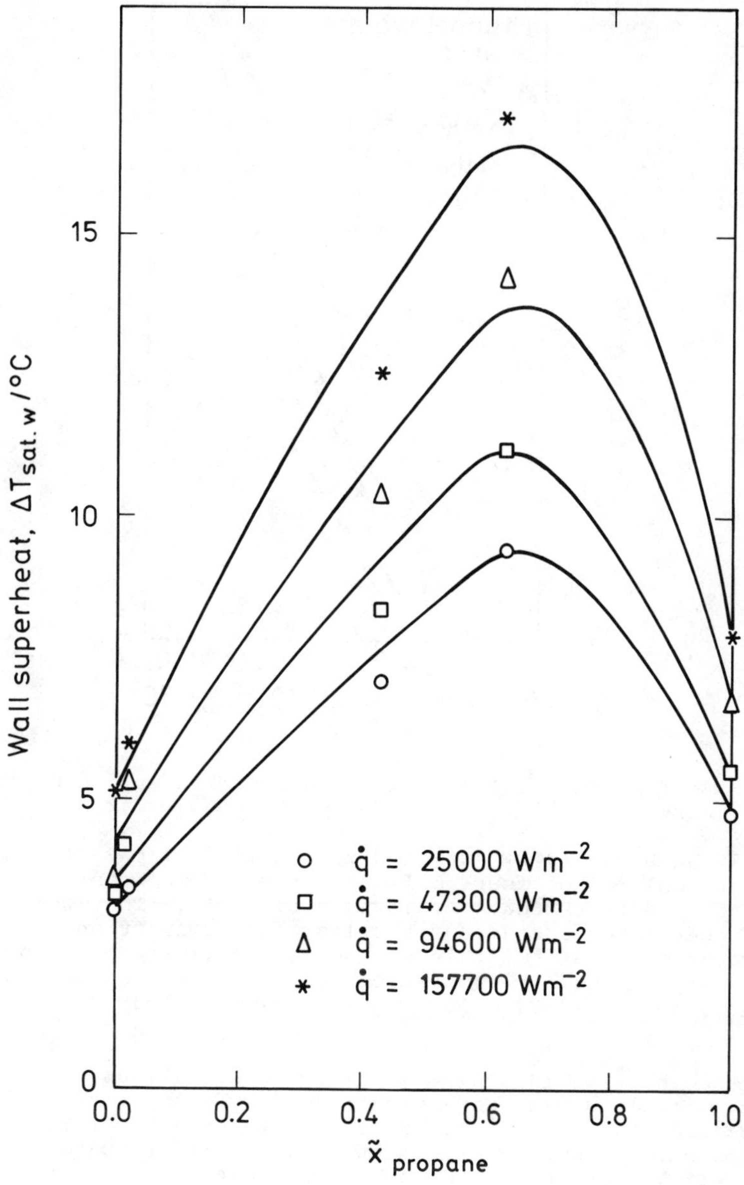

Fig. 19. Boiling data for propane/*n* butane mixtures. Clements and Colver (1972).

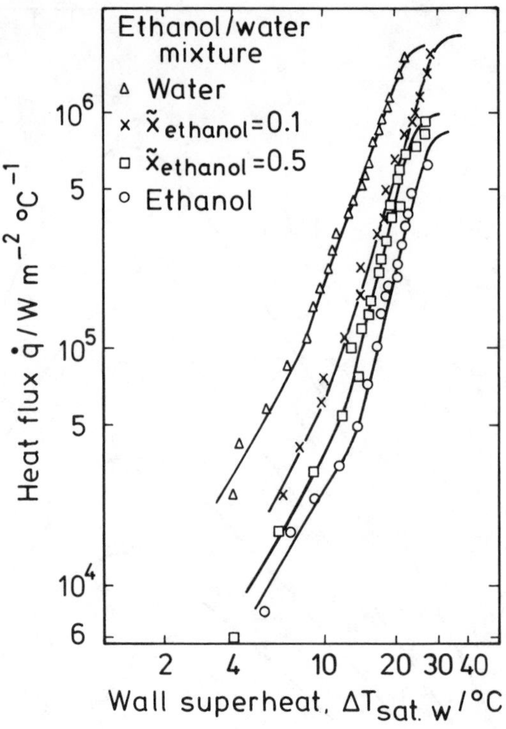

Fig. 20. Boiling curves for ethanol/water mixtures. Bonilla and Perry (1941).

Wall and Park (1978) examined forced flow of paraffin mixtures over a vertical row of four horizontal tubes. The coefficients for all the tubes were reduced for mixture coolants. They were higher for the upper tubes because of the increased convection caused by rising vapor. Because of this convection, nucleate boiling on the upper tubes is a smaller proportion of the whole and consequently the reduction in heat transfer was less significant on the upper tubes.

In-tube boiling in convective flow in situations dominated by nucleate boiling has been investigated by Rose et al. (1963) for subcooled boiling of aqueous alcohol solutions, by Bennett (1976) for water/ethylene glycol at flow qualities up to 26%, and by Toral (1979) for subcooled and saturated boiling of ethanol/cyclohexane. These studies showed marked decrease in heat transfer over the whole composition range—even for azeotrope-forming mixtures.

Nucleate pool boiling of ternary mixtures was investigated by Grigor'ev et al. (1968) using acetone/ethanol/water and acetone/methanol/water and by Stephan and Preusser (1978) using the latter mixture. Both groups found deterioration in heat transfer for these mixtures as illustrated in Fig. 21, which shows lines of constant heat transfer coefficient for acetone/ethanol/water.

Filatkin (1972) studied nucleate boiling heat transfer in water/ammonia mixtures and found that there too the loss of available superheat at the bubble boundaries leads to a decrease in heat transfer coefficient.

Boiling in cryogenic fluids is not, in principle, different from that in organic fluids. Several investigations into nucleate boiling of cryogenic mixtures have shown the same deterioration in heat transfer. The mixtures investigated have included nitrogen/oxygen (Lyon, 1964; Kosky and Lyon, 1968), nitrogen/methane (Ackerman et al., 1975), and nitrogen/argon (Thome, 1978).

Oil leakage in compressors has made the boiling of refrigerant/oil mixtures important. Several researchers, such as Dougherty and Sauer (1974) and Sauer et al. (1978), found that addition of up to 5% by weight of oil causes an increase in boiling heat transfer coefficient; further addition causes a decrease in the coefficient. Figure 22 illustrates such trends. Stephan (1963) and Henrici and Hesse (1971) found a decrease in heat transfer for

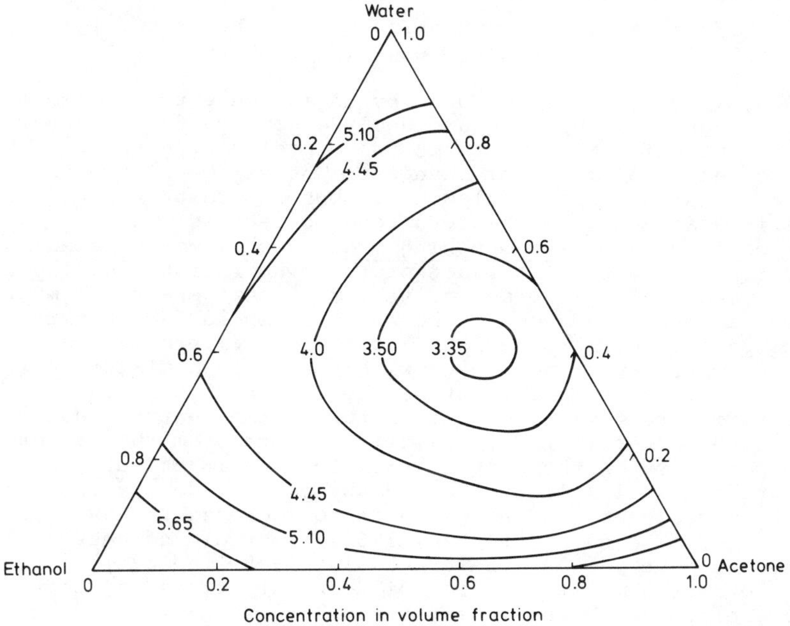

Fig. 21. Boiling heat transfer in a ternary mixture showing lines of constant heat transfer coefficient, kW m^{-2} $^{0}C^{-1}$, at 116 kW m^{-2}, Grigor'ev et al. (1968).

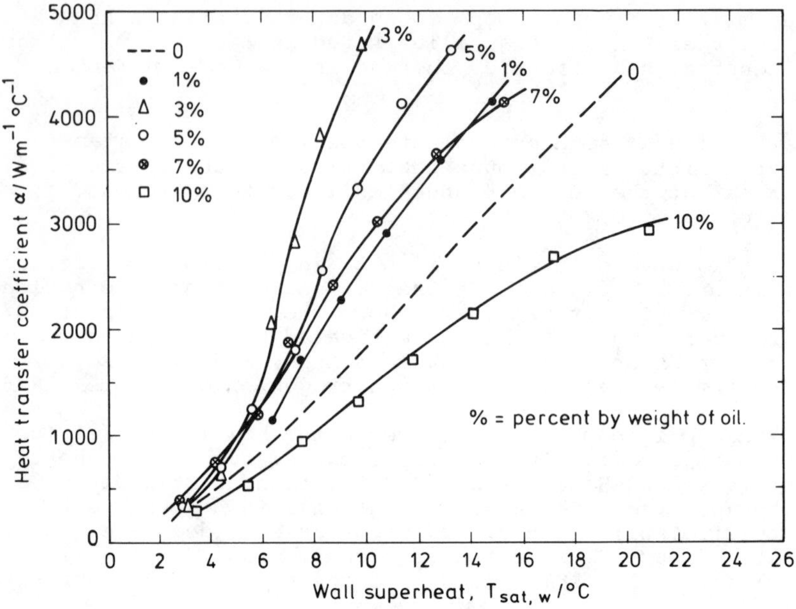

Fig. 22. Effect of oil concentration on boiling of refrigerant/5 GS oil mixtures, Sauer et al. (1978).

up to 1% by weight of oil followed by an increase to a maximum, at 3% by weight, and then a sharp decline with further increase in oil content. In these mixtures two opposing effects are acting to produce the complex set of changes. The effect of the volatility difference acts through the slowing of bubble growth via the decrease in available superheat to tend to decrease the heat transfer coefficient. The opposing trend occurs with the Marangoni effect acting along the interface of the evaporation microlayer. Figure 23a shows a typical σ-x curve for a refrigerant/oil mixture at ambient temperature showing the increase in surface tension that accompanies an increase in oil content. Figure 23b shows an evaporating microlayer under a growing bubble in a mixture. Since the film is thinner at A than at B, the instantaneous heat transfer and evaporation rate will be greater toward the center. Hence the light component (refrigerant) is stripped more from the center than from the periphery of the microlayer. The resultant surface tension gradient pulls liquid toward the center, delays the drying out of the microlayer, and hence tends to increase the overall heat transfer rate. The data (Stephan, 1963; Henrici and Hesse, 1971; Dougherty and Sauer, 1974; Sauer et al., 1978) indicate that the Marangoni effect dominates at low oil concentrations and the volatility effect at high concentrations.

Aqueous solutions of inorganic salts have also been found to show a decrease in heat transfer coefficient compared with pure water (Cryder and Gilliland, 1932; Minchenko and Firsova, 1972;

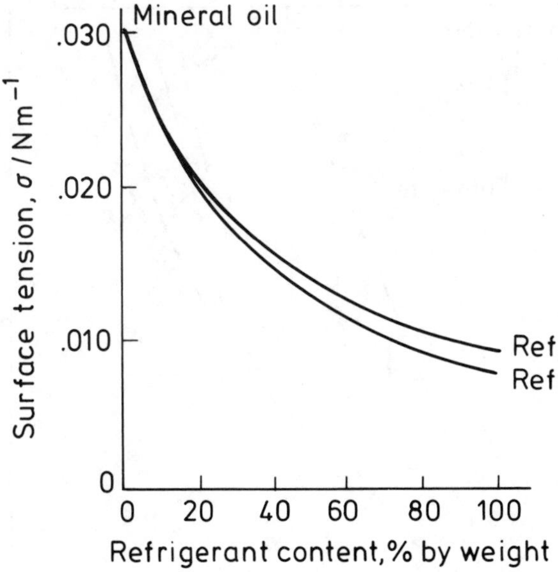

Fig. 23a. Surface tension of refrigerant/oil mixtures at 15°C.

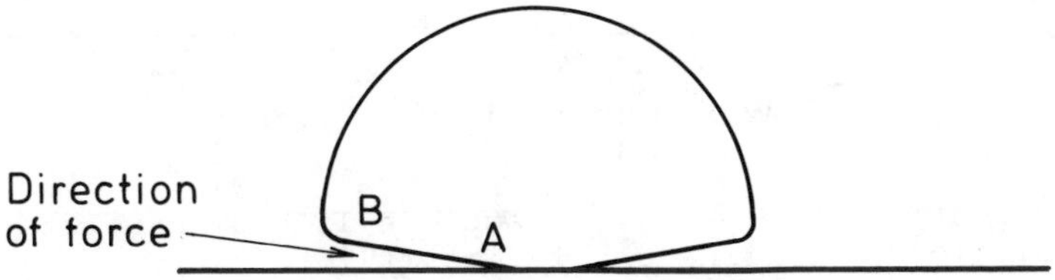

Fig. 23b. Marangoni effect acting at an evaporating microlayer.

Fig. 23. Boiling in refrigerant/oil mixtures.

Ohnishi and Tajima, 1975). This is clearly illustrated by the results of Ohnishi and Tajima for lithium bromide, reproduced in Fig. 24. The deterioration in heat transfer in solutions was found to be greatest at low pressures. Here too several effects may be occurring. As discussed by Van Stralen (1968a), an enrichment of the heavy component (salt) at the interface leads to a decrease in available superheat and contributes to the observed reduction in heat transfer coefficient. However, the increase in salt concentration near the interface can cause marked local changes in several physical properties. The increase in viscosity

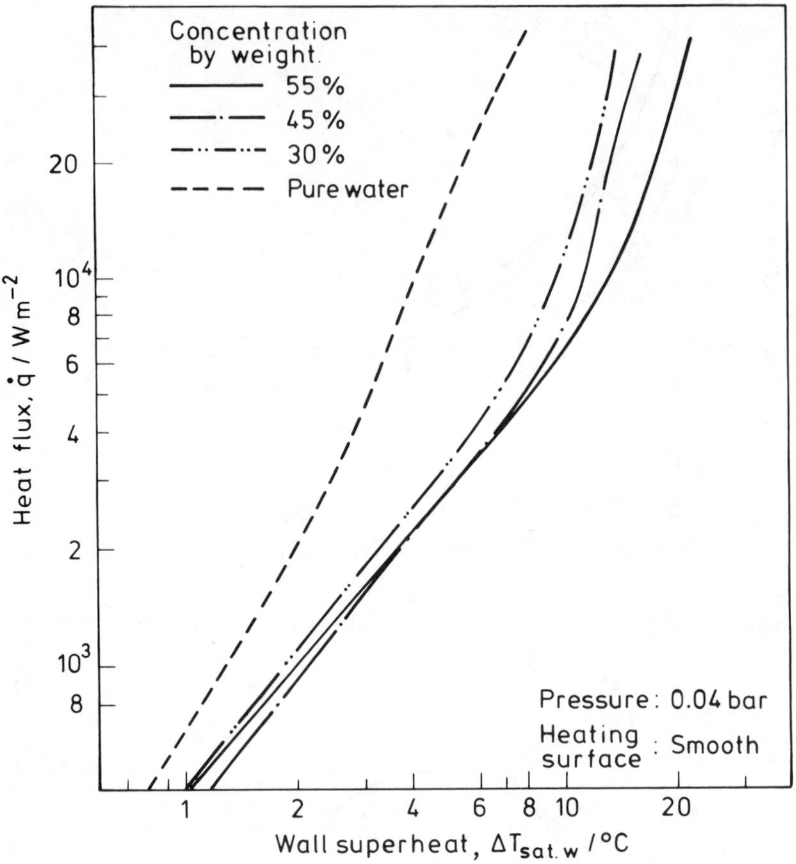

Fig. 24. Nucleate boiling curve of aqueous lithium bromide solutions. Ohnishi and Tajima (1975).

can reduce the bubble growth rate during the prolonged initial growth period as well as reducing the tendency of neighboring bubbles to coalesce.

Several investigators have examined mixtures where the primary effect of an additive is to affect the physical properties around the bubble—the volatility effect is small. Averin and Kruzhilin (1955) added *iso*-amyl alcohol to water primarily to reduce the surface tension, and they found that the heat transfer coefficient, α, varied according to

$$\alpha \propto \left(\frac{1}{\sigma}\right)^{1/3} \tag{50}$$

Hall and Hatton (1965) similarly found that addition of small amounts of Teepol to water could increase the pool-boiling heat transfer coefficient on electrically heated rods; the effect was seen on a rough surface but not on a smooth one. They noted that bubble coalescence was reduced. Other workers (Van Stralen, 1968b; Isshiki and Nikai, 1973) made the same observation. High-molecular-weight nonvolatile surfactants at low concentrations reduce coalescence, and encourage foaming by their hindrance of drainage of thin interbubble layers. They form an orientated monolayer at the interface with consequently reduced surface tension. When the thin film between the bubbles starts to drain, the exposed surface is temporarily pure water, until the monolayer has time to reestablish. The locally high surface tension causes a force that pulls liquid into the interbubble region and hence delays bubble coalescence. An alternative, related mechanism acts simultaneously for volatile surfactants in larger concentrations and is illustrated in Fig. 25. The hindrance of nearby bubbles causes mass transfer to be relatively less efficient in region *B* than in *A* or *C* and reduces replenishment of the light component. Hence the concentration of lights may be higher at *A* and *C* and there may be a surface tension gradient along the interface. Depending on whether the mixture is positive (increase in more volatile concentration increases surface tension) or negative (Zuiderweg and Harmens, 1958), liquid will either be drawn into the region between the bubbles, keeping them separate and encouraging foaming, or drawn away, encouraging coalescence. As we shall see, this can also influence the boiling crises.

Morgan et al. (1949) found that the increase in heat transfer coefficient depended on the concentration of the surfactant. A 1% by weight solution of Drene in water showed a considerably greater increase in α than did a 0.1% solution, although the static surface tension, as measured with a du Nouy tensiometer, for each solution at its boiling point is almost identical. Morgan et al. suggested that, for the more dilute solution, the surfactant monolayer cannot become established during the bubble lifetime, whereas for the 1% solution there are enough molecules near the interface to ensure that the dynamic surface tension is effectively reduced.

Averin and Kruzhilin (1955) investigated the effect of viscosity changes by dissolving beet sugar in water. The heat transfer coefficient varied according to

$$\alpha \propto \left(\frac{1}{\mu}\right)^{0.45} \tag{51}$$

Such trends agree qualitatively, though not quantitatively, with

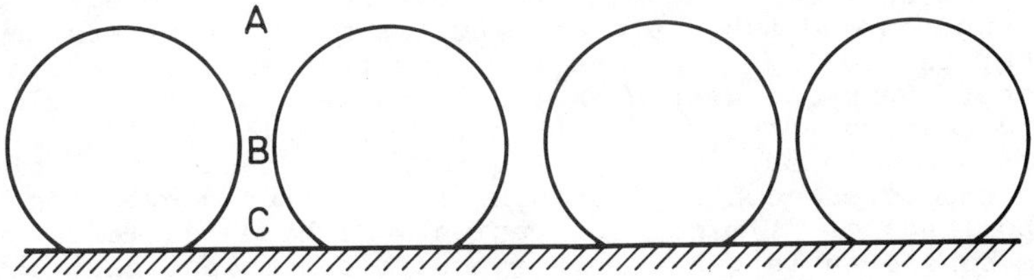

Fig. 25. Illustration of Marangoni effect and influence on bubble coalescence.

well-established nucleate boiling correlations for pure substances (Rohsenow, 1952; Forster and Zuber, 1955). By contrast, Kotchaphakdee and Williams (1970) found that polymeric additives ($\tilde{M} \approx 10^3$) in small concentrations in water gave an increase in heat transfer that was attributed to the effect of viscosity change. Similarly, Gannet and Williams (1971) found that addition of high-molecular-weight ($\tilde{M} \approx 10^6$) polymers to cyclohexane improved the nucleate boiling heat transfer. Dunskus and Westwater (1961) added low-molecular-weight polymers ($\tilde{M} \approx 2000$) to *i*-propanol and obtained small improvements or deteriorations in heat transfer depending on the additive and on the concentration. The improvements in heat transfer are contrary to Eq. (51), and it is suggested that the apparently anomalous results were produced by reduced bubble coalescence. This occurred because of increased viscosity around the bubble, which inhibits draining of the liquid separating the bubbles. Indeed, Dunskus and Westwater actually observed a decrease in bubble coalescence using films of the boiling process.

Another means by which small additions of large molecules can improve heat transfer was suggested by Lowery and Westwater (1957). They added to methanol cationic, nonionic, and anionic agents that were not surface active and measured the boiling curves. The molecular weights of the additives were in the range 340–466. They suggested that the measured increase in nucleate boiling heat transfer is due to the additive molecules increasing the active site population by acting as nuclei. Jontz and Myers (1960) investigated nucleate boiling of aqueous solutions of Aerosol and Tergitol. Both gave improvements in nucleate boiling heat transfer; in the former case it was attributed to nucleation on the additive molecules and in the latter to the decrease in surface tension at the bubble boundary. An interesting observation made by Jontz and Myers was that many forms of impurities such as bits of wood, rubber stoppers, and gasket cement could cause nonreproducibility when added to the liquid pool—as little as 10 ppm of foreign material was sufficient. This illustrates the need for scrupulous cleanliness in all such experiments, and it is quite possible that such care was not always taken in many of the experiments reported in this review. However, in view of the consistency of the general trends observed, we may assume that there is some "threshold level," usually exceeded, beyond which the results are independent of the impurity concentration. In other words, the absolute values of the coefficients may be in error, but the relative values provide a sound basis for comparison.

Thus far, only miscible mixtures have been considered here. The literature on the subject of immiscible mixtures is sparse, but interest is growing as a result of its relevance to spills of hydrocarbons. Sump and Westwater (1971) showed that the boiling curve could be much more complex than for miscible fluids. If the heater is a wire or tube in a liquid pool, it can be in either layer. Furthermore, each layer can be in the free convection, nucleate boiling, or film boiling region. Eighteen different situations can be envisaged from the possible combinations. Sump and Westwater investigated several of these situations using binary mixtures of water with Refrigerant-112, -113 and *n*-hexane.

They noted that the results varied considerably when the heater tube was moved from one liquid phase to another and that, for nucleate boiling, the presence of the second liquid phase caused only a small change in the heat flux at a constant metal wall temperature. This contrasts with the marked change in growth rate for a bubble at the interface (Van Stralen, 1966b) (see Sec. 4.3). They observed that bubbles rising from a tube in the lower phase were held up at the liquid-liquid interface and coalesced before continuing to rise. A theoretical study of the stability and shape of such bubbles was presented by Oktay (1971). He also noted that when these bubbles reach the upper surface they tend to condense and drop back to the lower liquid level.

5.2 Correlations

5.2.1 General Form of Correlations

Several correlations for mixture-boiling heat transfer coefficients have been published in recent years, and are listed in Table 5. These all allow for the volatility effects and are not valid for surface-active additives or those that greatly increase the viscosity; some of these correlations are merely data-fitting exercises to a particular set of data over a narrow range of conditions; others (Afgan, 1966; Stephan and Körner, 1969; Clements and Colver, 1972; Calus and Rice, 1972; Calus and Leonidopoulos, 1974; Happel, 1977; Stephan and Preusser, 1978) are of a more general nature, and some of these are discussed here. Before examining some suggested correlations, in Sec. 5.2.2, this section discusses the form that they should take and the trends that they should reproduce.

Most correlations are intended to be of use to equipment designers. Though, as we have seen, the interface conditions are the important factor underlying bubble growth and nucleate boiling heat transfer, the engineer normally only knows the bulk fluid condition and does not wish to evaluate those at the interface. Hence most correlations must use the bulk mixture physical properties and a temperature-driving force related to the bubble point of the bulk fluid. Since this will overestimate the heat transfer coefficient, we must apply a correction factor that depends on the boiling range or composition difference between liquid and vapor and on the mass diffusivity.

Most correlations for nucleate boiling in pure substances are of the form

$$q = P\Delta T^m_{\mathrm{sat},w} \tag{52}$$

where P and m are functions of the relevant physical properties and perhaps of the surface topography; m is usually of the order 3 to 5. This suggests that the q (or α) - $\Delta T_{\mathrm{sat},w}$ curves are straight parallel lines when plotted on logarithmic axes. Generally, however, such lines are neither straight for a given substance throughout the nucleate boiling region [due largely to the existence of the separate regions, isolated bubbles, and interference in the

Table 5. *Published Boiling Data and Correlations*

Authors	Fluids	Boiling Surface	Pressure	Additional Remarks
Bonilla and Perry (1941)	All binary combinations of water, ethanol, *n*-butanol, acetone	Horizontal chromium plate	0.2–1.3 bar	
Cichelli and Bonilla (1945)	Water/ethanol, propane/pentane	Horizontal chromium plate	1–60 bar	
Leppert et al. (1958)	*i*-propanol/water, methanol/water	Outside stainless steel tube, D = 28 mm	1 bar	Forced flow of coolant over heater
Sternling and Tichacek (1961)	14 binary mixtures	Outside stainless steel tube, D = 4.5 mm	1 bar	Boiling ranges of mixtures up to 100°C
Huber and Hoehne (1963)	Benzene/diphenyl	Outside stainless steel tube, D = 9.5 mm	1–3.5 bar	
Rose et al. (1963)	Aqueous solutions	Inside vertical stainless steel annulus, center rod D = 6.2 mm	2.1 bar	Subcooled flow boiling; subcooling 35–60°C; velocity 1.2 m s^{-1}
Afgan (1966)	Ethanol/benzene	Outside stainless steel tube, D = 5.1 mm	6–15 bar	Correlation proposed
Afgan (1968)	Ethanol/water, water/glycerine, ethanol/benzene	Not stated	1–9 bar	
Grigor'ev et al. (1968)	Acetone/methanol/water, acetone/ethanol/water	Outside stainless steel tube	1 bar	Ternary mixtures

Table 5. *Published Boiling Data and Correlations* (*Cont.*)

Stephan and Körner (1969)	17 systems	Used published data	1–10 bar	Generalized correlation proposed
Takeda et al. (1970)	Methanol/water, ethanol/water, acetone/water, methyl ethyl ketone/water	Copper plate, platinum wire, D = 0.2 mm	1 bar	
Wright et al. (1971)	Ethane/ethylene	Outside gold-plated tube, D = 20.6 mm	2.4–36.6 bar	
Clements and Colver (1972)	Propane/n-butane, propane/n-pentane	Outside gold-plated tube, D = 20.6 mm	1.6–29.3 bar	Correlation proposed
Calus and Rice (1972)	i-propanol/water, acetone/water	Nickel/aluminium wire, D = 0.31 mm	1 bar	Correlation proposed
Tolubinskiy et al. (1973)	Ethanol/water	Outside vertical stainless steel tube, D = 4.5 mm	1–15 bar	Correlation proposed
Isshiki and Nikai (1973)	Ethanol/water, water/ethylene glycol, water/n-butanol	Nickel wire, D = 0.3 mm	1 bar	
Calus and Leonidopoulos (1974)	n-propanol/water	Nickel/aluminium wire, D = 0.31 mm	1 bar	Correlation proposed
Tolubinskiy et al. (1975)	Ethanol/benzene	Outside vertical stainless steel tube, D = 4.5 mm	1–18 bar	Correlation proposed

Table 5. *Published Boiling Data and Correlations (Cont.)*

Happel (1977)	Benzene/toluene, ethanol/benzene, water/*i*-butanol	Outside nickel tube, D = 14 mm	0.5–2 bar	Correlation proposed
Bennett (1976)	Water/ethylene glycol	Insider vertical tube, D = 19 mm	2 bar	Forced flow of coolant; correlation proposed
Wall and Park (1978)	*n*-pentane/*n*-hexane	Outside gold-plated tubes, D = 26.6 mm in tube bundle	Not stated	
Stephan and Preusser (1978)	Acetone/methanol, acetone/water, water/methanol, acetone/methanol/water	Outside nickel tube, D = 14 mm	1 bar	Correlation proposed for multicomponent mixtures
Toral (1979)	Ethanol/cyclohexane	Inside vertical tube, D = 26 mm	2 bar	Forced flow of coolant
Cryder and Gilliland (1932)	Aqueous solutions of Na_2SO_4, NaCl	Outside brass tube, D = 26.8 mm	1 bar	
Filatkin (1972)	Water/ammonia	Outside tube, D = 28 mm	1 bar	Correlation proposed
Minchenko and Firsova (1972)	Aqueous solutions of LiCl, LiBr	Outside stainless steel tubes, D = 11, 12, 13 mm	0.05–1 bar	Correlation proposed
Ohnishi and Tajima (1975)	Aqueous solutions of LiBr	Outside copper tube, D = 19.9 mm	0.04–0.4 bar	correlation proposed
Lyon (1964)	Oxygen/nitrogen	Copper block	not stated	Varying orientation of heater

Table 5. *Published Boiling Data and Correlations (Cont.)*

Kosky and Lyon (1968)	Oxygen/nitrogen	Flat Platinum Disk	1–32.4 bar	
Ackermann et al. (1975)	Nitrogen/methane	Horizontal copper plate	2–35 bar	
Thome (1978)	Argon/nitrogen	Copper block, prepared sites	1–1.3 bar	Effect of surface finish investigated

nucleate boiling region, as described by Zuber (1963)], nor are they parallel for different substances. Corty and Foust (1955) showed that the slope and position of the boiling curves depend on the surface roughness. The effective surface roughness can vary greatly for pure substances and, as we have seen in Sec. 3, it can vary significantly with composition, especially for aqueous solutions. Nevertheless, the form of correlation in Eq. (52) is currently the best available.

The boiling curves for some mixtures are similar in slope to those for the pure fluids; for others definite trends with changes in composition can be seen. Table 6 shows the values of m for the data shown in Fig. 20, and a plot of m against x for

Table 6. *Slope of Boiling Curve: Variation with Composition*[a]

x_{ethanol}	m
0	2.69
0.1	3.12
0.5	4.05
1.0	3.97

[a]From Bonilla and Perry, 1941.

n-propanol/water (Calus and Leonidopoulos, 1974) is shown in Fig. 26.

Although the shape of the boiling curve can be related to the site distribution density (Mikic and Rohsenow, 1969; Singh et al. 1976; Lorenz et al., 1974), the variation of m with composition has not yet been studied in a systematic way and has not been included in any correlation. Therefore for mixtures the form of correlation that must be used at present is

$$q = CP\Delta T^{m}_{\text{sat},w} \tag{53}$$

where C is a correction factor that is not a function of q and tends to 1 as x tends to 0 or 1. Since C accounts for the decrease in heat transfer due to the reduction in available superheat, it must in some way be related to (1) the boiling range or to $|y^* - x|$, which expresses the tendency to strip the light component from the interface, and (2) the diffusivity, which expresses the tendency to replenish it. The modulus sign is used on y^*-x because boiling is reduced over the whole composition range. The difference ΔT_{sat} is given by $T_w - T_{\text{bub}}(x)$.

A further important parameter is the pressure. A number of

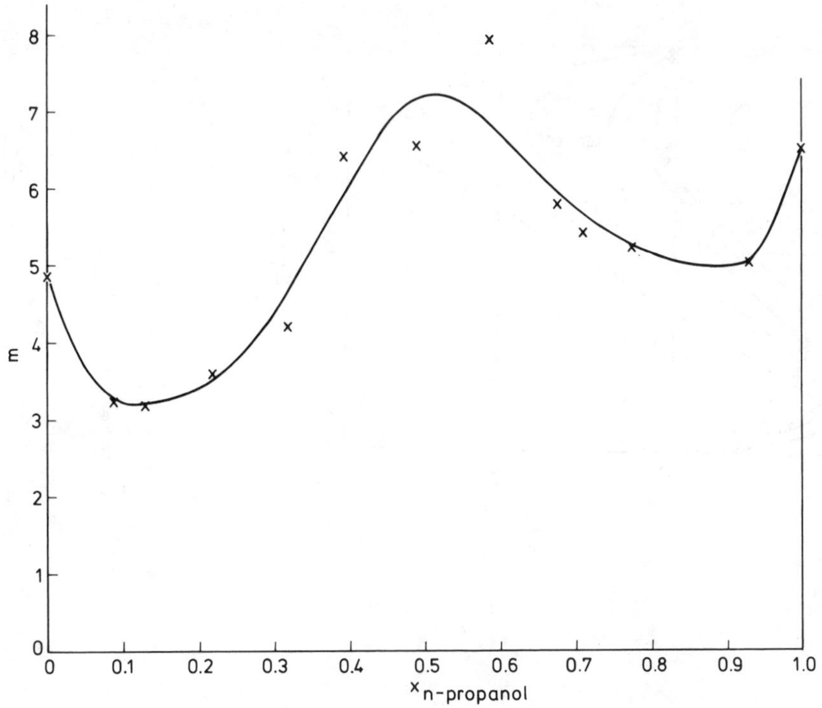

Fig. 26. Graph of m (Eq. 51) against $x_{n\text{-propanol}}$. Data of Calus and Leonidopoulos (1974).

authors (Afgan, 1966; Wright et al., 1971; Tolubinskiy et al., 1973; Tolubinskiy et al., 1975; Happel, 1977) have noted that the deterioration in heat transfer coefficient increases with pressure. Figure 27a shows data for ethanol/benzene mixtures in the range 1×10^5–13.6×10^5 Nm^{-2} presented by Tolubinskiy et al. (1975). Figure 27b shows the concentration difference y^*-x for the same range of conditions, and it can be seen that the absolute value of this difference decreases with pressure increase.

This effect of pressure has not yet been explained satisfactorily. Over the pressure ranges concerned, the difference y^*-x usually decreases with rising pressure, which suggests a lessening of the reduction in available superheat. The physical and transport properties do not change significantly. Hence changes in single-bubble dynamics cannot explain the effect. One possible explanation may lie in the work of Tolubinskiy et al. (1975) and of Happel (1977), who suggested that, for a mixture, the bubble population may not increase with pressure as much as it does for pure substances. This could serve as a satisfactory explanation of the observed effect of pressure on mixture boiling, but it in turn demands an explanation. To examine the effect of pressure on bubble population, we need to consider the terms in Eq. (5). Calculation shows that the form of the relationship $dp_v/dT - x$ changes little with pressure and does not indicate any increase in ONB superheat

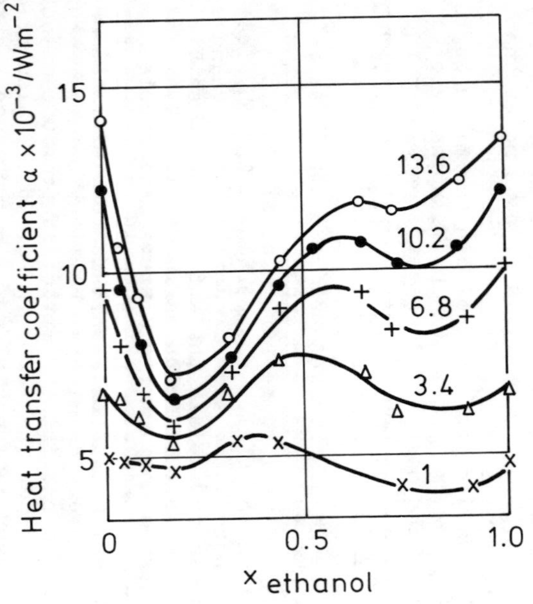

Fig. 27a. Heat transfer coefficient of ethanol/benzene mixtures at varying pressure $\dot{q} = 10^5$ Wm^{-2} figures denote P/bar.

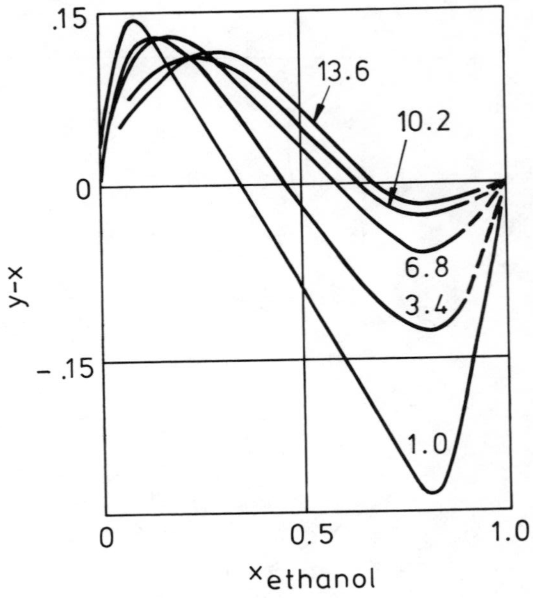

Fig. 27b. Phase equilibrium of ethanol/benzene mixtures. Figures denote P/bar.

Fig. 27. Boiling of ethanol/benzene mixtures at varying pressures. Tolubinskiy et al., (1975).

with increasing pressure. With a rise in pressure, and hence boiling temperature, the surface tension decreases, again indicating for all compositions an increase in bubble population. The effect of temperature change on contact angle is not known for metal surfaces, but it is known to decrease with increasing temperature for a number of organic fluids on PTFE (Yargin et al., 1972; Adamson, 1973); this would tend to decrease the active site density as the pressure increased. To the author's knowledge no experimental examination of the variation of active site density with pressure and composition has been conducted, but such a test is suggested.

An alternative line of argument is to note that as the pressure increases the departure diameter decreases and the bubble population increases (Séméria, 1962; Fath and Judd, 1978). Hence, with a larger number of smaller bubbles, it is tempting to suggest that the effectiveness of the mass transfer process in replenishing the light component at the bubble interface may be diminished due to the crowding of bubbles. Such an explanation would require that the increase in number density outweighs the decrease in departure size such that the proportion of the heater area covered by bubbles is increased. This, however, runs counter to three pieces of evidence: (1) As the pressure increases, there is an observed increase in the heat transfer area available for natural convection and entirely unaffected by bubble heat transfer (Fath and Judd, 1978), (2) At low pressures, below about one-third of the critical, q_{cr} increases with P, suggesting less interference of bubble columns if not of individual bubbles, (3) An increase in pressure diminishes bubble coalescence (Séméria, 1962), again suggesting reduced interference between neighboring bubbles at the surface.

An investigation of the effect of pressure on boiling of single bubbles at prepared sites and on boiling surfaces is also suggested in order to determine whether the changes in mixture boiling characteristics with pressure are due to the behavior of individual bubbles or to their mutual interference.

5.2.2 Empirical Correlations

Most of the available correlations relate only to binary mixtures. The first, and perhaps most widely used, is that of Stephan and Körner (1969), who suggested that the superheat at the heating surface is made up of two parts (see Fig. 28), an "ideal" given by

$$\Delta T_{sat,I} = \tilde{x}_A \, \Delta T_{sat,A} + \tilde{x}_B \, \Delta T_{sat,B} \tag{54}$$

and an excess $\Delta T_{sat,E}$. The superheats $\Delta T_{sat,A}$ and $\Delta T_{sat,B}$ are the superheats that would be found for each of the pure substances boiling on its own on the given surface and at the heat flux of interest. They must be calculated from an appropriate correlation (e.g., Rohsenow, 1952; or Forster and Zuber, 1955). Note that this correlation is not based on the expected coefficient from the EPF, but on the convenience of a linear ideal, Eq. (54).

Since the decrease in bubble growth depends on $\tilde{y}^* - \tilde{x}$, Stephan and Körner suggested that $\Delta T_{sat,E}$ would be a function of this

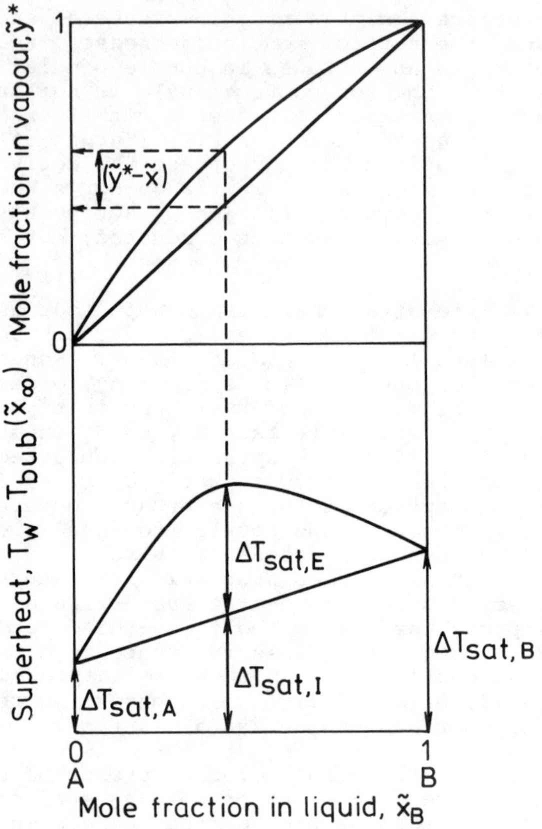

Fig. 28. The Stephan and Korner correlation for binary mixtures. Collier (1981).

concentration difference. Furthermore, they observed that it depends on the pressure; see Sec. 5.2.1. They proposed the following correlation scheme.

$$\Delta T_{sat,w} = \Delta T_{sat,I} (1 + |\Theta|) \tag{55}$$

where

$$\Theta = A(\tilde{y}^* - \tilde{x}) \tag{56}$$

The multiplier A depends on the pressure according to

$$A = B(0.88 + 0.12 \times 10^{-5} p) \tag{57}$$

and B depends only on the identity of the substances in the mixture. Using published data from a variety of sources, Stephan and Körner extracted values of B for various binary pairs, some of which are shown in Table 7.

Table 7. *Values of B for Various Binary Mixtures*

Mixture	B
Ethanol/acetone	0.75
Ethanol/benzene	0.42
Ethanol/water	1.21
i-Propanol/water	2.04
n-Propanol/water	3.29

They tested the correlation for 17 systems covering a wide range of concentration driving forces and found a root-mean-square (rms) error of 8.6%. Using a general value of B equal to 1.53, they found, for the same systems, a rms error of 15%, and they recommended this value if no data are available from which to extract B for a particular mixture.

Stephan and Preusser (1978) suggested an extension to the method in Eqs. (55) and (56) for ternary mixtures using the form

$$\Theta = A_{AC}\,(\tilde{y}_A - \tilde{x}_A) + A_{BC}\,(\tilde{y}_B - \tilde{x}_B) \tag{58}$$

or, in general terms for an n-component mixture,

$$\Theta = \sum_{i=1}^{n-1} A_{in}\,(\tilde{y}_i - \tilde{x}_i) \tag{59}$$

where the A_{in} refers to the binary mixture i-n and the ideal superheat is given by

$$\Delta T_{\text{sat},I} = \sum_{i=1}^{n} \Delta T_{\text{sat},i}\,\tilde{x}_i \tag{60}$$

In using Eqs. (59) and (60) the components must be numbered in order of increasing boiling point. In principle, then, the boiling behavior of a multicomponent mixture can be predicted knowing only the behavior of the binary permutations. Stephan and Preusser tested the method for the ternary mixture acetone/methanol/water at atmospheric pressure and obtained an average error of 14% on the predicted heat transfer coefficients. If instead of using the values of A that optimize the binary data ($A_{AC} = A_{\text{acetone/water}} = 0.81$, $A_{BC} = A_{\text{methanol/water}} = 0.56$), they optimized the values to give the best fit to the ternary data ($A_{AC} = 0.58$, $A_{BC} = 0.54$), the average error was reduced to 8.3%. Thus, if some data are available, the multicomponent extension to the original method can give comparable accuracy to the binary prediction.

In contrast to the method of Stephan and Körner (1969), the

scheme proposed by Happel suggested that the "ideal," for binary mixtures, is given by a heat transfer coefficient that varies linearly with composition. They postulated that

$$\frac{\alpha}{\alpha_I} = 1 - G(\tilde{y}^* - \tilde{x})^H \tag{61}$$

where $G = G$(components, p), $H = H$(components), and extracted values of G and H for three mixtures. A similar method but with H equal to 1 was proposed by Afgan (1966).

Although no particular theoretical reasons exist for preferring either of the schemes illustrated in Eqs. (55) and (61), the former, with only one adjustable constant, is simpler to use and to obtain the necessary data. Moreover, the data are available for more systems, and this scheme has already shown itself capable of extension to multicomponent mixtures.

5.2.3 Semiempirical Correlations

The correlation schemes described in Sec. 5.2.2 require at least one boiling heat transfer measurement using the mixture in question in order to extract empirical constants B or G and H. Calus and Rice (1972) suggested a method that circumvents this difficulty.

The bubble growth equations of Scriven and van Stralen, discussed in Sec. 4.2, indicate that the growth rate is reduced by a factor

$$1 - (y^* - x)\left(\frac{\kappa}{\delta}\right)^{1/2} \frac{c_l}{\Delta h}\frac{dT}{dx}$$

Calus and Rice examined whether such a factor might serve to correlate the reduction in heat transfer when applied to the coefficient for the equivalent pure fluid. They found that a satisfactory correlation was achieved by a modified, and simpler, form of the above factor, that is,

$$1 + |y^* - x|\left(\frac{\kappa}{\delta}\right)^{1/2}$$

the reciprocal of which is equivalent to C in Eq. (53). The heat transfer coefficient is given by

$$\alpha = \alpha_{EPF}/[1 + |y^* - x|(\kappa/\delta)^{1/2}]^{0.7} \tag{62}$$

The method agreed well with data gathered by Calus and Rice for i-propanol/water and acetone/water on a thin wire as well as by Sternling and Tichacek (1961) for glycol/water and glycerol/water on the outside of a tube.

Although in principle the method can be used knowing only the phase equilibrium and transport property data for the mixture, the variation of thermal and mass diffusivity (especially the latter) with composition and temperature is often difficult to find or to predict. Figure 29 shows that $(\kappa/\delta)^{1/2}$ can be a complex function of composition, at least in aqueous mixtures.

A suggested method combining the features of the Calus and Rice and the Stephan and Körner correlations was presented by Calus and Leonidopoulos (1974). The "ideal" superheat was taken to be linear with composition and, applying the correction factor, F (Eq. 38), they obtained

$$\Delta T_{sat} = \left(x_A\ \Delta T_{sat,A} + x_B\ \Delta T_{sat,B}\right)\left[1 + (x-y^*)\left(\frac{\kappa}{\delta}\right)^{1/2} \frac{C_l}{\Delta h}\frac{dT}{dx}\right] \quad (63)$$

This equation served to correlate well data for n-propanol/water mixtures boiling on a thin horizontal wire. Equation (63) is similar to Eq. (55) and suggests a comparison of Θ and $F - 1$, where

$$F - 1 = (x - y^*)\left(\frac{\kappa}{\delta}\right)^{1/2} \frac{C_l}{\Delta h}\frac{dT}{dx} \quad (64)$$

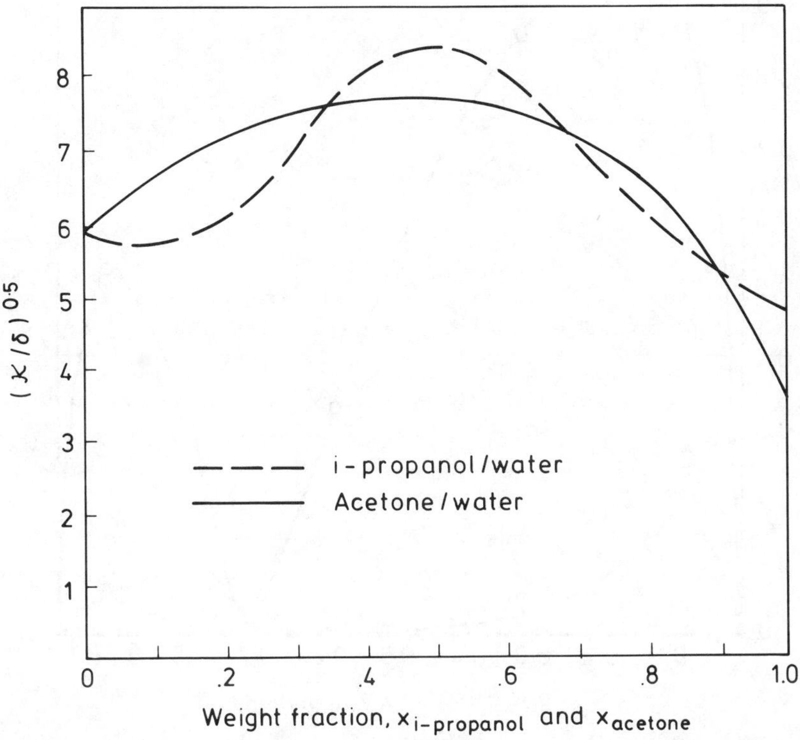

Fig. 29. Values of $(\kappa/\delta)^{1/2}$ as a function of concentration for two mixtures. Calus and Rice (1972).

Figure 30 shows the two correction factors, Θ being calculated with B = 3.29 (Table 7). The difference in values and also in trends with composition can be explained by the fact that, as noted earlier, the boiling behavior measured by Calus and Leonidopoulos, Fig. 31, is different from that more typically shown by the data used by Stephan and Körner, and therefore at present this method, though attractive, should be used with caution for other fluids and other geometries.

Although in this case direct comparison may not be valid, the approach of Calus and Leonidopoulos appears to offer possibilities. If an equation of the form of Eq. (63) could be shown to be valid generally, it might then be possible to evaluate the correction factor F at one composition, where the mass diffusivity was known, then the correction factor B (assumed to be composition-independent) could be calculated for use in Eqs. 55 to 57 from

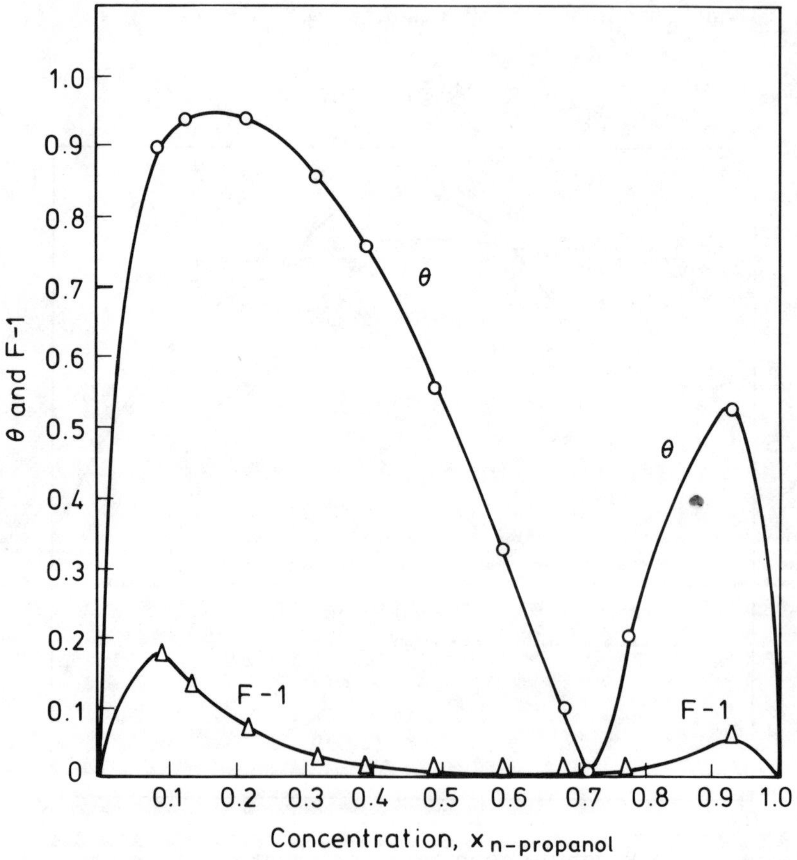

Fig. 30. Variation of correction factors Θ and F-1 with concentration of n-propanol in water. Calus and Leonidopoulos (1974).

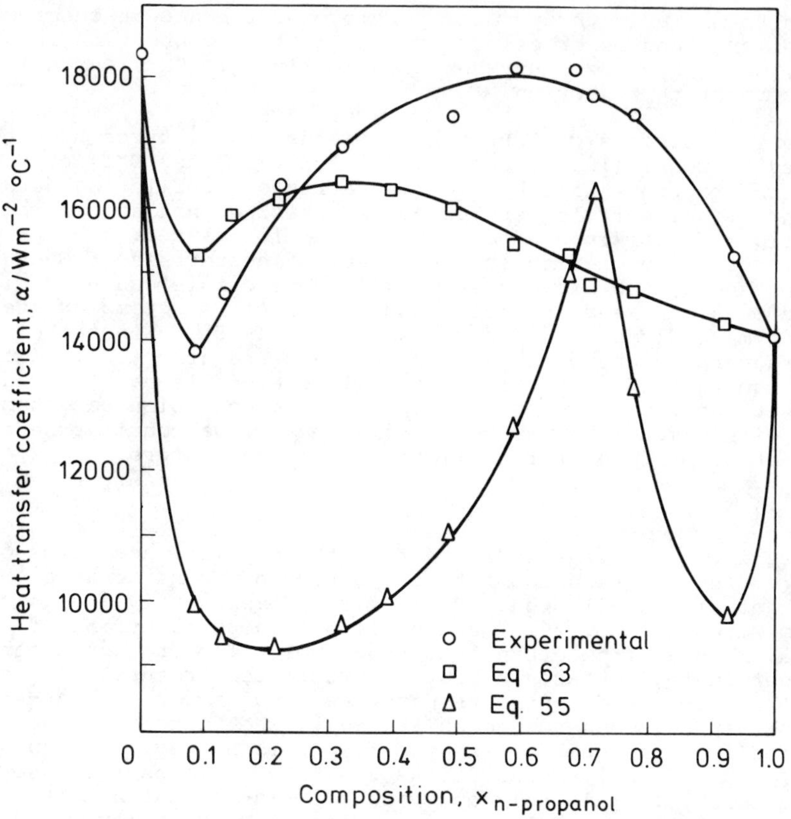

Fig. 31. Experimental and predicted boiling heat transfer coefficients for *n*-propanol/water mixtures at $\dot{q} = 4 \times 10^5$ Wm^{-2}. Calus and Leonidopoulos (1974).

$$B = \frac{F - 1}{(0.88 + 0.12 \times 10^{-5} p)\ |\tilde{y}^* - \tilde{x}|} \qquad (65)$$

Equation (65) is obtained by simultaneous solution of Eqs. (38), (55), and (63).

6 CRITICAL HEAT FLUX

6.1 Introduction

The term critical heat flux encompasses two phenomena: departure from nucleate boiling (DNB) and dryout. The former occurs in nucleate boiling-dominated situations, that is, pool boiling and subcooled and low-quality flow boiling, and is caused by a breakdown in the vapor-removal mechanism with consequent blanketing of the heater surface. The latter situation occurs in high-quality annular flow (Hewitt, 1978). Although dryout may be important in relation to mixtures through its possible occurrence in chemical plant boilers (and is briefly discussed in this connection in Sec. 8), most previous study has been confined to DNB, which is the concern of this section.

Previous surveys of DNB in pure fluids and mixtures have been presented by Owens (1964), Bergles (1967, 1975), and Collier (1981). The literature presents a bewildering array of conflicting data in the light of which it is perhaps not surprising that no general correlations for mixture effects have been formulated. The primary reason for this complexity is the effect of heater size. Even for pure substances, the processes governing the crisis can vary with heater dimensions, and the relationship $\dot{q}_{cr}(D)$ is a complex one depending markedly on D (Lienhard and Dhir, 1973). Since the various governing phenomena vary with D and are affected in different ways by the fact that the fluid is a mixture, it is easy to understand that the way in which $\dot{q}_{cr}$ depends on composition can change greatly with dimensions. Thus in this section we consider separately the effect of mixtures for small and for large heaters.

6.2 Thin Wires

Most investigations indicate that the critical heat flux on thin wires is increased in mixtures, often to a considerable extent. Van Wijk et al. (1956) produced the first comprehensive tests using platinum wires (D = 0.02 cm) and boiling aqueous solutions of alcohols, ranging from ethanol to n-octanol. The crisis is marked by an inordinate increase in heater surface temperature for a small increase in power and contrasts with the results for similar-sized heaters given by Lienhard and Dhir (1973), who found a gradual transition from the nucleate to the film boiling region. Van Wijk et al. found that all the solutions exhibited a large increase in $\dot{q}_{cr}$ at a fairly low alcohol concentration and that with further addition of solute the value of $\dot{q}_{cr}$ passed through a maximum. For most of the alcohols the critical heat flux at the maximum increased with increasing molar mass of alcohol and moved to a position of lower x_{alcohol}. Thus for ethanol the maximum value of the ratio R (= $\dot{q}_{cr}/\dot{q}_{cr,\text{water}}$) was 1.7 and occurred at x_{ethanol} = 0.19, whereas for n-pentanol the maximum value of R was 3.1 and occurred at $x_{n\text{-pentanol}}$ = 0.018. With further increases in molar mass, R was found to decrease, although remaining greater than unity. This was explained by the fact that the constant-boiling temperature approaches close to that of pure water and gives reduced boiling ranges for the heavier alcohols.

Van Stralen (1959b) continued this work by examining the effect of pressure, which is illustrated in Fig. 32. At subatmospheric conditions the critical heat flux is low—the first bubble to grow does so at high superheat and consequently grows so fast that it envelopes the wire, causing burnout. The crisis is determined not so much by the dynamics of bubble growth or the hydrodynamics of bubble departure and liquid replenishment but simply by the temperature at which the first bubble nucleates. Consequently, any effect of mixtures on bubble dynamics has little chance to assert itself; the effect of mixtures acts on ONB mainly through changes in contact angle (see Sec. 3). It appears that the net influence of composition on DNB is small, thus R tends to 1 at low pressures. With increasing pressure the effects of relative volatility come into

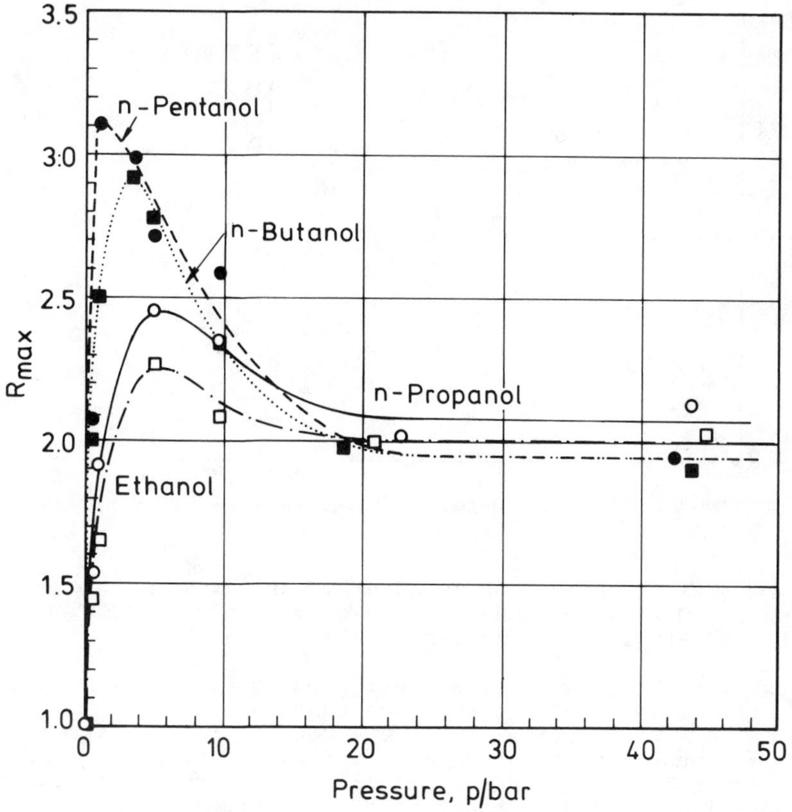

Fig. 32. Effect of pressure on R_{max} for several aqueous alcohol solutions. Van Stralen (1959b).

play and R_{max} increases. For further increase in pressure R_{max} decreases to a constant value of about 2, independent of solute molar mass. The presssure at which R_{max} occurs decreases with increase in molar mass. Experiments with aqueous methyl ethyl ketone, acetone, and ammonia gave similar results, showing that the effects were not limited to alcohol solutions. The effect of pressure was somewhat different when combined with subcooling for alcohol solutions (Van Stralen, 1962). The ratio R passed through a maximum with increasing pressure, but upon raising the pressure up to 130 bar, R continued to fall.

The complex effect of wire diameter and orientation was studied by Van Stralen and Sluyter (1969). Results for methyl ethyl ketone/water, using the concentration that gives maximum $\Delta T_{av}/G_d$, are shown in Fig. 33. For horizontal wires R is low, approximately 1, at very low wire diameters because few bubbles are needed for burnout—the values of $\dot{q}_{cr}$ are high because of the large convective

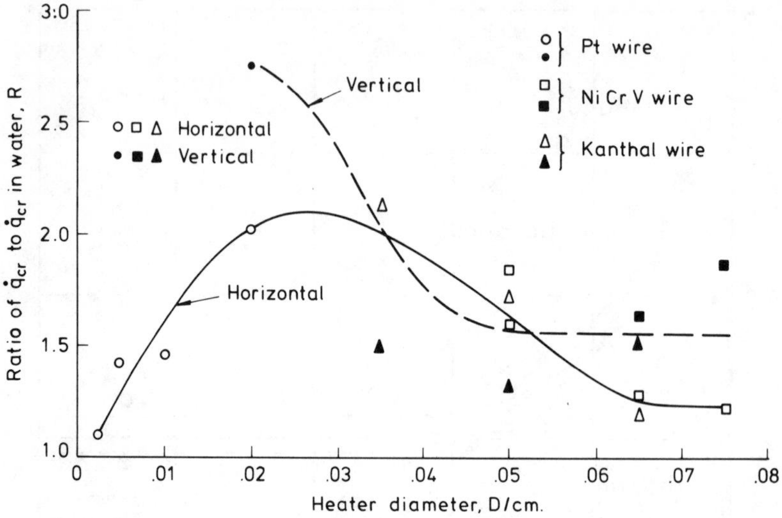

Fig. 33. Critical phenomenon in water/methyl ethyl ketone solutions, $x = 0.041$. Dependence of R on wire size and orientation. Van Stralen and Sluyter (1969).

heat flux. As the wire diameter increases to 0.025 cm, R increases to 2.1, after which it falls to a constant value at a diameter of about 0.06 cm. For vertical wires a somewhat similar pattern emerges, although R is greater for given wire diameter and fluid composition. Vertical wires show a lower $\dot{q}_{cr}$ for pure fluids than do horizontal wires due to the formation of slugs. Van Stralen and Sluyter suggest that orientation has less effect on the crisis for mixtures due to the Marangoni effect, which reduces coalescence as described in Sec. 5. Hence $\dot{q}_{cr}$ for mixtures is less affected by orientation for mixtures and R is greater for vertical than for horizontal wires.

Data given by Isshiki and Nikai (1973) for aqueous organic solutions support the general increase in $\dot{q}_{cr}$ found by Van Stralen. A contrasting set of data was given by the work of Pitts and Leppert (1966). They worked with aqueous solutions of methyl ethyl ketone on wires of Tophet A (80% Ni, 20% Cr), and their results are illustrated in Fig. 34. For the smallest wire (0.0087 cm) and the largest (0.13 cm) used, the critical heat flux decreased as the ketone was added. In each case a minimum was reached at about 2.3% by weight of methyl ethyl ketone, and $\dot{q}_{cr}$ then increased to little more than the value for pure water. For a wire of intermediate size, 0.0254 cm, $\dot{q}_{cr}$ increased but the maximum value of R was only 1.5, at a mass fraction of 0.07. For a 0.02 cm wire the corresponding R and x given by Van Stralen (1966b) are 2.6 and 0.041, respectively.

Although no general correlation is yet available, a comprehensive analysis of the critical phenomenon has been carried out

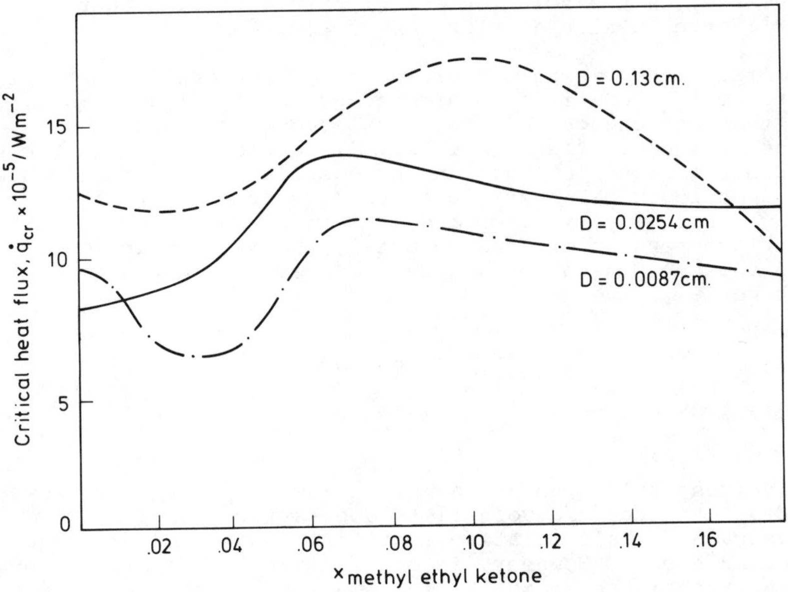

Fig. 34. Effect of wire diameter on critical heat flux in aqueous solutions of methyl ethyl ketone. Pitts and Leppert (1966).

by Van Stralen (1966b, 1968a) as part of his studies of bubble growth. The model of the critical phenomenon is purely thermal and assumes burnout to be caused by bubble packing. When the entire surface of the heater is covered by bubbles, the resulting vapor blanket produces the crisis. As discussed in Sec. 4, the total heat flux is the sum of two contributions

$$\dot{q} = \dot{q}_{nb} + \dot{q}_{\text{conv}} \tag{66}$$

Since the departure diameter in mixtures is reduced, a greater number of bubbles can be packed onto the surface before dryout occurs. Thus dryout is defined by a fixed value of the direct vaporization at the surface. Since the direct vaporization is decreased in mixtures but the convective contribution is more than correspondingly increased, the total heat flux is increased at the crisis. Van Stralen (1966b) gives data to confirm this point by comparing water and methyl ethyl ketone/water on a 0.02 cm wire. The bubble departure diameter at the crisis is lower by about 60% in the solution and the active nucleus density is roughly four times greater. The critical heat flux is increased by a factor 2.6; for water the direct vapor formation at the surface, measured at 37 x 10^4 W m^{-2}, is roughly half the value of $\dot{q}_{cr}$ (67 x 10^4 W m^{-2}); for the solution the direct contribution (26 x 10^4 W m^{-2}) is one-seventh the total (172 x 10^4 W m^{-2}). The convective contribution caused by the bubbles (increase in heat transfer above that due to natural convection and not due to direct vapor formation) is four times as large for the solution as for pure water and contributes about 60% of the total heat flux.

An alternative viewpoint for the trends in critical heat flux in mixture boiling on wires is offered by Hovestreijdt (1963), who proposed that the change in bubble coalescence characteristics, due to the Marangoni effect (see Sec. 5), could be as significant as the bubble growth slowing. An increase in bubble coalescence would produce lower DNB heat fluxes through an earlier vapor blanketing. For many mixtures the effects of relative volatility and of surface tension difference act in the same direction. Figure 35 shows $\dot{q}_{cr}$ data for n-butanol/water together with $\Delta T_{av}/G_d$ and M, where

$$\Delta T_{av}/G_d = (x - y^*) \frac{dT}{dx} \tag{67}$$

$$M = (x - y^*) \frac{d\sigma}{dx} \tag{68}$$

Positive values of $\Delta T_{av}/G_d$ and of M, at low $x_{n\text{-butanol}}$, suggest increased $\dot{q}_{cr}$ through the volatility and surface tension mechanisms. respectively, and their relative importance cannot be distinguished by this means alone. However, in one mixture, methanol in heptane, there is a large increase in $\dot{q}_{cr}$ for small increases in x_{methanol}, and this may be explained by the relative volatility effect alone because M is such that, acting alone, it would reduce $\dot{q}_{cr}$. Figure 35 shows the same effect for n-butanol-rich solutions. There is not yet evidence of any mixture where the surface tension effect is the primary agent in determining the crisis in any but the smallest concentration ranges.

The same effect that opposes coalescence can aid the stability of foams. Hovestreijt showed that the mere existence of a foam, achieved with a small addition of surface-active agent at the top of the liquid pool, is not necessarily accompanied by an increase in $\dot{q}_{cr}$. Van Stralen (1967) reached a similar conclusion with solutions of Teepol. Although the solutes can stabilize foams at the top of the pool, it may be assumed that there is no time to establish a monolayer at the bubble interface, thus there is no effective reduction of surface tension.

Van Stralen and Sluyter (1969) investigated ternary mixtures. They found similar increases in $\dot{q}_{cr}$ in line with the decrease in available superheat.

As we have seen, the surface superheat for mixture nucleate boiling at a given heat flux exceeds that for a pure substance. Since the critical heat flux is also increased, the surface temperature at the crisis tends to be doubly raised. The maximum reduction in bubble growth and the maximum increase in $\dot{q}_{cr}$ occur at the same, often low, solute concentration (Van Stralen, 1970); thus for pure water Van Stralen (1966b) found that the wire superheat at the crisis was 21.5°C, whereas for $x_{\text{methyl ethyl ketone}} = 0.041$ it was 34°C.

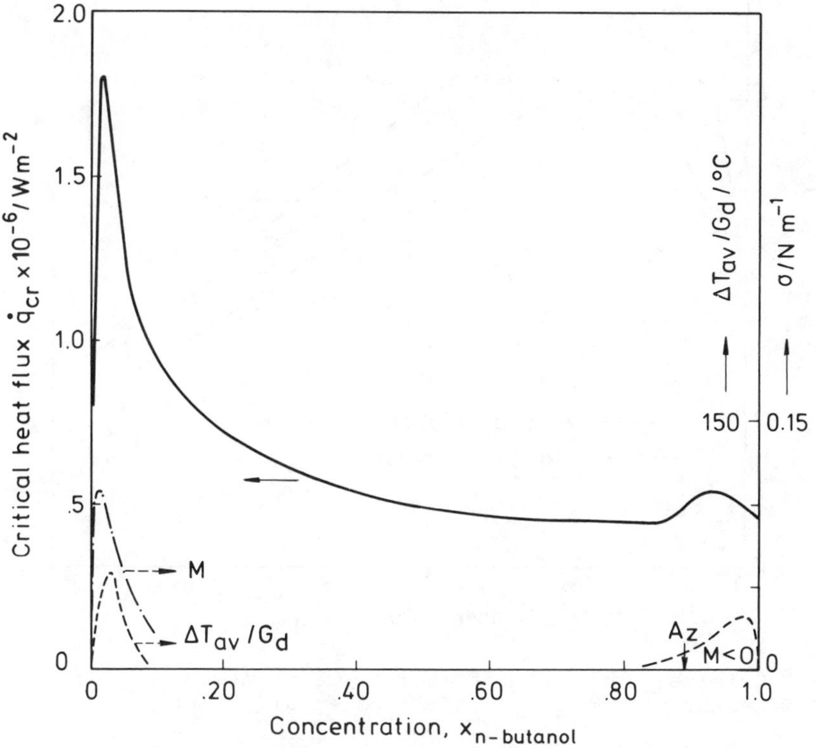

Fig. 35. Critical heat flux of n-butonal/water mixtures comparison with $\Delta T_{av}/G_d$ and M. Hovestreijdt (1963).

6.3 Large Heaters

A considerable amount of critical heat flux data is available for mixtures boiling on large tubes and flat surfaces. Bonilla and Perry (1941) measured $\dot{q}_{cr}$ for aqueous alcohol solutions on a flat copper surface and found increases of a few percent upon addition of small amounts of ethanol to water and a small decrease upon addition of butanol. Similarly, Owens (1964) studied boiling of water/methyl ethyl ketone and aqueous alcohol solutions on the outside of a tubular heater of diameter 0.5 cm and found that additions of up to 10% by weight of all the organics to water reduced $\dot{q}_{cr}$ by as much as 30%; Tobilevich et al. (1969) measured the $\dot{q}_{cr}$ for ethanol/water mixtures on the outside of a copper tube of o.d. 1 cm. They found a sharp minimum in $\dot{q}_{cr}$ at $x_{ethanol}$ = 0.088, with a critical heat flux some 30% of the value for pure water, and a maximum at $x_{ethanol}$ = 0.35, where the critical heat flux is about 75% of the value of pure water.

Kutateladze et al. (1966) illustrated the importance of the geometry using ethanol/water. For a wire they obtained a significant increase of $\dot{q}_{cr}$ with addition of ethanol and for a ribbon heater, on its side, they obtained a definite decrease in $\dot{q}_{cr}$ (see Fig. 36).

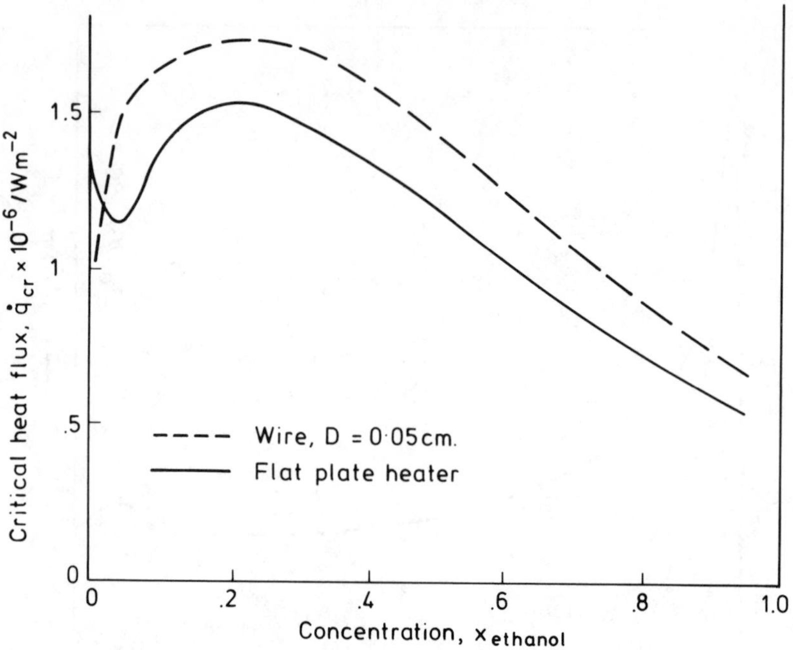

Fig. 36. Critical heat flux of ethanol/water mixtures. Effect of heater geometry. Kutateladze et al (1966).

A number of workers have studied nonaqueous mixtures. Wright and Colver (1969) used ethane/ethylene mixtures on a horizontal rod and concluded that the mixtures behaved almost identically to the pure fluids. Their results suggest, however, a moderate increase in $\dot{q}_{cr}$ at low pressures for low concentrations of ethylene and a decrease at high pressures for high ethylene concentrations. Bonilla and Perry (1941) studied butanol/acetone, butanol/ethanol and acetone/ethanol on flat heaters and found increases in $\dot{q}_{cr}$ of up to 17%. Cichelli and Bonilla (1945) found that n-pentane/propane mixtures at elevated pressures on a fouled surface showed $\dot{q}_{cr}$ about 50% greater than for pure propane and three times greater than for pure n-pentane. Benzene/diphenyl mixtures on a horizontal tube, o.d. = 0.95 cm, were studied by Huber and Hoehne (1963), who found a 100% increase in $\dot{q}_{cr}$ for a 6.3% by weight addition of benzene to diphenyl. Afgan (1966) studied ethanol/benzene mixtures boiling on the outside of a stainless steel tube (o.d. = 0.45 cm). Figure 37 shows the increase in $\dot{q}_{cr}$ that he found over a range of pressures on each side of the azeotrope.

Data for cryogenic mixtures on a vertical copper block are reported by Lyon (1964), whose results show a reduction in $\dot{q}_{cr}$ when nitrogen is added to oxygen (up to 20% molar) and a constant value for all compositions thereafter up to pure nitrogen. Kosky and Lyon (1968) report data for nitrogen/oxygen mixtures for pressures up to the critical. For high pressures they found only

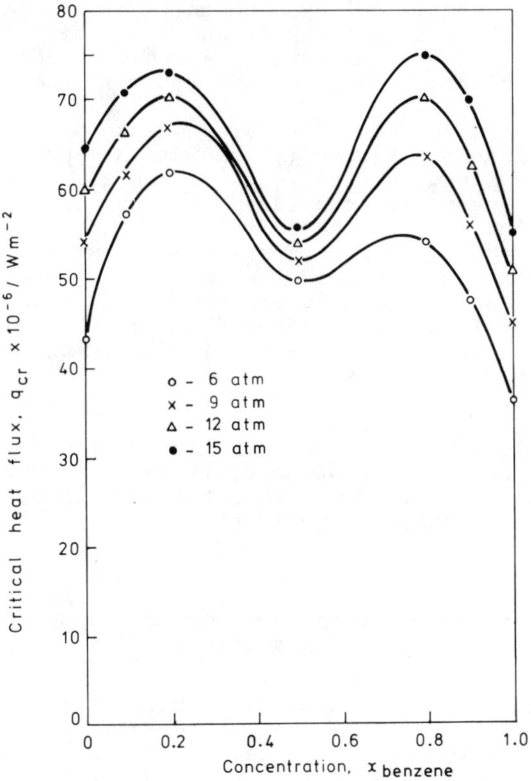

Fig. 37. Critical heat flux for ethanol/benzene. Afgan (1966).

gradual changes in $\dot{q}_{cr}$ with composition. Ackerman et al. (1975) investigated nitrogen/methane mixtures at moderate and high pressures and found a steep maximum in the critical heat flux.

The effect on $\dot{q}_{cr}$ of transverse flow past a heating tube was investigated by Leppert et al. (1958) for dilute aqueous alcohol solutions. The critical heat fluxes for the solutions were little different from those for pure water. Naboichenko et al. (1965) examined the crisis in monoisopropyl diphenyl/benzene mixtures in a similar geometry and obtained a maximum in the $\dot{q}_{cr}$-x curve at low concentrations of monoisopropyl diphenyl. The value of $\dot{q}_{cr}$ at this point was about twice that of pure benzene. They investigated the effect of subcooling and found that for high values a second, though smaller, maximum occurred at low benzene concentrations. In-tube DNB experiments for acetone/toluene and benzene/toluene with varying subcooling were conducted by Andrews et al. (1968) and by Carne (1963). They both obtained maxima in the $\dot{q}_{cr}$-x curves. The critical heat fluxes, though not the shape of the curves, varied with subcooling.

Several authors (Debbage et al., 1963; Sterman and Korychanek,

1970) have studied the critical phenomenon in high-molar-mass organic fluids (mostly polyterphenyls) and their mixtures because of their possibly attractive properties as reactor coolants. The behavior of such coolants is complex and not well understood.

Dunskus and Westwater (1961) studied dilute solutions (up to 1%) of high-molar-mass substances ($\tilde{M} \approx 1000$) in *i*-propanol (see also Sec. 5). They found critical heat fluxes as much as 60% above those for pure *i*-propanol and concluded that the increase was greatest as the molar mass of the additive increased. The increase in $\dot{q}_{cr}$ was attributed to the fact that locally increased viscosity at the bubble boundary prevents coalescence of adjacent bubbles. Conversely, Gannet and Williams (1971) found reductions in $\dot{q}_{cr}$ upon addition of very heavy substances ($\tilde{M} \approx 10^6$) to cyclohexane. The character of the maximum in the boiling curve changed, becoming broader. Such a result was also found by Gambill (1968) for dilute aqueous solutions of polyethylene oxide.

With the bewildering and often conflicting trends reported, even for solutions of simple compounds, it is perhaps not surprising that no general correlations have yet been proposed. However, several authors have considered their results in the light of the general relationship

$$\frac{\dot{q}_{cr}}{\Delta h\ \rho_v^{1/2}\ [\sigma\ (\rho_l - \rho_v)g]^{1/4}\ [(\rho_l + \rho_v)/\rho_l]^{1/2}} = K \qquad (69)$$

derived by Zuber (1958) and in a slightly different form by Kutateladze (1951), which is reasonably accurate for the crisis in pure substances. Kutateladze suggested that K is almost constant but varies slightly with geometry, being 0.12 for the outside of tubular heaters and 0.13 for flat plates. Zuber showed theoretically that K is 0.13 for a flat surface, although he also shows (Zuber, 1959) that it must be subject to some variation due to the spectrum of unstable wavelengths. Afgan (1966), Kutateladze et al. (1966), Gaidorov (1975), and Frea et al. (1977) have shown how K varies for their various mixtures. Figure 38 shows the variation with composition for the results of Afgan with ethanol/benzene.

7 FILM BOILING

In film boiling the heater surface is covered by a coherent layer of vapor, and the liquid is normally assumed to have no contact with the surface. Heat is conducted through the vapor to the neighboring liquid-vapor interface, where further vapor is generated. Wave disturbances may be present at this interface and, if they are of sufficient amplitude, vapor may be released in the form of bubbles at the nodes of the waves. This is more common for horizontal surfaces—vertical heaters are cooled by a vapor film that increases in thickness with height but with an interface that is usually smooth. Film boiling of pure substances is considered by Bromley (1950) for horizontal tubes of large diameter and by Lienhard and Wong (1964) for horizontal tubes of small diameter, that is, wires. In this context large and small usually

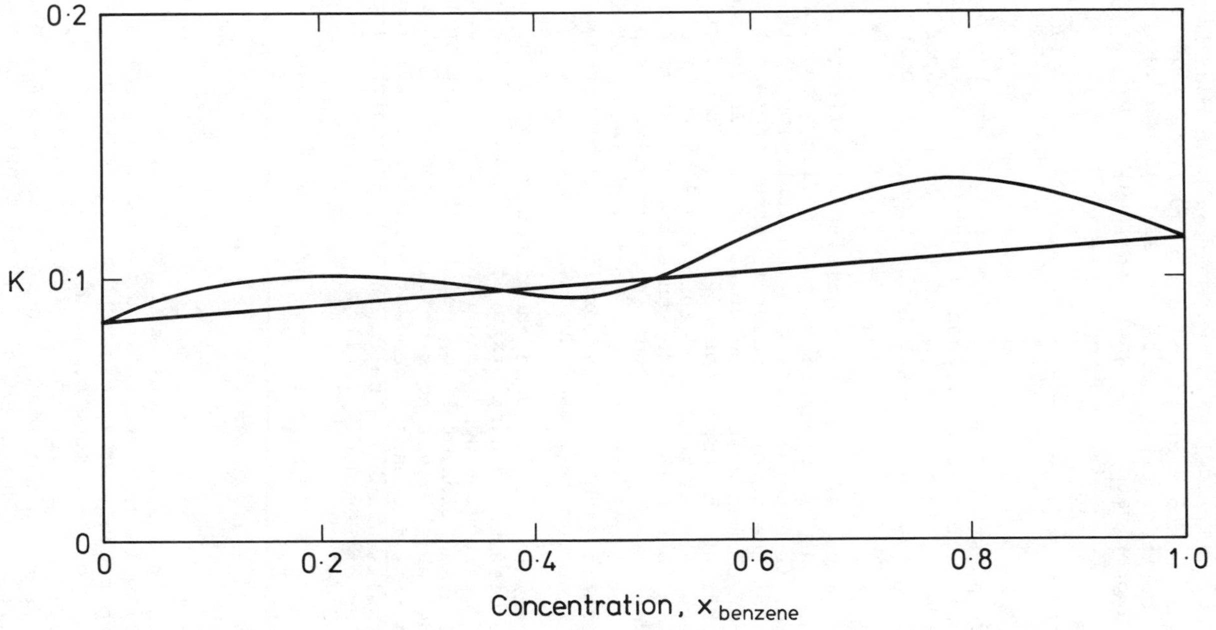

Fig. 38. Variation of K with concentration for ethanol/benzene. Afgan (1966)

mean greater or less than about 0.1 cm. Berenson (1961) considered film boiling on horizontal flat surfaces. His model, which considers the dominant wavelength λ_D gives a similar result to that of Bromley with the tube diameter replaced by λ_D. Lienhard and Wong extended the hydrodynamic instability theory to thin wires by accounting for the effect of surface tension across the curved periphery of the liquid-vapor interface—their experiments showed that the average bubble spacing was approximately λ_D, where λ_D is given by

$$\lambda_D = 2\pi \left[\frac{3\sigma}{g(\rho_l - \rho_v)} \right]^{1/2} \tag{70}$$

Film boiling of pure substances on a vertical flat plate has been studied by Hsu and Westwater (1960), whose model encompasses the case where the distance in the direction of flow or the vapor flow rate is sufficiently great to cause the transition to turbulent flow.

Film boiling in mixtures is again significantly affected by the volatility difference but, in contrast to pool boiling, an improvement in heat transfer is generally found. The stripping of the light component causes the interface saturation temperature to increase. Making the common assumption of saturation in the bulk thus indicates a gradient in temperature between the interface and the bulk fluid, hence heat is conducted into the liquid. The common explanation is that this provides an extra mechanism for heat removal and increases α for given T_w but an alternative viewpoint might be that with the liquid acting as a heat sink less heat is available for evaporation and so the vapor film is thinner and the resistance to heat transfer is thus reduced.

Kautzky and Westwater (1967) were the first to investigate the film boiling of miscible binary mixtures of volatile substances. They studied carbon tetrachloride/Refrigerant 113 on a horizontal flat plate at 1 bar. Addition of 1% by weight of Refrigerant to carbon tetrachloride (which gives no significant change in physical or transport properties) increased the heat transfer coefficient by some 20%. Their results are illustrated in Fig. 39, which also shows the correlation of Berenson (1961),

$$\alpha = 0.425 \left[\frac{\lambda_v^3 \; \rho_v \; (\rho_l - \rho_v) \; \Delta h' \; g}{\mu_v \; \Delta T \; [\sigma/g \; (\rho_l - \rho_v)]^{1/2}} \right]^{1/4} \tag{71}$$

where $\Delta h'$ is a specific latent heat modified to allow for the vapor superheating. Equation (71) is successful for the pure subtances but cannot predict the large increase in α when a small amount of solute is added to pure solvent at each end of the concentration range. Similar behavior was found by Basu (1975) for carbon tetrachloride/chloroform mixtures boiling on thin wires. By contrast, however, Basu did not find such behavior for carbon tetrachloride/acetone, and Wright et al. (1971) found no such effect for ethane/ethylene mixtures boiling on a tube. Ackerman et al. (1976) found little effect of volatility on film boiling of nitrogen/methane mixtures, except perhaps at low pressure and high heat fluxes, where a decreased heat flux is suggested.

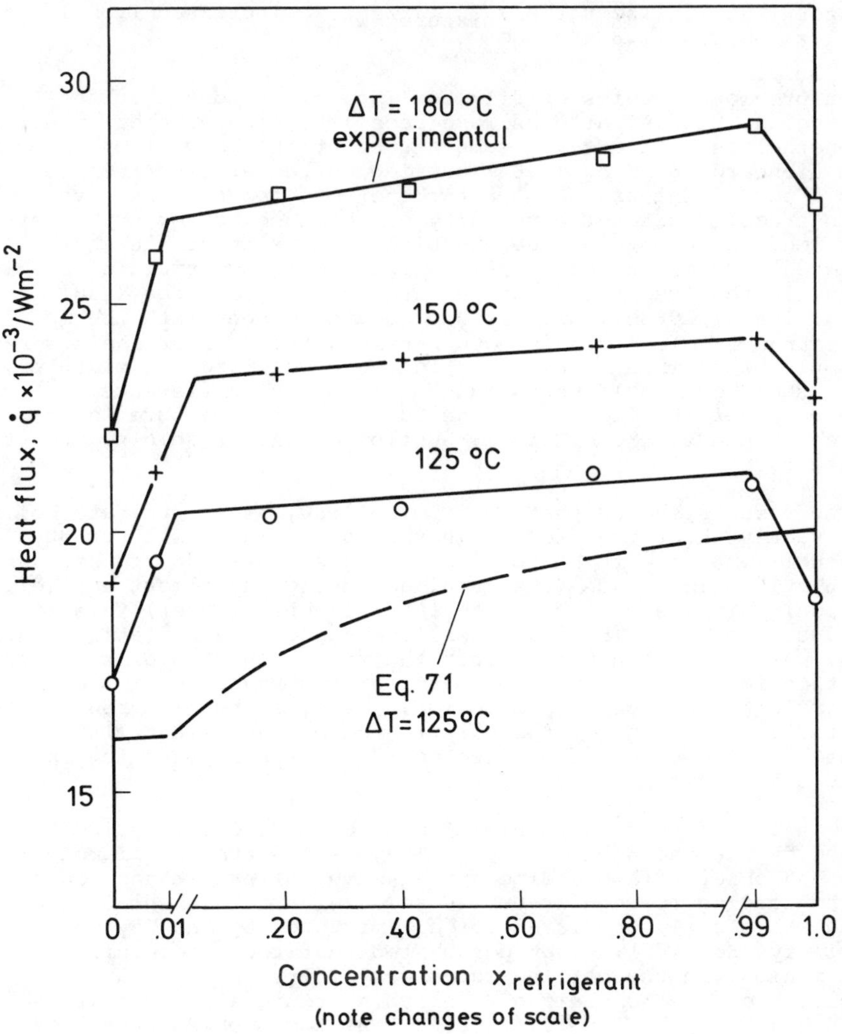

Fig. 39. Effect of composition on film boiling heat flux. Kautzky and Westwater (1967).

Van Stralen et al. (1972) examined film boiling from thin wires where a regular pattern of bubble columns can be seen. They used pure water and methyl ethyl ketone/water solutions, weight fraction 4.1% (which gives the maximum slowing of bubble growth in nucleate boiling). They noticed much reduced coalescence between adjacent bubbles for the solutions, which they attributed to Marangoni effects. Over a large range of superheats the heat transfer results showed the same trends as those of Kautzky and

Westwater (1967). The ratio $\alpha_{mixture}/\alpha_{water}$ increased from 1.2 at $\Delta T_{sat,w}$ = 400°C to 1.8 at 1100°C.

In previous studies of film boiling at horizontal surfaces, where regular bubble patterns were seen, the liquid-vapor interface was nonetheless implicitly assumed to be at the saturation temperature. Lienhard and Wong (1964) considered the excess pressure across a curved interface but not the excess temperature. Van Stralen et al. doubted the validity of this neglect of superheat. They filmed the growth of the "bubbles" at the interface and considered them in an identical fashion to the treatment of bubbles growing, in the asymptotic period, at a heating surface (Van Stralen, 1968a, 1968b). They found good agreement with the model, and inferred from measured growth rates that for water the superheat could be as much as 6°C. An increase occurred for mixtures, and the superheat could be as much as 11°C, the difference in bubble point of the liquid and the boiling point of pure water, which corresponds to complete exhaustion of the organic solute at the interface.

The results showed that in film boiling, as in nucleate boiling the convective heat transfer is increased in mixtures. For pure water or ethanol some 95% of the total heat transfer was due to net vaporization at the interface (the remainder being due almost entirely to radiation); in the mixture (methyl ethyl ketone/water) this figure was reduced to 53%. The remainder, apart from the radiation contribution, was due to conduction into the film. In view of the increase in α, this must more than compensate for the reduction in heat transfer due to net vapor production. Measured bubble growth rates indicated a reduction for the mixture in good agreement with the nucleate boiling case; they also noticed again a reduction in bubble departure diameter.

Heat transfer at the minimum point on horizontal cylinders was examined by Yue and Weber (1974). They used a tube of diameter 0.8 cm and observed a one-dimensional wave pattern at the top of the tube leading to regular bubble release. The average bubble spacing was closely predicted for all mixtures by the Taylor instability, Eq. (70). They postulated that the vapor removal rate at the minimum in the boiling curve was limited by the Taylor instability on the top half of the cylinder and assumed that the bubbles had a depth of $\pi D/2$. Noting that two bubbles break off for each wave oscillation, they showed that the heat flux at the minimum should be given by

$$\dot{q}_{min} = 0.15\rho_v\ \Delta h \left[\frac{\sigma g(\rho_l - \rho_v)}{(\rho_l + \rho_v)^2}\right]^{1/4} \tag{72}$$

Using Bromley's (1950) equation for the film boiling heat transfer coefficient (thus considering here only pure fluids),

$$\alpha = 0.62 \left[\frac{\lambda_v^3\ g\rho_v(\rho_l - \rho_v)\Delta h}{\mu_v\ D\Delta T}\right]^{1/4} \tag{73}$$

they showed that the temperature difference (wall to bulk liquid) at the minimum in the boiling curve should be

$$\Delta T_{min} = 0.334 \left[\frac{D}{\lambda_D}\right]^{1/3} \left[\frac{\rho_v \, \Delta h}{\lambda_v}\right] \left[\frac{g(\rho_l-\rho_v)}{(\rho_l+\rho_v)}\right]^{2/3} \left[\frac{\sigma}{g(\rho_l-\rho_v)}\right]^{1/2} \left[\frac{\mu_v}{g(\rho_l-\rho_v)}\right]^{1/3} \quad (74)$$

Data for pure substances and for mixtures of low volatility were predicted well by Eq. (74), if the constant 0.334 were replaced by 0.42, but it underpredicted the minimum superheat for mixtures of high relative volatility, $\alpha>4$. For such mixtures Yue and Weber proposed to correlate ΔT_{min} by writing it as the sum of two temperature differences, one through the vapor and one through the liquid (zero for coolants of low relative volatility):

$$\Delta T_{min} = (T_w - T_i) + (T_i - T_\infty) \quad (75)$$

The first term on the right-hand side is evaluated using Eq. (74) for all coolants. Yue and Weber showed that $(T_i - T_\infty)$ could be correlated against $[(y^*-x_\infty)/(y^*-x)_{max}](T^* - T_\infty)$. The various items of this correlating term are illustrated in Fig. 40; $(y^*-x)_{max}$ is the greatest value of $(y^* - x)$ for the mixture in question. They found that, for five mixtures, $(T_i - T_\infty)$ is equal to this correlating group, as shown in Fig. 41.

The problem of film boiling of binary mixtures on an isothermal vertical flat plate, with laminar flow in both phases, was comprehensively analyzed by Yue and Weber (1973) and by Marshall and

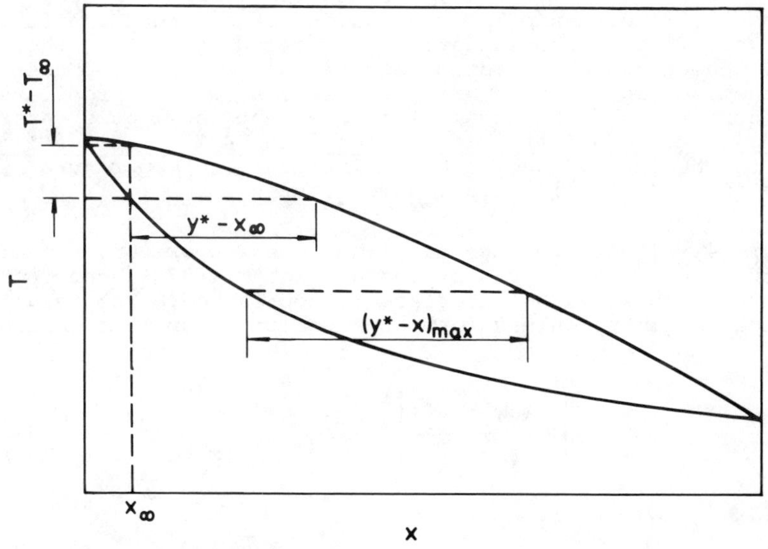

Fig. 40. Vapor liquid equilibrium diagram for film boiling correlation.

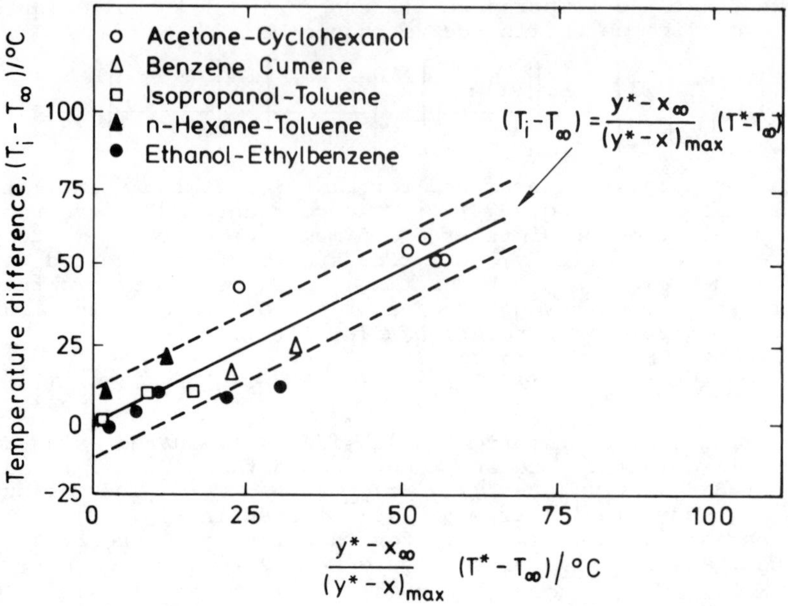

Fig. 41. Liquid film temperature difference for mixtures. Yue and Weber (1974).

Moresco (1977). The analogous situation on a horizontal plate, with net flow of liquid over it, was studied by Prakash and Seetharamu (1978) with similar conclusions regarding the effect of relative volatility. The situation is modeled by writing the mass, momentum, and energy conservation equations for the two phases and solving them, employing similarity transforms, with appropriate boundary conditions. The assumptions are that the liquid-vapor interface is smooth and free of bubbles (Fig. 42), that the liquid far from the heater surface is at the saturation temperature $T_{bub}(x_\infty)$, and that the interface temperature is $T_{bub}(x_i)$.

Figure 43 illustrates typical liquid and vapor concentration and temperature fields, given by Yue and Weber (1973), showing the large decrease in x as the interface is approached. The results are plotted in dimensionless form, the position parameter being given by

$$\eta = \left[\frac{g(\rho_l - \rho_v)}{4\nu_v^2 \rho_v}\right]^{1/4} yz^{-1/4} \qquad (76)$$

for the vapor and

$$\eta = \left[\frac{g\beta_l(T_w - T_\infty)}{4\nu_l^2}\right]^{1/4} yz^{-1/4} \qquad (77)$$

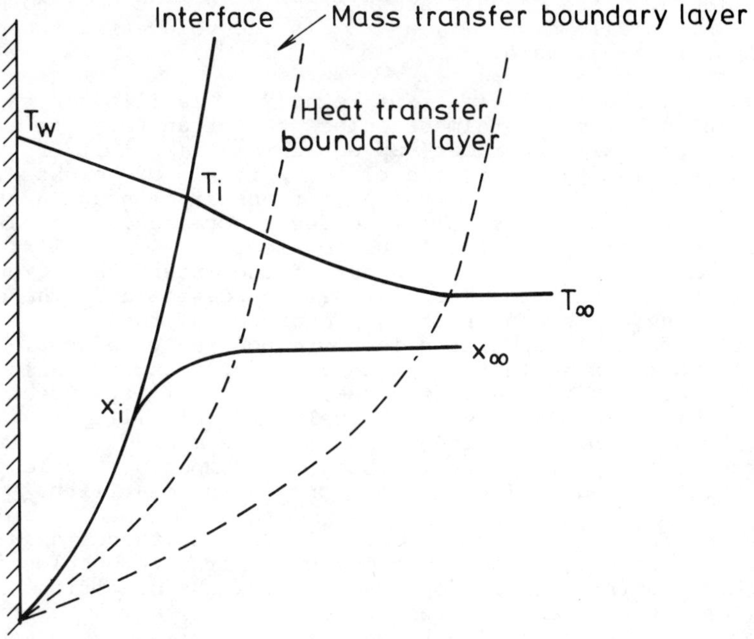

Fig. 42. Physical model for boiling on a vertical flat plate.

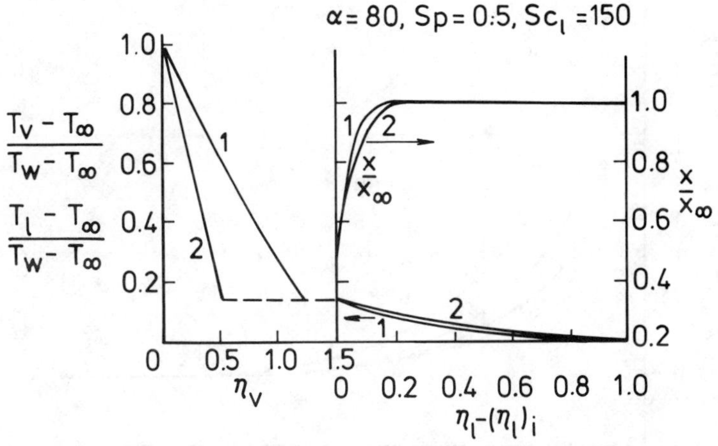

Fig. 43. Temperature profiles in liquid and vapor and concentration profiles in liquid for two film thicknesses. Yue and Weber (1973).

for the liquid, where y and z are respectively distances normal to the wall and axially along it, and β_l is the volumetric expansion coefficient in the liquid.

Figure 44 shows the effect of relative volatility on the heat transfer in the form of a dimensionless heat transfer rate against a dimensionless wall superheat, Sp equal to $c_v\ (T_w - T)/\mathrm{Pr}_v \Delta h$. The conditions are averaged over the plate, which is of height H. The general effect is to increase the heat transfer although as the superheat increases the volatility effect decreases. The maximum effect of conduction into the liquid phase occurs when all the light component has been stripped out of the interface region; $T_i - T_\infty$ is then maximized. For further increases in T_w the conduction heat transfer is a decreasing proportion of the total and the beneficial effect of conduction is correspondingly reduced. Figure 44 also shows the rise in total heat transfer with increasing relative volatility and indicates that the volatility effect can persist at higher superheats if the relative volatility is raised. If the Schmidt number is raised, that is, the mass diffusitivity is lowered, the replenishment of the light component is slowed, and the effect is equivalent to an increase in volatility.

Yue and Weber performed experiments to test these ideas with a horizontal carbon cylinder, D = 0.8 cm. Although bubbles formed in a regular pattern at the top of the tube, the conditions around most of the circumference correspond closely to the mathematical model described above for a vertical flat plate, and the results agreed well with the predictions over a wide range of volatilities and superheats, as illustrated in Fig. 45 for acetone/cyclohexanol

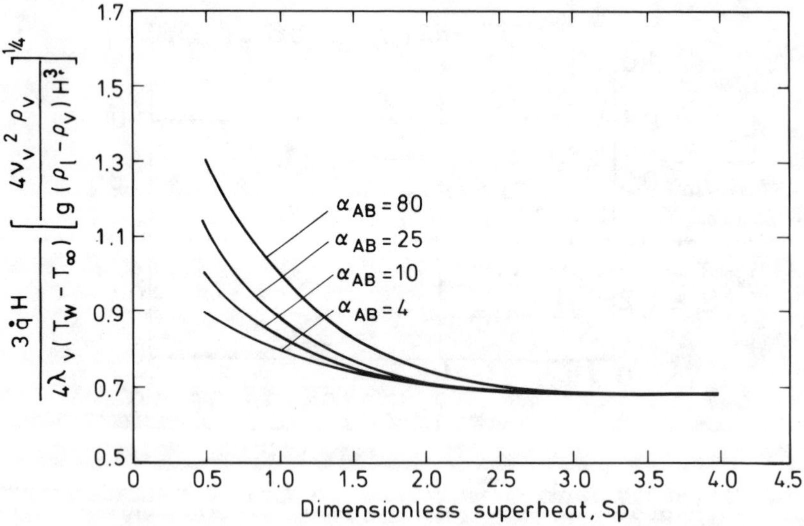

Fig. 44. Effect of relative volatility on film boiling heat transfer of mixtures on vertical flat plates. Yue and Weber (1973).

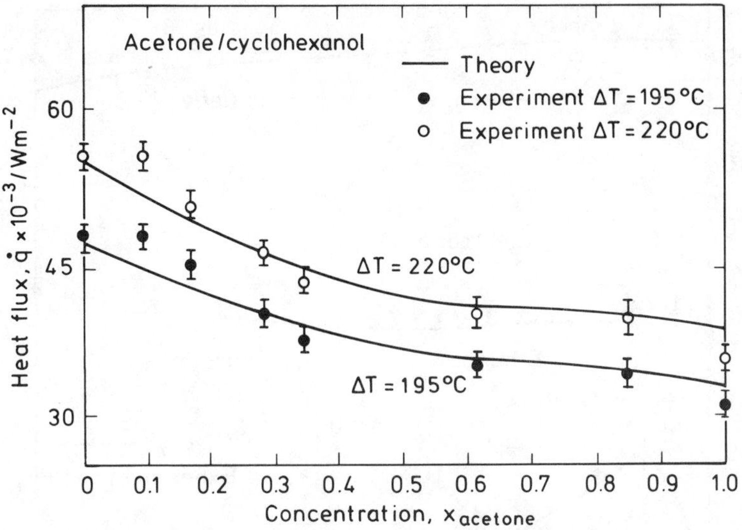

Fig. 45. Effect of composition on film boiling heat flux for acetone/cyclohexanol. Yue and Weber (1973).

(relative volatility 80).

This model for heat flux at given superheat can be combined with the model discussed earlier for wall superheat at the minimum (Yue and Weber, 1974) to give the $\dot{q}$ and ΔT_{sat} coordinates for the minimum in the boiling curve. Figure 46 shows excellent agreement for four mixtures over the whole composition range. Note the almost twofold increase in $\dot{q}_{min}$ at 9.5% acetone and also the absence of any mixture effect for carbon tetrachloride/Refrigerant 113 in contrast to the previously mentioned conclusion of Kautzky and Westwater (1967) for constant ΔT.

It should be noted that although the model proposed by Yue and Weber seems successful in predicting the significance of mixture effects for large heaters, it fails for small ones, that is, thin wires. As noted earlier, Basu (1975) found a significant increase in heat transfer for carbon tetrachloride/chloroform mixtures on a 0.004 cm-diameter horizontal platinum wire but not for carbon tetrachloride/acetone. The greatest mixture effect in the first case, a tripling of the heat flux for given $\Delta T_{sat,w}$, occurred for Sp ~ 0.8 and relative volatility about 1.7, whereas the work of Yue and Weber indicates that there should be an increase of only a few percent in heat flux. The carbon tetrachloride/acetone mixtures at the same Sp and α showed no appreciable mixture effect. This discrepancy is not perhaps surprising when it is realized that the work of Yue and Weber is based on a vertical surface and the above results were taken with thin wires. The modeling of the two situations has already been shown to be significantly different for pure substances (Bromley, 1950; Berenson, 1961; Lienhard and Wong, 1964).

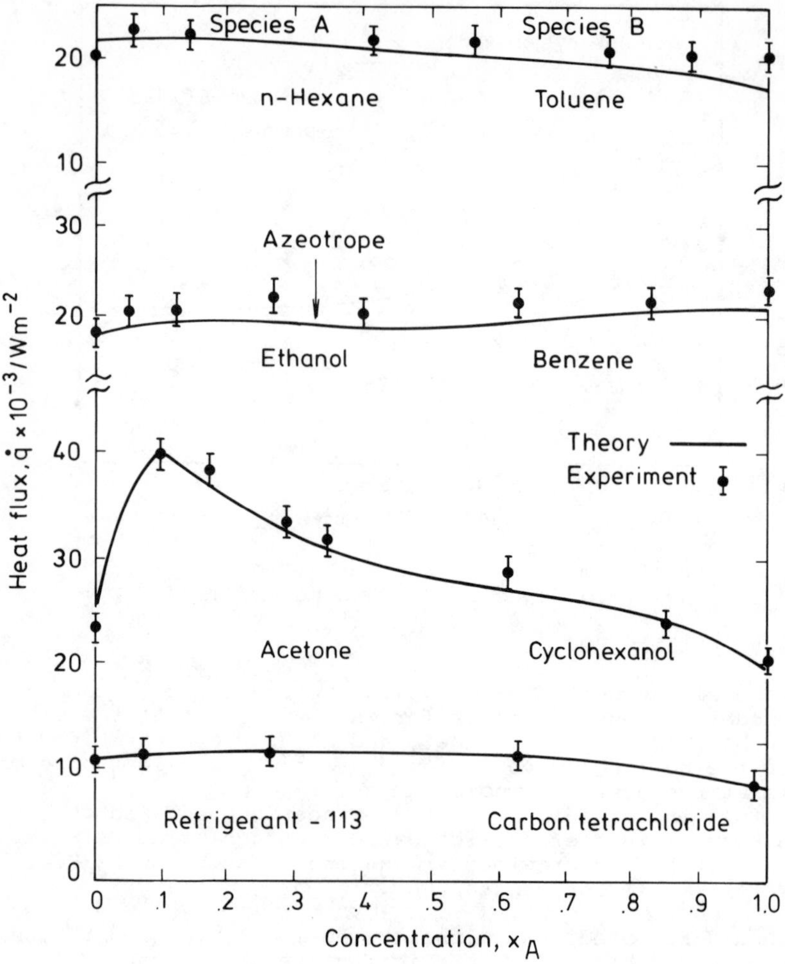

Fig. 46. Minimum film boiling heat flux of mixtures. Yue and Weber (1974).

Although the effect of turbulence has been discussed by Hsu and Westwater (1960) in relation to vertical film boiling of pure substances, its effect has not yet been considered with respect to mixtures. If the shear in the liquid is sufficient to cause turbulence within it, the interface will be replenished more quickly with the light component. The increase in heat transfer due to conduction into the liquid will therefore be diminished, but whether this will offset the effect of increased interface evaporation rates (tending to increase the stripping of the light component) remains a subject for examination.

Several workers have investigated mixtures other than those of volatile, miscible fluids. The effect of addition of high-molar-mass additives to organic substances was noted by Dunskus and Westwater (1961) and by Gannet and Williams (1971). Both investigations showed increases in film boiling heat transfer.

Film boiling of immiscible mixtures was studied by Bragg and Westwater (1970) and by Yilmaz et al. (1976). The former used water with both more and less dense organic additives, perchlorethylene, Refrigerant-113, and *n*-hexane. The lower liquid was in film boiling and its properties controlled the boiling process. The primary effect of the upper liquid was to subcool the lower one and thus to increase the heat flux. Yilmaz et al. used water, always the lower liquid, and a series of paraffins, *n*-hexane to tetradecane. The lighter additives increased the heat transfer. With increasing molar mass of additives, the volatility difference decreased and so did the enhancement of heat transfer—indeed, a deterioration was noted with the heavier additives. With these mixtures no subcooling of the water was found, and it may be assumed that the water at the liquid-liquid interface was superheated and that this causes the deterioration in heat transfer.

8 FLOW BOILING AND EVAPORATION

In many of the boilers used in the chemical and petroleum industries the feedstock is a mixture of many components, perhaps up to 40 in a typical petroleum feed. Few studies have been carried out in such systems, and interpretation of the available results is exceedingly complicated.

McAdams et al. (1942) studied vaporization of benzene/oil mixtures in a multipass horizontal-tube evaporator. The results were dominated by the heat used to maintain the increasing bubble point as the benzene was preferentially vaporized. This caused a progressive decrease in heat transfer coefficient, overcoming the tendency of the coefficient to increase with increasing flow quality.

Performance data for vertical thermosiphon reboilers operating with pure substances and with mixtures are given by Bonnett and Gerster (1951) and by Shellene et al. (1968). The latter show clearly that the overall heat transfer coefficient in *i*-propanol/water is reduced in a similar way to that in pool boiling (Fig. 47). Their results suggest that the maximum heat flux, due to DNB, is reduced in mixtures of these two substances. In contrast, Calus et al. (1973) obtained overall heat transfer data for a single tube, steam-heated natural-circulation boiler and suggest an <u>increase</u> in heat transfer for *n*-propanol/water mixtures. This is yet another area in multicomponent boiling where further data are needed before general conclusions can be drawn.

Since the quality, and hence the flow regime, changes along a boiler channel, the measured overall thermal performance can be the manifestation of several different heat transfer modes. A true insight into the effects involved can be gained only from local measurements. Toral (1979) presents such data for ethanol/cyclohexane mixtures in a single electrically heated vertical tube.

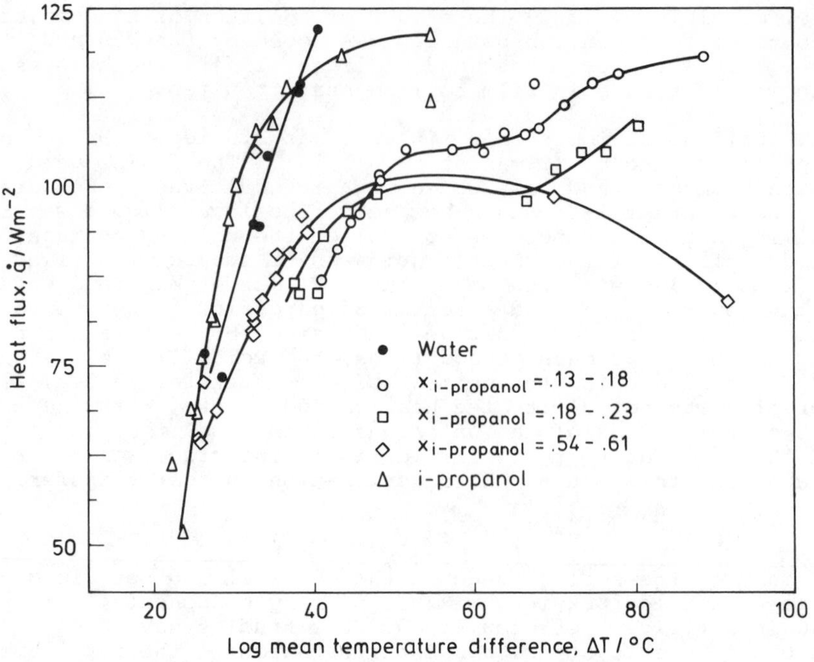

Fig. 47. Heat transfer data for *i*-propanol/water mixtures in a vertical thermosiphon reboiler. Shellene et al. (1968).

More than 50 thermocouples along the tube are used to measure local heat transfer coefficients. The results show the heat transfer to be dominated by nucleate boiling at qualities up to 40%, and in such cases the reduction in boiling coefficient is more than in the no-flow case. Bennett (1976) gives data for water/ethylene glycol mixtures in a vertical-tube system. Electrical heating was used to raise the flow quality to the required level and a further, independently controlled, short heater length was used to measure the local heat transfer coefficients. In this case low heat fluxes could be used at high qualities, and both convective and nucleate boiling modes were investigated. Further local data for flow boiling are given by Mishra et al. (1978), who measured heat transfer to Refrigerants-12 and -13 and their mixtures in forced convection in horizontal tubes, although few definite trends can be discerned.

Shock (1973) and Bennett (1976) have suggested correlation methods for the flow boiling of mixtures based on the well-known correlation of Chen (1966) for pure substances. In this the total heat transfer is divided into a nucleate boiling, "microconvective," contribution and a flow-induced "macroconvective" contribution.

The microconvective contribution is evaluated using the Forster-Zuber (1955) pool boiling method multiplied by a suppression factor $S_c(\mathrm{Re}_{TP})$ where Re_{TP}, the two-phase Reynolds number, is given by

$$\mathrm{Re}_{TP} = \frac{G(1 - \dot{x})D}{\mu_l} (F')^{1.25} \tag{78}$$

where S_c accounts for the steepening of the temperature gradient and F' for the increase in apparent liquid velocity; $\dot{x}$ is the flow quality. F' is an empirical function of the quality and the physical properties. For mixtures Bennett suggested that this value should in turn be multiplied by a factor $1/F^{0.99}$; see Fig. (38). (The apparently curious power arises by chance through a combination of physical property effects in the Forster-Zuber method.) Bennett's data appear to agree with this theoretically derived correction, although the effect on the overall heat transfer coefficient is stated to be slight since the data were dominated by the macroconvective contribution. Shock (1973) proposed a method similar in principle but different in detail. He suggested that the wall superheat should be calculated for each component in pool boiling for the microconvective heat flux. The Stephan and Körner (1969) method (Sec. 5.2.2) should then be used to obtain the pool boiling superheat at this heat flux for the mixture (at the local flash evaporation condition). Hence the local microconvective heat transfer coefficient could be found by dividing the microconvective heat flux by this pool boiling superheat. In view of the simultaneous macroconvective heat transfer, an iterative solution is necessary. Shock suggested that a further correction factor may be necessary to account for the effect of the fluid flow on mass transfer in the layer surrounding the bubbles. This is supported by the data of Toral (1979) in a nucleate boiling-dominated situation. Figure 48 shows the results plotted in the "boiling curve" form for pure cyclohexane, pure ethanol and an intermediate concentration of large y^*-x. Also shown is the curve predicted for the mixture using the method of Stephan and Körner (1969), with the value of the characteristic constant B obtained by Körner (1967) from experiments on a horizontal heater. The figure suggests a greater decrease in mixture boiling coefficient than in pool boiling. However, it could be that the difference is due to a dependence of B on the orientation, causing a change in mutual bubble interaction; hence a different value may be appropriate to the vertical tube. A further point is that B may also depend on the heater surface and its finish, as suggested by Stephan and Preusser (1978).

Shock (1973) and Bennett (1976) also discuss the macroconvective heat transfer in terms of an annular-flow mass transfer model. This flow regime is entered at qualities above a few percent where heat transfer is increasingly dominated by the thin annular film dragged up the tube walls by the shear exerted by the fast-moving gas core (Hewitt and Hall-Taylor, 1970). Bennett (1976) used water/ethylene glycol as a test fluid, and since the glycol is almost involatile, he considered mass transfer only in the liquid phase. He considered the mixture effects to alter the interface condition but not the resistance of the liquid film. A mass balance over a short elemental length of tube, assuming interface equilibrium and neglecting the change in enthalpy of the liquid, leads to the conclusion that the macroconvective heat transfer coefficient for the EPF should be multiplied by

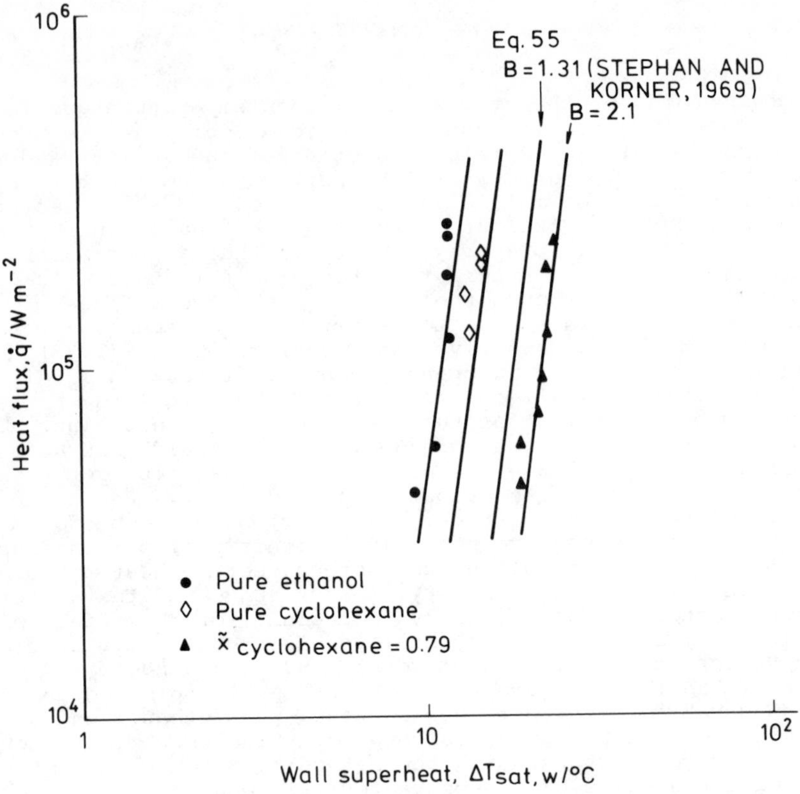

Fig. 48. Convective boiling of ethanol/cyclohexane mixtures in a vertical round tube, D = 26 cm. Toral (1969).

$$1 - \frac{(1 - y)\dot{q}}{\rho_{v,A}\Delta h \beta \Delta T_{sat,w}} \cdot \frac{dT}{dx}$$

where β, the mass transfer coefficient is given by

$$\beta = 0.023(\mathrm{Re}_{TP})^{0.8}(\mathrm{Sc})^{0.4}\,\frac{\delta}{D} \tag{79}$$

This relationship, when combined with the microconvective heat transfer, gives good agreement with the results of Bennett. Shock (1976) has considered the analogous situation where both components have significant volatilities. He drew up mass and energy balances for elemental lengths in each phase and solved them to give expressions for the axial and radial concentration gradients in the phases. The equations were solved numerically for one mixture, ethanol/water. It is shown that, contrary to expectation, the concentration gradient in the vapor is such as to cause diffusion of light component toward the interface; there is still a net evaporation of light component due to the proportion of light component in the net evaporating stream which outweighs this diffusion stream. The calculations by Shock suggest that for this mixture the effects of the mass transfer resistances on the concentration distribution are small (Fig. 49) and the effects on the wall temperature are insignificant. Such a model has yet to be examined for mixtures of higher relative volatility, including dissolved gases, and also remains to be tested experimentally.

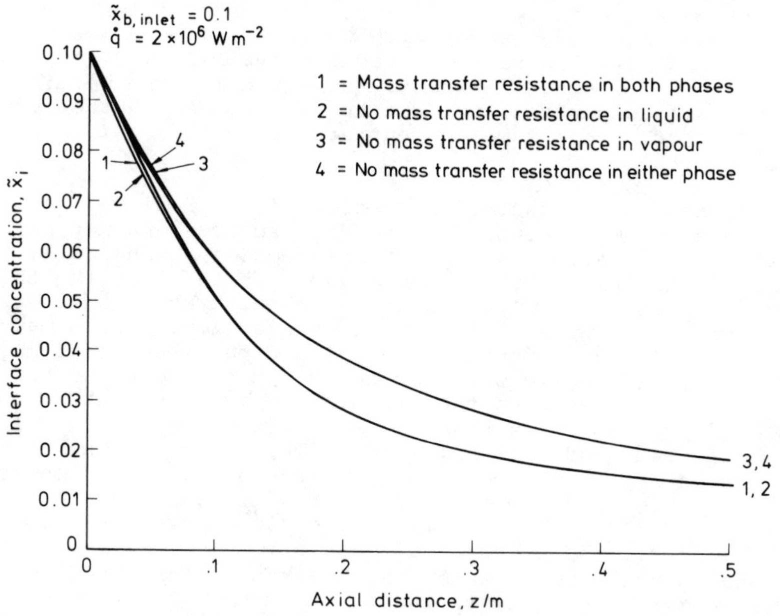

Fig. 49. Axial variation of interface concentration for annular flow heat transfer to ethanol/water. Shock (1976).

Detailed analyses of the concentration profiles in the two phases can be very useful. This is also true in the field of multicomponent condensation where equivalent-laminar-film models are now being used for design. A simpler approach which is also used in condensation, and which might perhaps be adopted for evaporation, is that of Bell and Ghaly (1973). This accounts for the heat release due to the fact that the dew point of the condensing mixture decreases as the quality decreases. Certain simplifying assumptions are made. The most important of these assumptions are that the liquid and vapor compositions are in equilibrium at the vapor bulk temperature, that the sensible heat of the vapor is transferred to the interface by a convective mechanism with a heat transfer coefficient calculated assuming only the vapor phase to be present, and that the total latent heat of condensation and the sensible heat of the condensate and vapor are transferred through the entire liquid film resistance. It is clear, however, that whatever method is to be used, or adopted from the condensation field, there is an urgent need for experimental data with mixtures over a wide range of volatilities.

The end of the annular regime is marked by the dryout point—when the liquid film is evaporated completely. At low pressures the subsequent heat transfer to droplet-laden gas is markedly reduced and, in heat flux-controlled situations, physical burnout can result. At high pressures the deterioration is less noticeable. The effect of mixture composition has not yet been identified, but Marangoni effects could be significant. An indication of possible effects can be deduced from work done by Norman and Binns (1960) on minimum wetting rates in flows of *i*-propanol/water films down an unheated rod surrounded by vapor with which the liquid was not in equilibrium. As the flow rate was reduced the film was seen to break down into rivulets. Figure 50 shows the minimum wetting rate as a function of concentration. The turning points in the curve are seen to occur at the same concentrations as the maximum and minimum in $y^* - x$.

The explanation of these observations is as follows. A film flowing down a rod will be subject to disturbances and may have a wavy profile. The parts that momentarily become thinner will tend to approach equilibrium with the bulk vapor faster than the thicker parts. Consequently, local gradients of concentration and thus of surface tension will be set up along the interface. These may cause flow either into or out of the thinned part of the film, hindering or aiding breakdown. For the mixture used by Norman and Binns, the result depends on the mixture composition. At concentrations of *i*-propanol less than the azeotropic, the mixture is positive (Zuiderweg and Harmens, 1958). Thin parts of the film are lower in concentration of *i*-propanol and the surface tension is increased; liquid is drawn in and the film is stablized against breakdown. A small flow rate will suffice to cover the rod without breaking down. The converse effect is found in mixtures rich in *i*-propanol, where the film is destabilized against breakdown and the minimum wetting rate is increased. Such effects may also be expected in evaporation heat transfer, especially for thin films approaching the dryout point, where the large disturbance waves, which may swamp the effect described, are not present. Destabilization of the film with negative mixtures could accelerate the breakdown of the film and the heat transfer crisis.

When the dryout point has been reached, heat transfer occurs direct to the gas and thence to the remaining liquid droplets in the gas core, and also to those liquid droplets that interact directly with the heater. Landis and Mills (1974) have examined the behavior of droplets of binary mixtures in high-temperature fields. The work was conducted with direct reference to furnace design, but their findings could have some relevance to two-phase mist flow. They calculated the predicted concentration and temperature fields within a droplet, initially of uniform concentration and temperature, which was suddenly exposed to a high gas temperature. Since they were concerned mainly with combustion, they considered gas temperatures up to 2000°C. They ignored liquid circulation within the droplet and assumed no resistance to heat or mass transfer in the gas. Their results suggest that thermal diffusion resistance is significant at short times and that mass diffusion resistance is significant at long times. Since the

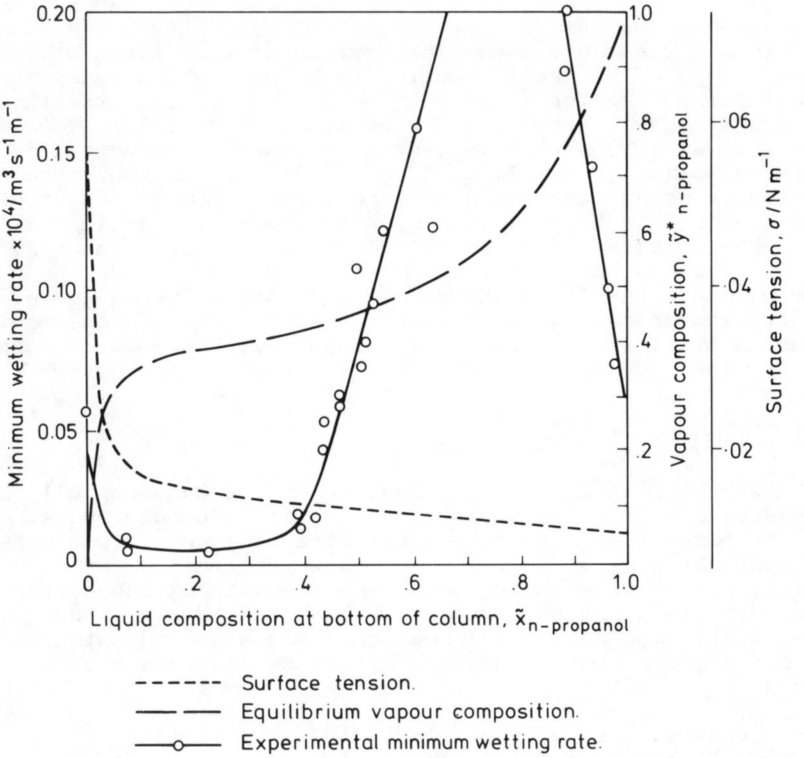

Fig. 50. Minimum wetting rates from rectification experiments. Norman and Binns (1960).

Lewis numbers are greater than 1 (about 20 in the cases considered), the center temperature changes more rapidly than does the center concentration, and they show that the temperatures within the droplet can reach high supersaturations for the local concentration conditions. Since droplets under such conditions have been observed to shatter spontaneously, it may be assumed that they have reached homogeneous nucleation conditions at the interior. Also of possible relevance to two-phase flow is the prediction that the mass transfer resistance causes the droplet lifetime to increase by a small amount—5% for a heptane/octane droplet of initial diameter 0.1 cm exposed to 2000°C at 1 atm. Law (1978) has extended the arguments and shown that the likelihood of significant superheat occurring within the droplets is increased as both temperature and pressure are raised.

Yang (1976) has studied the evaporation of a droplet on a heated surface where the heat transfer mode into the droplet is dominated by nucleate boiling. Since the nucleate boiling heat transfer into the droplets is reduced for mixtures, the evaporation rate is slower. He models the evaporation rate by assuming uniform concentration and temperature fields within the droplets.

These works suggest a complex picture of mist flow in binary mixtures. Some of the heat transfer is being used to evaporate the droplets and reaches them by three routes: first, via the gas to the droplets near the channel center; second, to droplets that contact the wall; and third, to droplets that approach but do not contact the wall. In the latter case, analogous to the bouncing water droplet on an oven, film boiling occurs between the droplet and the heater. We have already seen in Sec. 7 that film boiling can be more efficient in mixtures than in pure substances. By contrast, as shown above, the evaporation rate of a droplet in a high-temperature gas field and of one in contact with a heater is less efficient in mixtures. Thus the evaporation of droplets in two-phase mist flow of mixtures will change with composition in different ways depending on which of the above heat transfer routes is dominant.

9 CONCLUSIONS

The behavior of bubbles in binary mixtures can generally be modeled with almost as much confidence as in pure substances. A number of correlation methods for nucleate boiling heat transfer coefficients have been suggested and show varying degrees of generality. Further work is needed, however, in extending the methods, especially those based on fundamental bubble dynamics, to the true multicomponent case. Section 5 has also indicated that the effect of pressure changes on mixture boiling has yet to be explained.

The greatest area of uncertainty in pool boiling, arising from conflicting data even for the same surface geometry, is the understanding of the departure from nucleate boiling. This is especially true on surfaces where the characteristic dimension is larger than the bubble size. It is important to understand the phenomenon in relation both to DNB in chemical plant boilers and to the possible use of mixtures in cooling high-power-density electronic components. The advantage of mixtures lies in the possible increased heat loading available with the increase in $\dot{q}_{cr}$, which may outweigh the disadvantages of reduced heat transfer coefficients, but the geometric limits within which $\dot{q}_{cr}$ is increased must be elucidated.

Another important aim for the study of mixture boiling is the prediction of flow boiling heat transfer for application in the chemical and petrochemical industries. Although there has been some work on this, there is clearly a lack of data from which to evolve generalized correlation methods. The apparatus and techniques involved are complex, but the economic advantages that would accrue from acceptable mixture boiling correlations are considerable.

Generally the area of mixture boiling is one where great progress has been made in recent years and yet one where important and challenging problems remain.

NOMENCLATURE

a	dummy integration variable
A	parameter defined by Eq. (57)
b^*	relaxation microlayer height as fraction of bubble diameter
B	binary mixture characteristic constant in Stephan and Körner (1969) correlation (Eq. 57)
c	liquid concentration, kg m^{-3}
c_l	liquid specific heat capacity, J kg^{-1} K^{-1}
c_v	vapor specific heat capacity, J kg^{-1} K^{-1}
C	reduction in heat transfer (Eq. 53)
C_d	drag coefficient
D	heater diameter, m
F	bubble growth rate correction factor (Eq. 38)
F	force, N
F'	correction factor in Chen correlation
g	acceleration due to gravity, m s^{-2}
g_c	conversion factor
G	parameter in Eq. (58)
G	total mass flux, kg m^{-2} s^{-1}
G_d	mass vaporization fraction
h	specific enthalpy, J kg^{-1}
Δh	specific latent heat, J kg^{-1}
$\Delta\tilde{h}$	molar latent heat, J $kmol^{-1}$
$\Delta h'$	modified specific latent heat $(\Delta h+0.5c_v\Delta T_{sat,w})$, J kg^{-1}
H	parameter in Eq. (58)
H	height of heating surface, m
J	nucleation rate, m^{-3} s^{-1}
Ja	Jakob number; dimensionless superheat (Eq. 40)
k	Boltzmann constant, J K^{-1}
K	equilibrium constant, y^*/x
K	dimensionless critical heat flux (Eq. 69)
l	$(\Delta h_2/\Delta h_1)$ $(1 - x_\infty)/(x_\infty)$ (Eq. 31)
m	index in boiling curve equation
M	coalescence parameter (Eq. 68), N m^{-1}
M	molar mass, kg $kmol^{-1}$

n	exponent in Eq. (44)
N	number density of molecules, m^{-3}
p	pressure, N m^{-2}
p_i	partial pressure of inert gas, N m^{-2}
p_l	liquid pressure, N m^{-2}
p_v	vapor pressure, N m^{-2}
P	property group in nucleate boiling (Eq. 52), W m^{-2} K^{-m}
Pr	Prandtl number
$\dot{q}$	heat flux, W m^{-2}
$\dot{q}_{bi}$	heat flux for direct vaporization at the interface, W m^{-2}
$\dot{q}_{\mathrm{conv}}$	natural convection heat flux, W m^{-2}
$\dot{q}_{cr}$	critical heat flux, W m^{-2}
$\dot{q}_{nb}$	nucleate boiling heat flux, W m^{-2}
$\dot{q}_{\mathrm{min}}$	heat flux at the minimum in the boiling curve, W m^{-2}
r	radial coordinate, m
R	bubble radius, m
R	ratio of critical heat flux to that of water
$\tilde{R}$	universal gas constant, J kmol^{-1} K^{-1}
$\dot{R}$	velocity of liquid-vapor interface relative to bubble center, m s^{-1}
$\ddot{R}$	acceleration of liquid-vapor interface, m s^{-2}
R_c	cavity radius, m
R_D	bubble departure diameter, m
R_e	effective nucleus radius, m
Re_{TP}	two-phase Reynolds number (Eq. 78)
R_{max}	maximum value of critical heat flux ratio
R_s	radius of dry patch, *m*
S_c	suppression factor in Chen correlation
Sc	Schmidt number
Sp	dimensionless temperature difference, $c_v(T_w - T_\infty)/(\mathrm{Pr}_v\ \Delta h)$
t	time, s
t_d	delay time, s
t_g	bubble growth time, s
T	temperature, K
T_{bub}	bubble point temperature, K

T_{dew}	dew point temperature, K
ΔT_{av}	loss of available superheat, K
ΔT_{min}	temperature difference at the minimum in the boiling curve, K
ΔT_{sat}	superheat, K
$\Delta T_{sat,E}$	excess superheat, K
$\Delta T_{sat,I}$	ideal superheat (Eq. 54), K
u	velocity, m s^{-1}
$\tilde{V}$	molar volume, m^3 $kmol^{-1}$
V_D	departure volume of bubble, m^3
x	mass fraction in liquid
$\tilde{x}$	mole fraction in liquid
$\dot{x}$	flow quality
y	distance from wall, m
y	mass fraction in vapor
$\tilde{y}$	mole fraction in vapor
z	axial distance, m

Greek Letters

α	heat transfer coefficient, W m^{-2} K^{-1}
α	relative volatility, $\tilde{y}(1-\tilde{x})/\tilde{x}(1-\tilde{y})$
β	bubble growth constant (Eq. 23)
β	mass transfer coefficient, m s^{-1}
β_l	volumetric expansion coefficient in the liquid, K^{-1}
γ	characteristic time (Eq. 42), s
Γ	κ/δ
δ	diffusivity, m^2 s^{-1}
δ	correction factor for effect of pressure on vapor pressure
ε	$(\rho_l-\rho_v)/\rho_l$
ζ	ρ_v/c_∞
η	dimensionless position parameter (Eqs. 76, 77)
Θ	superheat increase factor (Eq. 56)
κ	thermal diffusivity, m^2 s^{-1}
λ	thermal conductivity, W m^{-1} K^{-1}
λ_D	most dangerous wavelength, m
μ	viscosity, kg m^{-1} s^{-1}

μ	ratio defined by Eq. (26), kg m^{-3}
ν	kinematic viscosity, m^2 s^{-1}
ν	$(c_l - c_v)/c_l$
ξ	$\rho_v \Delta h / \rho_l c_l T_\infty$
ρ	density, kg m^{-3}
σ	surface tension, N m^{-1}
τ	dimensionless superheat, $[T_\infty - T_{bub}(x)]/T_{bub}(x)$
ϕ	function defined by Eq. (27)
ϕ	contact angle, rad
ψ	function defined by Eq. (28)
ω	ρ_v/ρ_l

Superscripts

*	equilibrium value
-	average

Subscripts

A, B, C, D	species A, B, C, and D
b	bulk mean value
exp	experimental
E	excess
i	interface
I	ideal
l	liquid
ONB	at onset of nucleate boiling
v	vapour
w	wall
∞	far from heating surface

REFERENCES

Ackermann, H., L. Bewilogua, A. Jahn, R. Knoňer, and H. Vinzelberg 1976, Heat Transfer in Nitrogen-Methane Mixtures Under Pressure with Film Boiling *Cryogenics* vol. 16, pp. 497–499.

Ackermann, H., L. Bewilogua, R. Knöner, B. Kretzschmar, I. P. Usyugin, and H. Vinselberg 1975, Heat Transfer in Liquid Nitrogen-Methane Mixtures Under Pressure *Cryogenics*, vol. 15, pp. 657–659.

Adamson, A. W. 1973, Potential Distortion Model for Contact Angle and Spreading. II. Temperature Dependent Effects. *J. Colloid Interface Sci.* vol. 44, pp. 273–281.

Afgan, N. H. 1966, Boiling Heat Transfer and Burnout Heat Flux of Ethyl-Alcohol/Benzene Mixtures. *Third International Heat Transfer Conference, Chicago, Vol. 3,* pp. 175–185.

Afgan, N. H. 1968, Boiling Heat Transfer of the Binary Mixtures. *International Summer School of Heat and Mass Transfer, Boris Kidric Institute of Nuclear Sciences, Beograd, Yugoslavia (9–12 Sept.).*

Andrews, D. G., F. C. Hooper, and P. Butt 1968, Velocity, Subcooling and Surface Effects in the Departure from Nucleate Boiling of Organic Binaries. *Can. J. Chem. Eng.* vol. 46, pp. 194–199.

Averin, E. K., and G. N. Kruzhilin 1955, The Influence of Surface Tension and Viscosity on the Condition of Heat Exchange in the Boiling of Water. *Izvest. Akad. Nauk. S.S.S.R., Otdel, Tekh. Nauk.* vol. 10, pp. 131–137.

Bankoff, S. G. 1958, Entrapment of Gas in the Spreading of a Liquid over a Rough Surface. *AIChE J.* vol. 4, pp. 24–26.

Basu, D. K. 1975, Film Boiling of Organic Binaries on Horizontal Wire at Atmospheric Pressure. *Israel J. Technol.* vol. 13, pp. 386–390.

Beer, H., P. Burrow, and R. Best 1977, Nucleate Boiling; Bubble Growth and Dynamics. *Heat Transfer in Boiling,* H. Hahne, and V. Grigull, Eds. New York: Academic Press, pp. 21–52

Béhar, M., M. Courtaud, R. Ricque, and R. Séméria 1966, Fundamental Aspects of Subcooled Boiling with and Without Dissolved Gases. *Third International Heat Transfer Conference, Chicago, Vol. 4,* pp. 1–11.

Bell, K. J., and M. A. Ghaly 1973, An approximate Generalized Design Method for Multicomponent/Partial Condensers. *AIChE Symp. Ser.* vol. 69, pp. 72–79.

Benjamin, J. E., and J. W. Westwater 1961, Bubble Growth in Nucleate Boiling of a Binary Mixture. *International Developments in Heat Transfer, Part II. International Heat Transfer Conference, University of Colorado,* pp. 212–218.

Bennett, D. L. 1976, A Study of Internal Forced Convective Boiling Heat Transfer for Binary Mixtures. Ph.D. thesis, Lehigh University.

Berenson, P. J. 1961, Film-Boiling Heat Transfer from a Horizontal Surface. *J. Heat Transfer,* vol. 83, pp. 351–358.

Bergles, A. E. 1967, Augmentation of Boiling Heat Transfer. Two-Phase Gas-Liquid Flow and Heat Transfer, Special Summer Program, Massachusetts Institute of Technology.

Bergles, A. E. 1975, Burnout in Boiling Heat Transfer, Part 1: Pool Boiling-Systems. *Nuclear Safety* vol. 16, pp. 29–42.

Blander, M., and J. L. Katz 1975, Bubble Nucleation in Liquids. *AIChE J.* vol. 21, pp. 833–848.

Bonilla, C. F., and C. W. Perry 1941, Heat transmission to Boiling Binary Liquid Mixtures. *Trans. Am. Inst. Chem. Eng.* vol. 37, pp. 685–705.

Bonnet, W. E., and J. A. Gerster 1951, Boiling Coefficients of Heat Transfer—C_4 Hydrocarbon/Furfural Mixtures Inside Vertical Tubes. *Chem. Eng. Prog.* vol. 47, pp. 151–158.

Bragg, J. R., and J. W. Westwater 1970, Film Boiling of Immiscible Liquid Mixtures on a Horizontal Plate. *Fourth International Heat Transfer Conference, Paris*, Paper B7.1.

Bromley, L. A. 1950, Heat Transfer in Stable Film Boiling. *Chem. Eng. Prog.* vol. 46, pp. 221–227.

Bruijn, P. J. 1960, On the Asymptotic Growth Rate of Vapour Bubbles in Superheated Binary Liquid Mixtures. *Physica* Vol. 26, pp. 326–334.

Calus, W. F., R. K. Denning, A. di Montegnacco, and J. Gadsdon 1973, Heat Transfer in a Natural Circulation Single Tube Reboiler—Part II, Binary Liquid Mixtures. *Chem. Eng. J.* vol. 6, pp. 252–264.

Calus, W. F., and D. J. Leonidopoulos 1974, Pool Boiling—Binary Liquid Mixtures. *Int. J. Heat Transfer* vol. 17, pp. 249–256.

Calus, W. F., and P. Rice 1972, Pool Boiling—Binary Liquid Mixtures. *Chem. Eng. Sci.* vol. 27, pp. 1687–1697.

Carne, M. 1963, Studies of the Critical Heat Flux for Some Binary Mixtures and Their Components, *Can. J. Chem. Eng.* vol. 41, pp. 235–241.

Chen. J. C. 1966, Correlation for Boiling Heat Transfer to Saturated Fluids in Convective Flow. *I. and E. C. Proc. Des. and Dev.* vol. 5, pp. 322–329.

Cichelli, M. T., and C. F. Bonnilla 1945, Heat Transfer to Liquids Boiling Under Pressure. *AIChE J.* vol. 41, pp. 755–787.

Clements, L. D., and C. P. Colver 1972, Nucleate Boiling of Light Hydrocarbons and Their Mixtures. *Proceedings of the 1972 Heat Transfer and Fluid Mechanics Institute*, R. B. Landis and G. J. Hordemann, Eds. Stanford, Calif.: Stanford University Press, pp. 417–430.

Collier, J. G. 1981, *Convective Boiling and Condensation.* 2nd Ed. New York: McGraw-Hill.

Cooper, M. G., A. M. Judd, and R. A. Pike 1978, Shape and Departure of Single Bubbles Growing at a Wall. *Sixth International Heat Transfer Conference, Toronto*, Paper PB-1.

Cooper, M. G., and A. J. P. Lloyd 1969, The Microlayer in Nucleate Pool Boiling. *Int. J. Heat Mass Transfer* vol. 12, pp. 895–913.

Corty, C., and A. S. Foust 1955, Surface Variables in Nucleate Boiling. *Chem. Eng. Symp. Ser.* vol. 51, pp. 1–12a.

Cryder, D. S., and E. R. Gilliland 1932, Heat Transmission from Metal Surfaces to Boiling Liquids. *Ind. Eng. Chem.* vol. 24, pp. 1382–1387.

Debbage, A. G., M. Driver, and P. R. Waller 1963, Heat Transfer Properties of Organic Coolants Containing High Boiling Residues. AEEW Report. R261.

Denbigh, K. 1966, *The Principles of Chemical Equilibrium.* Cambridge: Cambridge University Press.

Dougherty, R. L., and H. J. Sauer 1974, Nucleate Pool Boiling of Refrigerant-Oil Mixtures from Tubes. *ASHRAE Trans.*vol. 80, pp. 175–192.

Dunskus, T., and J. W. Westwater 1961, The Effect of Trace Additives on the Heat Transfer to Boiling Isopropanol. *Chem. Eng. Prog. Symp. Ser.* vol. 57, pp. 173–181.

Eddington, R. I., and D. B. R. Kenning 1978, The Effect of Contact Angle on Bubble Nucleation. *Int. J. Heat Mass Transfer* vol. 22, pp. 1231–1236.

Fath, H. S. and R. L. Judd 1978, Influence of System Pressure on Microlayer Evaporation Heat Transfer. *J. Heat Trans.* vol. 100, pp. 49–55.

Filatkin, V. N. 1972, Boiling Heat Transfer to Water-Ammonia Mixtures. *Symposium on Problems of Heat Transfer and Hydraulics of Two-Phase Media.* S. S. Kutateladze, Ed. Paper 9.

Florschuetz, L. W., C. L. Henry, and A. Rashid-Khan 1969, Growth Rates of Free Vapor Bubbles in Liquids at Uniform Superheats Under Normal and Zero Gravity Conditions. *Int. J. Heat Mass Transfer* vol. 12, pp. 1465–1489.

Florschuetz, L. W., and A. Rashid-Khan 1970, Growth Rate of Free Vapor Bubbles in Binary Liquid Mixtures at Uniform Superheats. *Fourth International Heat Transfer Conference, Paris*, Paper B7.3.

Forster, H. K., and N. Zuber 1955, Dynamics of Vapor Bubbles and Boiling Heat Transfer. *AIChE J.* vol. 1, pp. 531–535.

Frea, W. J., R. Knapp, and T. D. Taggart 1977, Flow Boiling and Pool Boiling Critical Heat Flux in Water and Ethylene Glycol Mixtures. *Can. J. Chem. Eng.* vol. 55, pp. 37–42.

Gaidarov, S. A. 1975, Evaluation of Critical Heat Flow in the Case of a Boiling Mixture of Large Volume. *J. Appl. Mech. Tech. Phys.* vol. 16, pp. 601–603.

Gambill, W. R. 1968, Pool-Boiling Critical Heat Fluxes for Non-Newtonian Aqueous Solutions of Polyethylene Oxide. ORNL Report TM 2445.

Gannett, H. J., and M. C. Williams 1971, Pool Boiling in Dilute Non-aqueous Polymer Solutions. *Int. J. Heat Mass Transfer* vol. 14, pp. 1001–1005.

Grigor'ev, L. N., L. A. Sarkisyan, and A. G. Usmanov 1968, An Experimental Study of Heat Transfer in the Boiling of Three-Component Mixtures. *Int. Chem. Eng.* vol. 8, pp. 76–78.

Hall, I. S., and A. P. Hatton 1965, The Influence of Surface Roughness and a Wetting Additive on Pool Boiling on Horizontal Rods. *Proc. Inst. Mech. Eng.* vol. 180, pp. 160–179.

Happel, O. 1977, Heat Transfer During Boiling of Binary Mixtures in the Nucleate and Film Boiling Ranges. *Heat Transfer in Boiling.* E. Hahne and V. Grigull, Eds. New York: Academic Press. Also *Fifth International Heat Transfer Conference Tokyo, Vol. 6,* Paper B7.8, 1974.

Henrici, H., and G. Hesse 1971, Examinations Concerning the Heat Transfer When Evaporating R114 and R114 Oil Mixtures on a Horizontal Smooth Pipe. *Kaltetechnik* vol. 23, pp. 54–58.

Hewitt, G. F. 1978, Critical Heat Flux in Flow Boiling. *Sixth International Heat Transfer Conference, Toronto, Vol. 6,* Keynote Paper 13, pp. 143–172.

Hewitt, G. F., and N. S. Hall-Taylor 1970, *Annular Two Phase Flow.* Elmsford, N.Y.: Pergamon Press.

Holden, B. S., and J. L. Katz 1978, The Homogeneous Nucleation of Bubbles in Superheated Binary Liquid Mixtures. *AIChE J.* vol. 24, pp. 260–267.

Hovestreijdt, J. 1963, The Influence of the Surface Tension Difference on the Boiling of Mixtures. *Chem. Eng. Sci.* vol. 18, pp. 631–639.

Hsu, Y. Y., and J. W. Westwater 1960, Approximate Theory for Film Boiling on Vertical Surfaces. *Chem. Eng. Prog. Symp. Ser.* vol. 56, pp. 15–24.

Huber, D. A., and J. C. Hoehne 1963, Pool Boiling of Benzene, Diphenyl and Benzene-Diphenyl Mixtures Under Pressure. *Trans. Am. Soc. Mech. Eng., J. Heat Transfer* vol. 85, pp. 215–220.

Isshiki, N., and I. Nikai 1973, Boiling of Binary Mixtures. *Heat Transfer Jap. Res.* vol. 4. pp 56–66.

Jontz, P. D., and J. E. Myers 1960, The Effect of Dynamic Surface Tension on Nucleate Boiling Coefficients. *AIChE J.* vol. 6, pp. 34–38.

Judd, R. L., and K. S. Hwang 1976, A Comprehensive Model for Nucleate Pool Boiling Heat Transfer Including Microlayer Evaporation. *J. Heat Trans.* vol. 98, pp. 623–629.

Kautzky, D. E., and J. W. Westwater 1967, Film Boiling of a Mixture on a Horizontal Plate. *Int. J. Heat Mass Transfer* vol. 10, pp. 253–256.

Körner, M. 1967, Beitrag zum Warmenubergang bei der Blasenvertampfung binarer Gemische. Diss. th., Aachen.

Kosky, P. G., and D. N. Lyon 1968, Pool Boiling Heat Transfer to Cryogenic Liquids, III. Nucleate Boiling Data and Peak Nucleate Boiling Fluxes for Nitrogen-Oxygen Mixtures. *AIChE J.* vol. 14, pp. 383–387.

Kotchaphakdee, P., and M. C. Williams 1970, Enhancement of Nucleate Pool Boiling with Polymeric Additives. *Int. J. Heat Mass Transfer* vol. 13, pp. 835–848.

Kutateladze, S. S. 1951, A Hydrodynamic Theory of Changes in the Boiling Process Under Free Convective Conditions. *Izv. Akad. Nauk, USSR, Otd. Tekh. Nauk* no. 4, p. 529.

Kutateladze, S. S., G. I. Bobrovich, I. I. Gogonin, N. N. Mamontova, and V. N. Moskvicheva 1966, The Critical Heat Flux at the Pool Boiling of Some Binary Liquid Mixtures. *Third International Heat Transfer Conference, Chicago, Vol. 3*, pp. 149–159.

Landis, R. B., and A. F. Mills 1974, Effect of Internal Diffusional Resistance on the Evaporation of Binary Droplets. *Fifth International Heat Transfer Conference, Vol. 4*, Paper B7.9.

Law, C. K. 1978, Internal Boiling and Superheating in Vaporizing Multicomponent Droplets. *AIChE J.* vol. 24, pp. 626–632.

Leppert, G., C. P. Costello, and B. M. Hoglund 1958, Boiling Heat Transfer to Water Containing a Volatile Additive. *Trans. Am. Soc. Mech. Eng.* vol. 80, pp. 1395–1404.

Lienhard, J. H., and V. K. Dhir July 1973, Extended Hydrodynamic Theory of the Peak and Minimum Pool Boiling Heat Fluxes. NASA Report CR-2270.

Lienhard, J. H., and P. T. Y. Wong 1964, The dominant unstable wave-Length and Minimum Heat Flux During Film Boiling on a Horizontal Cylinder. *J. Heat Transfer* vol. 86, pp. 220–226.

Lorenz, J. J., B. B. Mikic, and W. M. Rohsenow 1974, The Effect of Surface Conditions on Boiling Characteristics. *Fifth International Conference, Tokyo,* Paper B2.1.

Lowery, A. J., and J. W. Westwater 1957, Heat Transfer to Boiling Methanol—Effect of Added Agents. *Ind. Eng. Chem.* vol. 49, pp. 1445–1448.

Lyon, D. N. 1964, Peak Nucleate-Boiling Heat Fluxes and Nucleate Boiling Heat-Transfer Coefficients for Liquid N_2, Liquid O_2 and their Mixtures in Pool Boiling at Atmospheric Pressure. *Int. J. Heat Mass Transfer* vol. 7, pp. 1097–1116.

McAdams, W. H., W. E. Kennel, C. S. Minden, R. Carl, P. M. Picornell, and J. E. Dew 1949, Heat Transfer at High Rates to Water with Surface Boiling. *Ind. Eng. Chem.* vol. 41, pp. 1945–1953.

McAdams, W. H., W. K. Woods, and L. C. Heroman 1942, Vaporization Inside Horizontal Tubes—II. Benzene-Oil Mixtures. *Trans. ASME*, pp. 193–200.

Marschall, E., and L. L. Moresco 1977, Analysis of Binary Film Boiling. *Int. J. Heat Mass Transfer*. vol. 20, pp. 1013–1018.

Mikic, B. B., and W. M. Rohsenow 1969, A New Correlation of Pool-Boiling Data Including the Effect of Heating Surface Characteristics. *J. Heat Transfer* vol. 91, pp. 245–250.

Minchenko, F. P., and E. V. Firsova 1972, Heat Transfer to Water and Water-Lithium Salt Solutions in Nucleate Pool Boiling. *Symposium on Problems of Heat Transfer and Hydraulics of Two-Phase Media*. S. S. Kutateladze, Ed., Paper 10.

Mishra, M. P., H. K. Varma, and C. P. Sharma 1978, Heat Transfer Coefficients in Forced Convection Evaporation of Refrigerant Mixtures. *International Seminar on Momentum Heat and Mass Transfer in Two-Phase Energy and Chemical Systems, Dubrovnik,* Session 5.2.

Moalem, D., W. Zijl, and S. J. D. Van Stralen 1977, Nucleate boiling at a Liquid-Liquid Interface. *Lett. Heat Mass Transfer* vol. 4, pp. 319–329.

Moore, F. D., and R. B. Mesler 1961, The measurement of Rapid Surface Temperature Fluctuations During Nucleate Boiling of Water. *AIChE J.* vol. 7, pp. 620–624.

Morgan, A. I., L. A. Bromley, and C. R. Wilke 1949, Effect of Surface Tension on Heat Transfer in Boiling. *Ind. Eng. Chem.* vol. 41, pp. 2767–2769.

Murphy, R. W., and A. E. Bergles 1972, Subcooled Flow Boiling of Fluorocarbons—Hysteresis and Dissolved Gas Effects on Heat Transfer. *Proceedings of the 1972 Heat Transfer and Fluid Mechanics Institute*, R. B. Landis and G. J. Hordemann, Eds. Stanford, Calif.: Stanford University Press, pp. 400–416.

Naboichenko, K. V., A. A. Kiryutin, and B. S. Gribov 1965, A Study of Critical Heat Flux with Forced Flow of Mono-isopropyl Diphenyl/Benzene Mixture. *Teploenergetika* vol. 12, pp. 81–86..

Norman, W. S., and D. T. Binns 1960, The Effect of Surface Tension Changes on the Minimum Wetting Rates in a Wetted-Rod Distillation Column. *Trans. Inst. Chem. Eng.* vol. 38, pp. 294–300.

Ohnishi, M., and O. Tajima 1975, Pool Boiling Heat Transfer to Lithium Bromide Water Solution. *Heat Transfer Jap. Res.* vol. 4, pp. 67–77.

Oktay, S. 1971, Multi-Fluid Subdued Boiling: Theoretical Analysis of Multi-Fluid Interface Bubbles. *IBM J. Res. Dev.* vol. 15, pp. 342–354.

Owens, W. L. 1964, An Analytical and Experimental Study of Pool Boiling with Particular Reference to Additives. AEEW Report no. R180.

Pitts, C. C., and G. Leppert 1966, The Critical Heat Flux for Electrically Heated Wires in Saturated Pool Boiling. *Int. J. Heat Mass Transfer* vol. 9, pp. 365–377.

Porteous, W., and M. Blander 1975, Limits of Superheat and Explosive Boiling of Light Hydrocarbons, Halocarbons. and Hydrocarbon Mixtures. *AIChE J.* vol. 21, pp. 560–566.

Prakash, M. V., and K. N. Seetharamu 1978, Interface Conditions in Boiling of Saturated Binary Mixtures. *International Seminar on Momentum Heat and Mass Transfer in Two-Phase Energy and Chemical Systems, Dubrovnik,* Session 5.2.

Prausnitz, J. M. 1969, *Molecular Thermodynamics of Fluid Phase Equilibria.* Englewood Cliffs, N.J.: Prentice-Hall.

Robb, W. M., and R. Cole 1970, A Study of Incipient Vapor Nucleation Within Liquid Filled Conical Cavities. *Fourth International Heat Transfer Conference, Paris,* Paper B2.7.

Rohsenow, W. M. 1952, A Method of Correlating Heat-Transfer Data for Surface Boiling of Liquids. *Trans. ASME* vol. 74, pp. 969–976.

Rose, W. J., H. L. Gilles, and V. W. Uhl 1963, Subcooled Boiling Heat Transfer to Aqueous Binary Mixtures. *Chem. Eng. Prog. Symp. Ser.* vol. 59, pp. 62–70.

Ruckenstein, E. 1964, Influence of the Marangoni Effect on the Mass Transfer Coefficient. *Chem. Eng. Sci.* vol. 19, pp. 505–506.

Sauer, H. J., R. K. Gibson and S. Chongrungreong 1978, Influence of Oil on the Nucleate Boiling of Refrigerants. *Sixth International Heat Transfer Conference, Toronto,* Paper PB-12.

Scriven, L. E. 1959, On the Dynamics of Phase Growth. *Chem. Eng. Sci.* vol. 10, pp. 1–13.

Semeria, R. L. 1962, An Experimental Study of the Characteristics of Vapour Bubbles. *J. Mech. E. Symposium on Two-Phase Flow, Session 2,* Paper 7, pp. 26–34.

Shah, V. L., and W. T. Sha 1978, Growth of a Sodium Vapor Bubble Rising in the Superheated Liquid. *Nuclear Eng. Design* vol. 45, pp. 81–91.

Shellene, K. R., C. V. Sternling, N. H. Snyder, and D. M. Church 1968, Experimental Study of a Vertical Thermosiphon Reboiler. *Chem. Eng. Prog. Symp. Ser.* vol. 64, pp. 102–113.

Shock, R. A. W. 1973, The Evaporation of Binary Mixtures in Forced Convection. AERE Report R7593.

Shock, R. A. W. 1976, Evaporation of Binary Mixtures in Annular Flow. *Int. J. Multiphase Flow* vol. 2, pp. 411–433.

Shock, R. A. W. 1977, Nucleate Boiling in Binary Mixtures. *Int. J. Heat Mass Transfer* vol. 20, pp. 701–709.

Singer, R. M. 1970, An Experimental Demonstration of Inert Gas Effects upon the Incipient Pool Boiling of Sodium. *Nucl. Sci. Eng.* vol. 42, pp. 427–428.

Singh, A., B. B. Mikic, and W. M. Rohsenow 1976, Active Sites in Boiling. *J. Heat Transfer* vol. 98, pp. 401–406.

Skinner, L. A., and Bankoff, S. G. 1964a, Dynamics of Vapor Bubbles in Spherically Symmetric Temperature Fields of General Variation. *Phys. Fluids* vol. 7, pp. 1–6.

Skinner, L. A., and S. G. Bankoff 1964b, Dynamics of Vapor Bubbles in Binary Liquids with Spherically Symmetric Initial Conditions. *Phys. Fluids* vol. 7, pp. 643–648.

Stephan, K. 1973, Influence of Oil on Heat Transfer of Boiling Refrigerant-12 and Refrigerant-22 *Proceedings of the 15th International Congress on Regrigeration,* Paper II-6.

Stephan, K., and M. Körner 1969, Calculation of Heat Transfer in Evaporating Binary Liquid Mixtures. *Chem. Ing. Tech.* vol. 41, pp. 409–417.

Stephan, K., and P. Preusser 1978, Heat Transfer in Natural Convection Boiling of Polynary Mixtures. *Sixth International Heat Transfer Conference, Toronto, Vol. 1,* Paper PB-13.

Sterman, L. S., and J. Korychanek 1970, Critical Heat Flux During Boiling of High Boiling Heat Carriers. *Soviet Atomic Energy* vol. 29, pp. 1124–1125.

Sternling, C. V., and L. J. Tichacek 1961, Heat Transfer Coefficients for Boiling Mixtures. *Chem. Eng. Sci.* vol. 16, pp. 297–337.

Stewart, J. K., and R. Cole 1972, Bubble Growth Rates During Nucleate Boiling at High Jakob Numbers. *Int. J. Heat Mass Transfer* vol. 15, pp. 655–663.

Stone, C. R. 1980, Boiling: Bubble Growth in Pure and Binary Liquids. D. Phil. thesis, Oxford University.

Sump, G. D., and J. W. Westwater 1971, Boiling Heat Transfer from a Tube to Immiscible Liquid-Liquid Mixtures. *Int. J. Heat Mass Transfer* vol. 14, pp. 767–779.

Takeda, H., T. Hayakawa, and S. Fujita 1970, Nucleate Boiling Heat Transfer in Binary Liquid Mixtures. *Kagaku Kogaku* vol. 34, pp. 751–755.

Thome, J. R. 1978, Bubble Growth and Nucleate Pool Boiling in Liquid Nitrogen, Liquid Argon and Their Mixtures. D. Phil. thesis, Oxford University.

Tobilevich, N. Y., I. I. Sagan, and N. A. Pryadko 1969, The Transition Regime from Bubble Boiling to Film Boiling for Water and Water-Alcohol Mixtures with Vapor Heating of the Heat-Transfer Surface. *Inzhenerno-Fizicheskii Z.* vol. 16, pp. 610–616.

Tokuda, N. 1970, Dynamics of Vapor Bubbles in Binary Liquid Mixtures with Translatory Motion. *Fourth International Heat Transfer Conference, Paris*, Paper B7.5.

Tolubinskiy, V. I., A. A. Kriveshko, J. N. Ostrovskiy, and V. Y. Pisarev 1973, Effect of Pressure on the Boiling Heat Transfer Rate in Water-Alcohol Mixtures. *Heat Transfer Sov. Res.* vol. 5, pp. 66–68.

Tolubinskiy, V. J., and J. N. Ostrovskiy 1966, On the Mechanism of Boiling Heat Transfer (Vapour Bubbles Growth Rate in the Process of Boiling of Liquids, Solutions, and Binary Mixtures). *Int. J. Heat Mass Transfer* vol. 9, pp. 1463–1470.

Tolubinskiy, V. J., and J. N. Ostrovskiy 1969, Mechanism of Heat Transfer in Boiling of Binary Mixtures. *Heat Transfer Sov. Res.* vol. 1, pp. 6–11.

Tolubinskiy, V. J., J. N. Ostrovskiy, and A. A. Kriveshko 1970, Heat Transfer to Boiling Water-Glycerine Mixtures. *Heat Transfer Sov. Res.* vol. 2, pp. 22–24.

Tolubinskiy, V. J., J. N. Ostroviskiy, V. Y. Pisarev, A. A. Kriveshko and D. M. Konstanchuk 1975, Boiling Heat Transfer Rate from a Benzene-Ethanol Mixture as a Function of Pressure. *Heat Transfer Sov. Res.* vol. 7, pp. 118–121.

Toral, H. 1979, Flow Boiling Heat Transfer in Mixtures. D. Phil. thesis, Oxford University.

Van Ouwerkerk, H. J. 1971, The Rapid Growth of a Vapour Bubble at a Liquid-Solid Interface. *Int. J. Heat Mass Transfer* vol. 14, pp. 1415–1431.

Van Ouwerkerk, H. J. 1972, Hemispherical Bubble Growth in a Binary Mixture. *Chem. Eng. Sci.* vol. 27, pp. 1957–1967.

Van Stralen, S. J. D. 1959a, Heat Transfer to Boiling Binary Liquid Mixtures. Part I. *Br. Chem. Eng.* vol. 4, pp. 8–17.

Van Stralen, S. J. D. 1959b, Heat Transfer to Boiling Binary Liquid Mixtures—Part II. *Br. Chem. Eng.* vol. 4, pp. 78–82.

Van Stralen, S. J. D. 1962, Heat Transfer to Boiling Binary Liquid Mixtures—Part IV. *Br. Chem. Eng.* vol. 7, pp. 90–97.

Van Stralen, S. J. D. 1966a, The Mechanism of Nucleate Boiling in Pure Liquids and in Binary Mixtures—Part I. *Int. J. Heat Mass Transfer* vol. 9, pp. 995–1020.

Van Stralen, S. J. D. 1966b, The Mechanism of Nucleate Boiling in Pure Liquids and in Binary Mixtures—Part II. *Int. J. Heat Mass Transfer* vol. 9, pp. 1021–1046.

Van Stralen, S. J. D. 1967, The Mechanism of Nucleate Boiling in Pure Liquids and in Binary Mixtures—Part III. *Int. J. Heat Mass Transfer* vol. 10, pp. 1469–1484.

Van Stralen, S. J. D. 1968a, The Growth Rate of Vapour Bubbles in Superheated Pure Liquids and Binary Mixtures, Part I: Theory. *Int. J. Heat Mass Transfer* vol. 11, pp. 1467–1489.

Van Stralen, S. J. D. 1968b, The Growth Rate of Vapour Bubbles in Superheated Pure Liquids and in Binary Mixtures, Part II: Experimental Results. *Int. J. Heat Mass Transfer* vol. 11, pp. 1491–1512.

Van Stralen, S. J. D. 1970, The Boiling Paradox in Binary Systems. *Fourth International Heat Transfer Conference, Paris,* Paper B7.6.

Van Stralen, S. J. D., C. J. J. Joosen, and W. M. Sluyter 1972, Film Boiling of Water and an Aqueous Binary Mixture. *Int. J. Heat Mass Transfer* vol. 15, pp. 2427–2445.

Van Stralen, S. J. D., and W. M. Sluyter 1969, Investigations on the Critical Heat Flux of Pure Liquids and Mixtures Under Various Conditions. *Int. J. Heat Mass Transfer* vol. 12, pp. 1353–1384.

Van Stralen, S. J. D., M. S. Sohal, R. Cole, and W. M. Sluyter 1975, Bubble Growth Rates in Pure and Binary Systems: Combined Effect of Relaxation and Evaporation Microlayer. *Int. J. Heat Mass Transfer* vol. 18, pp. 453–467.

Van Stralen, S. J. D., W. Zijl, and D. A. de Vries 1977, The Behaviour of Vapour Bubbles During Growth at Subatmospheric Pressures. *Chem. Eng. Sci.* vol. 32, pp. 1189–1195.

Van Wijk, W. R., A. S. Vos, and S. J. D. Van Stralen 1956, Heat Transfer to Boiling Binary Liquid Mixtures. *Chem. Eng. Sci.* vol. 5, pp. 65–80.

Wall, K. W., and E. L. Park 1978, Nucleate Boiling of *n*-pentane, *n*-hexane, and Several Mixtures of the Two from Various Tube Arrays. *Int. J. Heat Mass Transfer* vol. 21, pp. 73–75.

Wright, R. D., L. D. Clements, and C. P. Colver 1971, Nucleate and Film Boiling of Ethane-Ethylene Mixtures. *AIChE J.* vol. 17, pp. 626–630.

Wright, R. D., and C. P. Colver 1969, Saturated Pool Boiling Burnout of Ethane-Ethylene Mixtures. *Chem. Eng. Prog. Symp. Ser.* vol. 65, pp. 204–210.

Yang, W-J. 1976, Nucleate-Boiling Type Evaporation of Binary-Liquid Drops on Heated Surfaces. *Lett. Heat Mass Transfer* vol. 3, pp. 467–474.

Yargin, V. S., L. D. Volyak, Ju. V. Tarlakov, and V. G. Stepanov 1972, The Effect of Temperature on the Wetting Angle. *Russian J. Phys. Chem.* vol. 46, pp. 1046–1048.

Yatabe, J. M., and J.W. Westwater 1966, Bubble Growth Rates for Ethanol-Water and Ethanol-Isopropanol Mixtures. *Chem. Eng. Prog. Symp. Ser.* vol. 62, pp. 17–23.

Yilmaz, B. S., S. F. Clarke, and J. W. Westwater 1976, Heat Transfer from Water in Film Boiling to an Upper Layer of Paraffinic Hydrocarbon. *ASME*, Paper no. 76-HT-24.

Yue, P. L., and M. E. Weber 1973, Film Boiling of Saturated Binary Mixtures. *Int. J. Heat Mass Transfer* vol. 16, pp. 1877–1887.

Yue, P. L., and M. E. Weber 1974, Minimum Film Boiling Flux of Binary Mixtures. *Trans. Inst. Chem. Eng.* vol. 52, pp. 217–221.

Zeugin, L., J. Donovan, and R. B. Mesler 1975, A Study of Microlayer Evaporation for Three Binary Mixtures During Nucleate Boiling. *Chem. Eng. Sci.* vol. 30, pp. 679–683.

Zijl, W., D. Moalem, and S. J. D. Van Stralen 1977, Inertia and diffusion Controlled Bubble Growth and Implosion in Initially Uniform Pure and Binary Systems. *Lett. Heat Mass Transfer* vol. 4, pp. 331–339.

Zuber, N. 1958, On the Stability of Boiling Heat Transfer. *Trans. ASME* vol. 80, pp. 711–720.

Zuber, N. 1959, Hydrodynamic Aspects of Boiling Heat Transfer. AEC Report AECU-4439.

Zuber, N. 1963, Nucleate Boiling. The region of Isolated Bubbles and the Similarity with Natural Convection. *Int. J. Heat Mass Transfer* vol. 6, pp. 53–78.

Zuiderweg, F. J., and A. Harmens 1958, The Influence of Surface Phenomena on the Performance of Distillation Columns. *Chem. Eng. Sci.* vol. 9, pp. 89–108.

Appendix

Beer et al. (1967) show that bubble departure is determined by a balance of forces (Fig. 17).

drag + surface tension = inertia + excess pressure + buoyancy (A-1)

or, in mathematical term,

$$\frac{1}{2}\, C_d \rho_l \dot{R}^2 \pi R_s^2 + 2\pi R_s \sin\phi = -V_D \rho_l \ddot{R} + \left(\frac{2\sigma}{R} + \Delta p\right) \pi R_s^2 + V_D(\rho_l - \rho_v) g \qquad \text{(A-2)}$$

In this appendix the net force on the bubble is evaluated as a function of time and, when the force is zero, departure is presumed to occur.

The radius-time relationship is evaluated from Eqs. (33) and (38).

$$R = \frac{1}{F} \sqrt{\frac{12}{\pi}} \frac{\rho_l c_l \ \Delta T_{sat} \sqrt{\kappa t}}{\rho_v \Delta h} \tag{A-3}$$

Beer et al. (1977) suggest that the drag coefficient can be calculated from

$$C_d = \frac{5360}{(R\dot{R} \ \rho_l/\mu_l)^{79}} \tag{A-4}$$

and that the base radius R_s is given by

$$R_s = \frac{7}{11} R \tag{A-5}$$

The latter relationship will not hold for thin wires, and so the numerical values calculated in this appendix are strictly valid for surfaces of large characteristic dimension.

The excess pressure inside a bubble was shown by Beer et al. (1977), for water, to drop quickly to about 15 N m^{-2}. Although this is not true in the early stages of bubble growth, it is used here, for all fluids, as a reasonable approximation at all times. It will be approximately true at departure, which is the only time at which strict accuracy is required.

Calculations were carried out in the form of two parametric studies. The first is based on dilute aqueous solutions at normal pressures with F ranging from 1 to 5, and with a superheat of 10.0°C. It is assumed that all the physical properties are those of pure water except that σ is reduced from 0.060 N m^{-1} to 0.040 N m^{-1} for the solutions. A reasonable estimate of typical contact angles is thought to be 40° for pure water and 20° for solutions. It is almost certain that Eq. (A-5) will not hold for large changes in contact angle, but it is used here for want of a better relationship. Figure A-1 illustrates for pure water the variation of the individual forces with time, as well as the net upward and downward forces. Departure is seen to occur at 12.0×10^{-3} s, at which point the radius is 2.65 mm. Similar calculations, with F ranging from 2 to 5 to illustrate the effect of slowing down the bubble growth rate, are shown in Table A-1.

Table A-1. *Departure Time and Size for Bubbles in Aqueous Solutions*

F	$t_d \times 10^3$ (s)	R_D (mm)
1	12.0	2.65
2	13.2	1.40
3	11.4	0.87
4	9.2	0.58
5	6.8	0.40

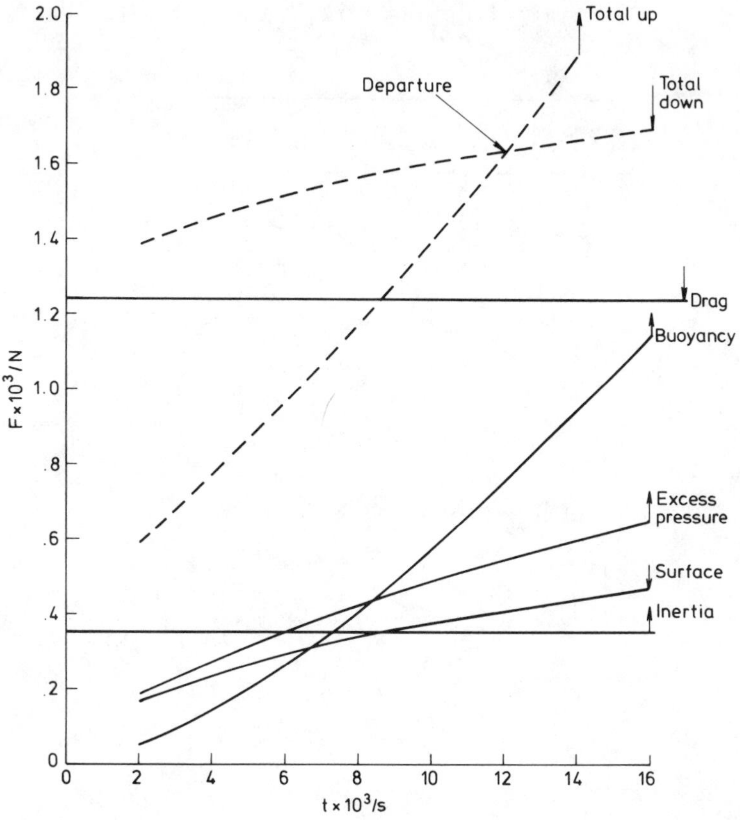

Fig. A-1. Forces on a bubble growing in water, $\Delta T_{sat,w} = 10.0^0C$, p = 1 bar.

It is seen that the departure time increases at first before decreasing, but it should be noted that a step change in σ and in ϕ has been introduced into the calculations between $F = 1$ and the other values and this causes the apparent anomaly. Nevertheless, R_D is predicted to decrease monotonically as F increases.

A second example, to illustrate binary organic mixtures, is based on the physical properties of ethanol at its normal boiling point. The contact angle has been taken as 20^0. The results in Table A-2 show a steady decrease in growth time and departure radius.

Table A-2. *Departure Time and Size for Organic (Ethanol) Solutions*

F	$t_d \times 10^3$/ (s)	R_D/ (mm)
1	22.2	1.20
2	16.5	0.55
3	10.0	0.28
4	5.8	0.16
5	3.5	0.10

Chapter 4

Contact Angles

Jacques Chappuis
Laboratoire de Technologie des Surfaces,
Ecole Centrale de Lyon, Ecully, France

Present address: Department
of Mechanical Engineering, University
of Toronto, Toronto, Ontario, M5S 1A4,
Canada

1 INTRODUCTION

1.1 Behavior of Liquids Relative to Solid Surfaces

A drop of a liquid, placed on the surface of a solid, can exhibit two different behaviors:

1. The drop may spread completely over the solid, and after a certain length of time, form a very thin film.

2. The drop may become distorted, and finally take an equilibrium configuration.

In the second case, at one point on the edge of the drop (also called the three-phase line), the tangential plane to the liquid surface forms with the plane of the solid surface an angle θ, different from zero and called the "contact angle" (Figs. 1 and 2).

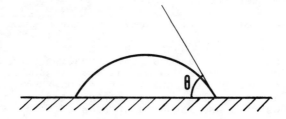

Fig. 1. A liquid drop on a solid with contact angle θ less than 90^0.

For most liquid-solid systems, the contact angle θ is about the same at each point of the edge of the drop, but can take different values corresponding to different stable equilibrium configurations of the drop. In this case, the contact angle values are included between two extreme values called "advancing contact angle θ_A" and "receding contact angle θ_R".

$$\theta_R \leqslant \theta \leqslant \theta_A \tag{1}$$

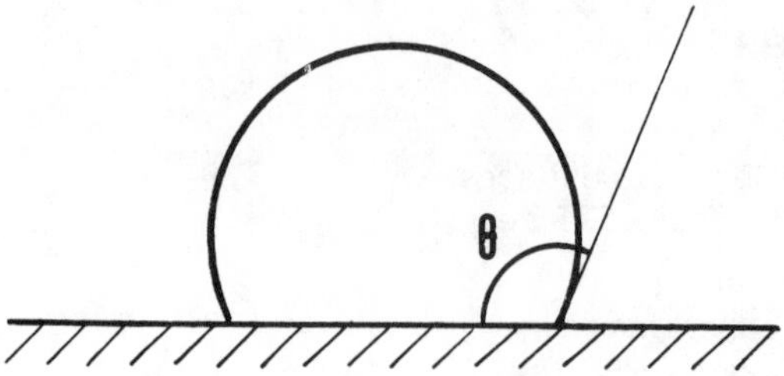

Fig. 2. A liquid drop on a solid with contact angle θ greater than 90^0.

The difference between θ_A and θ_R can be very large (more than 50^0). This phenomenon, as well as the actual difference between advancing and receding contact angle, is called "contact angle hysteresis."

On perfectly smooth and homogeneous solid surfaces, the contact angle would take a single value. In this theoretical case, a first interpretation of contact angles will be given: When the liquid completely spreads on the solid surface (contact angle zero), the attraction of the molecules of the liquid by the atoms of the solid is larger than the mutual attraction of the molecules of the liquid; when there is not complete spreading, the larger the contact angle is, the smaller the affinity of the liquid's molecules for the solid, in proportion to the mutual affinity of the liquid molecules. This very relevant reasoning was first expressed by Thomas Young in 1805 (see Young, 1855). It shows, first, that it will be possible to explain contact angles in terms of intermolecular forces, and second, that contact angles are correlated with solid-liquid adhesion and liquid cohesion.

Finally, one of the most important features of contact angles is the adsorption phenomenon: The solid surfaces are mostly covered by foreign molecules, which considerably modify their power of attraction relative to liquids.

1.2 Applications of Contact Angles

Contact angles play a major role in a number of technological processes. In various situations, the spreading of liquids on solids is desired: insecticides on leaves, oils on metallic surfaces, paints on solid supports, and so on. Dyeing, waterproofing, and cleaning of fabrics are closely related to contact angles, and textile industries use large quantities of tensio-active products. Hydrophobic coatings are largely employed to make porous materials water repellent, or to prevent adhesion (e.g., in frying pans). Optimum adhesion by adhesives depends on contact angle values. Flotation is a rather complex application in the mining industry that permits the separation of various types of solid particles from each other. Boiling and condensation phenomena are other examples of contact angle applications: For instance, the mode of condensation of steam (filmwise or dropwise) on a cooled metal surface is important in connection with the efficiency of industrial heat exchange plants. In fact, all the domains of applied surface chemistry (including environmental science) are more or less related to contact angles.

More recently, the importance of contact angles in biology has been made evident. Many instances are found in the animal and vegetable kingdoms, at the human and microscopic levels. The structure of ducks' feathers closely conforms to theoretical requirements for water repellency. For insects, complex systems containing hydrophobic hair permit some of them to stand on the surface of the water (*Podura aquatica*), or to crawl on the underside of a water surface (*Octhebius quadricollis*). Wettability is a very important parameter in the study of cell surfaces and biological surfaces. For example, it has been shown that there is a diminished hydrophilicity of granulocytes in children who are prone to bacterial infections (Van Oss et al., 1977).

Finally, the measure of contact angles is an extremely sensible technical investigation process of solid surfaces, which will probably be used more and more in the future. This is illustrated by the two following examples.

In experiments about contact angles and their temperature dependance, Neumann (1974) has registered sharp changes of contact angle values, corresponding to phase changes of polymeric solids. His investigations with Teflon showed that not only crystalline transitions, but also glass transitions could be detected (for thin films, glass temperatures are difficult to determine by conventional techniques).

Contact angles are often used to detect and study monomolecular layers on solids. The quantity of a substance necessary to make a monolayer is as small as 10^{-7} g for each square centimeter of covered surface. The presence of the monolayer not only considerably increases the contact angles of the testing liquids, but these angles are sensitive to small differences in orientation, packing, or composition of the adsorbed molecules (Zisman, 1959). Of course, experimental results are not always easy to expound.

1.3 Theoretical Complexity and Present Standards of Understanding

Contact angle phenomena are complex for the following reasons:

1. Real solid surfaces are nonhomogeneous, noncontinuous, nonisotropic regions. They are rough and chemically transformed (oxides).

2. The adsorption phenomenon is itself complex. On a metallic sample exposed to the air, there is simultaneous adsorption of water, oxygen, nitrogen, and grease vapors. There is little or no adsorption only on low-energy surfaces (polymers). There is also adsorption at liquid interfaces: Impurities in liquids, at mass concentrations as low as 10^{-4}, sometimes greatly decrease the surface tension of liquids and consequently modify the contact angles.

3. The liquid molecular structure is little known in the bulk of liquids, and we are not aware of the arrangement of liquid molecules near the different liquid interfaces. This will be a limitation for molecular interpretations.

At the present time, we do not have quantitative laws to predict contact angles in every case. The theories treat the simplest cases (pure liquids, smooth surfaces, controlled adsorption) and are then corrected by taking into consideration supplementary parameters taken separately (roughness, for instance). To establish and verify these theories, correct experiments must satisfy very severe conditions (very carefully purified liquids, well-known solid surfaces, controlled atmosphere, etc.), and some results found in the literature are doubtful.

Practically, and without a preliminary experimental study, it is impossible to predict the advancing and receding contact angle values of a given liquid on a given solid. But knowledge of existant theories permits us to explain contact angle variations in a systematic study, or to know how to increase or decrease contact angles in practical applications.

1.4 Purpose and Scheme of the Chapter

Contact angles are presented here in a tutorial manner. The text should be easily read even by nonspecialists. We have tried:

To explain all necessary notions (liquid surfaces, solid surfaces, adsorption).

To clarify all controversial points and to reexamine every theory that is not completely justified.

To present the simplest models. To this end, the force concept has been deliberately chosen instead of the energetic one. Many relations about wettability have been, until now, established thermodynamically, and will be established here by pure mechanical considerations.

The next part of this text (Sec. 2) is about liquid surfaces. A firm knowledge of liquid interfaces is indispensable for the study of contact angles. Interfacial tensions are introduced in this simple case and are discussed in relation to molecular structures. The equations of the profiles of liquid meniscii are established, in cases so demanding, in the following sections of the text. Some notions about adsorption at liquid surfaces permit us to obtain information about the surface tension of solutions.

The surfaces of solids are studied in Sec. 3. Proofs are given for solid surface tensions and some generalities concerning heterogeneity and roughness of real solid surfaces are exposed.

The fourth section is about adsorption on solid surfaces: the mechanism, laws of adsorption at the solid-liquid and at the solid-gas interfaces, and decrease of the solid surface tension related to adsorption.

Sections 2, 3, and 4 supply only the knowledge necessary to begin the study of contact angles. All the following sections treat contact angles exclusively.

The simplest case (ideal homogeneous and smooth solid surfaces) is discussed in Sec. 5. Young's equation and its correct vectorial model, including the reaction of the solid, is presented first. Then, without using thermodynamics, all usual useful quantities are established: adhesion tension, work of wetting, work of adhesion, and the spreading coefficient. An important paragraph closes this fifth section: It explains the notion of reversibility and irreversibility in wetting phenomena. This concept, which will be used a lot in the following section, often allows for the solution of contact angle problems more easily than would the use of thermodynamics.

Section 6 demonstrates that hysteresis is due mostly to surface heterogeneity and roughness. For heterogeneity, the demonstration is done with strip-type and patch-type heterogeneous model surfaces. For roughness, there is a transposition of the preceding models. The other possible causes of hysteresis are mentioned. The results of the present analysis of contact angle hysteresis are compared to those in the literature.

Section 7 is the complete description and analysis of the measurement of contact angles by use of a wetting tensiometer. Rather than quickly describing all methods of contact angle measurement, it has been thought more advisable to analyze one of them fully. After discussing the advantages of the chosen method, the immersion and emersion curves are completely explained by taking account of edge effects.

An interpretation of contact angle values, in the case of low-energy surfaces, is proposed in Sec. 8. This interpretation is quite different from those that have been given thus far. The differences in the hypotheses and in the interpretations are carefully demonstrated. The new interpretation leads to a new equation of state, which has important consequences with regard to solid-liquid adhesion and measurement of adsorption.

Sections 5 through 8 concern equilibrium contact angles. In most cases, however, contact angles are not in thermodynamic equilibrium. Section 9 outlines some aspects of this large and often undiscovered area (surface-driven phenomena, the nonequilibrium of a sessile drop, and systems with permanent pulsations).

We have chosen to use the following customary units:

$dyn\ cm^{-1}$ for surface tensions

$erg\ cm^{-2}$ for surface energies

kcal for bond energies

We give here the conversion factors to SI units:

$1\ dyn\ cm^{-1} = 10^{-3}\ N\ m^{-1}$

$1\ erg\ cm^{-2} = 10^{-3}\ J\ m^{-2}$

$1\ kcal = 4.18\ 10^{3}\ J$

2 LIQUID SURFACES

The physical chemistry of surfaces concerns the study of physical and chemical phenomena produced at the interfaces separating two phases. The different types of interfaces can be classified by taking account of the three states of matter: solid, liquid, and gas. The following interfaces may be distinguished:

Liquid-gas

Liquid-liquid (nonmiscible)

Solid-gas

Solid-liquid

Solid-solid

The interfaces concerning two miscible liquids or two gases are not stable interfaces. Among the different types of stable interfaces, the first two are fluid interfaces. They separate phases that are sufficiently mobile for the interface to take on a configuration of mechanical equilibrium. All that will be said in this section is valid not only for liquid-gas interfaces but also for liquid-liquid interfaces.

2.1 Surface and Interfacial Tensions of Liquids, and Other Surface Thermodynamic Quantities

Increasing the surface area of a liquid necessarily requires bringing to the surface molecules that had previously been elsewhere in the liquid. Experience shows that it is always necessary to furnish work. For a given liquid, at a given constant temperature, the work that corresponds to a surface increase of one area unit is always the same, on condition that the transformation is made reversibly. To increase the extent of a surface reversibly, one must exert exactly the necessary force f' on the plane of the surface, equally distributed along the length L' of the line of discontinuity of the surface, and perpendicular to it. By displacing this line of $d\ell'$, the surface is increased by $dA = L'd\ell'$. The work $d\tau'$ associated with a displacement $d\ell'$ of the point of application of f' is

$$d\tau' = f'\, d\ell' = \frac{f'}{L'}\, dA \qquad (2)$$

The force, of which the value is f'/L' is called superficial tension γ. It represents the force necessary to maintain in equilibrium the unit of length of the line of discontinuity. It acts in the plane of the liquid's surface, perpendicular to the line of discontinuity of the surface, and on each unit of length of this line. Its orientation corresponds to that which would bring about a reduction of the liquid surface. γ is expressed in dyn cm^{-1} or in N m^{-1}. γ can also be called "work of surface formation". γ represents, in effect, the work necessary to form a new unit of surface: If we use $\gamma = f'/L'$ from eq. (2), we arrive at

$$\gamma = \frac{d\tau'}{dA} \tag{3}$$

γ is expressed here in erg cm^{-2} or in J m^{-2} (units equivalent to the preceding). Let us call G, the free energy of a system constituted by the surface and the phases that it separates. Certain authors assimilate γ to the ratio $\Delta G/\Delta A$, when there is a finite variation of the interfacial area. This definition would not be valid unless the system was such that the variation of A remained independent of other variables. Only the relation

$$\gamma = \left(\frac{\delta G}{\delta A}\right)_{v_1, v_2, \ldots, v_n} \tag{4}$$

in which A, v_1, v_2, . . . , v_n are the variables defining the system, is thermodynamically correct. The temperature T of the system is one of the variables v_i. But, when there is an increase of surface area, the system absorbs heat called "heat of surface extension." In order to maintain the temperature of the system, it must be furnished with a certain quantity of heat equal to $- T\, d\gamma/dT$ (according to the first law of thermodynamics applied to fluid interfaces). This quantity ($- T\, d\gamma/dT$) is positive because the surface tension of pure liquids decreases, linearly in general, as a function of the temperature, until it becomes null at the critical temperature. The total energy e_s of a square unit of fluid surface area is thus equal to

$$e_s = \gamma - T\frac{d\gamma}{dT} \tag{5}$$

Let us consider a practical example, that of water at 20°C:

$$\gamma = 72 \text{ dyn cm}^{-1} \ (0.072 \text{ N m}^{-1})$$
$$-\frac{T\,d\gamma}{dT} = 48 \text{ dyn cm}^{-1} \ (0.048 \text{ N m}^{-1})$$

The quantity of heat necessary upon an increase of 1 cm^2 of surface is thus equal to

$$\frac{48}{4.18}\, 10^{-7} = 1.15\text{x}10^{-6} \text{ cal } (4.8\text{x}10^{-6} \text{ J})$$

Let us imagine a system constituted by 100 g of water. If we furnish only the mechanical work necessary to spread out its surface (72.8 erg cm^{-2}), the temperature decrease of the system corresponds to $1.15\text{x}10^{-8}$ degrees for each square centimeter of surface formed under reversible conditions. This temperature variation is weak enough not to modify greatly the quantities (such as γ) that depend on the temperature.

The above example shows us that the work put into play is practically independent of possible thermal transfers; from an experimental point of view, it is possible to measure the work of surface extension; on the other hand, a quantity of heat on the order of 10^{-6} cal is much more difficult to measure.

In what follows, so as to avoid any confusion, we shall never use the word "energy." We shall speak only of "work" furnished or received by a system upon variations of interface areas during reversible transformations, and we shall not worry about possible thermal transfers. This work is the work carried by the superficial tension, which is a macroscopic force.

Figure 3 represents a frame (made of iron wire, for example), of which one edge *AB* is mobile. By plunging this frame into a soap solution and carefully withdrawing it, a liquid film is obtained, bounded by the four edges of the frame. This film has a nonnegligible thickness and the liquid-vapor interface area is double that of the frame. To keep the barrier *AB* in equilibrium, a force that is double that of the soap solution's superficial tension must be exerted on each of its units of length. Let us now place a knotted wire delicately on the liquid film (Fig. 3); and let us pierce the liquid film with a pin at the interior of the loop formed by the wire (Fig. 4); the wire takes the form of a circumference that helps reduce the remaining interfacial surface maximally. This experimental result is the proof that the surface tension is exerted equally in all directions. This experiment shows that the interface between two nonmiscible fluids is similar to a very thin elastic membrane stretched equally in all directions. As early as 1805, Young stated: "It is as if the two homogeneous fluids were separated by a uniformly stretched membrane without thickness." The hypothesis of the existence of a membrane that tends to contract at a liquid surface was foreseen and discussed, in particular by Adam (1941). Such a membrane does not really exist: One cannot separate it from the liquid; it is nonetheless an extremely useful image to explain the phenomena of capillarity.

Macroscopically, interfacial tension is thus a force that is exerted on all points of an interface, in the plane of the interface, and in all directions. Its value is the same at every point

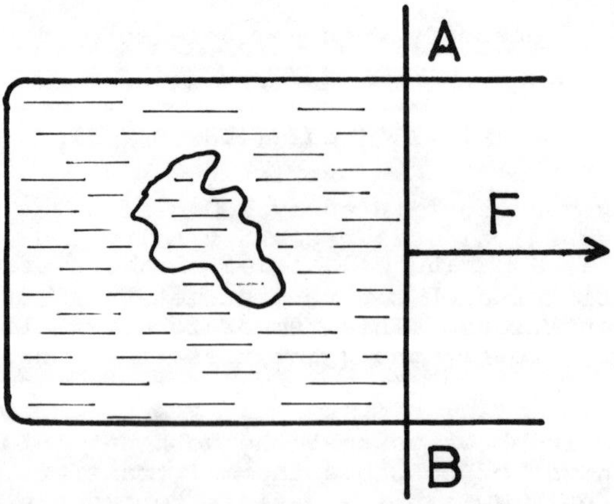

Fig. 3. Soap film stretched over a wire frame. For equilibrium of the movable edge *AB*, a force *F* is necessary to counterbalance surface forces.

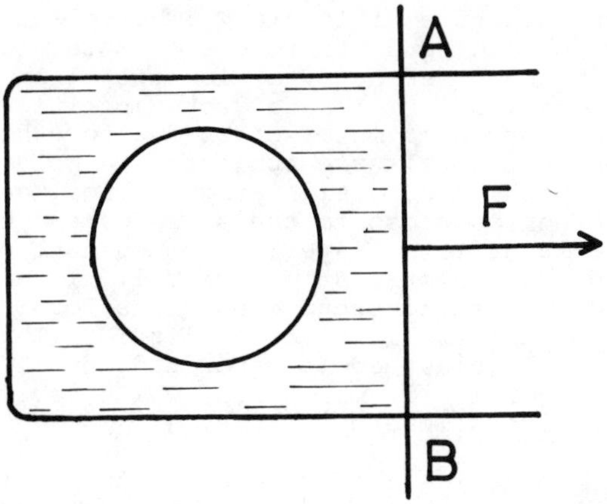

Fig. 4. After the breaking of the soap film inside a knotted wire, this latter takes the form of a circumference.

of the surface of a given liquid. It can be measured along a line of discontinuity of the interface. It is, in such a case, tangent to the interfacial surface, normal to the line of discontinuity, and aimed in the direction that tends to diminish the interfacial area.

2.2 Surface Tension Related to the Molecular Structure of Liquids

Superficial tension is the macroscopic consequence of forces that are mutually exerted among the molecules of a liquid near the surface. Among the intermolecular forces, some are strong (ionic and metallic bonds), others medium (hydrogen bonds), and others weak (van der Waals forces). The corresponding liquids have surface tensions that are respectively high, medium, and low. The molecular interpretation that follows explains the relationship between molecular forces and superficial tension. The molecules of the superficial layer have fewer neighbors than the molecules in the bulk. When the area of a liquid surface is increased, it is necessary to bring to the surface molecules from a lower depth, and thus to break molecular bonds. For this, work must be furnished that is none other than the work of surface formation (equal to the superficial tension). The heat that must be furnished at the same time (entropic term equal to $- T\, d\gamma/dT$), indicates that the arrangement of molecules in the superficial layers is not as good as it is at the center of the liquid. Also, the intermolecular forces are short-range forces; it is generally admitted that the disturbed zone constituted by the liquid surface has a very minute thickness, on the order of 10 Å. In this zone, the arrangement of molecules differs from that which exists in the bulk of the liquid.

The aim of the model below is to show that the surface molecules are more widely spaced than those found at the liquid's center. This model is not based on the information given by current theories about the molecular structure of liquids. The molecules of a liquid are subject, on one hand, to forces of molecular interaction, and they possess, on the other hand, a vibrational energy. We shall consider that they continually vibrate around positions for which the resultant of the forces of interaction due to other molecules is zero: Molecules in the bulk are subjected to molecular forces by surrounding molecules; the field is symmetrical and has no effect. For the molecules close to the surface, and even for those that constitute the superficial layer, the resultant of intermolecular forces should also be statistically null. If not, either the molecules will be displaced to occupy positions for which this resultant is null, or they will leave the liquid. This last case occurs when certain molecules become vaporized, but it concerns only a very small proportion of molecules, which have, moreover, at a given moment, the highest energy levels.

Let us interest ourselves, therefore, in the surface molecules that have an insufficient level of energy to become vaporized. These molecules vibrate around positions of equilibrium for which the forces of repulsion are equal to the forces of attraction. Let us consider, for example, that the forces of interaction are van der Waals forces. These forces are additive, and Fig. 5 represents the force curve for two molecules as a function of separation. The repulsive force operates only when the molecules come close together. For two molecules that are not close neighbors, there are only attractive forces between one and the other.

To simplify the model we shall assume that the equilibrium positions of the surface molecules are regrouped on successive layers, layer number 1 being the superficial layer (Fig. 6). Let us consider a molecule I from among all those (1, 1, 1, . . .) of

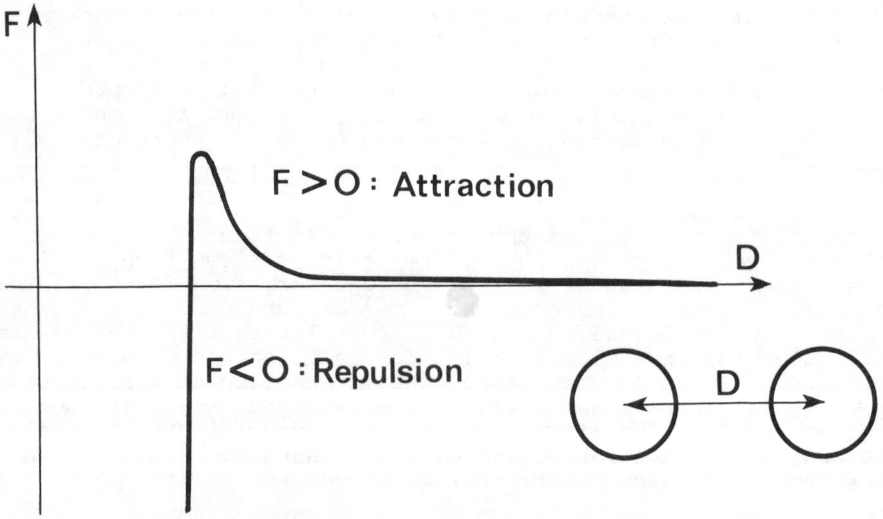

Fig. 5. Force curve for two molecules as a function of separation (van der Waals forces).

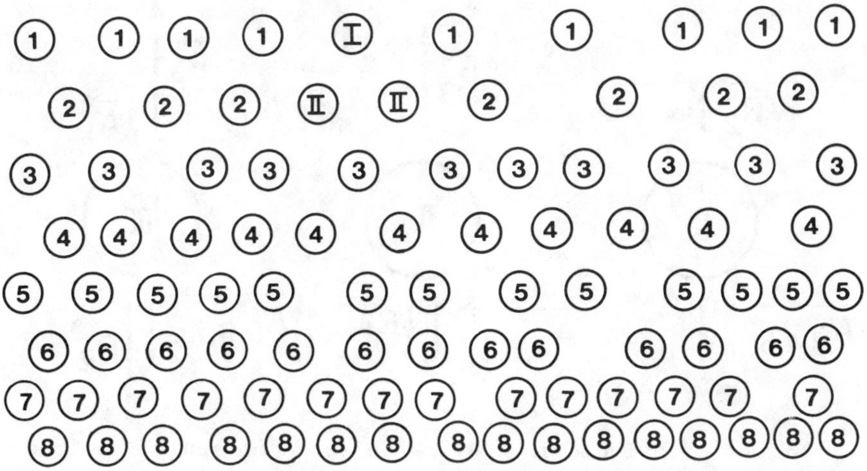

Fig. 6. Model in which surface molecules are regrouped on successive layers.

layer number 1. When this molecule occupies its position of equilibrium, the resulting effect of the molecules from layer 1 (molecules 1, 1, 1, . . .) is null. The molecule is attracted by the molecules of the second layer (molecules 2, 2, 2, . . .) except those with which it is in contact (molecules II, II, . . .). It is attracted by the molecules of layers 3, 4, 5, ... The resultant of all the attractive forces is vertical, and directed toward the interior of the liquid. It is equilibrated by the resultant of the forces of repulsion, which are exerted by a small number of molecules of layer 2 (molecules II, II, . . .). When, during its movements, molecule I begins to move toward the exterior of the liquid's surface, it is subject to forces of attraction including those of the molecules II, II and of the molecules (1, 1, 1, . . .). This explains that molecule I is kept close to the liquid's surface if its energy is less than a necessary given value for evaporation. To summarize, molecule I, when it is in its position of equilibrium, is subject to:

A force of attraction $\overrightarrow{A_{3,4,5,...}}$ coming from the molecules of layers 3, 4, 5, . . .

A resultant R_2, which is thus a force of repulsion coming from the molecules of layer 2 (combined effect of repulsion of II, II, . . . and of attraction of 2, 2, 2, . . .)

We have thus

$$\overrightarrow{R_2} + \overrightarrow{A_{3,4,5...}} = 0 \qquad (6)$$

(see Fig. 7).

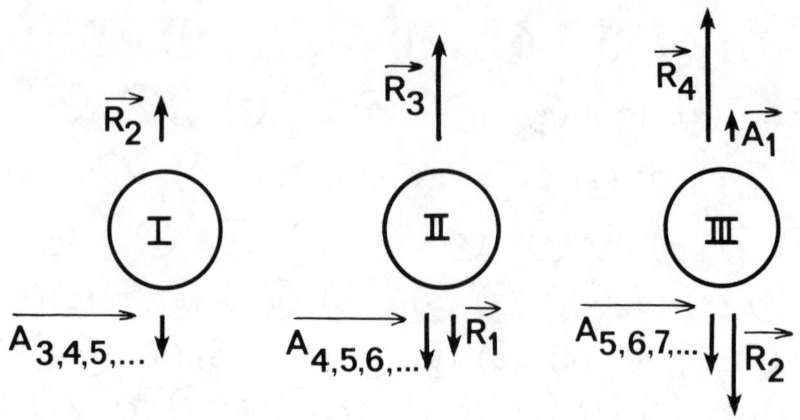

Fig. 7. Effect of neighboring layers on (a) a molecule I belonging to layer 1, (b) a molecule II belonging to layer 2, and (c) a molecule III belonging to layer 3.

Let us now consider a molecule II of layer 2. The effect $\overrightarrow{R_1}$ of all the molecules from layer 1 is, approximately, equal and opposed to the resultant of the forces that all the molecules from layer 2 exert on a molecule 1.

$$\overrightarrow{R_1} = -\overrightarrow{R_2} \tag{7}$$

$\overrightarrow{R_1}$ is thus a force of repulsion directed downwards. Molecule II is also subjected to:

A null force from the other molecules of layer 2.

A force of attraction $\overrightarrow{A_{4,5,6,...}}$ from the molecules of layers 4, 5, 6, . . .

A force of repulsion $\overrightarrow{R_3}$ from the molecules of layer 3 (combined effect of repulsion of III, III and of attraction of 3, 3, 3, . . .)

The whole of the forces acting on molecule II should have a null resultant. Figure 7b shows clearly that $|\overrightarrow{R_3}|$ is larger than $|\overrightarrow{R_2}|$ by a quantity equal to $|\overrightarrow{A_{4,5,6,...}}|$. This signifies that the forces of repulsion between II and III, III, . . . are larger than those between I and II, II,. . . Thus the spacing between layers 2 and 3 is smaller than that between 1 and 2.

This reasoning can be reproduced for a molecule III of layer 3, for a molecule IV of layer 4, and so on. For the molecule III, there appears a force of attraction A_1 directed upwards, due to layer 1. For the molecule IV, there appears a force of attraction $A_{1,2}$ directed upwards, due to layers 1 and 2. We can consider that these forces of attraction are respectively less than the forces of attraction $\overrightarrow{A_{5,6,7,..}}$ and $\overrightarrow{A_{6,7,8,...}}$ and can deduce from this that $|\overrightarrow{R_4}| > |\overrightarrow{R_3}|$ and that $|\overrightarrow{R_5}| > |\overrightarrow{R_6}|$. Consequently, the spacing between the layers diminishes gradually with the depth of the liquid. This effect will disappear for the molecule N of layer n, when the effect of attraction of layers 1 2, . . ., $n - 2$, is equal and opposed to the effect of attraction of layers $n + 2$, $n + 3$, $n + 4$, . . . , that is, when the distance from layer 1 to layer n becomes equal or superior to the range of the molecular forces. Van der Waals attractive forces vary as the inverse of the seventh power of the distance that separates the molecules ($F = K/D^7$). These are thus short-range forces for which the effect is completely negligible for more than 10 molecular thicknesses. Our model shows, thus, that the zone in which the arrangement of molecules differs from that at the liquid's interior has a thickness on the order of 10 Å.

We shall keep in mind that, in the upper molecular layers, the spacing between molecules increases gradually as we near the surface. However, the molecules remain close to each other, because they exert forces of repulsion between close neighbors when they are in their mean positions. The above model is valid for solids as well as liquids. In both cases, and because of their vibrational energy, molecules of the superficial layer can distance themselves from their position of equilibrium toward the liquid's exterior. The author thinks that in the case of liquids, and contrary to that of solids, the amplitude of vibration is large enough for a superficial molecule to evacuate its equilibrium site when it is in its most distant position. This site can then be occupied by a molecule of layer 2, whereby a "hole" is formed that can spread to the liquid's interior.

The above model permitted us to conclude that, in a direction perpendicular to the surface, there is a slight, gradual increase in the spacing between the molecules as we near the surface. Given the great mobility of liquid molecules, we cannot imagine a liquid surface structure in which the molecular concentration would vary in a direction (perpendicular to the surface), and would remain constant in the different planes parallel to the surface. Consequently, the molecular density n_1 in layer 1 must be less than the molecular density n_2 in layer 2, and so on up to layer n above which the molecular density is constant.

$$n_1 < n_2 < n_3 < \ldots < n_n \qquad (8)$$

This result slightly modifies some of the equations used in the course of reasoning. Equation (7), for example, $\overrightarrow{R_1} = -\overrightarrow{R_2}$, should have been written as

$$\overrightarrow{R_1} = -\frac{n_1}{n_2}\overrightarrow{R_2} \qquad (7a)$$

Nonetheless, since n_1/n_2 is little different than 1, the entire demonstration remains correct. Figure 6, which represents molecules of the superficial layers, takes into account the decrease of molecular densities in the layers as one approaches the surface.

Thus this figure is not valid in the case of solid surfaces.

From our model we can propose an interpretation for superficial tension: In the bulk of the liquid, the molecules are subject, in all directions, to forces of repulsion from their close neighbors, and to forces of attraction from all others. The repulsive forces are stronger, but since the field is symmetrical, the resultant on each molecule is null. Molecules of the superficial layers are subject to much weaker repulsion forces from their close neighbors, and to attractive forces from others. We have seen that the resultant of the vertical components of these forces is null for each molecule. As for the horizontal components, their resultant is null because of symmetry. However, if we consider that the attractive forces are greater than the repulsive forces, there results a stretch strain showing all the characteristics of superficial tension. Thus, superficial tension results from the fact that the molecules of the superficial layers are attracted equally in all directions, in planes parallel to the surface.

We have constructed a molecular model of liquid surfaces to justify a number of affirmations that are generally not proven in the literature. However, this model differs greatly from the actual structure of liquids: First, the model is static although molecules are in perpetual motion; second there is no reason for the superficial molecules to be regrouped on successive layers.

2.3 Interfacial Tension Between Two Nonmiscible Liquids

Let us imagine the interface between two liquids 1 and 2. The molecules of liquids 1 and 2 exert forces of interaction among themselves. Using the same reasoning as in the preceding model, we can conclude that the arrangement of molecules in liquid 1, near the interface, is less disturbed than in the case of a liquid-vapor interface. Liquid 1 will thus be the seat of an interfacial tension $\gamma_{1(2)}$ weaker than γ_1; $\gamma_{1(2)}$ could be called interfacial tension of liquid 1, modified by liquid 2. In the same way, liquid 2 is the seat of an interfacial tension $\gamma_{2(1)} < \gamma_2$.

The interfacial tension of the interface, γ_{12}, is equal to the sum $\gamma_{1(2)} + \gamma_{2(1)}$:

$$\gamma_{12} = \gamma_{1(2)} + \gamma_{2(1)} < \gamma_1 + \gamma_2 \tag{9}$$

Some authors have tried to connect interfacial tension γ_{12} with interfacial tensions γ_1 and γ_2 of liquids 1 and 2. We have listed below the results of their work.

Antonow's (1907) rule indicates that the interfacial tension γ_{12} between two mutually saturated regular solutions is equal to the difference of the superficial tensions of the two phases:

$$\gamma_{12} = |\gamma_1 - \gamma_2| \tag{10}$$

In fact, as Defay et al. (1966) have shown, this rule is only an approximation.

Girifalco and Good (1957) have proposed the relation

$$\gamma_{12} = \gamma_1 + \gamma_2 - 2\phi\sqrt{\gamma_1\gamma_2} \tag{11}$$

in which ϕ is a characteristic number of the studied system. Its expression brings into action dipolar moment, polarizability, ionization energy, and molecular radius of the two phases' molecules. Its value, generally included between 0.5 and 1.1, tends toward unity when the interaction forces are of the same type on each side of the interface.

Fowkes (1962) distinguishes, in the liquid superficial tension, between the part due to London dispersion forces (which he denotes γ^d), and the part due to other molecular forces (which he denotes $\Sigma\gamma^h$), like those that result from hydrogen bonds, for example. So he writes for any liquid:

$$\gamma_1 = \gamma_1^d + \Sigma\gamma_1^h \tag{12}$$

In the case of the interfacial tension between two liquids, Fowkes obtains

$$\gamma_{12} = \gamma_1 + \gamma_2 - 2\sqrt{\gamma_1^d\gamma_2^d} \tag{13}$$

To use this relation, it is necessary to evaluate the contributions γ_1^d and γ_2^d of the dispersion forces to the superficial tensions γ_1 and γ_2. The method consists of measuring the interfacial tension between the considered liquid and a reference liquid in which the intermolecular forces are only London forces, because, in this case $\gamma = \gamma^d$. Saturated hydrocarbons, which fill this condition, were used by Fowkes to measure the γ^d of diverse liquids.

The relations (11) of Girifalco and Good and (13) of Fowkes generally lead to results that agree with the experimental values.

2.4 Adsorption at Liquid Surfaces

Let us consider a solution, that is, a solute dissolved in a solvant. The molecules of the solute are spread out, regularly, among the molecules of the solvent. However, in the superficial layer, the concentration of the solute is generally superior or inferior to the mean. We call "adsorption" and denote by Γ, the algebraic excess of solute in the superficial layer (Γ is expressed in moles of solute by area unit). Gibbs' (1961) law allows that, for a constant given temperature T, the algebraic variation of the superficial tension γ is equal to

$$d\gamma = -\Gamma RT \ell n\, a_s \tag{14}$$

in which R is the universal gas constant and a_s is the activity of the solution (or its concentration for very dilute solutions); a positive adsorption (excess of solute in the superficial layer) leads to a decrease in superficial tension. It is the case for hexanol in water; for example, a solution of 0.2 g of hexanol per liter of water has a superficial tension of 67 dyn cm^{-1}, instead of 72 dyn cm^{-1} for pure water at 20°C.

Some substances considerably decrease the superficial tension of water when they are added at very weak concentrations. These substances, called tensio-actives, have numerous industrial and household uses. For some of them, Fig. 8 represents the superficial tension variations of water as a function of the concentration at which they are added. Tensio-active molecules usually present a hydrophilic polar portion and a hydrocarbonated hydrophobic tail.

In order to get exact experimental measurements with pure liquids, great care must be taken to have liquids that are really pure. One must furnish oneself with the purest liquids available. Also, one must avoid the presence, in nonpolar liquids, of small amounts of highly adsorbable impurities, such as polar molecules. To ensure the absence of this type of impurity, every liquid has to be passed through a column of an appropriate adsorbent (activated alumina, silica gel, florisil). High-boiling liquids have to be stripped of volatile impurities.

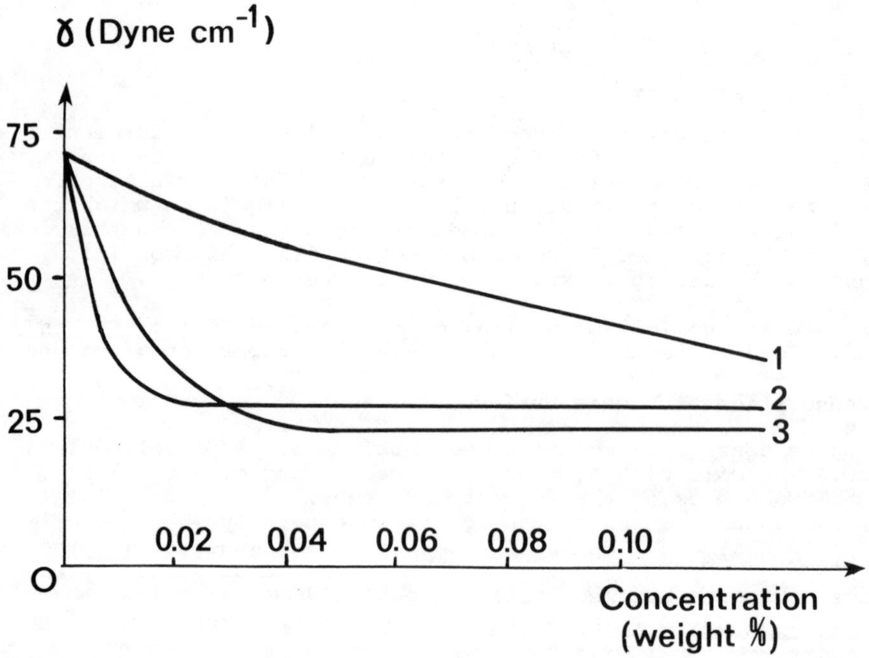

Fig. 8. Surface tension of tensio-actives in water as a function of the concentration at which they are added. 1, Sodium lauryl sulfate. 2, Cetyltrimethylammonium bromide. 3, Sodium di-*N*-octylsulfosuccinate.

2.5 Laplace's Law; The Condition of Mechanical Equilibrium of a Surface

Let us consider a point P, belonging to any surface, represented in Fig. 9. Let us trace a curve at a constant distance ρ' from point P, on this surface. This curve defines a spherical cap for which we propose to determine the equilibrium conditions by considering ρ' to be extremely small. Two orthogonal planes containing the normal line to the surface at point P cut the spherical cap following arcs AB and CD, of which the radii of curvature are r_1' and r_2'. By virtue of Euler's theorem, r_1' and r_2' verify the relation

$$\frac{1}{r_1'} + \frac{1}{r_2'} = \frac{1}{R_1'} + \frac{1}{R_2'} \tag{15}$$

in which R_1' and R_2' are the principal radii of curvature.

To maintain an element $d\ell$ passing through A and belonging to the circumference in equilibrium, a force must be exerted on it. The projection of this force on the normal line PN corresponds to

$$\gamma\ d\ell\ \sin\phi_1 = \gamma\ d\ell\ \phi_1 = \gamma\ d\ell\ \frac{\rho'}{r_1'} \tag{16}$$

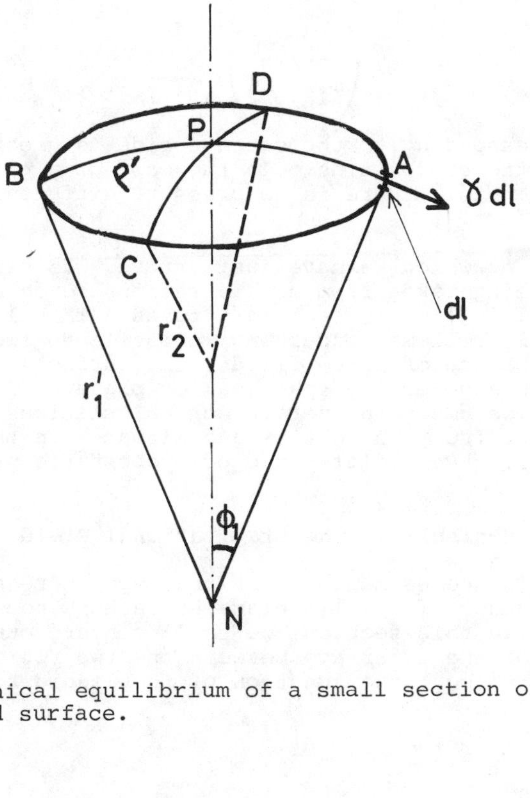

Fig. 9. Mechanical equilibrium of a small section of an arbitrary curved surface.

ϕ_1 being extremely small angle $\widehat{PNA}$. A force perpendicular to the spherical cap, and equal to

$$\gamma\, d\ell \left(\frac{2\rho'}{r'_1} + \frac{2\rho'}{r'_2}\right) = 2\,\rho'\,\gamma\, d\ell \left(\frac{1}{R'_1} + \frac{1}{R'_2}\right) \qquad (17)$$

is associated with the group of four elements $d\ell$ on the circumference, situated at A, B, C, and D.

The integration spread over the whole circumference gives a perpendicular resultant four times too big, since each element would have been counted four times. The perpendicular resultant is thus equal to

$$\rho'\,\frac{\gamma}{2}\left(\frac{1}{R'_1} + \frac{1}{R'_2}\right) 2\,\pi\,\rho' = \pi\,\rho'^2\,\gamma \left(\frac{1}{R'_1} + \frac{1}{R'_2}\right) \qquad (18)$$

It can only be equilibrated by a difference of pressure ΔP on all parts of the surface. Hence the relations:

$$\Delta P\,\pi\,\rho'^2 = \pi\,\rho'^2\,\gamma \left(\frac{1}{R'_1} + \frac{1}{R'_2}\right) \qquad (19)$$

and

$$\Delta P = \gamma \left(\frac{1}{R'_1} + \frac{1}{R'_2}\right) \qquad (20)$$

The pressure corresponding to the concave side is greater than that corresponding to the convex side. In the case of a plane surface, R'_1 and R'_2 are infinite; there is no pressure difference on either side of the interface.

The relation at which we have just arrived was stated by de Laplace (1806), being taken from considerations concerning the upward movement of liquids in capillary tubes. This law allows us to confront all problems concerning meniscii profiles and drop shapes. In the absence of force fields, the surfaces delimiting different phases are generally spherical or planar. In the gravitational field g, we need consider in our calculations only that two points, distant from a height z_1 and situated in the same fluid of density ρ, have a difference of hydrostatic pressure equal to $\rho g z_1$.

2.6 Cylindrical Meniscii in the Gravitational Field

This section concerns meniscii of the type represented in Fig. 10. The presence of a solid plane surface is necessary for their existence. In this section, we shall concern ourselves only with the profile of the interface between the two fluid phases, and shall not worry about the junction of the liquid to the solid.

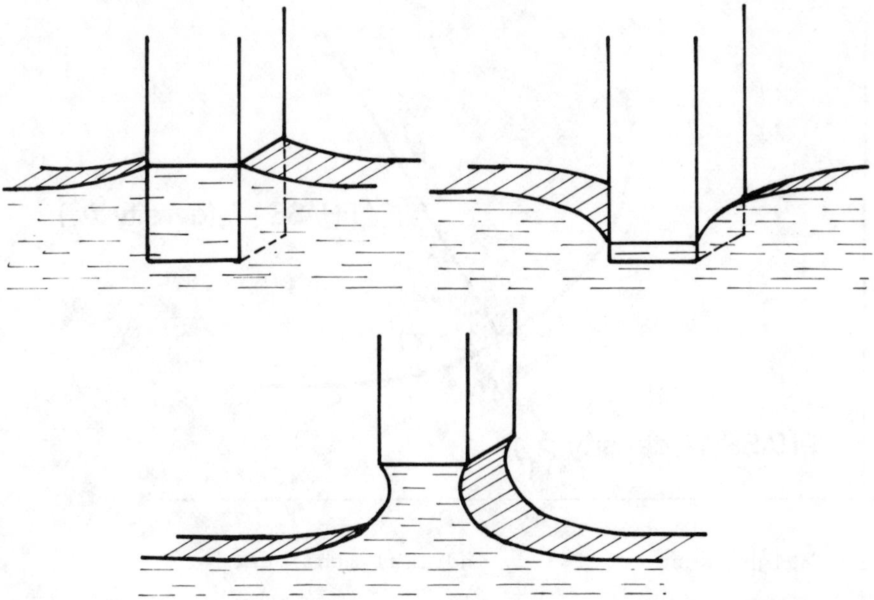

Fig. 10. Different possible shapes of cylindrical meniscii in the gravitational field.

Figure 11 represents the profile of a right-angle section of the meniscus. We shall take as surface radii of curvature, respectively, radius R' situated in the plane of the figure, and the infinite radius perpendicular to it. Laplace's law gives, in this case,

$$\Delta P = \gamma \left(\frac{1}{R'}\right) \tag{21}$$

On Fig. 11, M represents any point of the profile. Suppose that M' is a point close to M in the fluid phase of greatest density ρ_1, and that M'' is a point close to M in the fluid phase of weakest density ρ_2. The phase of density ρ_1 is always a liquid. That of density ρ_2 can be a liquid that is nonmiscible with the preceding, or a gas. M is located at a distance z above the plane portion of the interface where pressure P_0 prevail. The pressures at M' and M'' are, respectively,

$$P_{M'} = P_0 - \rho_1 gz \quad \text{and} \quad P_{M''} = P_0 - \rho_2 gz \tag{22}$$

whence

$$\Delta P = P_{M''} - P_{M'} = (\rho_1 - \rho_2)gz = \Delta\rho gz \tag{23}$$

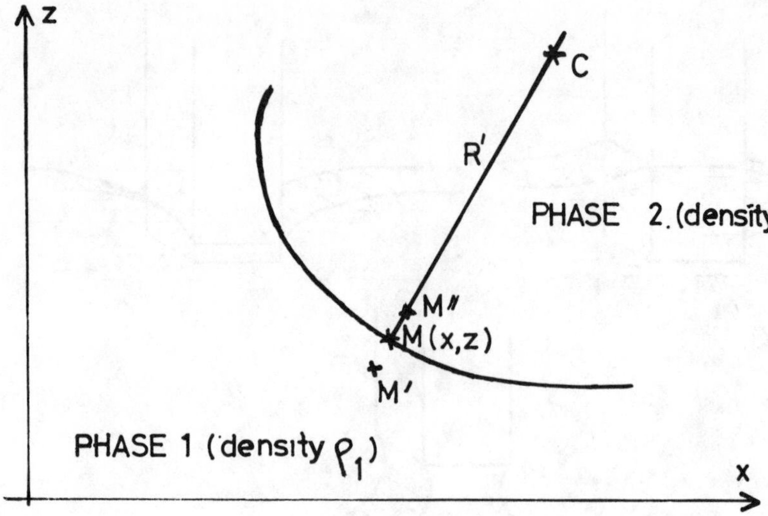

Fig. 11. Right section of a cylindrical meniscus.

ΔP represents the pressure variation due to curvature of the interface. By combining (21) and (23), we get;

$$\Delta\rho g z = \gamma \frac{1}{R'} \tag{24}$$

2.6.1 Cartesian Equation of the Curve $z = f(x)$

The radius of curvature R' is equal to

$$R' = \frac{[1 + (dz/dx)^2]^{3/2}}{d^2z/dx^2} \tag{25}$$

Equation (24) becomes

$$\Delta\rho g z = \gamma \frac{d^2z/dx^2}{[1 + (dz/dx)^2]^{3/2}} \tag{26}$$

When the liquid's surface is planar, at an infinite distance from the meniscus, the resolution of the differential equation (26) leads us to

$$x = (2a^2 - z^2)^{1/2} - \frac{a}{\sqrt{2}} \ln\left[\frac{|z|}{a\sqrt{2} - (2a^2 - z^2)^{1/2}}\right] + C \tag{27}$$

where a represents the capillary constant, defined by

$$a = \sqrt{\frac{2\gamma}{\Delta\rho g}} \tag{28}$$

and C represents a constant that depends on the origin of the axis of the abscissa. The graphic representation (Fig. 12) of this equation constitutes the profile of the cylindrical meniscii. This curve is composed of two parts that are symmetrical in relation to Ox. The part $z>0$ corresponds to the meniscii situated above the plane portion of the interface, and the part $z<0$ corresponds to those situated below.

The importance of Eq. (28) and of its graphic representation resides in the fact that a cylindrical meniscus is completely governed by the capillary constant a, which brings into play only those parameters dependent on the liquid. Thus, the curve is unique for a given liquid at a given temperature. If we use the same liquid, the different right sections of the meniscii from Fig. 10 have profiles corresponding to portions of the same curve, of which we have given the equation. The solid governs only the contact angle, which implies the choice of the right portion of this curve.

2.6.2 Parametric Equation of the Curve

Angle φ between the horizontal and the tangent at a point M of the curve is a very useful parameter: If θ is the contact angle of the meniscus to the solid;

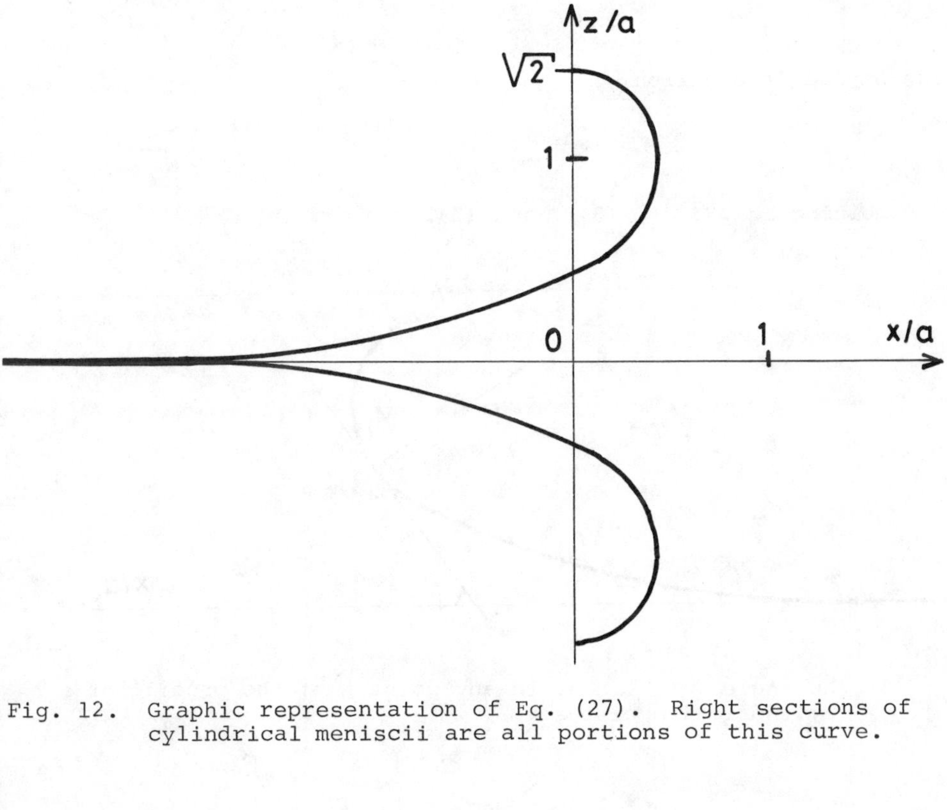

Fig. 12. Graphic representation of Eq. (27). Right sections of cylindrical meniscii are all portions of this curve.

$$\varphi = \frac{\pi}{2} - \theta \tag{29}$$

Suppose that M_1 is a point in the neighborhood of M on the profile (Fig. 13). We have

$$MM_1 = ds \tag{30}$$

M_1 is located by angle $\varphi + d\varphi$. The normal lines to the curve at M and M_1 intersect at a point P. Let us note that

$$PM_1 = PM = R' \tag{31}$$

(R' is the radius of curvature of the profile at M), and

$$\widehat{MPM}_1 = d\varphi \tag{32}$$

from which

$$R' \, d\varphi = ds \tag{33}$$

also

$$\sin \varphi = \frac{dz}{ds} \tag{34}$$

Using Laplace's equation,

$$\Delta\rho g z = \frac{\gamma}{R'} \tag{24}$$

and combining Eqs. (33), (34), and (24), we obtain

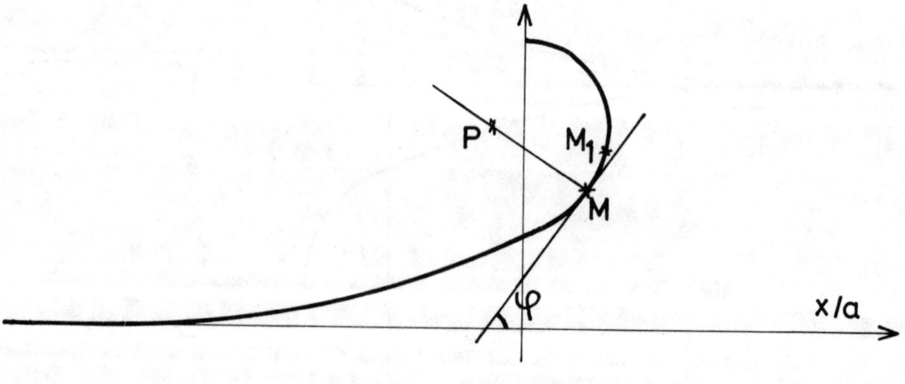

Fig. 13. An angle φ is linked to any point M of the profile of a cylindrical meniscus.

$$\Delta\rho g z \; dz = \gamma \; d\varphi \; \sin\varphi \tag{35}$$

of which the integration

$$\frac{\Delta\rho g}{2}\left[z^2\right]_0^z = \gamma\left[-\cos\varphi\right]_0^\varphi \tag{36}$$

leads to

$$z = \pm\sqrt{\frac{2\gamma}{\Delta\rho g}}\sqrt{1-\cos\varphi} = \pm\, a\sqrt{1-\cos\varphi} \tag{37}$$

The utilization of

$$dx = \frac{ds}{\cos\varphi} \tag{38}$$

permits us to express $x = f(\varphi)$ by

$$x = a\left[(1+\cos\varphi)^{1/2} - \frac{1}{\sqrt{2}}\left(\frac{(1-\cos\varphi)^{1/2}}{\sqrt{2}-(1+\cos\varphi)^{1/2}}\right)\right] \tag{39}$$

2.6.3 Volume of the Meniscus

Let us calculate the volume of a section of meniscus of width w, and limited by a vertical plane passing through point M located by angle φ (Fig. 14). The volume of a slice of thickness dx is

$$dV = zw \; dx \tag{40}$$

Since

$$z = \pm a\sqrt{1-\cos\varphi} \tag{37}$$

and

$$dx = \frac{ds}{\cos\varphi} \tag{38}$$

according to the geometric considerations of Fig. 14. Also

$$\Delta\rho g z = \gamma\,\frac{1}{R'} \tag{24}$$

$$R' \; d\varphi = ds \tag{33}$$

Taking account of these relations, expression of the elementary volume takes the form

$$dV = \frac{\gamma w}{\Delta\rho g}\cos\varphi \; d\varphi \tag{41}$$

and an integration between the limits $\varphi = 0$ and $\varphi = \varphi$ leads to

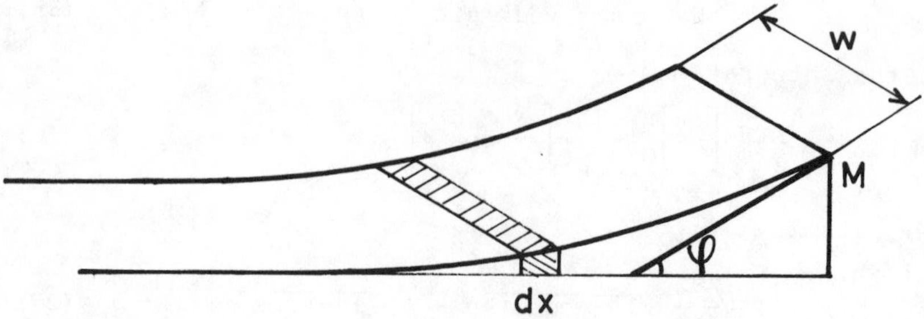

Fig. 14. Section of a cylindrical meniscus of width w, and limited by a vertical plane passing through point M located by angle φ.

$$V = \frac{\gamma\ w}{\Delta\rho g} \sin\varphi \tag{42}$$

2.6.4 Elementary Work of Formation of a Cylindrical Meniscus

Given a cylindrical meniscus of height z, connected with a vertical solid surface along a length w. If, during a reversible transformation, the meniscus evolves, and if its height becomes $z + dz$, we may assume that the meniscus has been subjected to a translation dx toward the exterior of the solid, and that it has been completed by an elementary slice of thickness dx situated between the transferred meniscus and the solid. The work that must be performed to form a meniscus of height $z + dz$ from a meniscus of height z corresponds simply to the work of formation of an elementary slice of thickness dx, taking into account a disappearance of free horizontal liquid surface of area $w\ dx$. (We shall not worry ourselves in this section with work due to variations of areas of solid-liquid and solid-vapor interfacial surfaces. We assume that this work is not included in the work of formation of the cylindrical meniscus. Of course, in a global treatment we shall take this work into account; see Secs. 5.5 and 5.9.) According to Fig. 15,

$$dx = \frac{dz}{tg\varphi} \tag{43}$$

The work of formation dW_f of the elementary slice includes a work of surface extension dW_s and a work against gravity dW_g:

$$dW_f = dW_s + dW_g \tag{44}$$

The mass of the elementary slice $\Delta\rho\ wz\ dx$ is raised to a medium height $z/2$. Whence

$$dW_g = wz\ dx\ \Delta\rho g\ \frac{z}{2} \tag{45}$$

The area of the liquid-vapor interface formed is $w\ ds$, but an area $w\ dx$ of the liquid's horizontal surface disappears at the same time. Thus

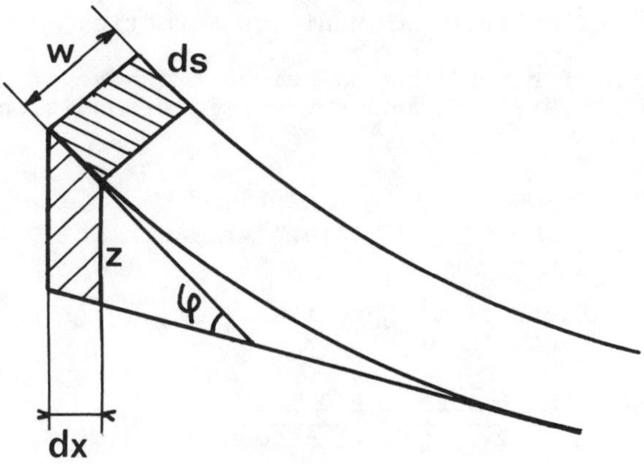

Fig. 15. Elementary slice of a cylindrical meniscus.

$$dW_s = w(ds - dx)\gamma \tag{46}$$

An examination of Fig. 15 leads to

$$\sin \varphi = \frac{dz}{ds} \tag{47}$$

Combining relations (45) and (46) with relations (37), (43), and (47), we obtain

$$dW_g = w \ \gamma \left(\frac{1 - \cos \varphi}{\tan \varphi}\right) dz \tag{48}$$

$$dW_s = w \ \gamma \left(\frac{1}{\sin \varphi} - \frac{1}{\tan \varphi}\right) dz \tag{49}$$

Equations (44), (48), and (49) lead to

$$dW_f = w \ \gamma \ dz \ \sin \varphi \tag{50}$$

or, considering θ, the angle formed by the profile of the meniscus and the vertical, that is, to say the contact angle,

$$dW_f = w \ \gamma \ dz \ \cos \theta \tag{51}$$

This expression is valid in the case of a meniscus situated below the surface. In effect, the term dW_s of work of surface extension has the same value, but the term dW_g could be considered to have a value opposite that of the preceding case. Actually it is not true at all: During the formation of the elementary slice, the liquid that was found at the site of the meniscus had to be raised to the level of the liquid's free surface. dW_g is positive in this case too.

2.6.5 Total Work of Cylindrical Meniscus Formation

The formation of a meniscus of height h (above or below the liquid's horizontal surface), under reversible conditions, requires a total work

$$W_f = \int_{h=0}^{h=h} \omega \gamma \sin \varphi \, dz \tag{52}$$

The parametric equations of the profile, (35) and (37), permit us to express dz as a function of φ:

$$dz = \frac{\sqrt{\gamma}}{\sqrt{2 \Delta\rho g}} \cdot \frac{\sin \varphi \, d\varphi}{\sqrt{1 - \cos \varphi}} \tag{53}$$

Under these conditions, Eq. (52) becomes

$$W_f = \frac{\omega \gamma^{3/2}}{\sqrt{2 \Delta\rho g}} \int_{\varphi=0}^{\varphi=\varphi} \frac{\sin^2 \varphi \, d\varphi}{\sqrt{1 - \cos \varphi}}$$

and after integration, with the condition that $0 < \varphi < \pi$,

$$W_f = \omega \gamma \sqrt{\frac{2\gamma}{\Delta\rho g}} \frac{2\sqrt{2}}{3} \left(1 - \cos^3 \frac{\varphi}{2}\right) \tag{54}$$

and, using the contact angle $\theta = \pi/2 - \varphi$,

$$W_f = \frac{\omega \gamma}{3} \sqrt{\frac{2\gamma}{\Delta\rho g}} \left[2 \sqrt{2} - (1 + \sin \theta)^{3/2}\right] \tag{55}$$

2.7 Axially Symmetrical Meniscii in the Gravitational Field

A drop placed on a plane surface, or a liquid column contained in a capillary tube, constitute usual examples of axially symmetrical meniscii. Such meniscii possess, on every point of their surfaces, two finite radii of curvature, and they can be described by a rotation of the profile around the vertical axis. Figure 16 represents several real cases of such meniscii. The cross-hatched sections correspond to a liquid of density ρ_1, and the non-cross-hatched sections to a gas, or a liquid of density ρ_2, which is less than ρ_1. These meniscii are connected with a horizontal solid surface, but, here again, the solid governs the contact angle, and in no way the shape of the fluid interface. It is impossible to associate an equation expressed in cartesian coordinates with each of these profiles. Nonetheless, in 1892, Bashforth and Adams gave a satisfactory mathematical solution to the problem. They established that any point S belonging to the surface of the meniscus (Fig. 17) verifies the relation

$$2 + \beta \frac{z}{b'} = \frac{1}{R'/b'} + \frac{\sin \phi'}{x/b'} \tag{56}$$

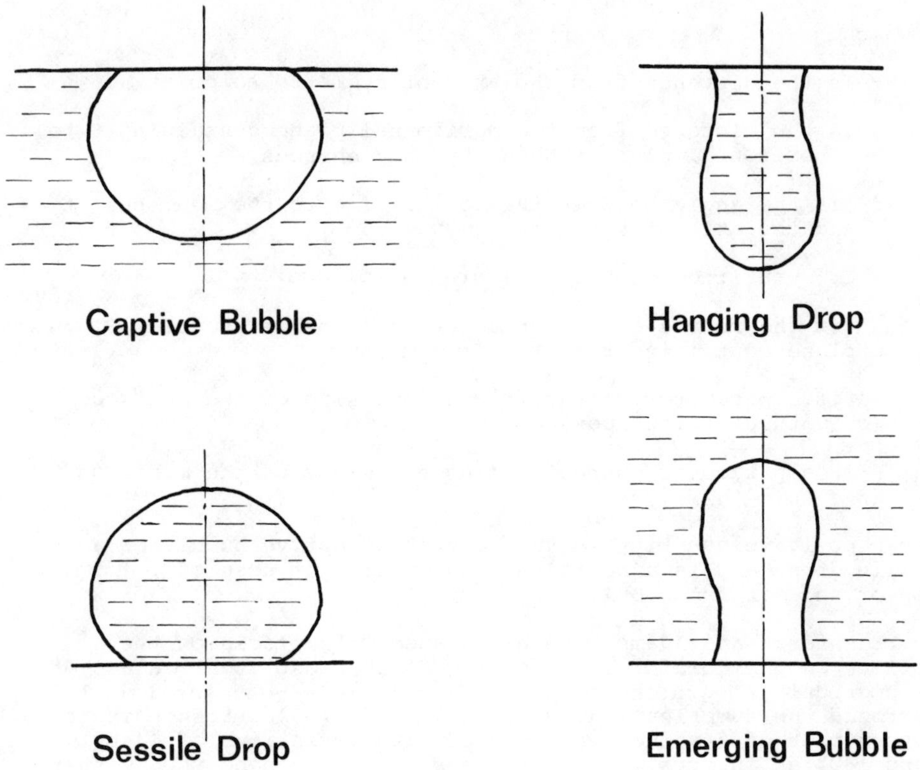

Fig. 16. Different shapes of axially symmetrical meniscii in the gravitational field.

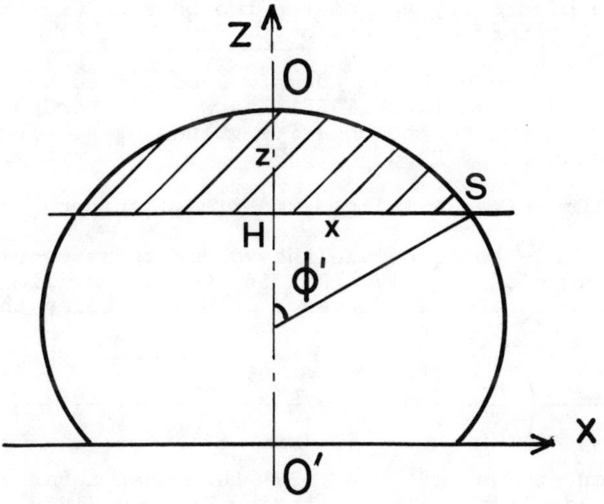

Fig. 17. Axially symmetrical meniscus. The volume of the cross-hatched zone is given in Eq. (57).

in which

x is the distance from the axis of symmetry to point S

z is the distance from the horizontal plane containing S to the horizontal plane tangent to the meniscus

ϕ' is the angle between the normal line to the meniscus at S and the axis of symmetry

The other quantities in Eq. (56) are as follows:

R' is the radius of curvature at the profile at point S, in a plane containing the axis of symmetry

b' is a parameter characterizing the size of the meniscus (curvature at its apex)

β is a parameter characterizing the shape of the meniscus ($\beta = \Delta\rho\, b'^2 g/\gamma$)

β is a positive quantity in the case of a captive bubble or a sessile drop, and is a negative quantity in the case of a hanging drop or an emerging bubble.

Bashforth and Adams have calculated with precision the profiles in the domain corresponding to the most usual values of β, by numerical integration of Eq. (56). In this monumental work are found the numerical values of x/b' and z/b' corresponding to a great number of β values, associated with variations of ϕ' between 0 and 180° at intervals of 5°. These results can be used either to reproduce the shape of the profile for known values of β and b', or, more often, to estimate the values of β and b' from measured values of x/b' and z/b' relative to a given meniscus.

Knowledge of gravity and of the two phases' densities permits us to calculate interfacial tensions.

The work of Bashforth and Adams was completed by Blaisdell (1940), Tawde and Parvatikar (1958), as well as Fordham (1948) and Mills (1953) relative to the negative values of β. The different tables were gathered by Padday (1969).

2.7.1 The Volume of an Axially Symmetrical Meniscus

It is possible to calculate the volume corresponding to the cross-hatched zone of Fig. 17, that is, the volume limited by a horizontal plane located at distance $OH = z$ (O being the apex of the meniscus).

$$V = \frac{\pi x^2 b'^2}{\beta}\left(\frac{1}{R'} - \frac{\sin\phi'}{x}\right) = \frac{\pi x^2 \gamma}{\Delta\rho g}\left(\frac{1}{R'} - \frac{\sin\phi'}{x}\right) \tag{57}$$

The combination of Eqs. (56) and (57) furnishes a variant of Bashforth and Adam's equation, cited by Princen (1969):

$$2 + \beta \frac{z}{b'} = \frac{\beta V}{\pi x^2 b'} + \frac{2 \sin \phi'}{x/b'} \qquad (58)$$

2.8 Influence of a Liquid's Curvature on Its Vapor Tension

At a given temperature, a plane liquid surface is in equilibrium with its vapor when the partial pressure of this latter is equal to the pressure of the liquid's saturating vapor, written p_o.

At the same temperature, a curved surface has a saturating vapor pressure of equilibrium p, different from p_o. p and p_o are linked by the relation

$$\log \frac{p}{p_o} = \gamma \left(\frac{1}{R_1'} + \frac{1}{R_2'} \right) \frac{V_m}{RT} \qquad (59)$$

in which

V_m is the liquid's molar volume

R is the universal gas constant

T is the temperature of the system

γ is the liquid's surface tension

R_1' and R_2' are the principal radii of curvature

For the case of a spherical surface of radius r, we have

$$\log \frac{p}{p_o} = \frac{2\gamma}{r} \frac{V_m}{RT} \qquad (60)$$

Equation (60) is frequently called the Kelvin equation (Thomson, 1871). Let us remember that r is positive if the liquid is on the concave side of the surface, and negative in the contrary case. Relation (60) shows that the vapor pressure of a drop of liquid is greater than that of the same liquid having a plane surface, and is all the greater as the drop is smaller. In the same way, the vapor pressure of a bubble enclosed by a liquid of the same nature as the bubble is lower than the saturating vapor pressure of the plane liquid, and is all the lower as the bubble is smaller. Relation (60) allows us to explain the phenomenon of capillary condensation in the pores of a solid, for a vapor pressure lower than p_o.

3 SOLID SURFACES

3.1 Surface and Volume Stresses

A solid is a piece of matter whose shape is determined. However, any solid becomes deformed by the action of forces applied to it.

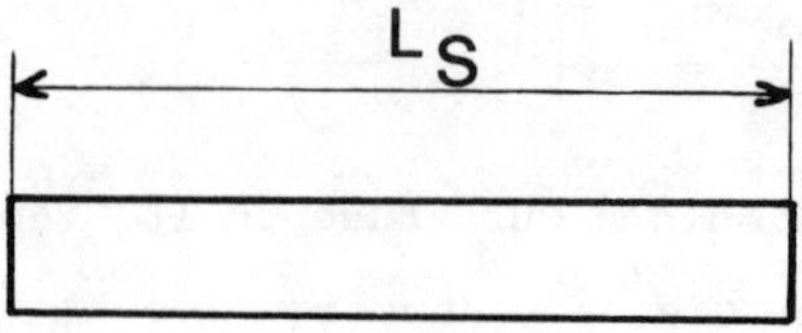

Fig. 18. Thin solid plate of length L_S.

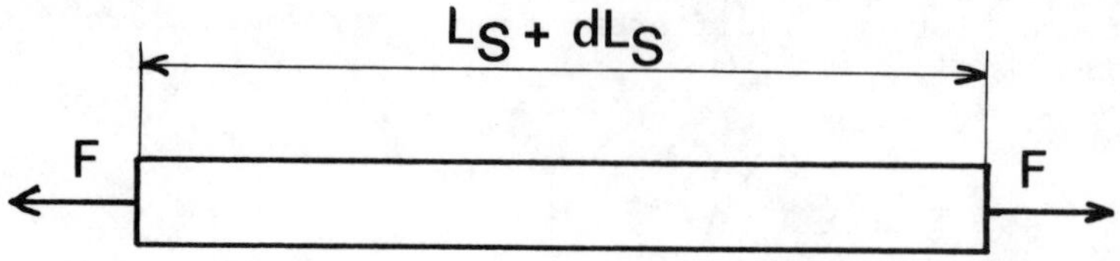

Fig. 19. When subjected to a traction force F, the plate lengthens by dL_S.

Let us consider a solid, thin plate of length L_S and of section S' (Fig. 18). We can subject it to a traction test by pulling at each of its extremities with a force F. The plate is thus subject to a "traction stress" equal to F/S'. The plate now has a length equal to $L_S + dL_S$ (Fig. 19). When the forces applied to the plate are removed, it does not regain its original length unless the exterior traction stress is less than a given value called "yield stress." In that case, we call the deformation purely elastic. One can then explain the plate's deformation by considering that each atom is slightly distanced from its neighbors in a direction parallel to the axis of force. In this direction, repulsive forces diminish between neighboring atoms. If we cut the plate with an imaginary plane, perpendicular to the axis of force, the interatomic forces that are exerted between the two parts are attractive forces for which the resultant is equal to force F.

These interatomic attractive forces are of two kinds:

1. There are those that are exerted between the atoms originally situated at the plate's interior. They are proportional to the section of the plate. At the interior of the solid we thus find volume stresses.

2. There are those forces that are exerted between the atoms originally situated in the superficial layers of the plate. They are proportional to the plate's perimeter. At the surface of the solid, there are thus surface stresses.

In general, the forces resulting from surface stresses are negligible when compared to those resulting from volume stresses. By contrast, in the case of extremely thin plates, the influence of surface stresses is measurable.

Surface stresses can be justified only by the existance of

superficial tensions at solid surfaces. The surface tension of a solid, γ, is also the work of formation, at a constant intensive state, of a unit of new surface. As such, it is a scalar. The surface stress measures the excess force required to stretch a solid when a surface is present. Since this excess force is dependant on direction. the general surface stress is a tensor. It has been shown (Shuttleworth, 1950; Herring, 1951; Mullins, 1963) that each component of surface stress is related to the surface tension γ and to strains.

Surface stresses of solids (of the order of magnitude 10^3 dyn cm^{-1}) are very small compared to volume stresses of solids (on the order of magnitude 10^{10} dyn cm^{-2}), and have little visual effect on solids: They do not influence the shape in the same way that surface tension does for liquids. With very minute particles, however, some such effects may be observed.

3.2 Superficial Tension of Solids

Some authors reject the notion of superficial tension for solids. In fact, it is difficult to obtain direct experimental evidence of the superficial tension of a solid. However, the existence of surface stresses at the surface of solids is one proof of their existance. What is more, Udin et al. (1949) have attributed the shortening of a thin copper wire, brought to a temperature slightly below the fusion point, to the superficial tension of the solid. Their estimated value of this tension is based on surface stresses deduced from the experiment. The result obtained, 1350 dyn cm^{-1}, is in agreement with that of the surface tension of liquid copper at a temperature slightly above the melting point (1300 dyn cm^{-1}).

For their part, Defay and Prigogine (1951) justify the existence of surface tensions at solid surfaces in the following way: A melted substance (for example, liquid glass), which cools progressively, manifests a very clear surface tension as long as it is in liquid form. Nothing shows that this surface tension disappears during the hardening process. This would therefore lead us to the belief that the solid glass has a surface tension.

The surface tension of a solid in vacuum is different from the surface tension of the same solid in a gaseous atmosphere. In the second case, we have the adsorption phenomenon (which we shall study in the following section). This signifies that there is a fixing of foreign molecules at the solid's surface. These molecules are subject to interaction forces from the atoms in the solid's superficial layers, and the forces between these atoms are themselves modified. The surface tension, being the macroscopic resultant of these forces, thus has a different value.

For the same reason, the surface tension of a solid in contact with a liquid will have a different value, called γ_{SL}, the interfacial tension of the solid-liquid interface.

In a crystal, the relative positions of atoms are different, according to the different crystalline planes. For the same crystal, the differently oriented faces thus have markedly different surface tensions. For example, Nicholas (1968) has shown that for latticed c.f.c. metals, $\gamma_{110} > \gamma_{100} \simeq \gamma_{111}$. However, the superficial tension is independent of the surface orientation in the case of amorphous compounds.

3.3 Measurement of Surface Tensions of Solids

The experimental obtainment of solid surface tensions requires us to vary the interfacial areas experimentally.

When the solid is in contact with a phase (for example, liquid) on one portion of its surface, and with another (for example, gaseous) on another portion, it is easy to vary the same quantity, simultaneously, of the areas of solid-liquid and solid-vapor interfaces. This possibility lets us experimentally obtain the difference $\gamma_{SV} - \gamma_{SL}$, in which γ_{SV} represents the solid's surface tension in the presence of vapor, and γ_{SL}, the solid-liquid interfacial tension. However, it does not let us obtain γ_{SV} or γ_{SL} separately. Most of that which follows in this text is devoted to the study of this kind of problem.

When the solid is in contact with only one phase, liquid or gaseous, its surface area variation necessitates a modification of its form. This implies bringing into play volume and surface stresses of the solid. The experimental methods for measuring the surface tension of solids are rather complex, and we shall not discuss them in detail in this chapter. We shall mention the following however:

The cleavage method is one of the most interesting, because what is truly measured is the surface tension γ. There are, however, numerous theoretical and experimental difficulties.

The shrinkage method consists of measuring the shortening of thin metal wires at temperatures between 80 and 100% of the melting temperature. This is the most precise method, but there are still theoretical difficulties in determining exactly what is measured.

The methods we have just cited are the most important ones. Linford (1973) presents and discusses them with great care. In the same text, other methods are mentioned. The numerical results obtained by any method have only an exactitude on the order of 20 to 50%.

3.4 Heterogeneity of Solid Surfaces

We shall see in Sec. 3.5 that solid surfaces are rough. A crystal thus presents different crystalline planes at different parts of its surface, with different surface tension values.

Metal alloys are of two kinds: There are "unlimited solid solutions" in which the grains have a continuous composition variation from their center toward the exterior; there are also "limited solid solutions" in which there are several kinds of grains of markedly different composition. In both cases, there exists a surface heterogeneity of the material.

Relatively pure metals (99—99.9%) contain grains of metal that are almost pure, whereas the impurities are assembled in the joints of the grains. Their surface is also heterogeneous.

Plastic substances are heterogeneous materials made up of a polymer and numerous additives. But even plastic substances made up only of a polymer have surface heterogeneity, because the structure presents crystalline zones in some areas and amorphous zones in others.

Moreover, the machine finishing of metallic surfaces considerably modifies the initial structure of the material. From the center of the piece up to its surface, one can distinguish three different zones in the case of a metallic surface obtained by machine finishing and polishing:

The undisturbed region at the core

A layer of metal (whose thickness is of the order of ten micrometers) made up of tiny, very deformed grains (work-hardened layer). In this layer, the deformation increases as one approaches the surface.

A completely disorganized zone (whose thickness is of the order of 1/100th of a micrometer), comprised of metal, oxyde fragments, abrasives, lubricants, etc. This layer is called "Beilby's layer."

In addition, all solid surfaces are covered by a layer of adsorbed molecules due to the adsorption phenomenon. This layer is usually heterogeneous.

3.5 Roughness of Solid Surfaces

Let us imagine a square of 1 cm per side, traced on the surface of a solid sample. The apparent area is equal to 1 cm^2, but the actual area is larger because there are hollows and asperities at the surface. Profiles of the samples can be obtained experimentally by different optical or mechanical methods, along chosen directions. Different criteria of roughness have been defined to obtain numerical values of roughness from these profiles, but statistical studies of the points of the profiles allow us to characterize the roughness of solids better.

Most machine-finished samples are nonergodic; this means that the profiles have different aspects according to the different directions at the surface of the sample. This is what Fig. 20 shows in the case of a ground sample.

Some kinds of machine finishing (sanding, for example) let us obtain ergodic surfaces. In this case the roughness criteria have the same values for profiles measured in any direction at the surface of the sample.

We present below the order of dimensions of distance h' separating the bottom of the hollows and the summit of the asperities,

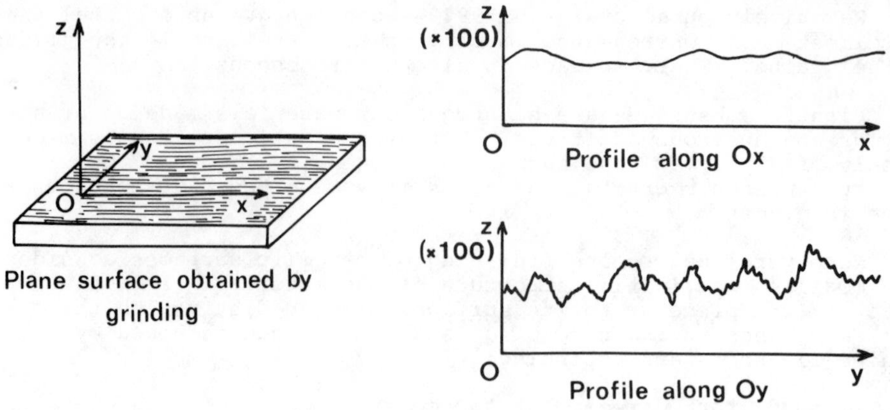

Fig. 20. Profiles along two perpendicular directions in the case of a ground sample.

according to the method of machine finishing.

Method of Machine Finishing	h' (μm)
Turning and drilling operations	5—50
Grinding	1—5
Polishing	about 1
Super finishing	0.05—0.5
Electrolytic polishing	0.01—0.1

Only cleavage of crystals following the crystalline planes lets us obtain rigorously smooth, homogeneous solid surfaces. Mica is one of the materials that can be cleaved the most easily; it is, however, difficult to get truly molecular planes.

4 ADSORPTION ON SOLID SURFACES

4.1 General Characteristics of Adsorption

4.1.1 Definition of Adsorption

Adsorption is the fixation of molecules, belonging to the phase with which the solid is in contact (gas or liquid solution), at the surface of the solid. Adsorbed molecules partially or completely cover the solid surface, and the adsorbed layer is always very thin; it is often a monomolecular layer (i.e., a covering of the solid by a single layer of foreign molecules), but it can also be multimolecular (of several molecular thicknesses only). The amount of adsorbed product on a massive sample is always extremely minute, and is measurable only if the absorbed product is radioactive. For example, if we consider an adsorbed layer with a uniform thickness of 10 Å on a sample with 10 cm^2 of real, total surface, the adsorbed product will occupy a volume equal to $10 \times 10^{-7} = 10^{-6}$ cm^3. Considering a volumic mass of 1 g cm^{-3}, the mass of adsorbed product is equal to one-millionth of a gram. The elucidation of these adsorption phenomena and their study were carried out with porous samples or solids in powder form (which represent a much greater surface than the massive samples).

4.1.2 Specific Surface Area of Solids

The specific surface area of a solid is the area of the solid surface corresponding to one unit of mass of material. It is generally expressed in $m^2\ g^{-1}$.

In the case of powders, the specific surface area is greater as the grains are smaller. If we imagine a powder composed of cubical grains, 1 mm per side, with a density equal to 1, the specific surface area is equal to $6 \times 10^{-3}\ m^2\ g^{-1}$. If the powder is composed of cubes of 1 μm per side, the specific surface area is then equal to $6\ m^2\ g^{-1}$. In the case of porous solids, the solid surface at the interior of the pores in communication with the exterior belongs to the real surface. Some porous solids have extremely large specific surface areas: between 100 and 1000 $m^2\ g^{-1}$ for activated carbon (that is charcoal that has had a particular treatment).

4.1.3 Demonstration of Adsorption Phenomena at the Solid-Gas Interface

Let us take a certain volume of carbonic gas, for example, in a test tube placed over a vat of mercury. We introduce a certain volume of granulated "active carbon" into the tube. We shall observe a rise of the mercury in the test tube and a more-or-less total disappearance of the carbonic gas (according to the respective quantities of gas and carbon present). The molecules of the gas have become fixed at the surface of the pores. We can adsorb on the order of 300 cm^3 of CO_2 per gram of carbon at ordinary temperature and under atmospheric pressure. In order to desorb the carbonic gas completely, we must combine the action of the vacuum and temperature. The same experiment can be executed with any other gas and any very porous solid (silicagel, bentonite, zeolite, etc.).

4.1.4 Demonstration of Adsorption Phenomena at the Solid-Liquid Interface

A solution of water containing fushine, a red dye, is placed in contact with a given quantity of activated carbon. After intimate mixing, the solution becomes colorless, indicating that the fushine has been completely adsorbed at the surface of the activated carbon. If decoloration is only partial, a supplementary quantity of activated carbon must be added.

4.1.5 The Adsorption Mechanism

When a molecule strikes a solid surface, it does not always rebound elastically. It may be kept at the surface. In fact, forces are exerted between the solid atoms and one molecule that is in the proximity of the solid surface. These forces, of van der Waals type, have a resultant that is a repulsive force when the molecule is extremely near the solid's superficial atoms, and an attractive force when the molecule distances itself from this position. In order to escape from the attractive forces due to the solid's atoms, the molecule must have an energy greater than a given value ε, which represents the interaction energy of the molecule with the solid.

In the adsorbed state, the molecule vibrates with a pseudo-frequency ν. During its movements, the molecule may acquire a supplementary energy resulting from collisions with its neighbors, and if its energy becomes greater than the value ε, it may return to its original phase. The molecule will then be called "desorbed."

With each oscillation, an adsorbed molecule has an energy whose distribution is aleatory and may be considered Maxwellian (De Boer, 1953). The probability of desorption with each oscillation is equal to

$$e^{-\varepsilon/kT} \tag{61}$$

where k = Boltzmann's constant

T = temperature of the system in K

There are ν oscillations per unit of time. The chance of desorption during a unit of time is thus

$$\nu e^{-\varepsilon/kT} \tag{62}$$

The mean duration of adsorption of the molecule is thus expressed by the relation

$$\tau_a = \frac{1}{\nu}\, e^{+\varepsilon/kT} = \tau_o\, e^{E/RT} \tag{63}$$

where

τ_o is the pseudo-period of vibration of the molecule in the adsorbed state

E is the energy of desorption per mole ($E = N\,\varepsilon$), N being Avogadro's number

R is the universal gas constant ($R = N\,k$)

τ_a will be called the "adsorption time."

4.1.6 Physical Adsorption or Physisorption

When the interaction forces between the molecules and the solid surface are solely electrostatic forces (van der Waals forces), there is physical adsorption. Desorption energy equals approximately 10 kcal mole^{-1} if the molecule is very polar, and about 1 kcal mole^{-1} if the molecule is not polar. Using the relation

$$\tau_a = \tau_0 \ e^{E/RT} \qquad (63)$$

and taking $\tau_0 = 10^{-13}$ s the calculation gives us, in the case of a system at 25°C, adsorption times equal to 10^{-6} s if the molecule is polar, and 10^{-12} s if it is not polar.

These adsorption times seem very small on a human scale, but they are nonetheless 10 to 10^7 times longer than the value of 10^{-13} s corresponding to the staying time at the solid's surface of a nonadsorbed molecule during its specular reflection. This explains why there is an accumulation of molecules of the phase in contact at the solid's surface. The adsorbed layer is made up of molecules that interchange continually with those of this phase. At equilibrium, during a given time, as many molecules get adsorbed as are desorbed, and the number of adsorbed molecules on each surface unit of the solid is statistically constant. Let us note that the mean adsorption time of adsorbed molecules increases if the temperature of the system diminishes. Physical adsorption is promoted by a temperature diminution of the system.

For a given product, adsorption (that is, the quantity of molecules adsorbed) is larger as the concentration of this product in the phase in contact is larger. If this concentration decreases, adsorption decreases. Physical adsorption is a reversible phenomenon.

4.1.7 Chemical Adsorption or Chemisorption

We know that nickel and oxygen interact chemically to form nickel oxide, NiO. The energy of the bond is high (116 kcal $mole^{-1}$), but the reaction does not take place unless the molecules' initial energy is greater than the activation energy of the reaction. That is to say, the reaction is promoted by an increase in temperature.

Let us now consider an oxygen molecule that is physically adsorbed on a clean nickel surface. If the oxygen molecule has, at a given moment, enough energy, there may take place the formation of true chemical bonds (with electron sharing) between the two oxygen atoms of the molecule and the nickel atoms of the solid's surface. We then say that the oxygen molecule is chemically adsorbed. Chemical adsorption is in no way a general phenomenon. It can take place only if the solid's atoms and those of the phase in contact are capable of reacting chemically between themselves. Chemical adsorption is promoted by an increase of the system's temperature, and it is always limited to a monolayer. The bonding energy is high (20 to 100 kcal $mole^{-1}$). Calculation thus gives an adsorption time greater than 10^{13} s. Chemical adsorption is thus considered to be an irreversible phenomenon. The molecules are very difficult to remove and desorption, if it takes place, is often accompanied by chemical changes.

4.2 Adsorption of Gases on Solids

Because the specific surface area of porous or finely divided solids is so high, it is relatively easy to measure the amount of gas adsorbed at different pressures, by gas-volume measurements. If the specific surface of the solid used is known, we can calculate the adsorption Γ, that is, the number of adsorbed molecules on each unit of area of the solid surface.

By plotting on a graph the experimental values of Γ, corresponding to measurements taken at the same temperature and at different pressures, we can get the curve $\Gamma = f(P)$, which is called the adsorption isotherm.

These isotherms are always limited by the saturating vapor pressure value of the gas at the temperature at which the experiment is performed. In fact, if we compress the gas at a pressure greater than the value p_o, there will be a condensation of this latter on the solid and on the sides of the vessel.

The isotherms of adsorption of a pure gas on a solid usually have the aspects represented in Fig. 21 and 22. We consider the isotherms of the type in Fig. 21 to be roughly characterized by a monotonic approach, to a limiting adsorption that presumably corresponds to a complete monolayer. In isotherms of the type shown in Fig. 22, multilayer adsorption occurs. In some rarer cases, the experimental isotherms have different forms from those of Figs. 21 and 22. All the forms of isotherms have been cataloged by Brunauer (1945), who enumerated five kinds.

By supposing that adsorption was limited to a single layer of molecules, Langmuir (1918) showed that the isotherm of adsorption had the equation

$$\theta_a = \frac{BP}{1 + BP} \tag{64}$$

in which B is a constant, and θ_a represents the fraction of solid surface covered by an adsorbed layer. This equation corresponds

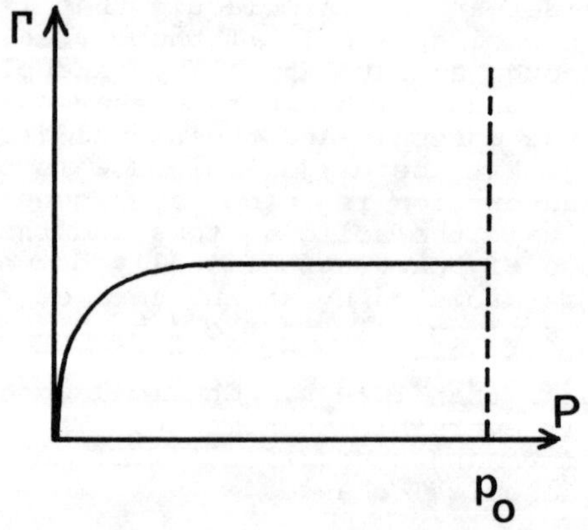

Fig. 21. Typical shape of a Langmuir isotherm (monolayer adsorption).

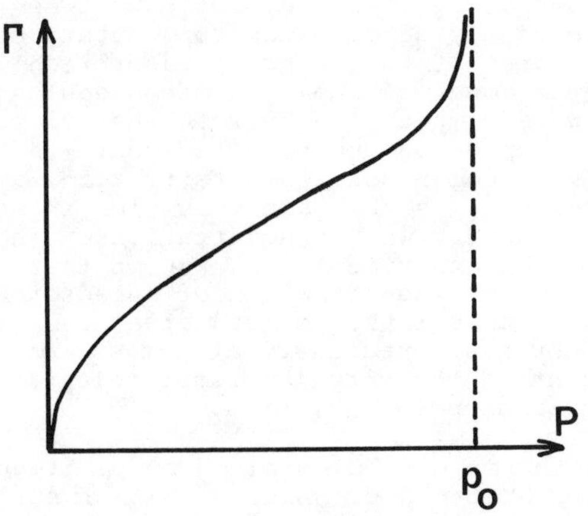

Fig. 22. Typical shape of a B.E.T. isotherm (multilayer adsorption).

to the isotherm of Fig. 21 and is representative of homogeneous surfaces. Usually, however, the surface shows sufficient heterogeneity to cause the heat of adsorption to decrease as the coverage increases. The Freundlich (1926) isotherm is obtained if an exponential distribution of adsorption heats is assumed. The Freundlich isotherm can be written in the form

$$\theta_a = K'P^{1/n} \tag{65}$$

where K' and n are constants, with $n>1$. This last equation has been found to represent many cases of chemisorption rather well.

Brunauer, Emmet, and Teller (1938) showed how to extend Langmuir's approach to multilayer adsorption, and their equation has come to be known as the B.E.T. equation. The isotherm of Fig. 22 agrees with the results of the B.E.T. theory.

4.3 Adsorption from Solutions

An adsorbed layer is often described as the coverage of the surface by joined spheres of one or several layers. An adsorbed layer corresponds to a compact arrangement of molecules as in liquids or solids, because the effect of molecular interaction forces prevails over that of the kinetic energy of molecules. If the phase in contact with the solid is itself in a dense state, the volume variation due to the adsorption phenomenon will be extremely small and impossible to measure in practice. No results exist concerning the adsorption of a pure liquid while this latter is in contact with the solid. We can only bring the adsorbed layer into evidence after having separated the liquid from the solid, in the case where the adsorption time of the liquid molecules is sufficiently long, that is, if there has been chemical adsorption.

With a solution containing solutes at low concentrations,

however, changes in the degree of adsorption between the components of the solution are easily detected as concentration changes in the solution. It is clear that this kind of adsorption is always equivalent to a replacement of one kind of molecules by another kind: There is always competition between the solutes and solvant. With a solution made up of one nonpolar solvent (an alkane, for example), and one very polar solute (a fatty acid, amine, etc.), one can form adsorbed layers that are very rich in solute, even when this latter is in very weak concentration in the solution. Bigelow et al.'s (1946) experiments have shown that only a few seconds are necessary for the formation of an adsorbed film of solute molecules when this latter exists at a mass concentration of 10^{-3}, and several hours are necessary at a mass concentration of 10^{-6}. We now understand the very important role played by the polar impurities of a nonpolar solution.

Adsorption isotherms can be obtained by plotting adsorption (number of molecules adsorbed per square unit of surface area) as a function of the concentration of solute in the solution. The adsorption of substances in solution does not usually seem to lead to layers whose thickness is more than one molecule. In many cases, the given experimental quantities can be easily represented by an empirical isotherm proposed by Freundlich in 1909 (see Freundlich, 1926):

$$\frac{m}{m_a} = Kc^{1/n} \tag{66}$$

where m and m_a are, respectively, the masses of adsorbed substances and adsorbant, c is the concentration of the solution when equilibrium is reached, and n is an empirical constant generally greater than 1.

When the solute is an electrolyte, the above equation is no longer valid, because the adsorption of ions is linked with more complex electrical phenomena.

Let us note, finally, that in some cases, the solute concentration of the adsorbed layer is weaker than the concentration of solute in the solution (for example, in the case of a solute that is less polar than the solvent). We then say that we have a negative adsorption.

It is also important to note that in most cases we never make an adsorbed film containing one kind of molecules only. Adsorption is a factor of heterogeneity.

4.4 Modification of the Surface Tension of Solids by Adsorption

The fixing of foreign molecules on a solid modifies its surface tension. The surface tension γ_{SV} of a solid in equilibrium with a vapor is different from the surface tension γ_{SO} of the same solid surface placed in a vacuum.

Bangham and Razouk (1937) have extended Gibbs' adsorption

equation to solid-gas interfaces. By likening the pressure and the fugacity of gas (the domain of weak pressures), they finally get the relation

$$\gamma_{SO} - \gamma_{SV} = RT\int_0^P \Gamma(P)\, d(\ell n\, P) \tag{67}$$

in which

R is the universal gas constant

T is the constant temperature of the system

P is the pressure of the gas

$\Gamma(P)$ is the quantity adsorbed at pressure *P*, expressed in moles per unit of surface area

Equation (67) was also demonstrated by Neumann (1974).

Since all the quantities located in the second member of Eq. (67) are positive, the quantity $\gamma_{SO} - \gamma_{SV}$ should also be positive. This signifies that the surface tension of a solid in equilibrium with its vapor (γ_{SV}) is weaker than the surface tension of the same solid in vacuum (γ_{SO}) when there is adsorption of the vapor. There is no vapor adsorption in the cases where such might lead to an increase in the solid's surface tension.

We note that the decrease of surface tension due to the adsorption phenomenon can be calculated when the adsorption isotherm of the gas on the surface $\Gamma = f(P)$ is known.

The quantity $\gamma_{SO} - \gamma_{SV}$ is usually called "spreading film pressure" and is then written Π_e:

$$\Pi_e = \gamma_{SO} - \gamma_{SV} \tag{68}$$

The idea of spreading film pressure was introduced at the time of the study of films made of insoluble compounds at the surface of water (that is, a water surface separated into two parts by a mobile barrier; at the side where the film is located, the barrier is subject to a measurable force that is the surface pressure of the film). Also, even though it is an image, one can consider that the adsorbed film at the solid's surface has a surface pressure Π_e.

5 CONTACT ANGLES ON SMOOTH AND HOMOGENEOUS SOLID SURFACES

We shall limit ourselves, in this section, to the study of wettability problems for "ideal" solid surfaces, that is perfectly planar solid surfaces for which the superficial tension is the same at all points.

5.1 Young's Model

In 1805, Young (1855) proposed a relation that is now classic to explain the behavior of a drop of liquid on an ideal solid surface. The first demonstration, based on a mechanical analysis of the problem, supposes that the three-phase separating line (solid, liquid, and vapor phases) is in equilibrium because of the action of the three interfacial tensions γ_{LV}, γ_{SV}, and γ_{SL}, which are exerted respectively in the planes of the liquid-vapor, solid-vapor, and solid-liquid interfaces (Fig. 23). The projection of these tensions upon the plane of the solid's surface leads to the classic relation

$$\gamma_{SV} - \gamma_{SL} = \gamma_{LV} \cos \theta \tag{69}$$

The projection on an axis perpendicular to the surface, makes a nonnull component appear, equal to $\gamma_{LV} \sin \theta$. Because of the existence of this component, the validity of Young's equation has often been doubted (Bikerman, 1967; Pethica and Pethica, 1967).

5.1.1 Thermodynamic Justification of Young's Equation

The thermodynamic demonstration of Young's equation was done by Johnson (1959). His treatment, which we shall not produce here, is a rearrangement of Gibbs' treatment (relative to the internal equilibrium of heterogeneous fluid masses in the gravitational field) to a sessile drop at equilibrium.

The mechanical equilibrium of the system involving a drop on a plane surface in the gravitational field leads to three equations: The first expresses Laplace's law for all points of the liquid-solid interface; the second expresses the difference of hydrostatic pressure between two points of different heights in the same phase; and the third is none other than Young's equation, expressed by means of the thermodynamic quantities γ_{SV}, γ_{SL}, and γ_{LV} and corresponding to tensions. Two other equations should come into play when the plane is not horizontal.

This demonstration imposes the fact that, even if Young's vectorial model was not well founded, his conclusions, written in the form of equation (69), was exact.

5.1.2 Quantities that May Come into Play in Young's Equation

In a general manner, Young's equation can be expressed equally

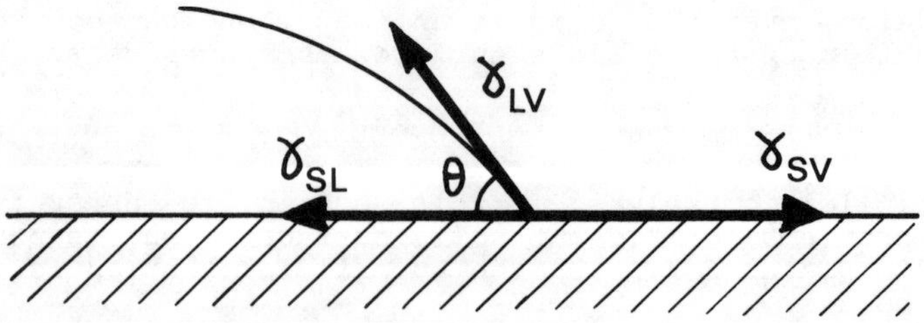

Fig. 23. A liquid drop on a solid. Each unit of length of the three-phase line is subjected to three interfacial tensions.

by means of the following quantities:

Surface tensions or interfacial tensions

Reversible surface formation works

Quantities of the type

$$\gamma_{SV} = \left(\frac{\partial G}{\partial A_{SV}}\right)_{T,V_i,\mu_i} \quad (70)$$

which, written in this form, constitutes the equation known as Young-Dupré's equation

On the other hand, in Young's equation, we cannot use quantities different from those defined before, such as the following quantities:

Surface specific energies: $e_s = Ts + \gamma + \Sigma\ \mu_i\ \Gamma_i$ (71)

Surface specific free energies: $g_s = \gamma + \Sigma\ \mu_i\ \Gamma_i$ (72)

Equations 71 and 72 are taken from Defay and Prigogine (1951).

5.2 Vectorial Justification of Young's Equation (Chappuis, 1974, 1977)

If it were possible to find a force that would equilibrate the vertical componant γ_{LV} sin θ, Young's vectorial model would be perfectly correct, and would cause no more controversy. To achieve this result, we suggest consideration not only of the edge's equilibrium, but that of the entire system: a sessile drop on a supposedly ideal and horizontal surface. Thus the drop consists of an axially symmetrical meniscus, and, if it is subject to the action of a gravitational field, its profile is described by Bashforth and Adams' equations (see Sec. 2.7).

Let us imagine a drop, in equilibrium on a solid horizontal surface, with a contact angle θ. The contact surface is a circle of radius x'. Let us take γ as the liquid's surface tension, ρ_L and ρ_V as the densities of the liquid and the vapor, and Δρ as the quantity $\Delta\rho = \rho_L - \rho_V$.

Figure 24 represents a diametrical section of the drop. Let R' be the radius of curvature of the profile in the neighborhood of the solid. Equation (57) gives us the drop's volume:

$$V = \frac{\Pi\ x'^2\ \gamma}{\Delta\rho\ g}\left(\frac{1}{R'} - \frac{\sin\ \theta}{x'}\right) \quad (57)$$

The apparent weight of the drop is

$$w_a = \Delta\rho\ g\ V \quad (73)$$

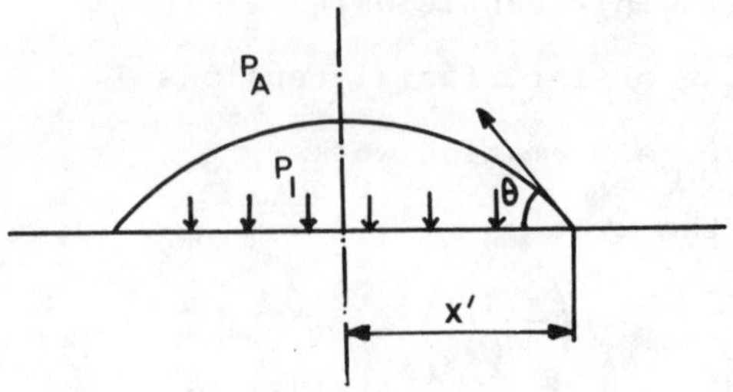

Fig. 24. Diametrical section of a sessile drop.

that is; $w_a = \pi x'^2 \gamma \left(\frac{1}{R'} - \frac{\sin\theta}{x'}\right)$ (74)

Laplace's equation lets us express the pressure P_I at the interior of the drop, in the neighborhood of the solid surface; if P_A is the atmospheric pressure;

$$\Delta P = P_I - P_A = \gamma \left(\frac{1}{R'} + \frac{\sin\theta}{x'}\right) \quad (75)$$

Over the entire surface that is common to the liquid and the solid, the solid opposes itself to the pressure due to the presence of the drop, by exerting a force F equal to

$$F = \pi x'^2 \Delta P = \pi x'^2 (P_I - P_A) \quad (76)$$

that is;

$$F = \pi x'^2 \gamma \left(\frac{1}{R'} + \frac{\sin\theta}{x'}\right) \quad (77)$$

F is greater than the weight of the drop by a quantity equal to $F - w_a$:

$$F - w_a = \pi x'^2 \gamma \left(\frac{2\sin\theta}{x'}\right) = 2\pi x' \gamma \sin\theta \quad (78)$$

The total reaction exerted by the solid on the drop should be equal to the apparent weight of the drop. In fact, F represents the reaction that a solid exerts on the entire surface common to the drop and the solid, but the solid exerts another reaction that can be applied only on the three-phase line. This new reaction, which, by reason of symmetry, should be spread equally along the line joining the liquid to the solid, is directed toward the interior of the solid, and has a vertical component equal to $2\pi x' \gamma \sin\theta$, that is

$$\frac{2\pi x' \gamma \sin\theta}{2\pi x'} = \gamma_{LV} \sin\theta \quad (79)$$

by unit of length of the joining line. It is this component that counterbalances the nonnull component of Young's model. The equilibrium vectorial model of the edge of a sessile drop, with which we shall be concerned hereafter, is represented in Fig. 25. Reaction R_S is delineated there perpendicular to the surface, as it should be in all cases. If it were not so, the projection in the horizontal plane of all the forces exerted on the edge of the drop could not lead to Young's equation, which has also been demonstrated by thermodynamic considerations.

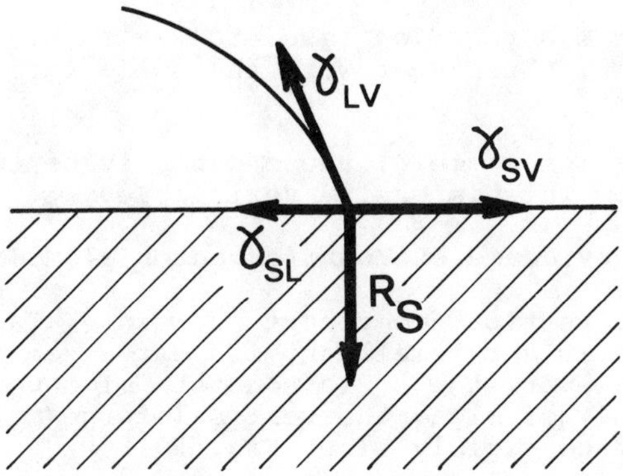

Fig. 25. Vectorial equilibrium of the edge of a drop.

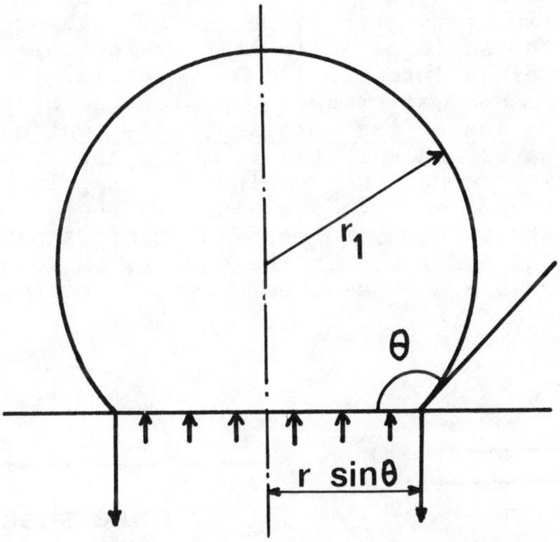

Fig. 26. Diametrical section of a spherical drop (in the absence of gravity).

5.2.1 Validity of Young's Vectorial Equation in the Absence of Gravity

In the absence of gravity, the drop is part of a sphere of of radius r_1. If it forms an angle θ with the solid surface, the area of the surface common to the solid and liquid equals $\pi r_1^2 \sin^2 \theta$ (by geometric considerations of Fig. 26):

The solid exerts a reaction equal to $\pi r_1^2 \sin^2 \theta \, \Delta P$, that is $\pi r_1^2 \sin^2 \theta \, 2 \gamma / r_1$, across the solid-liquid interface.

The solid exerts a reaction equal to $2 \pi r_1 \sin \theta . \ \gamma \sin \theta = 2 \pi r_1 \gamma \sin^2 \theta$, directed toward the solid interior, on the edge of the drop.

In the absence of a force field, these can only be two equal and opposite reactions. This is easily verifiable.

5.3 Experimental Evidence of Young's Vectorial Model

Concerning the vertical component $\gamma_{LV} \sin \theta$, Tabor (1969) states: "The force is very small and its effect on a solid is negligible. If, however, the solid consists of a very thin flexible sheet, e.g., of mica, the vertical force may be sufficient to distort the surface visibly" (see Fig. 27).

Of course, to give experimental proofs of this model, we cannot use nondeformable solids, for we would see nothing. For our part (Chappuis, 1977), we have chosen, as a very deformable solid surface, the thin, solid skin that appears on solidifying melted paraffin, and we have placed on it a sessile drop of water. This method has two advantages: First, we are sure to get a solid-liquid interface; when the solid paraffin skin is too thin, it breaks and the drop of water sinks into the liquid paraffin, the density of which is lower. Second, after having placed the drop, we wait for complete solidification of the paraffin, after which we can withdraw the drop of water and carefully observe the solid surface, possibly with a microscope. Concerning the experimental proceedure, we place the water droplet with a syringe. The temperature of the water must be slightly below the melting point of paraffin to avoid a local rapid solidification. In the cavity, there is a very circular ridge, visible to the naked eye, and located at the previous

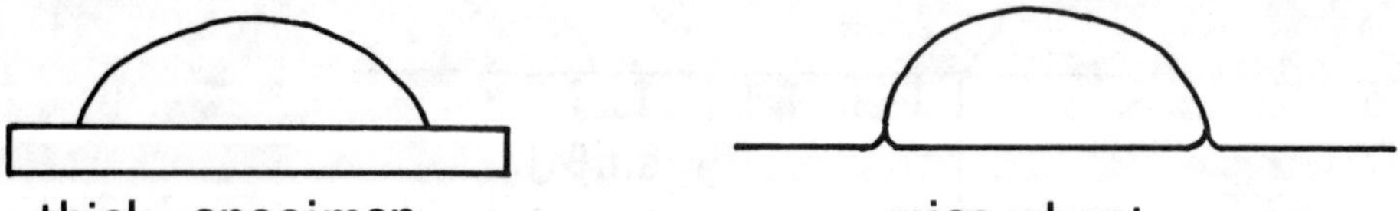

Fig. 27. The vertical component of the tensions has a negligible effect on a thick solid but distorts the shape of a thin mica sheet.

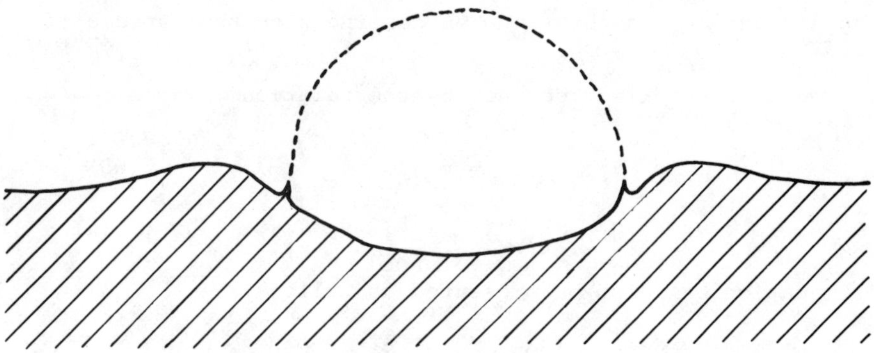

Fig. 28. A sessile drop of water is placed on the thin solid skin that appears on solidifying melted paraffin. After complete solidification of the paraffin and withdrawal of the drop, there is a circular ridge at the previous site of the edge of the drop.

site of the edge of the drop (see Fig. 28). We also obtained sectional views by quickly cutting the paraffin with a razor blade. The height of the ridge was about 0.1 mm. We have made a mathematical study, and the general profile of the paraffin can be explained if we take as a model a thin, elastic, permanent layer on a viscous, noncompressible fluid. Under such conditions, the ridge can be explained only as a local reaction of the solid.

In the case of highly deformable solids, Lester (1961) proposed a vectorial equation that can be substituted for Young's equation. In his model and because of the ridge, the solid-liquid and solid-vapor interfacial tensions are no longer in the initial plane of the solid.

By contrast, each time that the solid deformation is negligible, Young's equation is perfectly correct.

5.4 Contact Angle of a Drop of Liquid on a Clean, Solid Surface, in a Vessel Containing No Other Gas Than the Liquid's Vapor

At equilibrium, the contact angle is governed by Young's equation:

$$\gamma_{SV} - \gamma_{SL} = \gamma_{LV} \cos \theta \qquad (69)$$

Hence:

$$\cos \theta = \frac{\gamma_{SV} - \gamma_{SL}}{\gamma_{LV}} \qquad (80)$$

However, in that the solid surface in contact with the vapor is covered by an adsorbed layer of this vapor, we must consider γ_{SV}

in Young's equation and not γ_{SO} when dealing with the surface of the solid.

γ_{SV} and γ_{SO} are tied together by the relation

$$\gamma_{SO} - \gamma_{SV} = \pi_e \tag{68}$$

Thus: $\gamma_{SO} = \pi_e + \gamma_{SV}$ (81)

Young's equation can also be presented as follows:

$$\cos\theta = \frac{\gamma_{SO} - \gamma_{SL} - \pi_e}{\gamma_{LV}} \tag{82}$$

The presence of an adsorbed film raises the value of the contact angle. The existence of a film of gaseous structure (very mobile molecules at the solid surface), in two dimensions, gives us a fairly concrete image of the spreading pressure π_e that the adsorbed film exerts, and thus of its tendency to repel the drop and to increase the contact angle. Figure 29 represents the vectorial equilibrium of the edge, in relation with Eq. (82). Whether or not an adsorbed film exists at the solid's surface, γ_{SV}, γ_{SL}, and γ_{LV} should have determined values at equilibrium. Consequently, θ, defined by Eq. (82), can take only one value, whatever the volume of the drop.

5.5 Adhesion Tension; Work of Wetting

Although it is very difficult to measure γ_{SV} and γ_{SL} separately, it is very easy to measure their difference $\gamma_{SV} - \gamma_{SL}$ experimentally. Let us consider a vertical plate, partially immersed in a liquid. When equilibrium is reached, the different surface tensions γ_{SV}, γ_{SL}, and γ_{LV} impose an angle θ, verifying Young's equation.

Let us suppose that $\theta < \pi/2$, as it is in Fig. 30. The different exterior forces being exerted on the plate are forces of gravity, buoyancy forces, and surface forces. By surface forces we mean the forces situated at the interior of the liquid's surface, and by which this latter exerts a traction on each of the solid's vertical faces. Their resultant is vertical, directed downwards when $\theta < \pi/2$, and equal to $\ell\,\gamma_{LV}\cos\theta$ (ℓ being the perimeter of the plate).

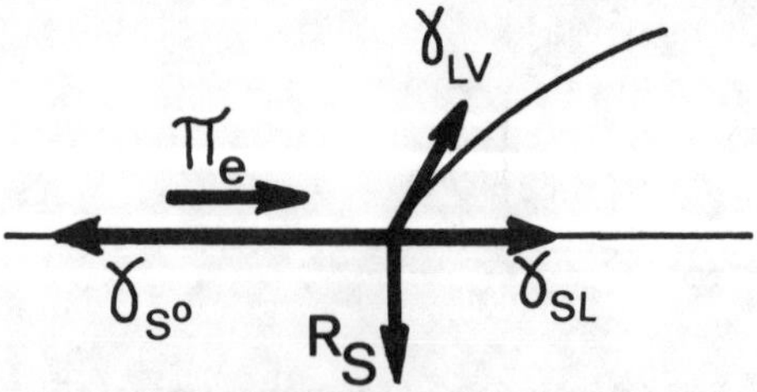

Fig. 29. Vectorial equilibrium of the edge of a drop when there is an adsorbed film at the solid-vapor interface.

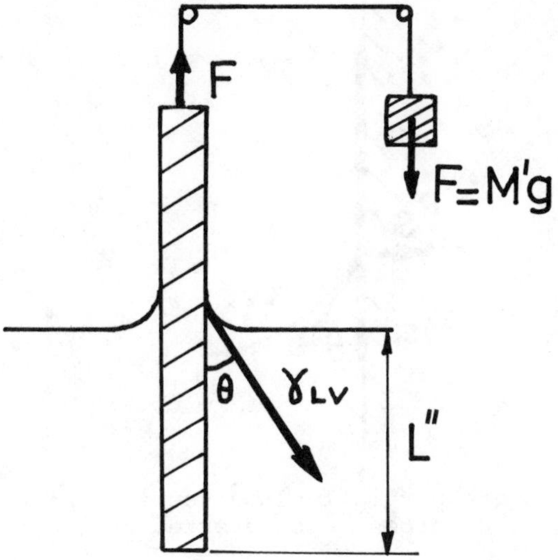

Fig. 30. Vertical plate partially immersed in a liquid: mechanical equilibrium.

If the plate is extremely thin, the gravitational forces and buoyancy forces are negligible, relative to surface forces. The plate equilibrium requires the intervention of the operator; he should exert a force

$$F = \gamma_{LV} \cos \theta \, \ell \quad (= M'g, \text{ for example}) \tag{83}$$

The forces being exerted on the three-phase line are represented in Fig. 31. The plate plunges into the liquid, but the liquid meniscus remains identical to itself. The line joining the liquid to the solid has not moved in relation to the liquid. The liquid's surface tension γ_{LV} has not furnished any work. The line joining the liquid to the solid, however, has been displaced from $dL"$ in relation to the surface of the solid. Since the point of application of the surface tensions γ_{SV} and γ_{SL} and of reaction R_S has moved, these forces have furnished work. That of R_S is null because its direction is perpendicular to the displacement. That of the solid-liquid interfacial tension is resistant work: Its direction is opposed to that of the displacement of the point of application. It is equal to $-\ell \, \gamma_{SL} \, dL"$. That of the solid-vapor interfacial tension is motor work, since its direction is that of the displacement of the point of application; it is equal to $\ell \, \gamma_{SV} \, dL"$.

Upon a reversible displacement $dL"$ the work corresponding to the wetting of the plate (that is, the work of wetting) is equal to $\ell(\gamma_{SV} - \gamma_{SL})dL"$. It serves to raise the mass M' by a height $dL"$, which represents work against gravity equal to $M'g \, dL"$. So we

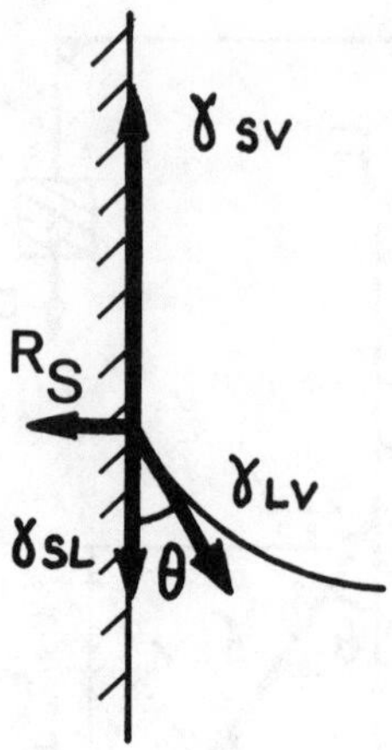

Fig. 31. Vertical plate partially immersed in a liquid: representation of the forces being exerted on the three-phase line.

obtain

$$\ell\ (\gamma_{SV} - \gamma_{SL})\ dL = M'g\ dL'' \tag{84}$$

Using the value of $M'g$ given in Eq. (83):

$$\ell\ (\gamma_{SV} - \gamma_{SL})\ dL'' = \ell\ \gamma_{LV} \cos\theta\ dL'' \tag{85}$$

which, after simplication, leads to Young's equation. It is thus possible to measure the work of wetting or dewetting dW of an area dA of the solid surface. This work is equal to

$$(\gamma_{SV} - \gamma_{SL})\ dA$$

We can thus obtain the quantity

$$\frac{dW}{dA} = \gamma_{SV} - \gamma_{SL} \tag{86}$$

This quantity is usually represented by the symbol τ, and is designated the "tension of adhesion,"

$$\tau = \gamma_{SV} - \gamma_{SL} \tag{87}$$

and, according to Eq. (69),

$$\tau = \gamma_{LV} \cos\theta \tag{88}$$

Authors rejecting the notion of superficial tension of solids cannot interpret this experiment: In that they do not accept the existence of superficial forces, they do not know to what force they should attribute the work associated with dewetting of a surface; they admit, however, that it cannot be the liquid's superficial tension that furnishes work, because its point of application is not displaced.

5.6 Work of Adhesion; Work of Cohesion

Given the following experiment: Let us consider a liquid whose free surface slightly exceeds the edges of the vat containing it. Let us take a solid, flat plate (a thin sheet of aluminium, for example), and let us put it on the liquid's surface (Fig. 32). If the plate is no longer held back, it begins to move spontaneously in the direction that tends to bring about a contact of the total inferior surface of the plate with the liquid. Whatever the solid and liquid used, the observed displacement always takes place in the same direction.

It is possible to operate under conditions fairly close to reversibility by using the apparatus represented in Fig. 33. A weight of mass M' equilibrates the resultant of forces applied to the plate by means of a frictionless pulley.

Let us take w' as the width of the plate, L as its length in

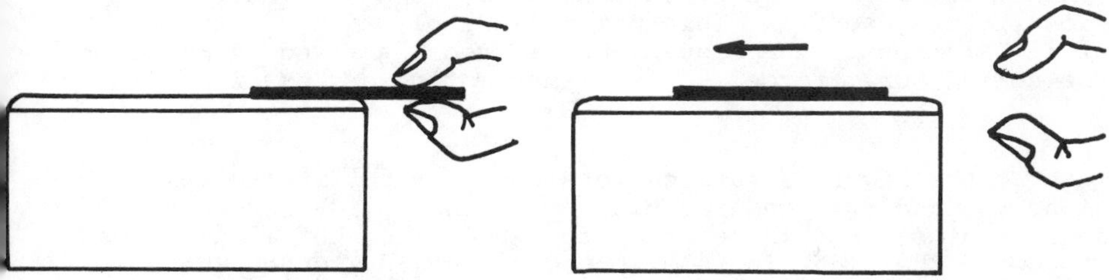

Fig. 32. A solid thin plate partially in contact with a horizontal liquid surface moves spontaneously in the direction that increases the area of the solid-liquid interface.

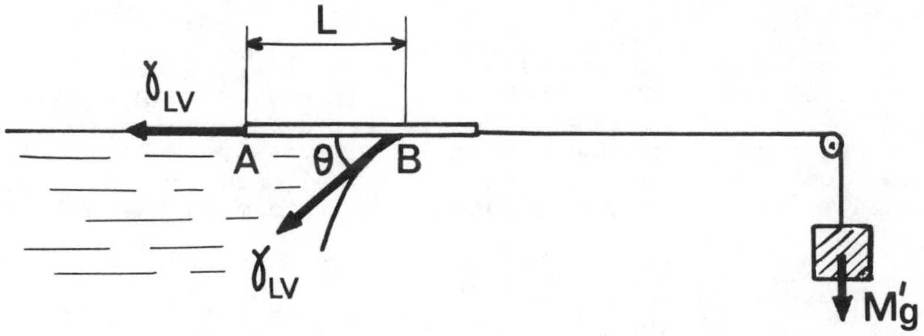

Fig. 33. Thin horizontal plate with one of its faces partially in contact with a liquid surface: mechanical equilibrium.

contact with the liquid, and θ as the angle joining the liquid to the solid on the inferior face of the plane. The exterior forces that are exerted on the plate are the liquid's surface tension along the line of junction labeled B in Fig. 33; and the liquid's surface tension along the extremity of the plate labeled A in Fig. 33. We shall disregard the gravity and buoyancy forces.

Weight $M'g$, which should equilibrate the horizontal resultant of the exterior forces, is

$$M'g = w' \gamma_{LV} (1 + \cos \theta) \tag{89}$$

Let us consider a reversible horizontal displacement of the plate, which increases by dL, the length of contact of the plate and the liquid. At the time of this displacement, the work dW furnished to the exterior (work against gravity for the weight $M'g$) is equal to $M'g\ dL$; hence

$$dW = w'dL\ \gamma_{LV} (1 + \cos \theta) \tag{90}$$

We can make use of Fig. 34 to determine the forces that actually come into action in accomplishing this work:

At A, the liquid's surface tension γ_{LV} and the interfacial tension γ_{SL} are exerted. This point is displaced by dL toward the left, in relation to the liquid, but it remains fixed in relation to the solid. Only force γ_{LV} furnishes motor work equal to $\gamma_{LV}\ w'dL$.

At B, the liquid's surface tension γ_{LV}, the interfacial tensions γ_{SV} and γ_{SL}, and reaction R_S of the solid are exerted. Point B remains fixed in relation to the liquid, but is displaced by dL toward the right in relation to the solid. Under these conditions γ_{LV} furnishes no work; γ_{SV} furnishes motor work equal to $\gamma_{SV}w'dL$; γ_{SL} furnishes resistant work equal to $\gamma_{SL}w'dL$ and R_S is perpendicular to the displacement. The assessment of the work furnished by all surface forces is

$$dW = \gamma_{LV}\ w'dL + \gamma_{SV}\ w'dL - \gamma_{SL}\ w'dL \tag{91}$$

Using Young's relation, we can verify the equality of relations (90) and (91). Dupré (1869) gives the name "work of adhesion W_A of a liquid to a solid" to work received (or furnished) upon contact (or separation)—brought about in a reversible manner—of one unit of liquid area with one unit of solid area (in the presence of the liquid's vapor), with the intention of forming one unit of solid-liquid area.

Let us use relations (90) and (91):

$$W_A = \gamma_{LV} (1 + \cos \theta) \tag{92}$$

$$W_A = \gamma_{SV} + \gamma_{LV} - \gamma_{SL} \tag{93}$$

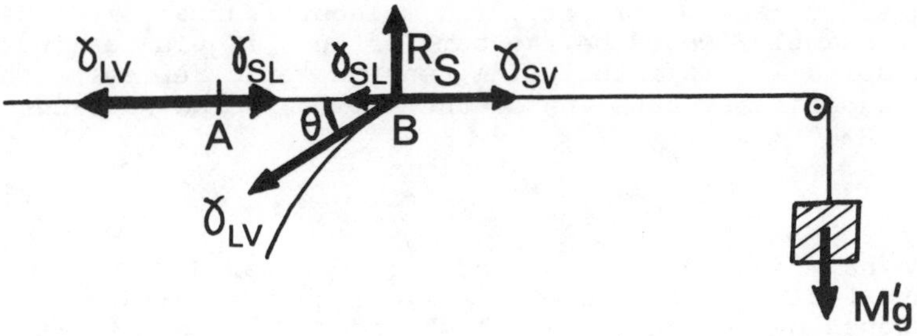

Fig. 34. Thin horizontal plate with one of its faces partially in contact with a liquid surface: representation of interfacial tensions acting on the plate.

If we place two units of liquid area in contact, there is a disappearance of these surfaces, and the work received, called "cohesion work of a liquid" is equal [according to Eq. (93)] to

$$W_C = \gamma_{LV} + \gamma_{LV} - \gamma_{LL}$$

In that the surface formed is located in the mass of the liquid, it cannot be the seat of any tension. Thus $\gamma_{LL} = 0$, and

$$W_C = 2\,\gamma_{LV} \tag{94}$$

The work of adhesion W_A, defined earlier, should not be confused with W_{AO}, defined by Harkins. To specify the part due to adsorption, Harkins and collaborators (Harkins and Loeser, 1950; Harkins and Livingstone, 1952) proposed an identical definition, but they estimate that the solid initial state is in equilibrium in vacuum. W_{AO} is thus expressed by

$$W_{AO} = \gamma_{SO} + \gamma_{LV} - \gamma_{SL} \tag{95}$$

Let us use relation (68):

$$\pi_e = \gamma_{SO} - \gamma_{SV} \tag{68}$$

We arrive at

$$W_{AO} = W_A + \pi_e \tag{96}$$

and

$$W_{AO} = \gamma_{LV}\,(1 + \cos\theta) + \pi_e \tag{97}$$

5.7 Physical Significance of W_{AO} and W_A

If there were no liquid-solid interaction, that is, if the

liquid and solid had no affinity, the molecular arrangement of the liquid (for example) would be the same as in a liquid-gas interface. The solid-liquid interfacial tension would represent the sum of the superficial tensions of the two connected surfaces, We would then have

$$\gamma_{SL} = \gamma_{LV} + \gamma_{SO} \tag{98}$$

The case we have just envisaged does not correspond to any reality and the interfacial tension of the solid-liquid interface is always lessened by a quantity A_O which indicates the affinity of the liquid for the solid. Under these conditions;

$$\gamma_{SL} = \gamma_{LV} + \gamma_{SO} - A_O \tag{99}$$

The rapprochement of (95) and (99) establishes the equality of quantities W_{AO} and A_O. W_{AO} expresses the affinity between the bare solid and the liquid.

By the same type of reasoning, we could prove that W_A expresses the affinity of the liquid for the solid surface in equilibrium with the vapor of the liquid, and knowledge thereof permits us to foresee the behavior of the solid-liquid pair studied. At the level of the solid-liquid interface, W_A has no significance (see examples 1 and 2 below).

W_{AO} does not let us predict the contact angle but is linked to the solid-liquid interfacial tension. Thus, depending on the case, one of the quantities is more interesting than the other.

Example 1. If, at the moment when the liquid is placed in contact with the solid, the solid surface is covered only by adsorbed molecules of the compound constituting the liquid, these molecules are no longer distinguishable from the molecules of the liquid phase when the solid-liquid interface is formed. In a more general manner, if the solid surface were covered with physically adsorbed molecules (of any compound), most of these molecules would have passed into the liquid phase at equilibrium. Hence, the initial number of physically adsorbed molecules has no influence on γ_{SL}, and it is preferable to link γ_{SL} to γ_{SO} by Eq. (95) than to γ_{SV} by Eq. (93).

Example 2. If the solid surface is initally covered with a uniform concentration of molecules (of a constituent analogous to or differing from that of the liquid), so soundly fixed to the solid that they will not mix with the liquid-phase molecules (the case of films with a particular structure and orientation, or the case of chemically adsorbed molecules), we shall assume that these molecules are an integral part of the solid. We shall take as "state of reference" a surface covered with such molecules, and call $\gamma_{SO'}$ its superficial tension; here again we shall define γ_{SL} by relation (95), written as $W_{AO} = \gamma_{SO'} + \gamma_{LV} - \gamma_{SL}$. Let us make it clear that, on such a surface, the saturating vapor of the drop's liquid may cause a supplementary adsorption, so that γ_{SV} will be different from $\gamma_{SO'}$.

5.8 Spreading Coefficient *S*

Spreading coefficient S was defined by Cooper and Nuttall (1915) in order to foresee the complete or partial spreading of a liquid on a given solid. S is defined by

$$S = W_A - W_C \tag{100}$$

or by

$$S = W_{AO} - W_C - \pi_e \tag{101}$$

Both of these equations lead to

$$S = \gamma_{LV} (\cos\theta - 1) \tag{102}$$

If S is negative, we can foresee the existence of a finite contact angle of the liquid on the solid; if S is positive, a complete spreading of the liquid over the solid will take place.

5.9 Reversible and Irreversible Processes in the Case of a Moving Three-Phase Line

5.9.1 An Irreversible Process

Let us consider a liquid and a solid plate of which the superficial tensions are such that

$$0 < \gamma_{SV} - \gamma_{SL} < \gamma_{LV}$$

The contact angle of the liquid will satisfy

$$\cos\theta = \frac{\gamma_{SV} - \gamma_{SL}}{\gamma_{LV}} \tag{80}$$

and θ will be less than $\pi/2$.

We are going to study the immersion process of the plate represented by stages a, b, c, and d of Fig. 35, while the plate is held back by the operator. In a, the plate is above the liquid and it is slowly lowered. In b, it is found in the neighborhood of the liquid. From the moment of contact with the liquid, a meniscus forms spontaneously (Fig. 35c) and makes an angle of junction θ, which is preserved during the plate's immersion (Fig. 35d).

Upon the plate's contact with the liquid, the line of junction, subject to different surface forces, finds itself in a position of disequilibrium. The resultant of forces applied to this line is not null. An irreversible movement of the line results, whose object is to make it assume a position of equilibrium corresponding to the value of the contact angle that verifies Young's equation. We notice that the work furnished by the wetting of the plate is more than that necessary for the formation of the meniscus under reversible conditions. The calculation could be done from Eqs. (55), (37), and (85). The difference is dissipated in caloric

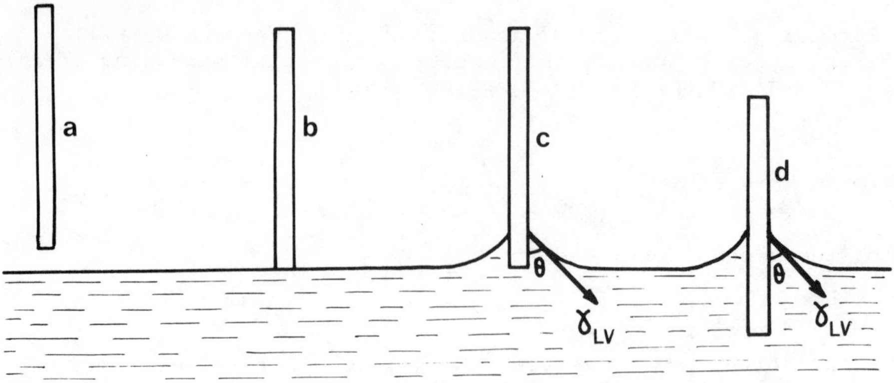

Fig. 35. Example of an irreversible process: the immersion of a vertical solid plate when the contact angle θ is less than 90°.

form. Upon an ulterior immersion of the plate, the process becomes reversible again, and the plate furnishes work of wetting (see Sec. 5.5).

5.9.2 A Reversible Process

Let us consider the same plate, and the same liquid. Thus θ is less than $\pi/2$. The plate is now completely immersed in the liquid, and we shall now consider its emersion process represented by stages a, b, c, and d of Fig. 36. During this emersion the real microscopic angle of junction always remains fixed at value θ imposed by Young's equation, and the apparent angle of junction α varies from $\pi/2$ to θ (for more details, see Sec. 7.2.4).

As long as the apparent angle of junction α has not reached the value θ, the meniscus remains attached to the upper edge of the plate (Fig. 36a and 36b). From the moment when angle θ is reached (Fig. 36c), the meniscus no longer deforms itself upon any further movement of the plate.

Let us consider stage b. Given α, the apparent angle formed by the tangent to the liquid's profile with the vertical. To keep the plate (of negligible weight) sustained, the operator must exert an upward traction equal to γ_{LV} cos α, by unit of length of the plate's perimeter. That is ℓ, this perimeter. Traction F, exerted by the operator, is equal to $F = \ell\ \gamma_{LV} \cos\alpha$. Whenever the operator raises the plate by dz, he furnishes work equal to

$$dW' = \ell\ \gamma_{LV} \cos\alpha\ dz \qquad (103)$$

Comparing this work with expression (51) establishes the fact that the work furnished by the operator exactly corresponds to the

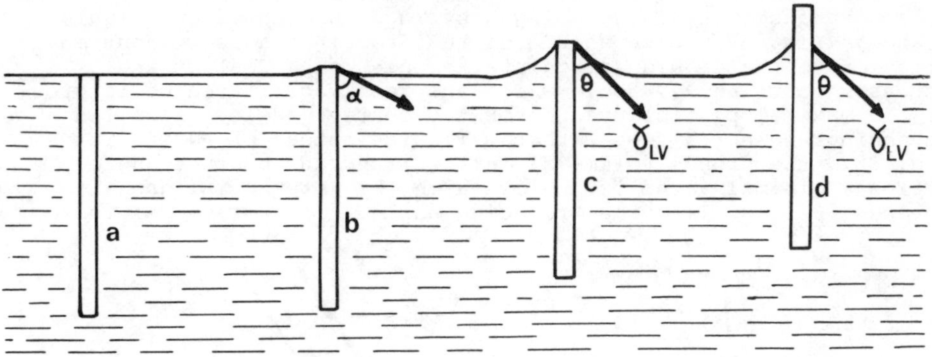

Fig. 36. Example of a reversible process: the emersion of a vertical solid plate when the contact angle θ is less than 90°.

elementary work of formation of the meniscus. For identical elementary raisings dz, the work furnished by the operator is very little at the beginning of emersion ($\cos \alpha \simeq 0$) and becomes greater and greater (α diminishes and cos α increases). The ultimate deformation of the meniscus (Fig. 36c) necessitates work equal to $\ell\ \gamma_{LV} \cos \theta\ dz$. Afterwards, because the meniscus remains identical to itself (Fig. 36d), the work furnished by the operator represents work of plate dewetting. According to what was said in Sec. 5.5, the dewetting of a plate of perimeter ℓ, during an ascension dz, requires work equal to

$$dW = (\gamma_{SV} - \gamma_{SL})\ \ell\ dz \tag{104}$$

Young's equation establishes the equality of quantities dW and dW' in relations (103) and (104) when $\alpha = \theta$. Hence, the following interpretation: Discontinuity in the phenomenon (limit stage c), comes about at the moment when the work that should be furnished to continue deforming the meniscus during an ascension dz becomes greater than the work that should be furnished to dewet the plate during the same ascension dz. Emersion of a plate on which the liquid forms an angle $\theta < \pi/2$ is an entirely reversible phenomenon. At every moment during this process, the resultant of forces applied on the three-phase line is null.

6 HYSTERESIS OF CONTACT ANGLES ON REAL SOLID SURFACES

6.1 Advancing and Receding Contact Angles

We noted in the preceding section that the contact angle of a liquid upon an ideally plane, homogeneous surface should have only one equilibrium value. Experimental results from real solid surfaces rarely lead to such a conclusion.

Let us consider the following experiment. A drop is placed on a plane surface. By introducing a syringe into the drop, liquid is added (Fig. 37). The volume of the drop increases without any variation of the liquid-solid interfacial area. Contact angle increases to value θ_A, above which the interfacial area would grow. θ_A, the maximal possible value for the contact angle, is called "advancing angle." If we withdraw liquid instead of adding it (Fig. 38), the drop's volume diminishes, and the contact angle attains a minimal value θ_R, below which the interfacial area recedes. θ_R is called the "receding angle." The origin of these

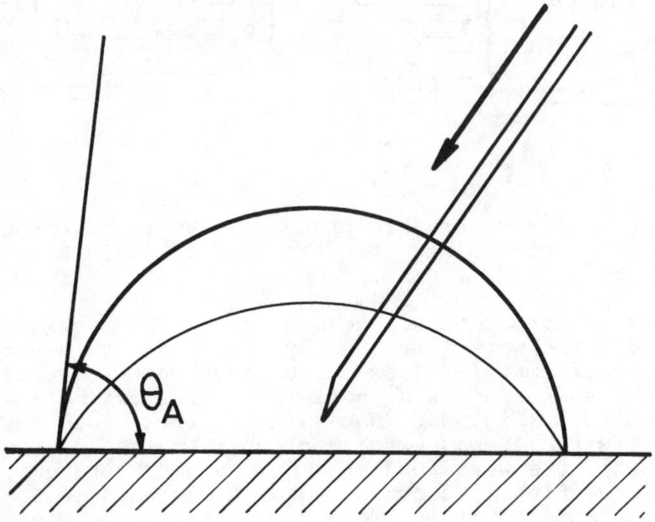

Fig. 37. Advancing contact angle.

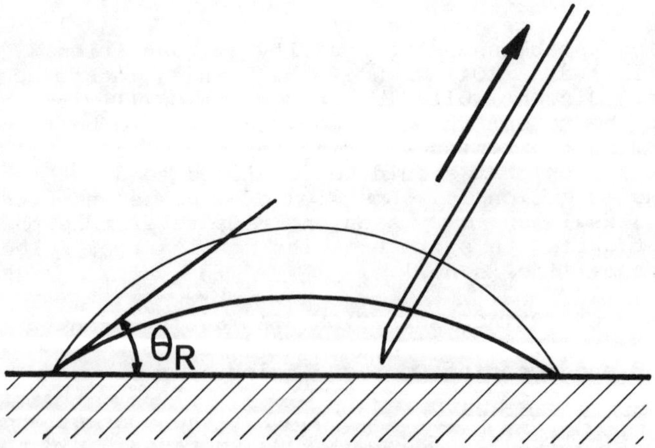

Fig. 38. Receding contact angle.

names is due to the fact that the advancing angle's measurement is realized by advancing the liquid's periphery over the solid surface; the receding contact angle is measured by withdrawing it.

A drop placed on a solid surface, and presenting a contact angle θ which is found between θ_A and θ_R, is in equilibrium. A "region of equilibrium" exists, often covering several dozen degrees. A measurement of the contact angle of a drop placed on a plane is thus of little significance, and it is much more important to know θ_A and θ_R. The difference between these two values is called "contact angle hysteresis."

In those applications where the contact angle should be as small as possible, it will be better to choose conditions such that the receding contact angle is established.

Three causes are mentioned in the literature to explain contact angle hysteresis:

1. The heterogeneity of solid surfaces
2. The roughness of solid surfaces
3. Modification of the solid surface resulting from its contact with the liquid.

We shall now study each of these factors so as to determine its influence on contact angle hysteresis.

6.2 Surface Heterogeneity and Hysteresis

The different facets of a crystal have different superficial tensions. With real materials (for example, alloys), one can expect greater fluctuations of the solid's superficial tensions. The adsorbed molecules on a surface modify its superficial tension. But these layers are rarely uniform; usually, adsorption is localized at a number of preferential sites on the surface, whereas the rest of the surface may be free of adsorbed molecules. A real surface, even a perfectly plane one, is never homogeneous as far as its tensions are concerned.

We shall try to understand, with the help of simple models, the influence of surface heterogeneity upon contact angle hysteresis.

6.2.1 Model 1: A Horizontally-Striped Plate

Given a flat plate whose surface is made up of very thin horizontal stripes endowed alternately with a strong and weak superficial tension (Fig. 39). The former (in black in the figure) will be labeled 1. They verify that

$$\gamma_{S_1V} - \gamma_{S_1L} = \gamma_{LV} \cos \theta_1 \qquad (105)$$

The latter (in white) will be labeled 2. They verify that

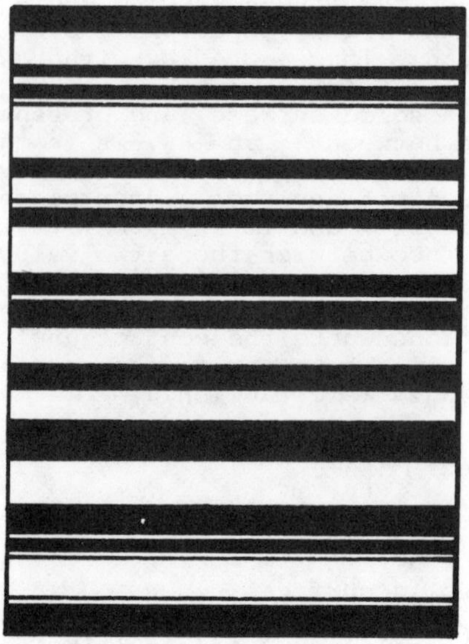

Fig. 39. Heterogeneous surface consisting of horizontal stripes.

$$\gamma_{S_2V} - \gamma_{S_2L} = \gamma_{LV} \cos \theta_2 \qquad (106)$$

θ_1 and θ_2 could be such that $\theta_1 = 30^0$ and $\theta_2 = 120^0$ (the case of Fig. 40 and 41). Let us proceed with a very slow immersion of the plate, held vertically, into the liquid.

If the first stripe corresponds to a zone of type 1, the line of junction will pass over it spontaneously, following an irreversible process of the type described in Sect. 5.9. But the line of junction is stopped as soon as it encounters the line below the first zone of type 2. Thus we immediately arrive at a state corresponding to the configuration represented in Fig. 40a. Zone 2 will not be wetted until the contact angle has attained a value equal to θ_2. At the time of the plate's penetration, the meniscus becomes deformed, but the line of junction remains fixed in relation to the solid until this value θ_2 is reached (Fig. 40b). From this moment on, the line of junction passes over zone 2 with the same speed as that of the plate's penetration. During this movement, the angle of junction remains constant and equal to θ_2 (Fig. 40c). As soon as the line of junction reaches a new zone of type 1, it abruptly bypasses it to stop at the lower limit of the following zone 2. During this spontaneous and irreversible process, angle θ decreases (Fig. 40d) by a value that is greater as the thickness of the zone 1 bypassed is larger. The line of junction will remain fixed at the lower limit of this zone 2 until the deformation of the meniscus, due to the plate's penetration, allows for reestablishement of angle θ_2. The system is once again in a state corresponding to that of Fig. 40b, and the preceding process reoccurs for

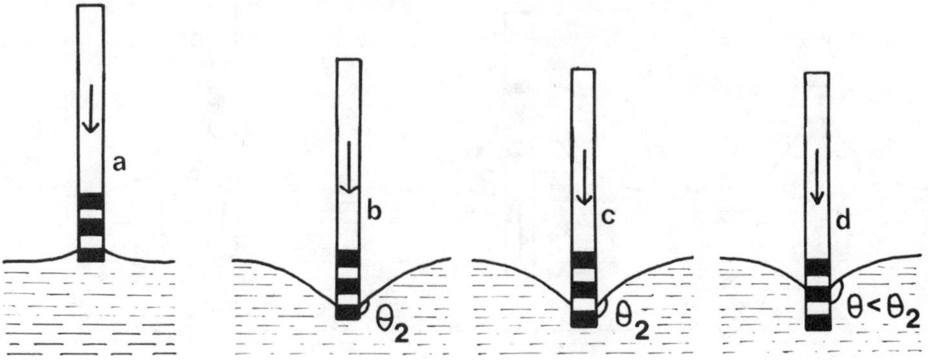

Fig. 40. Immersion process of a heterogeneous surface consisting of horizontal stripes.

each pair of stripes.

In conclusion, during a continuous penetration of the plate (Fig. 40b, c, d), the mean value of the contact angle is $\theta_2 - \Delta\theta_2$. Quantity $\Delta\theta_2$ is less as the thickness of these stripes is smaller in relation to the mean height of the meniscus [calculable by relation (37)]. We shall consider that $\theta_2 - \Delta\theta_2 \simeq \theta_2$ when the stripes are very thin.

We feel it necessary to summarize the earlier shown facts according to a new presentation. At the moment of a slow and continuous penetration of the plate into the liquid, we remark the appearance of the following phenomena depending on the zones passed over:

Reversible rises of the line of junction on the surface as long as the resultant of forces exerted on the line of junction is null (zones 2)

Irreversible and spontaneous rises of the line of junction on the surface beginning when the resultant of forces exerted on the line of junction stops being null and is directed upwards (Zones 1)

Absence of movement of the line of junction on the surface when the surface is in such a position that an elementary movement upwards or downwards would cause the appearance of a nonnull resultant of surface forces, aimed in the opposite direction from the displacement. (This stable position of the line of junction is situated (in the case of our model) at the level of the lines separating the upper portions of zones 1 from the lower portions of zones 2)

Without making such a detailed study, let us now consider the emersion of the same plate, initially immersed in the same liquid.

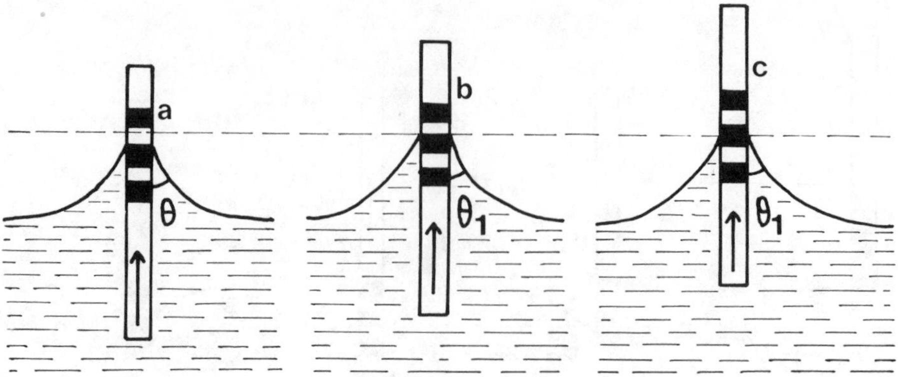

Fig. 41. Emersion process of a heterogeneous surface consisting of horizontal stripes.

At a given moment, the line of junction is in a stable position on a line of the solid separating a zone 2 (above) from a zone 1 (below), as in the Fig. 41a. While the plate is rising, the line of junction will not begin its displacement over zone 1 until the deformation of the meniscus has imposed a contact angle equal to θ_1 (Fig. 41b). We shall then have a reversible descent phase of the line of junction over the surface (zone 1), as in Fig. 41c. The act of bypassing the lower limit of zone 1 marks the end of this phase. An irreversible dewetting process of zone 2 immediately below occurs (the resultant of forces exerted on the line of junction at this moment is nonnull and directed downwards), which immediately causes the junction line to pass to the lower limit of the zone 2. The angle is now equal to a value θ, greater than θ_1. The process represented by Fig. 41a, b, c begins again.

In conclusion, during a continuous withdrawal of the plate, the receding contact angle corresponds to a mean value $\theta_1 + \Delta\theta_1$. Quantity $\Delta\theta_1$ is smaller as the stripes of type 2 are thinner. If the thickness of these stripes is negligible in relation to the mean height of the meniscus, then

$$\theta_1 + \Delta\theta_1 \simeq \theta_1$$

Hence, the advancing contact angle (at the time of immersion) is imposed by the zones of low superficial tension (zones 2), whereas the receding contact angle (at the time of emersion) is imposed by the zones of high superficial tension (zones 1). If the stripes are not very thick (less than 0.1 mm to give a definite idea), the contact angle upon immersion is θ_2, and upon emersion is θ_1; where θ_1 and θ_2 are Young's angles defined by Eqs. (105) and (106).

Finally, it is possible (for example, by inverting the direction of movement of the plate) to get any contact angle between

θ_1 and θ_2, except those corresponding to junctions on stripes of a high superficial tension. In fact, only stripes with a low superficial tension and limiting lines between two stripes of differing superficial tensions correspond to stable meniscal configurations.

6.2.2 Model 2: A Vertically Striped Plate

Using the plate described in the previous model, we shall perform a rotation of 90° (Fig. 42). When this plate is partially immersed, the junction of the liquid and solid takes place following a curve such that Young's equation is proven at all points. The equation of this curve (which cannot be a straight line) is of little importance here. Let us note only that it must be identical to itself, whatever the depth at which the plate is immersed. The same is true of the meniscus. While the plate is partially immersed, any immersion or emersion is thus a reversible process. During the course of such processes, the work of wetting or dewetting of the plate, respectively, comes into play.

Let ℓ be the perimeter of the plate. On a fraction $(1 - \alpha')$ of this perimeter there are zones of high surface tension (zones 1), and on a fraction α' there are zones of low surface tension (zones 2). Upon a penetration dz of the plate (without deformation of the meniscus), the work of wetting furnished by the plate is equal, according to Eq. (86), to

$$dW = \ell\ dz \left[\alpha' (\gamma_{S_2V} - \gamma_{S_2L}) + (1 - \alpha')(\gamma_{S_1V} - \gamma_{S_1L})\right] \qquad (107)$$

and applying relations (105) and (106) leads to

$$dW = \ell\ dz\ \gamma_{LV} \left[\alpha' \cos\theta_2 + (1 - \alpha') \cos\theta_1\right] \qquad (108)$$

Quantity (108) also represents the work that must be furnished to withdraw the plate by a displacement $d2$.

If the stripes are wide, the experimenter can measure angles of different values along the line of junction. The presence of very narrow stripes leads to the observation of a constant macroscopic contact angle, and of a nondeformed cylindrical meniscus of height z. This height is such that the work of deformation of the meniscus necessary to bring it to a value $z + dz$ is equal to the work of wetting of the plate related to an upward movement of the same height dz of the three-phase line.

Equation (51) expresses the value of the elementary work of formation of the meniscus:

$$dW_f = \ell\ \gamma_{LV}\ dz \cos\theta \qquad (51)$$

and Eq. (108) the value of the work of wetting of the plate. Condition $dW = dW_f$ leads to

$$\cos\theta = \alpha' \cos\theta_2 + (1 - \alpha') \cos\theta_1 \qquad (109)$$

Fig. 42. Heterogeneous surface consisting of vertical stripes.

or

$$\cos \theta = \alpha' \left(\frac{\gamma_{S_2V} - \gamma_{S_2L}}{\gamma_{LV}} \right) + (1 - \alpha') \left(\frac{\gamma_{S_1V} - \gamma_{S_1L}}{\gamma_{LV}} \right) \quad (110)$$

If the stripes are very thin (less than 0.1 mm), a constant contact angle θ, defined by relation (110), is associated with the model of the plate composed of vertical stripes. In such a case there is no contact angle hysteresis.

6.2.3 Model 3: A Surface of High Superficial Tension Interspersed with Small Patches of Low Tension

The advantage of such a model is its correspondence with a real case: that of adsorbed layers, such as fatty acids on metals, for example. The distribution of these layers of low superficial tension is not very uniform. If the quantity of adsorbed compound is much less than that corresponding to a complete monolayer, the compound is distributed over the surface in the shape of little islands. If the quantity of adsorbed compound is slightly less than a complete monolayer, the uncovered solid surface appears in the form of little islands spread over a surface of low superficial tension. This is shown in numerous photographs taken by electronic microscopy and published in the literature. For example, Ries and Walker (1961) have obtained electron micrographs of monolayers of n-hexatriacontanoic acid, transferred at different film pressures by the Langmuir-Blodgett method on collodion-covered supports. Figure 43a is the reprint, with the permission of the authors, of these photographs. Brockway and Jones (1964) obtained similar micrographs with monolayers of behenic acid adsorbed on glass by the retraction method (Bigelow et al., 1946).

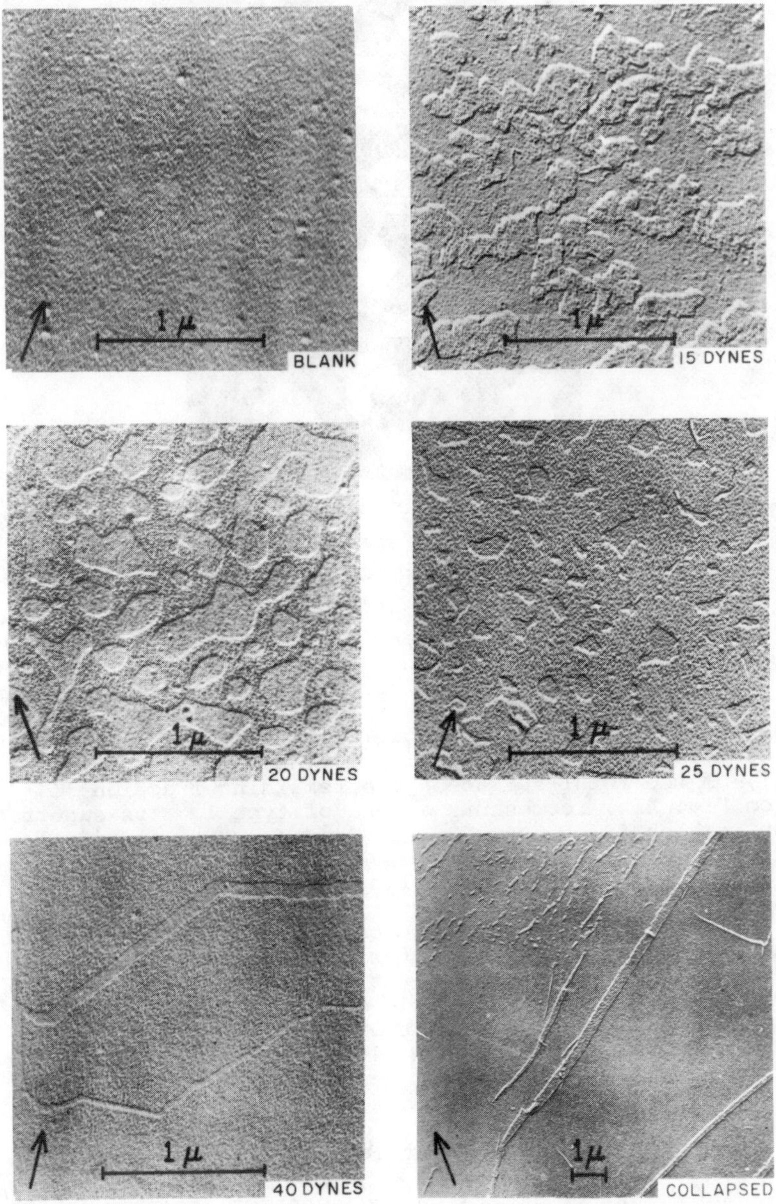

Fig. 43. (a) Electron micrographs of monolayers of n-hexatriacontanoic acid. From Reis and Walker (1961).

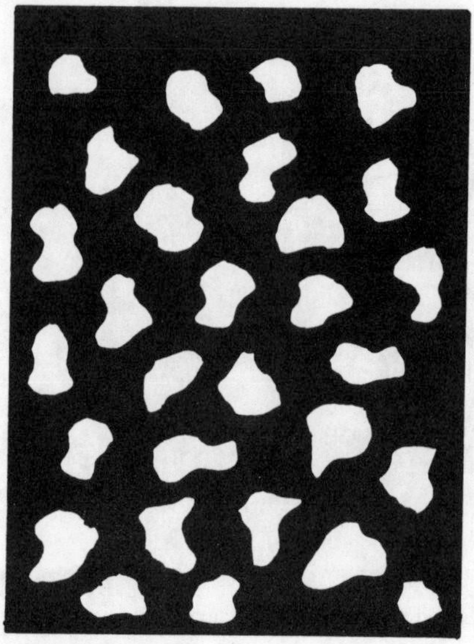

Fig 43(b): Heterogeneous surface consisting of patches of weak superficial tension spread over a surface of high superficial tension.

When a certain fraction of surface is covered (a fraction that may be quite variable), we proceed directly from a structure characterized by the presence of patches of low tension spread over a surface of high tension to one where the patches of high tension are spread over a surface of low superficial tension. In the model represented in Fig. 43b, very small patches (less than 0.01 mm^2) are zones of type 2 (of weak superficial tension) and they cover a fraction α', of the total surface. The remaining fraction $(1 - \alpha')$ in black on Fig. 43b, represents a zone of type 1. Its superficial tension is high.

Let us imagine a plate partially immersed in a liquid. Given the heterogeneity of the surface, the line joining liquid and solid is not a straight line. Because of the very small size of the patches, however, the experimenter will observe a macroscopically uniform meniscus joining itself to the solid. In order to use the results obtained during the study of the preceding models, we shall suppose that the patches consist of regularly distributed rectangles as indicated in Fig. 44. Such a model results from the combination of that composed of vertical stripes and that composed of horizontal stripes. On the same horizontal line we find either zones of high superficial tension only (line b on Fig. 44), or both zones of high and low superficial tension (line a on Fig. 44).

At the moment of immersion, the contact angle is imposed by the zones where the superficial tension is lowest, that is, those cut by lines of type a. We find ourselves now in the model of vertical stripes. Upon immersion, the meniscus attaches itself to the solid along a line of type a. The contact angle is equal, on slightly inferior to θ_A, such that

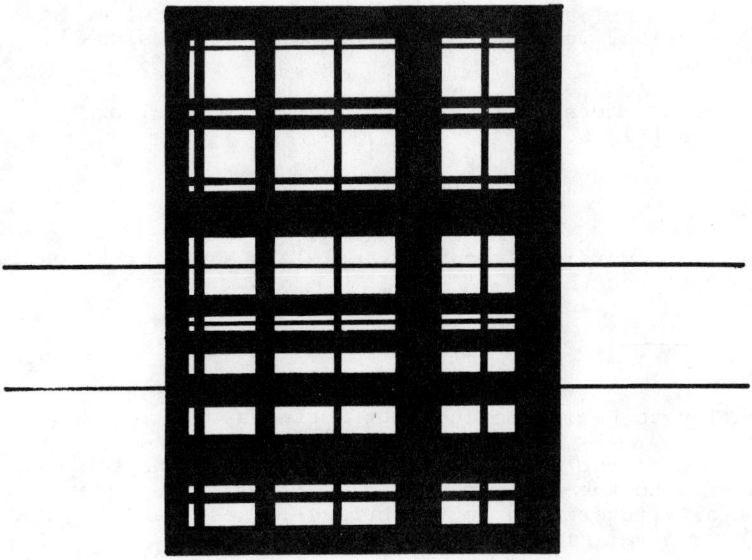

Fig. 44. Model of heterogeneous patchwise surface.

$$\cos\theta_A = \alpha'' \cos\theta_2 + (1 - \alpha'')\cos\theta_1 \qquad (111)$$

In relation (111), the coefficient α'', different from α', comes into play: α'' represents the percentage of zones of low superficial tension on a line of type a. If the distribution of zones of low superficial tension is the same horizontally and vertically, the total surface of these zones corresponds to $\alpha''^2 S$. Thus, according to the definition of α';

$$\alpha'' = \sqrt{\alpha'} \qquad (112)$$

and $$\cos\theta_A = \sqrt{\alpha'}\cos\theta_2 + (1 - \sqrt{\alpha'})\cos\theta_1 \qquad (113)$$

The angle upon immersion equals $\theta_A - \Delta\theta_A$ from the conclusions relative to the first model. The photographs of adsorbed layers, taken by electronic microscopy show that the average area of a patch is less than a millionth of a square millimeter. Consequently, $\Delta\theta_A$ can usually be neglected.

Let us now consider the emersion of the model plate. The model composed of horizontal stripes shows us that angle θ_R is governed by the zones of high superficial tension. They are those that are cut by lines of type b on Fig. 44. Thus the receding angle equals $\theta_1 + \Delta\theta_1$. But, given the size of the patches, $\Delta\theta_1 \ll \theta_1$.

$$\theta_R \simeq \theta_1 \qquad (114)$$

Considering the complementary model, corresponding to the

presence of rectangular patches of high surface tension over a surface of low tension, we come to the following conclusions by identical reasoning:

The angle upon immersion (advancing angle) is equal or slightly less than θ_2

$$\theta_A \simeq \theta_2 \tag{115}$$

The angle upon emersion (receding angle) is equal or slightly more than θ_R, such that

$$\cos \theta_R = \sqrt{1 - \alpha'} \cos \theta_1 + (1 - \sqrt{1 - \alpha'}) \cos \theta_2 \tag{116}$$

6.2.4 Application to Real Heterogeneous Surfaces

With the help of the above analysis, let us examine the real case corresponding to the diagram of Fig. 43b (which represents a monolayer partially covering a solid surface). Results (114) and (115) should remain valid; that is

The angle upon immersion (advancing contact angle) should be equal to or slightly less than θ_2 in the case corresponding to the presence of patches of high superficial tension over a surface of low tension.

The angle upon emersion (receding contact angle) should be equal to or slightly greater than θ_1 in the case corresponding to the presence of patches of low superficial tension over a surface of high tension.

For that which concerns relations (113) and (116), established from a particular model (particular because of the shape and distribution of patches), we have verified that they fairly well describe the aspect of experimental curves obtained with surfaces partially covered by an adsorbed layer of a polar-nonpolar compound.

For systems with three or more different zones (presenting different superficial tensions), one can only affirm that it is the zones of lowest superficial tension that impose the advancing contact angle, and the zones of highest superficial tension that impose the receding contact angle.

6.3 Surface Roughness and Hysteresis

We are going to show, with the help of simple models, how the roughness of a sample causes contact angle hysteresis. The imaginary models will be equivalent to those studied earlier.

6.3.1 Model 1: A Plate Presenting Inclined Horizontal Stripes

A plate representing inclined horizontal stripes, of supposedly uniform superficial tension, is shown in cross section in Fig. 45. We shall suppose that the stripes are narrow and invisible to the

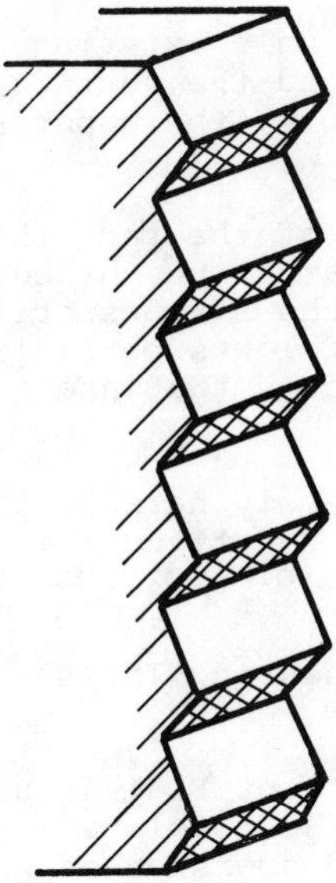

Fig. 45. Model rough surface: plate presenting inclined horizontal stripes.

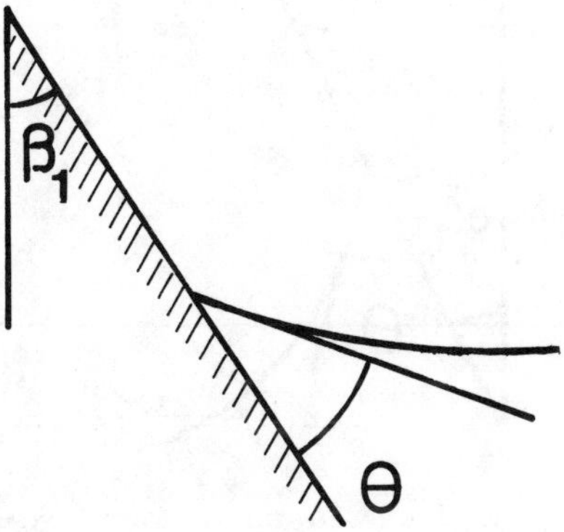

Fig. 46. Three-phase line located on an upward-inclined stripe.

naked eye. Those turned upward form an angle β_1 with the vertical, whereas those turned downward form an angle α_1 with the vertical. The real contact angle θ is the same on all points of the surface and proves that

$$\gamma_{LV} \cos \theta = \gamma_{SV} - \gamma_{SL}$$

Figure 46 takes account of the fact that when the line joining liquid to solid is found on a stripe turned upward, the apparent contact angle, that is, the angle existing between the tangent to the profile of the meniscus and the vertical, is equal to $\theta + \beta_1$. If the roughness is very minute, the apparent angle is the experimental measure.

Figure 47 takes account of the fact that when the line joining liquid to solid is found on a stripe turned downward, this same angle equals $\theta - \alpha_1$. Thus, the stripes oriented upward are equivalent to horizontal stripes of low superficial tension, and those turned downward constitute zones that are equivalent to stripes of high superficial tension.

If the stripes are extremely narrow, the apparent angle of the meniscus is equal to $\theta + \beta_1$ at the time of immersion (advancing angle) and $\theta - \alpha_1$ at the time of emersion (receding angle).

6.3.2 Model 2: A Plate Presenting Inclined Vertical Stripes (Fig. 48)

Let us consider the plate of Fig. 45 after a rotation of 90°. This model is equivalent to the model having heterogeneous vertical stripes in that immersion and emersion of such a plate are reversible phenomena. The real surface of the plate is equal to k' times its apparent area. The work of dewetting of this plate by a height dz is thus equal to

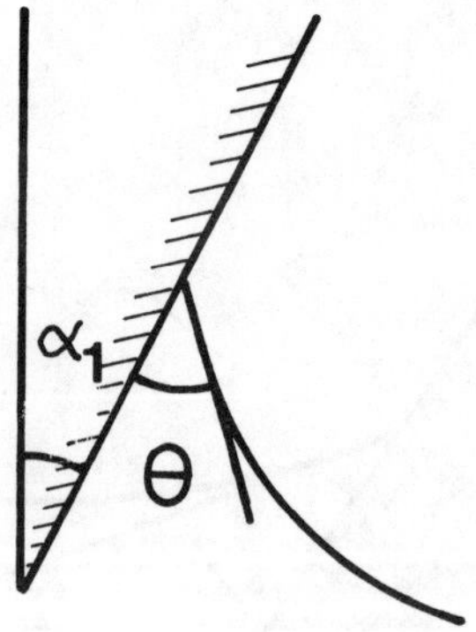

Fig. 47. Three-phase line located on a downward-inclined stripe.

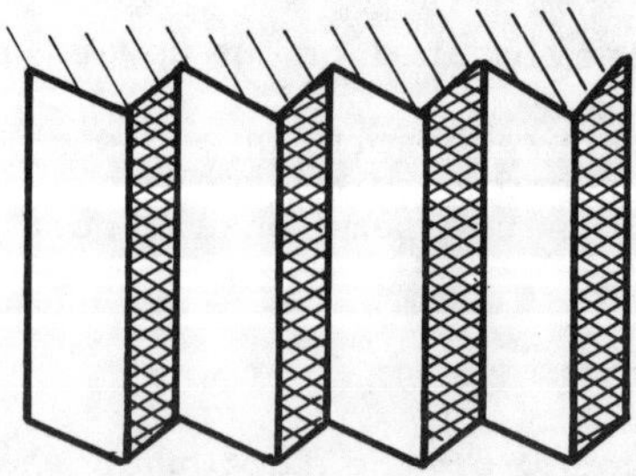

Fig. 48. Model rough surface: plate presenting inclined vertical stripes.

$$dW = \ell\ dz\ k'\ (\gamma_{SV} - \gamma_{SL}) = \ell\ dz\ k'\ \gamma_{LV} \cos\theta \qquad (117)$$

This work is equal to that of formation of the meniscus dW_f from height z to height $z + dz$:

From (51): $$dW_f = \ell\ dz\ \gamma_{LV} \cos\theta' \qquad (118)$$

θ' being the apparent macroscopic angle presented by the meniscus. Thus,

$$\cos\theta' = k' \cos\theta \quad \text{if } k' \cos\theta < 1 \qquad (119)$$

and $$\cos\theta' = 1 \quad \text{if } k' \cos\theta \geqslant 1 \qquad (120)$$

6.3.3 Application to Real Surfaces

Let us note first that a surface of the "corrugated iron" type (Fig. 49) leads to results similar to those established with the surface of models 1 and 2. For example, if α_1 and β_1 are the largest angles presented by the profile with the vertical on the plate in Fig. 49, the angle upon immersion should be equal to $\theta + \beta_1$ and the angle upon emersion equal to $\theta - \alpha_1$. Presented with its generatrices vertical, this same sample, upon immersion as upon emersion, presents an angle θ' such that $\cos\theta' = k' \cos\theta$, k' being the ratio of the real area to the apparent area.

The values of slopes of asperities for a real sample can take all values between 0 and the maximal slope corresponding to angle α_1 or β_1 of the preceding example.

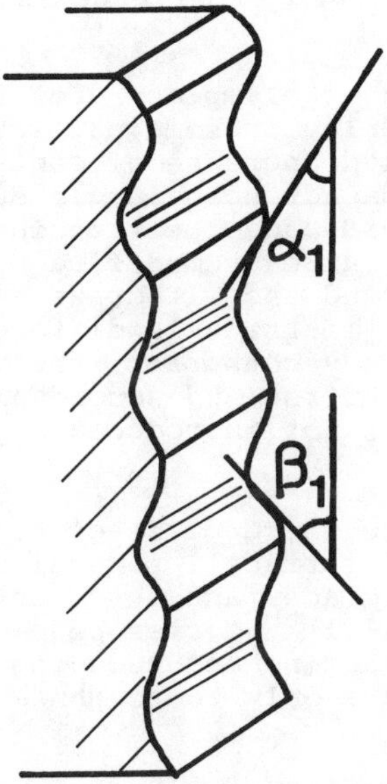

Fig. 49. Model rough surface of the "corrugated iron" type.

The reasoning we have used for heterogeneous surfaces, which we have transposed to the case of rough surfaces, lets us state that a rough sample is equivalent to a heterogeneous sample for which the superficial tension would be different at each point and would take all possible values between two extreme values. Furthermore, at the time of wetting or dewetting of a rough sample, we notice experimentally that air bubbles or water droplets may remain imprisoned at the bottom of cavities formed by the roughness. This complicates the phenomenon even more.

Therefore we shall not imagine a third model that could take account of the existence of roughness. We could obtain it by combining the first two. Unfortunately, it would not correspond to any reality. We shall only keep in mind that vertical streaks diminish the macroscopic contact angle, whereas horizontal streaks contribute to the phenomenon of hysteresis.

6.4 Other Causes of Variation in Contact Angles: The modifications of a Solid Surface After Its Contact with a Liquid

The surface of a solid can be different after having been in contact with a liquid. There are then changes in contact angles. For example, a solid surface covered by a layer of adsorbed fatty acid is plunged into a solvent. After withdrawal of the plate, some molecules of fatty acid have desorbed into the solvent, and are no longer in the adsorbed layer. Inverseley, a clean solid surface plunged into a solution containing a highly adsorbable substance will retain an adsorbed layer after immersion (which can, in addition, be in equilibrium with the vapor). Furthermore, a porous surface (a sheet of paper, for example) becomes impregnated with liquid upon immersion, and experiences a change in its superficial tension.

We can easily find other examples. Let us note, however, that these phenomena come into play in some particular cases in which they lead to an often marked increase of contact angle hysteresis. Generally, one measures the advancing angle during the immersion of the solid into the liquid, and the receding angle when the solid is emersed. If the plate, already modified by its contact with the liquid, is again immersed and withdrawn, the advancing angle at the time of the second immersion is different from that measured at the time of the first. By constrast, the values of receding angles obtained during the first and second emersions are identical if the second immersion has not introduced any further surface modifications.

In order to master the third cause of contact angle hysteresis (modification of the solid surface), several successive immersion-emersion cycles must take place, and this, until the angles measured during the nth and the $(n + 1)^{th}$ cycles become identical. The difference between the advancing and receding angle at the moment of the nth cycle is thus due only to roughness and heterogeneity of the modified surface.

Johnson and Dettre (1969) justly propose to separate the changes in contact angles due to kinetics of surface interaction from those due to surface roughness and heterogeneity. Only the latter present the general properties of hysteresis (particularly reversibility) that are observed in other fields. Kinetics of surface interaction modify the contact angle values, but the term hysteresis is unsuitable for these modifications.

6.5 Other Important Theories About Contact Angle Hysteresis

Although Lord Rayleigh in 1890, was the first to remark upon the phenomenon of contact angle hysteresis, the first equation bringing roughness into account as a parameter of contact angles is due to Wenzel (1936):

$$\cos \theta_W = \frac{A_a}{A_r} \cos \theta_O \tag{121}$$

In Eq. (121), A_a is the apparent area of the solid surface and A_r is the real area of the solid surface taking the peaks and valleys into account, θ_W is the angle observed on a surface having Young's intrinsic angle θ_O. Let us note that (121) and (119) are identical.

It has been shown (Shuttleworth and Bailey, 1948; Good, 1952) that Wenzel's contact angle is that which corresponds to the lowest free energy of the system. But the contact angles measured experimentally on a rough surface are found between the advancing and receding contact angles, and are usually different from Wenzel's contact angle.

Cassie and Baxter (1944) have extrapolated Wenzel's equation for the case where the liquid is not able to penetrate into cracks and crevices of very rough surfaces. Cassie and Baxter's angle is subject to the same commentaries as those concerning Wenzel's angle.

The first more quantitative results between contact angle hysteresis and heterogeneity as well as roughness are due to Johnson and Dettre (1969). To obtain these results they chose models with a drop of liquid centered on an array of alternating concentric rings. For the heterogeneous model surface, the concentric rings are of two types, one having an equilibrium contact angle θ_1 and the other θ_2. For the rough model surface, they combined grooves and hills in such a way that a normal section to the surface through the origin results in a sinusoidal wave form. They demonstrated the existence of a large number of metastable states by minimization of the overall free energy of the system, while neglecting gravity.

By minimization of the system's free energy, Neumann and Good (1972) studied heterogeneous surfaces consisting of horizontal strips, heterogeneous surfaces consisting of vertical strips, and patchwise heterogeneous surfaces. Their thermodynamic analysis, taking gravity into account, is rigorous, but for the mathematical calculations it is necessary to specify the numerical values corresponding to the model's parameters and to have access to a computer. In Secs. 6.2 and 6.3 we have presented a simplified discussion of Neumann and Good's model.

Johnson and Dettre's thermodynamic approach, on one hand, and that of Neumann and Good on the other, as well as the more physical analysis developed by the author (Chappuis, 1974) and reproduced in this text, all lead to the same general conclusions. It must be noted, however, that Eqs. (113) and (116) are original.

6.6 Conclusion and an Example of the Use of Results Obtained

The simple models presented above allow us to understand by what processes heterogeneity and roughness of solid surfaces bring about contact angle hysteresis. These results let us interpret a number of experimental data obtained during the course of systematic studies.

As an example, we shall consider the following experimental study (Chappuis et al., 1974). Radioactively marked stearic acid monolayers are placed on solid surfaces. Temporal evolution of the layer is studied while the samples are stocked in different environments. The radioactivity of the surface indicates the quantity of acid adsorbed. The measure of wetting of the solid by distilled water will give information concerning the layer.

6.6.1 Experimental Details

The samples are made of an unoxidizable alloy with cobalt as its base (stellite: Co 60%, Cr 30%, W 10%). The samples are polished with diamond paste (size of particles: 1 μm) and then cleaned in different solvents. The stearic acid monolayers are placed by the retraction method (Bigelow et al., 1946). The samples are plunged into a solution of cyclohexane containing 10^{-4} g cm^{-3} stearic acid during 10 min. After deposition of the monolayers, the samples are immediately tested. They are then divided into three groups, which are stored:

A: in a dry environment (dessicating apparatus)

B: in a dust-free environment (50% relative humidity at 21°C)

C: in water vapor at 100°C

The sample are then tested each day. The evolution of the quantity of stearic acid in an adsorbed layer is followed by measuring its activity with a scintillation counter. The measure of wetting by distilled water leads to the knowledge of the advancing and receding angles.

6.6.2 Experimental Results

The experimental results can be summarized as follows:

1. The quantity of fatty acid remains constant through time during the course of different trials. It is close to the quantity corresponding to a monomolecular oriented layer.

2. The values of advancing contact angles are almost constant through time. However, an evolution of receding contact angles

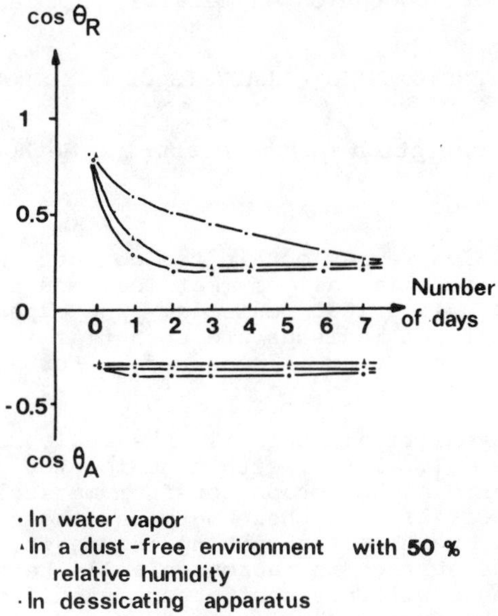

Fig. 50. Advancing and receding contact angles of water on a clean metallic sample, as a function of the elapsed time between cleaning and measurement, and following different storage conditions.

is observed. After some time, the receding contact angle stabilizes, and in all cases, tends toward the same equilibrium value (Fig. 50).

3. Equilibrium is obtained all the more rapidly when the percentage of water of the medium in which the samples are stored is greater.

6.6.3 Interpretation

Most of the surface is covered by a layer of fatty acid which imposes the advancing contact angle, whereas the receding angle is imposed by the islands of surface not covered by the film. The diminution of the cosine of the receding contact angle is linked to an increase of that angle, which signifies that there is a modification only of the naked surface. Since the rapidity of this modification is very sensitive to the percentage of water in the medium in which the samples are stored, it is probable that foreign molecules that modify the naked zones (by adsorption, for example) are water molecules. The interpretation of this experiment was very delicate to make, while we still had not considered that the receding angle was imposed by the highest superficial tension zones, that is, those not covered by a fatty acid film.

This experiment demonstrates as well the very important role played by water vapor adsorption on metallic samples, even in dry environments.

7 THE WETTING TENSIOMETER. ANALYSIS OF A METHOD OF MEASURING CONTACT ANGLES

7.1 Choice and Description of the Measuring Method

7.1.1 Choice of Method

The means by which we may obtain the contact angle of a liquid on a solid can involve an axially symmetrical meniscus (liquid drop or gas-bubble) or a cylindrical meniscus. The experimental conditions of formation of the meniscus are such that it is the advancing or the receding angle that is measured. (For a sessile drop, see Sec. 6.1.)

With axially symmetrical meniscii, the angle can be measured directly (Bigelow et al., 1946), with a goniometer (Fort and Patterson, 1963), from a photograph, or from measurements of the meniscus (Bikerman, 1941). In these methods, the solid surface is investigated only along the line joining the meniscus to the solid. What is more, if the surface is macroscopically heterogeneous, the measured contact angle will not be the same at different points along the edge of the drop. The last method gives a mean contact angle value.

On the other hand, immersion methods for a solid plate that is plunged more or less deeply into a liquid allow us to obtain a mean contact angle for each depth of immersion. The advancing and receding conditions are easily obtained respectively at the times of immersion and emersion of the plate. Three principal methods exist, which permit contact angle measurement when using a solid plate immerged in a liquid:

1. The "tilting plate method" attributed to Adam and Jessop (1863). The plate is inclined until the liquid surface appears to remain perfectly flat right up to the surface of the solid. The inclination of the plate in relation to the horizontal line then corresponds to the contact angle.

2. Neumann's (1974) method consists of plunging the plate vertically and measuring the height of the cylindrical meniscus. This height z is linked to the contact angle by relation (37):

$$z^2 = \frac{2\gamma}{\Delta\rho g}\,(1 - \sin\theta) \tag{37}$$

3. In the wetting tensiometer, the plate is also plunged vertically and we measure the weight of the cylindrical meniscus. This weight W_M is deduced from Eq. (42), and for a plate of perimeter ℓ equals

$$W_M = \gamma_{LV}\,\ell\,\cos\theta \tag{122}$$

First described by Guastalla (1956), the wetting tensiometer not only has advantages particular to immersion methods of a solid plate; it also lets us obtain recordings whose analysis (Chappuis and Georges, 1974) gives us the advancing and receding angles for each immersion depth of the plate.

We shall present and analyze only this latter method. Most experimental methods, with experimental procedures, have been recently reviewed and analyzed by Neumann and Good (1979).

7.1.2 Description of Method

The aim is to measure advancing and receding angles (θ_A and θ_R, respectively) on parallelepipedic samples of nonnegligible section, at different points of their surface. The dispersion of experimental results due to the nonobtention of advancing or receding conditions is eliminated, because the measurement is made during the movement of the plate. However, the displacement speed of the solid in relation to the liquid is so low that the measurements taken correspond to equilibrium states. The measured contact angles represent mean values along the perimeter encircling the solid.

The solid sample is suspended from the arm of an electromagnetic balance. During immersion and emersion of the sample (obtained by moving the liquid surface upward or downward), the force exerted on the arm of the balance is constantly recorded as a function of the height of the liquid surface. The recording obtained (on an *XY* recording table) constitutes the immersion and emersion curves.

The apparatus is assembled in the following way (Fig. 51). The plate is suspended vertically under the arm of an electromagnetic balance (Mettler H 20 E), which records the force variations around a given value corresponding to the sample's weight. The liquid is contained in a beaker of wide section, placed on a vertically mobile support. A small synchronous motor regulates the raising or lowering of the support at a very slow speed (less than 1 mm mn^{-1}). The motor also regulates a potentiometer that supplies a direct current, the tension of which is linear as a function of the height of the support. Thus, a raising of the support corresponds to a plunge of the plate into the liquid. We record the force applied under the balance scale on an *XY* recording table as a function of the distance from the solid's inferior face to the free surface of the liquid. The wetting experiment takes place in a laminar flow of filtered air.

We have verified the following points:

1. On account of the conception of the balance, the vertical position of the scale is not rigorously the same according to the value of the weight. However, with our experimental conditions, the total variation of the scale height is on the order of 0.01 mm, which is negligible.

2. The raising and lowering speed of the liquid (less than 1 mm mn^{-1}) is so low that we can always consider that the system is in equilibrium. For this, we have stopped the liquid displacement at different intervals and verified that the difference

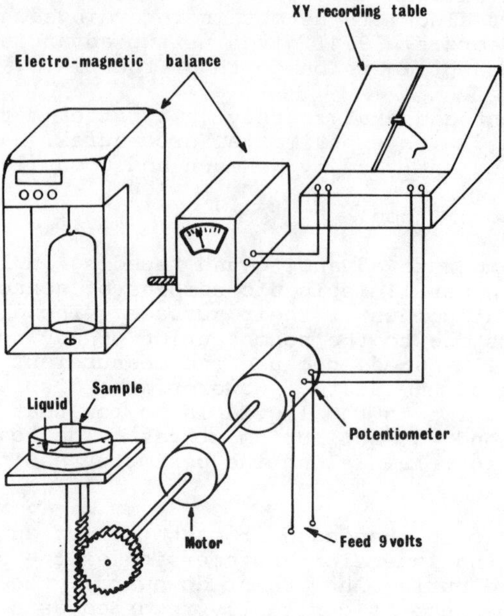

Fig. 51. Experimental apparatus for measuring advancing and receding contact angles.

between the measurement taken after a certain time and that obtained during movement is negligible.

7.2 Theoretical Analysis

7.2.1 Notation and Conventions

On a vertical axis (perpendicular to the water's surface), the forces and algebraic distances will always be counted positively when directed downward. Let us agree to the following:

S' = the section of the sample

ℓ = the perimeter of the sample

H = the height of the sample

M = the mass of the sample

P_A = the atmospheric pressure exerted on the free liquid surface

ρ_L = the volumic mass of the liquid

ρ_A = the volumic mass of air

$\Delta\rho = \rho_L - \rho_A$ the volumic mass of the liquid, less than of air

θ_I = the angle of junction upon immersion (advancing angle)

θ_E = the angle of junction upon emersion (receding angle); $d = \overline{OM}$ on Fig. 52

d = the algebraic distance from the free liquid surface to the inferior plane of the sample; d = OM on Fig. 52

h = the algebraic distance from the free liquid surface to the solid-liquid junction line; $h = \overline{OH}$ on Fig. 52

f = the resultant of forces exerted on the balance scale less the sample's weight in the air

F_S = the resultant of surface forces

F_B = the resultant of buoyancy forces

F_G = the gravitational force on the sample; $F_G = Mg$

On the curves representing the force exerted on the arm of the balance as a function of the sample's penetration into the liquid, we shall take the following origins:

The origin of the forces ($f = 0$) corresponds to the weight of the sample suspended in the air.

The origin of the penetration ($d = 0$) corresponds to the position where the sample's inferior plane becomes confused with the free surface of the liquid.

7.2.2 Nature of the Measured Forces

Three types of forces act on the arm of the balance: the gravitational force F_G, the buoyancy forces F_B, and the surface forces F_S. The force F acting on the arm of the scale is equal to

$$F = F_G + F_B + F_S \tag{123}$$

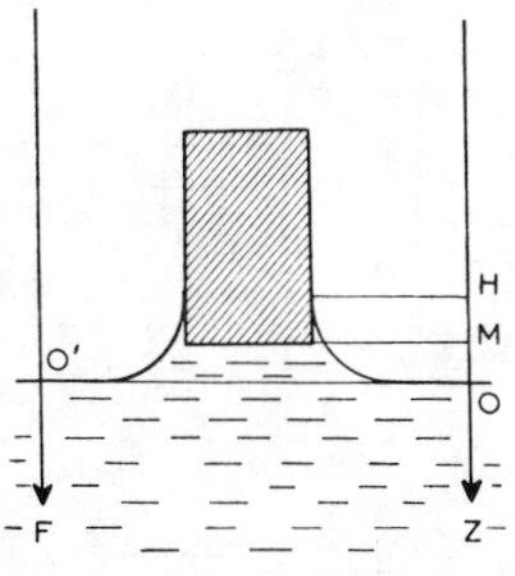

Fig. 52. Conventions

Buoyancy forces are due to different pressures that act on the faces of the solid (Fig. 53). When the sample is partially immersed, the vertical resultant on the superior face is equal to $[P_A - \rho_A g (H - d)]S'$, and on the inferior face $(P_A - \rho_L gd)S'$. Thus

$$F_B = [P_A - \rho_A g (H - d)]S' - (P_A + \rho_L gd)S'$$

$$= - \Delta\rho gdS' - \rho_A gHS' \quad (124)$$

Surface forces have a resultant equal to

$$F_S = \gamma \; \ell \; \cos \alpha \quad (125)$$

When the meniscus is above the free surface of the liquid, the resultant of the surface forces is directed downward (Fig. 54). When the meniscus is below the free surface of the liquid, the resultant of the surface forces in directed upward (Fig. 55). Acting on the scale is

$$F = F_G + F_B + F_S = Mg - \Delta\rho gdS' - \rho_A gHS' + \gamma \; \ell \; \cos \alpha \quad (126)$$

and f, defined as F diminished of the apparent weight of the sample

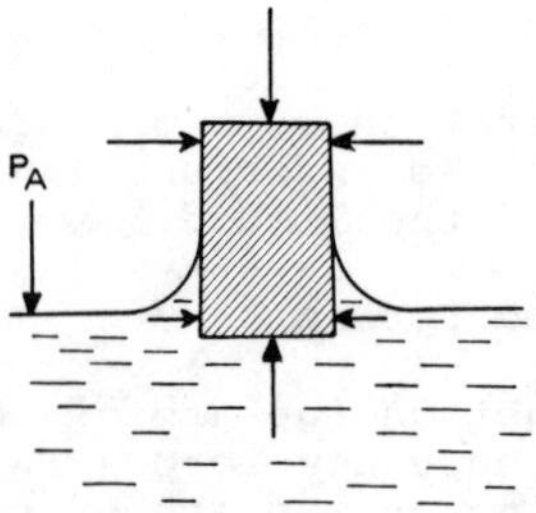

Fig. 53. Forces exerted on the faces of the sample.

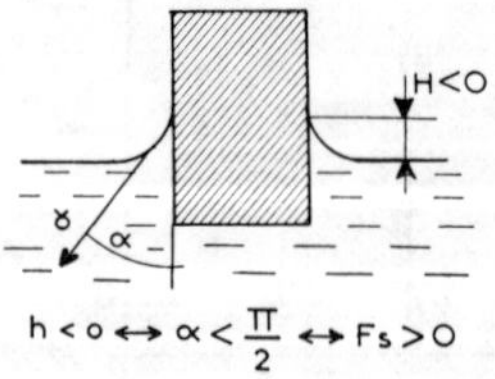

Fig. 54. When the meniscus is above the free surface of the liquid, the resultant of the surface forces is directed downward.

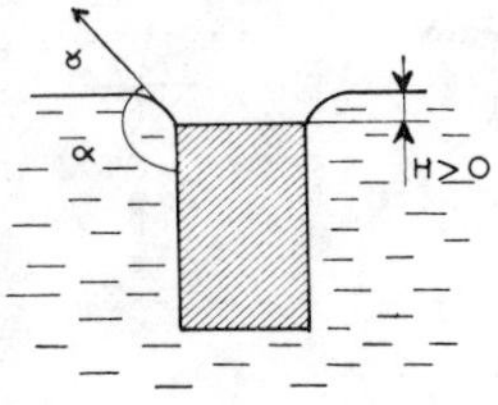

Fig. 55. When the meniscus is below the free surface of the liquid, the resultant of the surface forces is directed upward.

in the air, is written:

$$f = F - (Mg - \rho_A gHS')$$

$$= - \Delta\rho g S'd + \gamma\ \ell \cos\alpha \qquad (127)$$

7.2.3 Height of the meniscii

It has been shown [Eq. (37)] that

$$h^2 = \frac{2\gamma}{\Delta\rho g}(1 - \sin\alpha) = a^2 (1 - \sin\alpha)$$

With our conventions, it should be noted that

$h > 0$ when $\alpha > \pi/2$ (Fig. 55)

$h < 0$ when $\alpha < \pi/2$ (Fig. 54)

7.2.4 Case of Horizontal Edges

The samples have very acute edges. Their radius of curvature is on the order of several micrometers. At the microscopic level, they can be represented by Fig. 56. The penetration of the edge of a sample in a liquid takes place as shown in the succession of diagrams A', B', C', D', E' of Fig. 56 when the contact angle θ has a value between 0 and 90^0. This angle remains constant during the penetration of the edge. The experimenter can observe only the variation of the apparent macroscopic angle α (diagrams A, B, C, D, E of Fig. 57). As soon as the edge has been passed over, the meniscus attaches to the horizontal face of the sample (case D). An ulterior elementary penetration produces an irreversible phenomenon of slipping and rupture of the meniscus, leading abruptly to case E. This phenomenon takes place for the value $\alpha = \theta + \pi/2$. For the same reason, upon emersion of the sample,

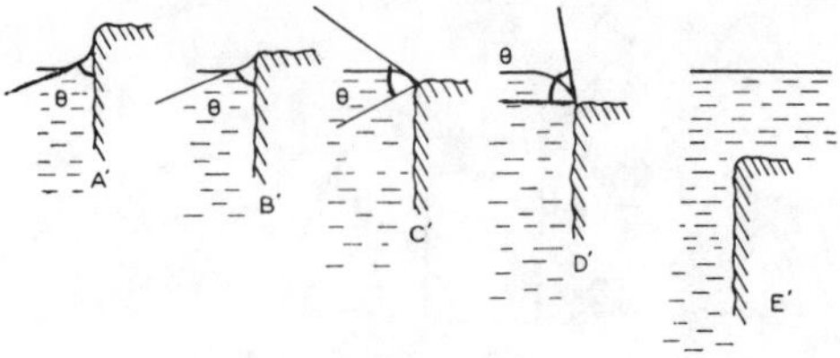

Fig. 56. Penetration of the edge of a sample in a liquid, presented at the microscopic level.

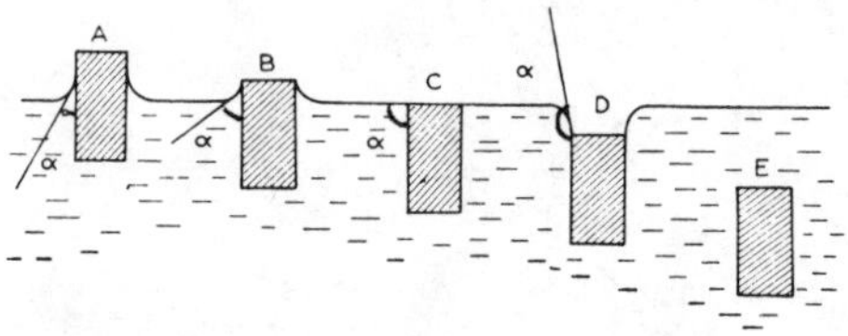

Fig. 57. Penetration of the edge of a sample in a liquid, presented at the macroscopic level.

there is a rupture of the meniscus for the value $\alpha = \theta - \pi/2$.

The length of the line joining the liquid to the solid varies at the time of the penetration of the edge. It can be admitted that if the width of the zone disturbed by the edge is negligible in relation to the dimensions of the samples, the line of junction has the same length as it would if the meniscus were attached to the vertical walls of the sample. In the following sections we shall represent macroscopic diagrams, but we shall always keep in mind equivalent, more realistic diagrams, like those of Fig. 56.

7.3 Physical Significance of the Recordings

7.3.1 Experimental Curves

All experimental curves obtained will be classed in two groups, I and II. Group I represents the immersion, then the emersion of a sample that forms contact angles $\theta < \pi/2$ with the liquid. The contact angles θ in group II are, by opposition, $> \pi/2$. The observation of different phases of wetting is paralleled to the recording of force-displacement curves, each letter corresponding to a particular stage. We thus find group I in Fig. 58, and group II in Fig. 59.

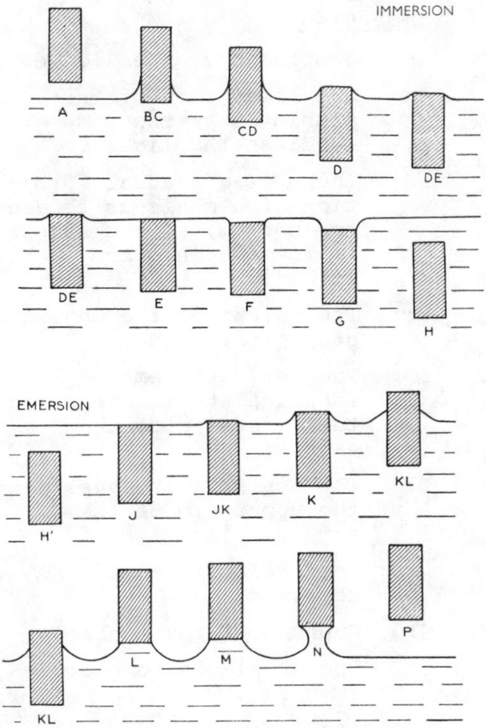

Figure 58A: Observation of different macroscopic aspects of the meniscus when θ<π/2.

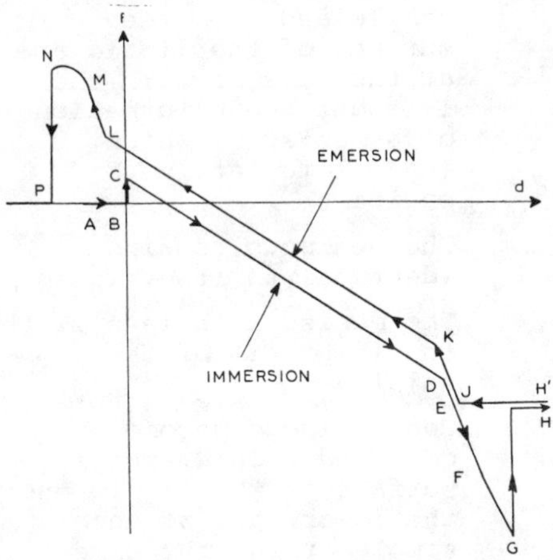

Figure 58B: Recorded curve when θ<π/2.

Immersion

A: Sample above the liquid $f = 0$.

B: The lower face of the sample is at the same level as the liquid. A meniscus of weight *BC* (on Fig. 58B) forms spontaneously.

CD: The meniscus remains identical to itself during the immersion of the sample.

D: The meniscus is level with the upper rim of the sample.

DEFG: The meniscus becomes deformed. In *E*, the free surface of the liquid and the upper face of the sample are at the same level.

G: Rupture of the meniscus.

H: The sample is completely immersed.

Emersion

H': The sample is completely immersed.

J: The upper face of the sample and the free surface of the liquid are at the same level.

JK: The meniscus is formed gradually as the sample is withdrawn.

K: The height of the meniscus stops increasing.

KL: The meniscus remains identical to itself during the emersion of the sample.

L: The meniscus is level with the lower rim of the sample.

LMN: Deformation of the meniscus.

N: Rupture of the meniscus.

P: The sample is completely emerged.

Fig. 58. Curve I. Meniscus above the free surface of the liquid. Macroscopic aspects of the meniscus, experimental curve, and experimental observations (θ<π/2).

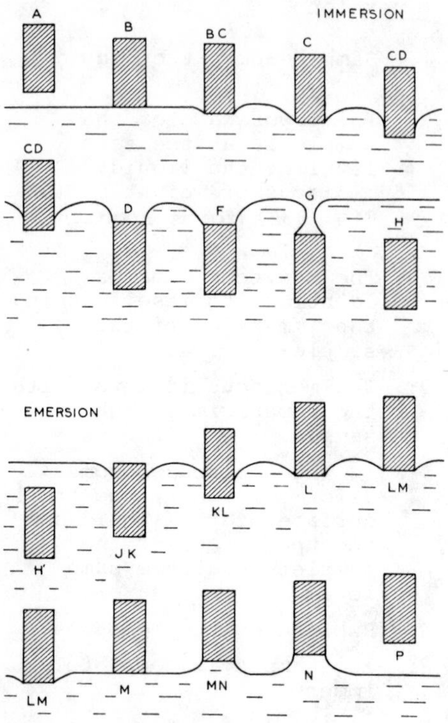

Figure 59A: Observation of different macroscopic aspects of the meniscus when $\theta>\pi/2$.

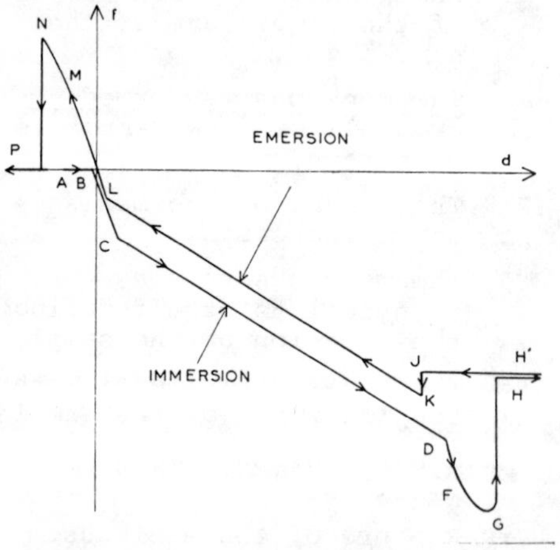

Figure 59B: Recorded curve when $\theta>\pi/2$.

<u>Immersion</u>

AB: Sample above the liquid.

B: The lower face of the sample is at the same level as the liquid.

BC: There is a gradual formation of a meniscus toward the bottom, as the sample is lowered into the liquid.

C: The height of the meniscus stops increasing.

CD: The meniscus remains identical to itself during the immersion of the sample.

D: The meniscus is level with the upper rim of the sample.

DFG: The meniscus become deformed.

G: Rupture of the meniscus.

H: The sample is completely immersed.

<u>Emersion</u>

H'J: The sample is completely immersed.

J: The upper face of the sample and the free surface of the liquid are at the same level; there is spontaneous formation of a meniscus, which leads immediately to position *K*.

KL: The meniscus remains identical to itself.

L: The meniscus is level with the lower rim of the sample.

LMN: The meniscus become deformed. In *M*, the free surface of the liquid and the lower face of the sample are at the same level.

N: Rupture of the meniscus.

P: The sample is completely emerged.

Fig. 59. Curve II. Meniscus below the free surface of the liquid. Macroscopic aspects of the meniscus, experimental curve, and experimental observations ($\theta>\pi/2$).

7.3.2 Theoretical Curves

The analysis shows four successive stages in both the immersion and emersion phenomena:

1. Formation of the meniscus

2. Displacement of the meniscus along the vertical walls of the sample

3. Evolution of the meniscus as it attaches to a horizontal rim

4. Rupture of the meniscus

We shall suppose that the sample presents a constant contant angle θ_I upon immersion, and a constant contact angle θ_E upon emersion.

Stage 1: Two cases are possible: In the first case, $\theta_E > \pi/2$ or $\theta_I < \pi/2$. A meniscus forms instantaneously and irreversibly when the lower (or upper) horizontal face of the sample attains the level of the free surface of the liquid. This is represented by the vertical parts *BC* and *JK* of Figs. 58B and 59B respectively.

$$\overline{BC} = \gamma \ \ell \cos \theta_I \tag{128}$$

$$\overline{JK} = \gamma \ \ell \cos \theta_E \tag{129}$$

In the second case $\theta_E < \pi/2$ or $\theta_I > \pi/2$. There is no abrupt variation of the angle θ, and the process is reversible. During this phase, the meniscus evolves, all the while attaching itself to a horizontal rim of the sample. The height of the meniscus varies, bringing about a variation of the angle α, thus of the surface force. We have two possible cases: $h = d$ when the meniscus attaches itself to the lower rim (Fig. 60), or $h = d - H$ when the meniscus attaches itself to the upper rim (Fig. 61). The equation of the curves $f(d)$ are then obtained by eliminating α and h between the following equations:

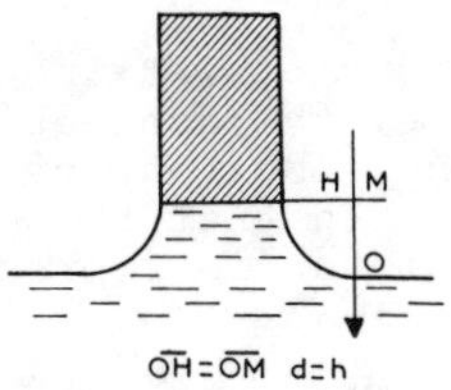

Fig. 60. The meniscus attaches itself to the lower rim of the sample.

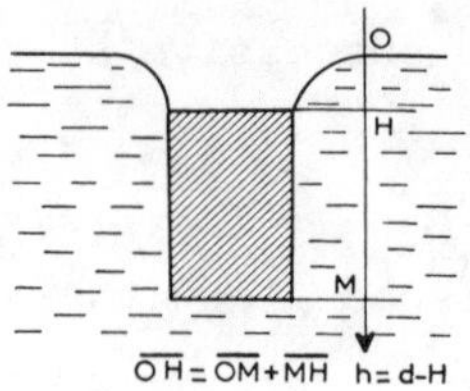

Fig. 61. The meniscus attaches itself to the upper rim of the sample.

$$h^2 = a^2 (1 - \sin\alpha) \qquad (37)$$

$$f = -\Delta\rho g S'd + \gamma\, \ell \cos\alpha \qquad (127)$$

$$\begin{cases} h = d \\ \text{or} \\ h = d - H \end{cases}$$

We obtain the equation of the curves C_1 (when $h = d$) and C_2 (when $h = d - H$).

$$C_1: \quad f = -\Delta\rho g S'd - \frac{\gamma}{a}\, \ell d \sqrt{2 - d^2/a} \qquad (130)$$

$$C_2: \quad f = -\Delta\rho g S'd - \frac{\gamma}{a}\, \ell (d - H) \sqrt{2 - (d - H)^2/a^2} \qquad (131)$$

The curves C_1 and C_2 are represented in Fig. 62. C_1 is centered in $B(0,0)$ and C_2 in $J(H, -\Delta\rho g S'H)$. C_2 is deduced from C_1 by a translation of vector $\overrightarrow{BJ}$. The dotted parts of the curves C_1 and C_2 do not have a physical significance. They are the parts corresponding to $d > a$ for C_1 (α would be $> \pi$), and to $d < (H - a)$ for C_2 (α would be < 0). In the case of immersion, the curve C_1 will be obtained experimentally between the ordinates $f = 0$ (point B) and $f = \gamma\, \ell \cos\theta_I$ (point C). In the case of emersion, the curve C_2 will be obtained experimentally between the ordinates $f = -\Delta\rho g H S'$ (point J) and $f = -\Delta\rho g H S' + \gamma\, \ell \cos\theta_E$ (point K).

Stage 2. If the conditions of wetting are uniform and identical all along the sample, the contact angle remains constant, and in consequence so does the resultant of the wetting forces F_S. Only buoyancy forces F_B vary. In that these forces are proportional to the penetration, the phenomenon is represented by a straight line, the slope of which is $-\Delta\rho g S'$.

This stage is represented by parts CD and KL of the experimental curves, which are straight lines of slope $-\Delta\rho g S'$, and intersecting the axis Of at the respective values $\gamma\, \ell \cos\theta_I$ and $\gamma\, \ell \cos\theta_E$.

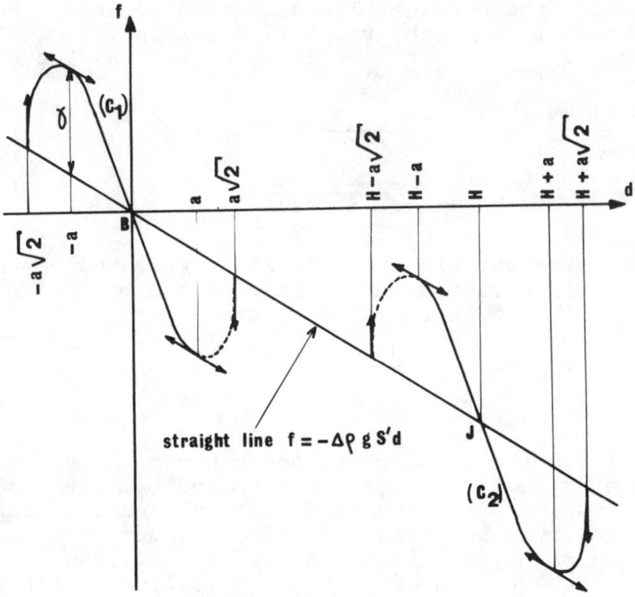

Fig. 62. Graphic representation of curves C_1 [Eq. (130)], and C_2 [Eq. (131)]. Portions of these curves are obtained experimentally when the meniscus attaches itself to the upper rim (C_1) or the lower rim (C_2) of the sample.

Stage 3. The parts *DFG* and *LMN* of the experimental curves indicate the phenomenon of junction of the meniscus to the horizontal rims of the sample. It is thus the same phenomenon as was already studied in stage 1 (second case). *DFG* and *LMN* are therefore portions of curves C_1 and C_2.

Stage 4. The rupture of the meniscus takes place upon immersion at point *G* for an angle $\alpha = \theta_I + \pi/2$, and upon emersion at point *N* for an angle $\alpha = \theta_E - \pi/2$ (from results of Sec. 7.2.4). We have calculated the abscissae of *G* and *N*:

$$\text{The abscissa of } G\text{:} \quad H + a\sqrt{1 - \cos\theta_I} \tag{132}$$

$$\text{The abscissa of } N\text{:} \quad -a\sqrt{1 + \cos\theta_E} \tag{133}$$

7.3.3 Theoretical Construction of a Curve

Knowing the dimensions of the sample and the constants of the liquid, the curve $f(d)$ can be constructed for a sample that offers a contact angle θ_I upon immersion and θ_E upon emersion. The different elements are parts of curves C_1 and C_2 (stage 1, second case, and stage 3; the straight lines Δ and Δ' of slope $-\Delta\rho g S'$ and intersecting the axis Of of the respective values $\gamma \ell \cos \theta_I$ and $\gamma \ell \cos \theta_E$ (stage 2); the straight lines d_1 and d_2, the equations of which are

$$d_1 = - a \sqrt{1 + \cos \theta_E}$$

$$d_2 = H + a \sqrt{1 - \cos \theta_I}$$

and possibly the straight lines $d = 0$, if $\theta_E > \pi/2$, and $d = H$ if $\theta_I < \pi/2$ (stage 1, first case). An example of a sketch with these elements is shown in Fig. 63.

7.4 Practical Use

7.4.1 Choice of Samples

Using parallelepipedic samples, one can get the values of θ_I and θ_E from every point of the lines *CD* and *KL*, but also from the abscissa of the points *G* and *N*. Using cylindrical samples, curves of the same type are obtained. The measure of θ_I and θ_E from lines *CD* and *KL* is done in the same fashion. The obtainment of θ_I and θ_E from the points of rupture is realized after the experimental measurements of the coordinates of *G* and *N*, and by using the relations

$$\gamma \ell \cos \theta_I = -\Delta\rho g S' d_G - f_G \tag{134}$$

$$\gamma \ell \sin \theta_E = \Delta\rho g S' d_N + f_N \tag{135}$$

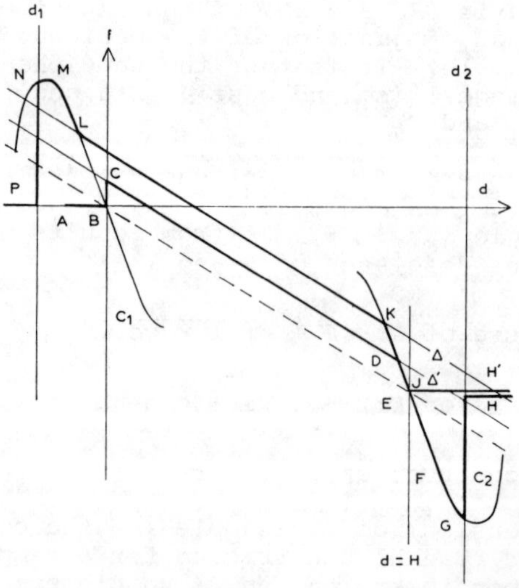

Fig. 63. Theoretical construction of a curve.

The method has also been extrapolated for spherical samples (Chappuis et al., 1977).

7.4.2 Choice and Influence of Liquid

It is preferable to choose a liquid of high surface tension (for example, distilled water, $\gamma = 72.8$ dyn cm^{-1}, or methylene iodide, $\gamma = 50.8$ dyn cm^{-1}). This offers several advantages:

A greater precision in the measurement of $\gamma \ell \cos \theta$.

With a given sample, the experiment most often shows a contact angle that is null with a liquid of low surface tension and a nonnull contact angle with a liquid of high surface tension. Thus, with a liquid of high surface tension it is generally possible to distinguish two solid surfaces that would offer the same behavior with a liquid of low surface tension.

So as to limit the adsorption of vapors of the selected liquid on the sample, it is necessary to choose a liquid possessing a low vapor tension. So as to make evident the modifications of the solid surface after its immersion in the liquid, it is a good idea to make two or more successive cycles of immersion and emersion (see Sec. 6.4).

7.4.3 Determination of Macroscopic Heterogeneity of Surface "Pollution"

For very smooth samples, the macroscopic heterogeneity of the surface pollution can be determined experimentally by observing the lines *CD* and *KL*, which may be straight lines (homogeneous surface) or, on the contrary, more or less deformed lines (heterogeneous surface). Figure 64 is the experimental curve obtained with a metallic sample covered with a monomolecular film of stearic acid. The angle upon immersion is $<\pi/2$, and the angle upon emersion

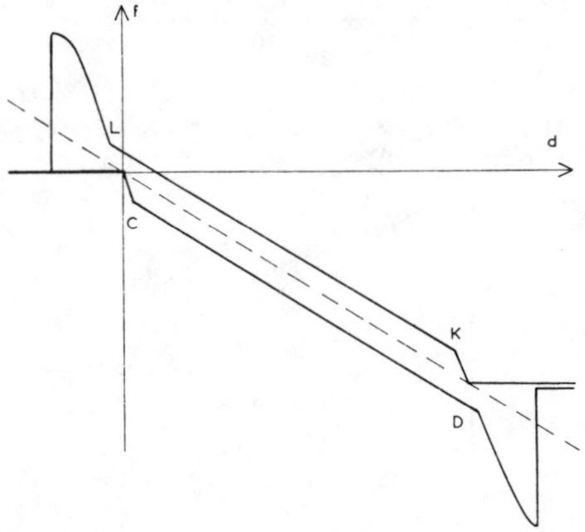

Fig. 64. Experimental curve obtained with a metallic sample covered with a monolayer of stearic acid.

is >π/2. The straightness of lines *CD* and *KL* proves the macroscopic homogeneity of the film on the surface.

Figure 65 is the experimental curve obtained with a metallic sample having gone through a cleansing cycle followed by drying in a stove. The deformed line *CD* shows the heterogeneity of the "cleanliness" of the surface. The line *KL* is not deformed because the contact angle upon emersion is null.

Figure 66 is the experimental curve obtained with a metallic sample that has been partially immersed in a solvent containing

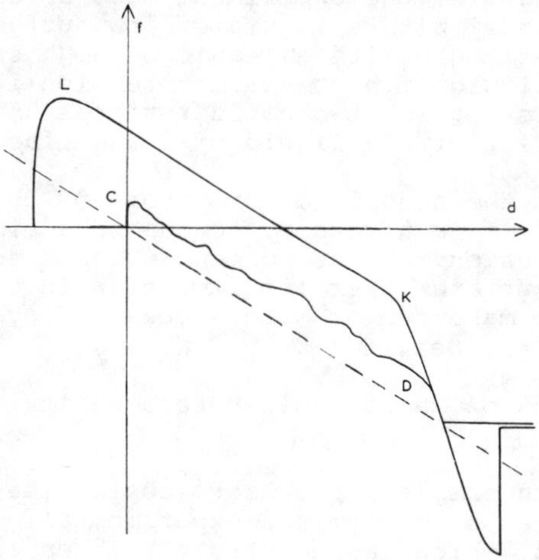

Fig. 65. Experimental curve obtained with a metallic sample that has been cleaned and dried.

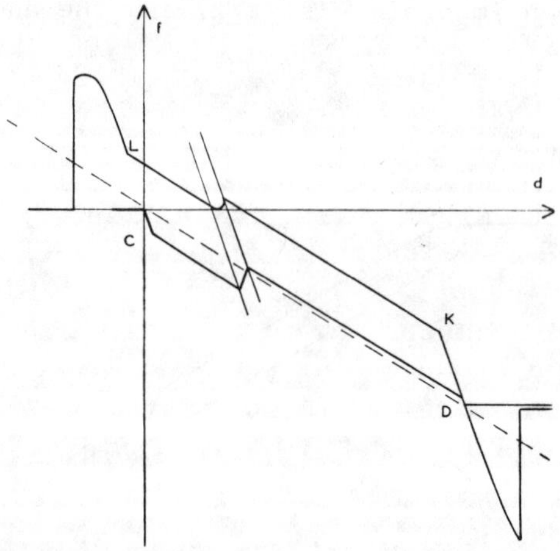

Fig. 66. Experimental curve obtained with a metallic sample that has been partially immersed in a solvent containing impurities.

impurities. A line situated at approximately on-third of the height from the base of the sample separates the cleanest upper zone (where the surface tension is the strongest) from the less - clean lower zone (where the surface tension is weaker).

7.4.4 Utilization of the Technique as a Test of Cleaning

It has been shown (Zisman, 1959) that the weakest values of θ characterize the best cleansings. In the laboratory, with nonporous metallic samples, and for different cleansing cycles, we have always obtained $\theta_E = 0^0$ and never $\theta_I = 0^0$, using water as the liquid of reference. The value of θ_I is thus a criterion for testing different cleansings. Also, the value of θ_I can be reproduced for analogous samples having gone through identical cleansing cycles.

The technique was also used to test lubricating films (monomolecular or not) voluntarily placed on the solids. The obtainment of values of θ_E and θ_I (never null) is very useful in gaining information about the films so placed. The reader may refer to Zisman's texts (1959, 1964) concerning this subject.

8 INTERPRETATION OF CONTACT ANGLE VALUES

8.1 Existence of an Equation of State

Suppose that we have use of an "ideal," perfectly smooth and homogeneous, horizontal solid surface, located in a vacuum. Its superficial tension is γ_{SO}. A drop of liquid, marked ('), having a superficial tension $\gamma_{L'V'}$, is introduced into the enclosure, and placed on the solid surface. The drop becomes partially vaporized so that the saturating vapor pressure that corresponds to equilibrium is established at all points of the enclosure. The presence of the liquid's vapor in the enclosure brings about the phenomenon of adsorption on the surface of the solid that is not in contact with the drop. The presence of the vapor's molecules, adsorbed at the solid's surface, diminishes its superficial tension by a value π'_e. Relation (68), established in Sec. 4.4, is now written

$$\gamma_{SO} - \gamma_{SV'} = \pi'_e \tag{136}$$

At equilibrium, the contact angle θ' of compound (') on the solid should prove Young's equation (69):

$$\gamma_{SV'} - \gamma_{SL'} = \gamma_{L'V'} \cos\theta' \tag{137}$$

or, by taking (136) into account;

$$\gamma_{SO} - \pi'_e - \gamma_{SL'} = \gamma_{L'V'} \cos\theta' \tag{138}$$

Superficial tension $\gamma_{L'V'}$ is known, and contact angle θ' can be measured. However, γ_{SO}, π'_e and $\gamma_{SL'}$ cannot be measured. We thus have at our disposition only one equation (138) for three unknowns (γ_{SO}, π'_e, and $\gamma_{SL'}$).

Repeating the operation with a second liquid, marked ("), leads to

$$\gamma_{SO} - \pi_e'' - \gamma_{SL''} = \gamma_{L''V''} \cos \theta'' \quad (139)$$

We shall now have a supplementary equation (139), but with two supplementary unknowns (π_e'' and $\gamma_{SL''}$).

One can thus imagine that the measurement of any liquid's contact angle, in equilibrium on an "ideal" solid surface, characterizes only the solid-liquid pair studied, and that it cannot give us any information concerning the behavior of another liquid on the same solid surface. But it has been verified that if we measure the contact angle θ of different liquids on the same real solid, and if we plot $\cos \theta$ as a function of γ_{LV} on a diagram, the experimental points are always assembled on a curve (see, for example, Fig. 67). This should signify that the quantities intervening in Eqs. (138) and (139) need not be independant, but may be connected by other relations. It is these relations that we propose to investigate by the interpretation of experimental curves of type $\cos \theta = f(\gamma_{LV})$. Of course, the choice of experimental results to be analyzed is extremely important, and we shall try to determine which criteria they should offer (see Sec. 8.3). Before doing this, we shall talk about the value of π_e according to the solid-liquid pair studied.

8.2 The Importance of π_e in Different Cases

Fowkes (1967) says that "materials having a high superficial energy cannot become adsorbed on materials having low energy in such a way to diminish the superficial tension of the latter." Neumann (1974) gets the same result by thermodynamic considerations. He thinks that adsorption will not play a major role for systems obeying $\gamma_{SO} < \gamma_{LV}$. He deduces from the few experimental data available at the present time (Whalen and Wade, 1967; Neumann and Sell, 1964) that the equilibrium spreading pressure π_e is normally less than approximately 1 erg cm^{-2} if the contact angle is not too low, say, above 20 or 30^0.

Thus, on a given solid, liquids of higher surface energy (or superficial tension) do not become adsorbed, and π_e must be null. By contrast, on the same solid, liquids of low superficial tension risk becoming adsorbed and forming a more or less dense film, which brings about the reduction of the solid's superficial tension. This reduction will be greatest for those liquids having a low superficial tension. In fact, these latter are the ones for which the saturating vapor pressure is greatest. Therefore, for two liquids having proximate affinities with regard to the solid's surface (two alkanes, for example), the one whose superficial tension is the lowest should produce the greatest spreading pressure π_e.

Moreover, with different liquids on the same solid, we proceed from the case of nonwetting (θ finite), to the case of complete wetting ($\theta = 0^0$) when the attraction of liquid molecules by the solid becomes greater than the attraction of liquid molecules among themselves (see Sec. 1.1). It is logical to think that when one takes liquids with lower and lower superficial tensions, and as one nears this transition, the attraction of molecules of the liquid's saturation vapor by the solid becomes great enough to form a dense layer cf adsorbed molecules, which has a nonnegligible spreading pressure π_e.

8.3 The Choice of Proper Experimental Results

We shall now investigate experimental results concerning contact angles of different liquids on a same solid. The criteria that will determine our choice are the following:

1. So as to have a general view of the phenomenon, the solid used should have a surface tension such that, in the range of liquids used, the vapor shows the adsorption phenomenon for some, and not for others. Since the mean superficial tension of solids is much higher than the mean superficial tension of liquids, the conclusions of Sec. 8.2 lead us to choose a solid with a low surface tension.

2. The measured contact angle should really correspond to the theoretical contact angle occurring in Young's equation. This requires control of the hysteresis phenomenon. Given that the solid should have a low superficial tension, it is possible that the heterogeneity of the solid's surface will convey itself by spots of higher tension (impurities, for example). On the other hand, when adsorption occurs we want to obtain the contact angle governed by the surface covered with adsorbed molecules. In the case, our conclusions from Sec. 6.2.4, let us state that it is the advancing angle that should be measured. In addition, we need very smooth solid surfaces, so that the hysteresis brought into play by roughness will remain negligible.

3. There should be a large number of liquids used, varying in chemical composition.

4. The experimental results should be precise and trustworthy.

These criteria led us to choose experimental results relative to the behavior of different liquids on polytetrafluorethylene (PTFE), mentioned by Fox and Zisman (1950). In fact:

1. PTFE is a solid with a very low superficial tension.

2. PTFE is a solid on which it is possible to obtain excellent surface conditions. Fox and Zisman proved that the roughness of their samples was not a parameter of contact angle hysteresis. Furthermore, and for experimental reasons, the measured contact angle was the advancing angle in all cases (it was the most easily reproducible one).

3. The list of liquid used by these authors is impressive. What

is more, the liquids were chosen from very different chemical families.

4. From an experimental point of view, the authors have taken great care regarding the preparation of the sample surfaces (smoothness), the cleansing of the PTFE, very extreme purification of liquids, determination of the liquid's superficial tensions, and contact angle measurements. We may have total confidence in these results. Besides, other authors (Doss and Rao, 1938) have used PTFE, and their results coincide perfectly with those of Fox and Zisman.

8.4 Analysis of Results

The curves $\cos\theta = f(\gamma_{LV})$ and $W_A - W_C = F(\gamma_{LV})$ figure in most books treating contact angles problems. They are reproduced in Figs. 67 and 68. Let us remember that W_A is the work of adhesion defined by (92), and that W_C is the cohesion work of the liquid defined by (94):

$$W_A = \gamma_{LV}\,(1 + \cos\theta) \qquad (92)$$

$$W_C = 2\,\gamma_{LV} \qquad (94)$$

Thus

$$W_A - W_C = \gamma_{LV}\,(\cos\theta - 1) \qquad (140)$$

All the liquids used have a surface tension between 12.8 and

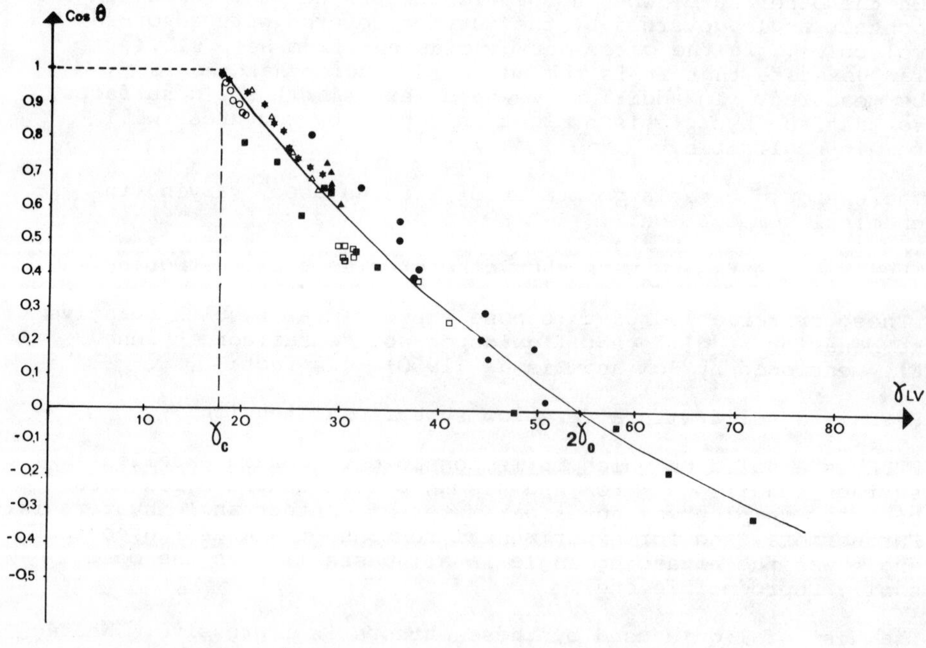

Fig. 67. Contact angles of liquids of different surface tensions on PTFE: cosine θ versus surface tension.

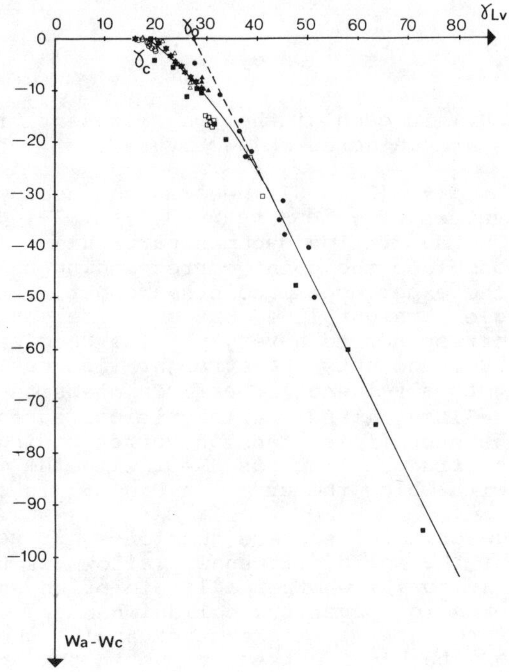

Fig. 68. Contact angles of liquids of different surface tensions on PTFE: $W_A - W_C$ versus surface tension.

72.8 dyn cm^{-1}, except mercury with γ_{LV} = 485 dyn cm^{-1}. This is due simply to the fact that, among liquids at a temperature of 20°C, mercury has the highest superficial tension (485 dyn cm^{-1}). This value is much greater than that relative to distilled water (72.8 dyn cm^{-1}), which constitutes the value immediately below it. Values corresponding to experiments with mercury thus have an exceptional interest for the validation of any given theory. We shall not include the results corresponding to mercury on our graphs, so that the experimental points are not all regrouped in an extremely narrow zone. Nevertheless, while interpreting these graphs we shall imagine the positions of points corresponding to mercury and defined by the following experimental values:

$$\gamma_{LV} = 485 \text{ dyn cm}^{-1}$$

$$\theta = 150^0$$

$$\gamma_{LV} \cos\theta = -420 \text{ dyn cm}^{-1}$$

$$W_A = 65 \text{ dyn cm}^{-1}$$

$$W_C \qquad = 970 \text{ dyn cm}^{-1}$$

$$W_A - W_C \qquad = - 905 \text{ dyn cm}^{-1}$$

As we said in Sec. 8.1 in each of the two representations, the experimental points are gathered on a curve.

The curve cos = $f(\gamma)$ (Fig. 67) becomes confused with a straight line for values of γ less than 40 dyn cm^{-1}, and distances itself later on. The separation becomes particularly large for high surface tensions (see the point corresponding to mercury). For low tensions, the experimental points are not, in fact, regrouped on a single straight line, but on different neighboring lines. Each one corresponds to a very precise chemical family (Figs. 69 through 72), and cuts the straight line $\cos \theta = 1$ at a point corresponding to a value of superficial tension γ between 17.5 and 20.5 dyn cm^{-1}. Zisman (1963) calls critical surface tension, γ_c, the value of the superficial tension corresponding to the intersection of the straight line $\cos \theta = 1$ with the mean straight line, of slope-b, with which the curve in Fig. 67 is confounded.

Even though the critical surface tension γ_c is not a characteristic quantity of the solid, it usually allows us to foresee (when γ_{LV} is not next to γ_c) whether a liquid of known superficial tension γ_{LV} will spread out over the solid (when $\gamma_{LV} < \gamma_c$) or, on the contrary, will present a finite contact angle (when $\gamma_{LV} > \gamma_c$). This concept is shown to be of extreme value in the analysis of wetting phenomena.

The curve $W_A - W_C = f(\gamma_{LV})$ cuts the straight line $W_A - W_C = 0$ at the point corresponding to the value $\gamma = \gamma_c$, which is a logical result. This curve includes a linear portion corresponding to surface tension values greater than 40 dyn cm^{-1} (see Fig. 68). The straight line on which there are points of superficial tension

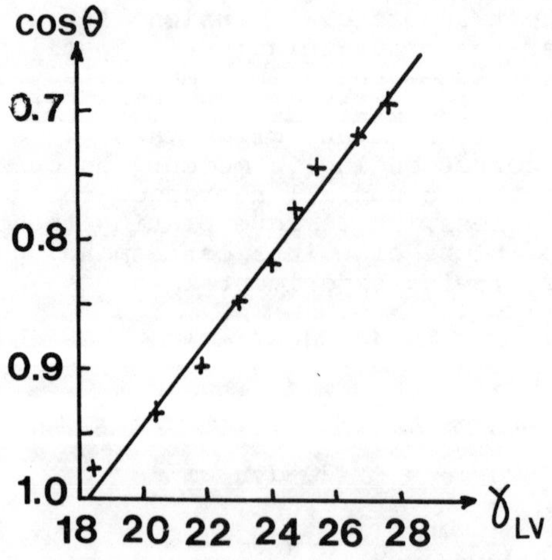

Fig. 69. Surface tension versus cos θ for the n-alkanes on PTFE.

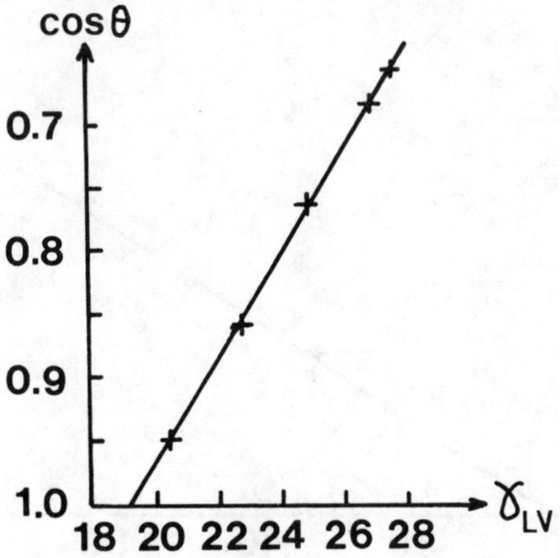

Fig. 70. Surface tension versus cos θ for the di(*n*-alkyl)ethers on PTFE.

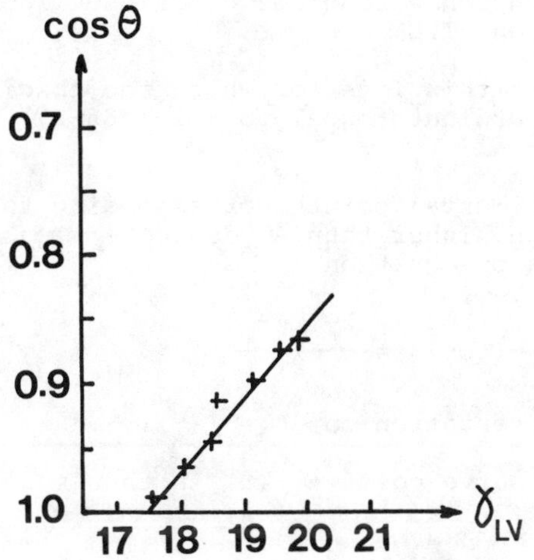

Fig. 71. Surface tension versus cos θ for the halocarbons on PTFE.

higher than 40 dyn cm^{-1} has a slope equal to -2, and cuts the axis of the abscissa ($W_A - W_C = 0$) at a point for which the abscissa is greater than γ_c. We shall call γ_O the abscissa of this point. The equation of this straight line is

$$W_A - W_C = -2(\gamma - \gamma_O) \quad (141)$$

In the case of PTFE $\gamma_O = 27$ dyn cm^{-1}.

The experimental point corresponding to mercury ($\gamma_{LV} = 485$ dyn cm^{-1}, $W_A - W_C = -905$ dyn cm^{-1}), is very near the line of Eq. (141).

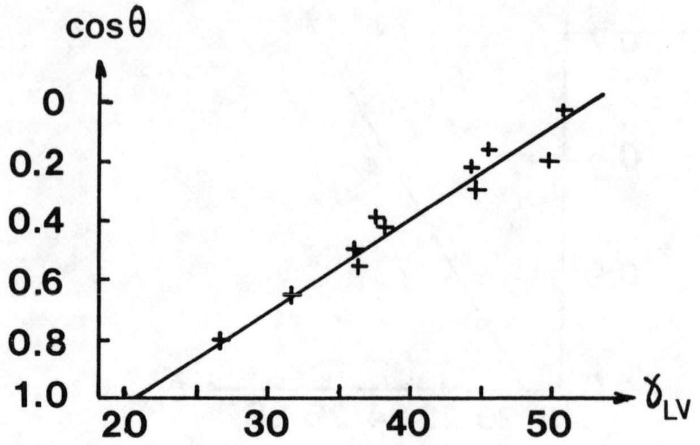

Fig. 72. Surface tension versus cos θ for the linear polymethylsiloxanes on PTFE.

In fact, the point of this line for which the abscissa is equal to 485 dyn cm^{-1} has an ordinate equal to - 2 (485-27), that is, - 916 dyn cm^{-1}.

Thus, all experimental points corresponding to liquids with a superficial tension higher than 40 dyn cm^{-1} satisfy relation (141), which is a hyperbola of equation.

$$\cos \theta = \frac{2\gamma_O - \gamma}{\gamma} \qquad (142)$$

in the graphic representation cos θ = $f(\gamma)$.

In conclusion, curve cos θ = $f(\gamma)$ is thus made of two distinct parts:

1. The points associated with abscissas less than 40 dyn cm^{-1} are regrouped on a network of neighboring straight lines. Each of them corresponds to a fixed chemical family.

2. The points corresponding to abscissas greater than 40 dyne cm^{-1} are regrouped on a single curve, which is a hyperbola.

8.5 Critical Analysis of Zisman's Interpretation

Zisman's interpretation is based on the curves analysis for values of γ_{LV} less than 40 dyn cm^{-1} with the hypothesis that π_e is negligible. For all liquids tested, the contact angle measured in an unlimited atmosphere was the same as that in a vessel containing the liquid's saturating vapor. Zisman used this fact as a basis to affirm that the pressure of adsorbed film is negligible, whatever the liquid used, on a surface of PTFE.

Let there be a system constituted by a drop of a given liquid in contact with a horizontal surface of PTFE. When this system exists in an unlimited atmosphere, there is a continual diminution of the partial pressure of the liquid's vapor, corresponding to the distance away from the drop. This phenomenon is governed by the laws of diffusion. Let us note that, at equilibrium, the saturating vapor pressure in the vapor layer in contact with the drop is imposed by the drop. Because of its curvature, the drop also imposes a saturating vapor pressure greater than that of a plane surface (see Sec. 2.8). Consequently, the vapor layer in proximity to the drop has a partial pressure of the liquid's vapor that is identical whether the measurement is taken in an open enclosure or in a closed one saturated by the liquid's vapor. Moreover, if the superficial tension of a solid is not uniform over its entire surface (because the adsorption is not everywhere the same), then the superficial tension of the solid in immediate proximity to the drop determines the liquid's behavior toward the solid.

It is thus normal to find identical contact angle values in both cases, but deducing from this that the adsorbed film's pressure π_e is negligible would be unjustified.

Later on, Zisman considers that the physical law between θ and γ in the absence of adsorption is characterized by the portions of curves corresponding to $\gamma<40$ dyn cm^{-1}. He is thus led to explain why liquids with a high superficial tension do not have contact angles verifying a linear relation $\cos\theta = f(\gamma)$, in the following way: If, for some liquids the linear law is not confirmed, this results from the formation of weak hydrogen bonds between the liquid's molecules and those of the solid surface. This is most likely in the case of liquids of high surface tension, which are always liquids with hydrogen bondings.

This last explanation is not satisfactory, because there exist liquids with a high superficial tension whose molecules contain no hydrogen atoms. Moreover, Zisman does not explain why the contact angles of liquids of high surface tension confirm a hyperbolic law of equation

$$\cos\theta = \frac{-\gamma + C}{\gamma}$$

in which C is a constant.

8.6 Other Important Theories in the Literature

Other studies have been carried out concerning the interpretation of contact angle values on low-energy surfaces. The most important, (Good, 1952, 1977, 1979), is based on the use of relation (11):

$$\gamma_{12} = \gamma_1 + \gamma_2 - 2\,\phi\,\sqrt{\gamma_1\gamma_2} \qquad (11)$$

established in the case of the interfacial tension between two nonmiscible liquids 1 and 2, and extrapolated to the solid-liquid interfacial tension γ_{SV}.

Fowkes (1964, 1967) slightly modified Good's model by using γ_{SV} extrapolated from Eq. (12):

$$\gamma_{12} = \gamma_1 + \gamma_2 - 2\sqrt{\gamma_1^d \gamma_2^d} \qquad (12)$$

Neumann (1974) proposed two methods to formulate an equation of state, both using Good's interaction parameter.

All these theories are now accepted in the field, and their knowledge is particularly important for a complete view of the subject. In each of them, however, it is always assumed, when using the experimental data, that π_e is negligible for all liquids on low-energy surfaces. As the author, for his part, refuses this assumption (see Sec. 8.2), his interpretation and conclusions will be quite different from those of Good, Fowkes, and Neumann.

8.7 The Author's New Interpretation (Chappuis, 1974)

Consequently, we shall stick with what was said in Sec. 8.2 concerning the estimation of the term π_e. Furthermore, given the solid we have chosen (see Sec. 8.3, first criterion), we can expect:

A null value of π_e for liquids having the highest superficial tensions

A nonnull value of π_e for liquids having the lowest superficial tensions

A larger and larger value of π_e for liquids with lower and lower superficial tensions

An examination of the diagrams (Fig. 67 and 68) shows a discontinuity in the outline of each of them, corresponding to an abscissa of the order of 40 dyn cm^{-1}. Our interpretation will be based on this single hypothesis:

$$\pi_e = 0 \quad \text{when } \gamma_{LV} > 40 \text{ dyn cm}^{-1}$$

$$\pi_e \neq 0 \quad \text{when } \gamma_{LV} < 40 \text{ dyn cm}^{-1}$$

It has been established in Secs. 5.6 and 5.7 that the work of adhesion of a liquid to a solid depends on the adsorption of the liquid's vapor at the solid surface. To separate the part due to adsorption, Harkins and Colleagues (Harkins and Loeser, 1950; Harkins and Livingstone, 1952) have introduced W_{AO} (work of adhesion of a liquid to the bare solid), which expresses the interaction between the molecules of the liquid and the atoms of the solid in the solid-liquid interface. W_{AO} is not influenced by the possible presence of an adsorbed film on the solid-vapor interface. The law $W_{AO} = f(\gamma_{LV})$ for different liquids on PTFE will then show no discontinuity whether γ_{LV} is greater or less than 40 dyn cm^{-1}. We are going to:

1. Establish the law $W_{AO} = f(\gamma_{LV})$ with liquids corresponding to

γ_{LV} greater than 40 dyn cm^{-1}. For them $\pi_e = 0$, so that W_{AO} is equal to the experimental quantity W_A.

2. Extrapolate the law $W_{AO} = f(\gamma_{LV})$ for liquids corresponding to γ_{LV} less than 40 dyn cm^{-1}.

3. Determine the theoretical equation of the complete experimental curve $W_A - W_C = f(\gamma_{LV})$ of Fig. 68.

4. Determine the theoretical equation of the complete experimental curve $\cos\theta = f(\gamma_{LV})$ of Fig. 67.

1. For liquids corresponding to γ_{LV}>40 dyn cm^{-1}, we have the relation expressing the nullity of π_e, Young's equation, and the experimental relation between $\cos\theta$ and γ, which is the equation of a hyperbola.

γ_{LV}>40 dyn cm^{-1}

$$\pi_e = 0 \qquad (143)$$

$$\gamma_{SO} - \pi_e - \gamma_{SL} = \gamma_{LV}\cos\theta \qquad (82)$$

$$\cos\theta = \frac{2\,\gamma_O - \gamma_{LV}}{\gamma_{LV}} \qquad (142)$$

Elimination of π_e and θ leads to the relation

$$\gamma_{SL} = \gamma_{SO} + \gamma_{LV} - 2\,\gamma_O \qquad (144)$$

Let us write again Eq. (95):

$$W_{AO} = \gamma_{SO} + \gamma_{LV} - \gamma_{SL} \qquad (95)$$

The combination of (144) and (95) leads to

$$W_{AO} = 2\,\gamma_O \quad (\text{when } \gamma_{LV} > 40 \text{ dyn cm}^{-1}) \qquad (145)$$

An analysis of the experimental curve corresponding to liquids having a superficial tension greater than 40 dyn cm^{-1} leads us to the conclusion that the adhesion work W_{AO} of any of these liquids on the bare solid is the same.

2. We have seen in Sec. 5.7 that, in opposition to W_A, W_{AO}, which represents the liquid's affinity for the bare solid, is not influenced by the possible presence of an adsorbed film on the solid. We can hereby extrapolate the relation we were able to establish in the case of liquids having high superficial tension to those with a low superficial tension. That is,

$$W_{AO} = 2\,\gamma_O \quad \forall\gamma \qquad (146)$$

3. We shall now try to determine the theoretical equation of

the complete curve $W_A - W_C = f(\gamma)$. Let us write (96) and (94):

$$W_{AO} = W_A + \pi_e \qquad (96)$$

$$W_C = 2\ \gamma_{LV} \qquad (94)$$

Combining (146), (96) and (94),

$$W_A - W_C = 2\ \gamma_O - 2\ \gamma_{LV} - \pi_e \qquad (147)$$

Equation (147) is thus the general equation of the experimental curve of Fig. 68: $W_A - W_C = f(\gamma)$. On Fig. 73 we have represented the experimental curve $W_A - W_C = f(\gamma)$ (reproduction of Fig. 68), and the straight line of equation

$$W_A - W_C = 2\ \gamma_O - 2\ \gamma \qquad (148)$$

with which the curve becomes confused when γ_{LV} is greater than 40 dyn cm^{-1}. Quantity π_e can be written by the combination of (147) and (148):

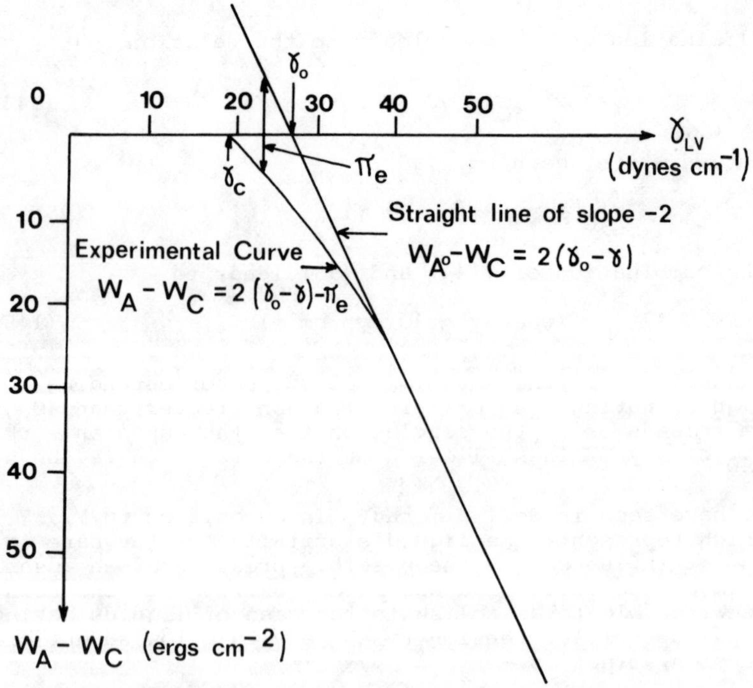

Fig. 73. Measurement of π_e from the representation $W_A - W_C$ versus surface tension.

$$\pi_e = (W_{AO} - W_C) - (W_A - W_C) \qquad (149)$$

π_e can thus be measured on Fig. 73 for a liquid having any superficial tension γ_{LV}. On a vertical line whose abscissa is γ_{LV}, π_e corresponds to the distance between the line of Eq. (148) and the experimental curve. Examination of Fig. 73 confirms that π_e is null when $\gamma_{LV}>40$ dyn cm^{-1}, π_e is nonnull when $\gamma_{LV}<40$ dyn cm^{-1}, and π_e is greater and greater for liquids with a weaker and weaker superficial tension. At point $\gamma_{LV} = \gamma_c$ (complete wetting), π_e is maximum and equals $2\,\gamma_0 - 2\,\gamma_c$.

For two compounds having the same low superficial tension, γ_{LV}, but belonging to different chemical families, nothing requires that the film spreading pressure π_e, due to adsorption of the vapor of each of the two compounds, be the same. We now understand the dispersion around a mean curve of experimental points having weak superficial tensions and corresponding to different chemical families.

The considerations concerning π_e from the examination of Fig. 73 are in perfect agreement with what we could have thought before.

4. We shall now determine the theoretical equation of the complete curve $\cos\theta = f(\gamma_{LV})$. In (144), γ_{SL} was established only in the case of liquids having high superficial tensions. A combination of the general relation

$$W_{AO} = 2\,\gamma_0 \qquad (146)$$

and Young's equation,

$$\gamma_{SO} - \pi_e - \gamma_{SV} = \gamma_{LV}\cos\theta \qquad (82)$$

permits us, by using Eq. (97) expressing W_{AO},

$$W_{AO} = \gamma_{LV}(1 + \cos\theta) + \pi_e \qquad (97)$$

to generalize the result expressed by (144):

$$\gamma_{SL} = \gamma_{SO} + \gamma_{LV} - 2\,\gamma_0 \qquad (150)$$

A combination of (82) and (150) leads to

$$\cos\theta = \frac{2\,\gamma_0 - \gamma_{LV} - \pi_e}{\gamma_{LV}} \qquad (151)$$

Equation (151) is the equation of the complete experimental curve $\cos\theta = f(\gamma_{LV})$ represented in Fig. 67.

8.8 Conclusions to Be Drawn from Our Interpretation

8.8.1 The Expression of W_{AO}

We have shown that the adhesion work of any liquid on a bare surface is the same. Its value $2\,\gamma_O$ can be measured experimentally. This result is in evident contradiction to the one established by Zisman (1964), which expresses W_{AO} in the form of a parabolic function of the liquid's surface tension. Our result should put the current theories concerning solid-liquid adhesion into question.

8.8.2 The expression of γ_{SL}

Let us recall Eq. (150) expressing γ_{SL}:

$$\gamma_{SL} = \gamma_{SO} + \gamma_{LV} - 2\,\gamma_O \tag{150}$$

γ_{LV} is relative to the liquid, γ_{SO} and γ_O are relative to the solid. Equation (150) is not symmetrical with regard to the liquid and the solid. Good's equation (11) and Fowkes' equation (12) established in the case of the interfacial tension γ_{12} between two nonmiscible liquids 1 and 2, are symmetrical relative to both liquids.

$$\gamma_{12} = \gamma_1 + \gamma_2 - 2\,\phi\,\sqrt{\gamma_1\gamma_2} \tag{11}$$

$$\gamma_{12} = {}_1 + \gamma_2 - 2\,\sqrt{\gamma_1^d\gamma_2^d} \tag{12}$$

In fact, when two liquids come into contact, each exerts actions on the other, leading to a mutual rearrangement of their molecular structures in the proximity of the interface. It is thus logical that the surface tensions of both liquids appear in a symmetrical manner in the expression of their interfacial tension. However, in the proximity of the surface common to a liquid and a solid, the liquid's molecules rearrange themselves to take the solid's presence into account, whereas the atomic superficial structure of the solid is not or only slightly modified (because of the volume stresses, the atoms occupy fixed positions inside the solid). Therefore the interfacial solid-liquid tension will be the sum of two quantities, γ_{SO}, which is the solid's superficial tension, and $\gamma_{LV} - 2\,\gamma_O$, which is the liquid's superficial tension modified by the presence of the solid. This sum cannot be symmetrical relative to γ_{SO} and γ_{LV}. These remarks are inconsistent with the extrapolation of relations (11) and (12) to solid-liquid interfaces.

8.8.3 The Significance of γ_O and γ_c

The quantity γ_O is rigorously constant for a given solid. γ_O is thus as useful as γ_{SO} in characterizing the solid.

The author believes that there should exist a simple relation between γ_O and γ_{SO}, but before trying to give a theoretical explanation of this relation, it would be necessary to have the values of γ_0 and γ_{SO} for a large number of solid surfaces with low energy.

On the contrary, γ_c depends on the solid, but also on the capacity of the liquids' vapors to form adsorbed films on the surface of the solid.

When $\gamma_{LV} = \gamma_c$, the liquid's affinity to its adsorbed layer is the same as that which a liquid of surface tension γ_0 would have with regard to the bare solid. The critical surface tension γ_c, defined by Zisman, remains very important in the analysis and experimental prediction of wetting phenomena.

8.8.4 The Measurement of π_e

Having obtained the experimental graph of Fig. 73, it is immediately possible to know the spreading pressure π_e corresponding to any liquid. This result is of the greatest importance, because until now, π_e constituted an experimental quantity that was extremely difficult to obtain. Even though the values of π_e for liquids having a weak surface tension is on the order of 10 to 20 dyn cm^{-1} (values taken from Fig. 73), only a few authors had questioned the "negligible value" of π_e for all liquids of PTFE. We can now say that <u>there is a nonnegligible adsorption of the vapor of liquids of low surface tension on PTFE</u>. On the other hand, the value of π_e corresponding to water is negligible or null: <u>On PTFE, water does not become adsorbed, or if so, very little</u>.

It should be noted that the studies carried out concerning a thorough analysis of curve $\cos\theta = f(\gamma)$ for $\gamma < 40$ dyn cm^{-1} should not be rejected, but rather reexamined and interpreted in terms of adsorption. The measurement of contact angles would show itself to be a very precise technique for measuring adsorption.

8.9 Recapitulation of Conclusions Drawn from PTFE Data

Table 1 recapitulates the hypothesis, the interpretations, and the conclusions of the theory of Zisman and that of the author. Figure 74 is the experimental curve $\cos\theta = f(\gamma_{LV})$ to which the table refers.

8.10 Extension of the Theory to Other Solids

Logically, the results we have established should be valid for low-energy solids other than PTFE. In effect, PTFE was chosen for its very low superficial tension, but there is no reason to believe that PTFE is a special solid.

In the case of solids with low superficial tensions, the curves obtained are of the same kind (Fox and Zisman, 1952a, 1952b; Shafrin and Zisman, 1952; Schulman and Zisman, 1952; Fox et al., 1953), and can be interpreted in the same way. However, if γ_c is greater than 30 dyn cm^{-1}, γ_0 may be greater than 45 dyn cm^{-1}, and a nonnegligible film pressure may be expected for all liquids except mercury. In that case we have only the point corresponding

Table 1

	Zisman's Interpretation	New Interpretation
Hypothesis	$\pi_e = 0, \quad \forall \gamma$	$\pi_e = 0$ when $\gamma > 40$ dyn cm^{-1}
The portion of the curve used to elaborate the theory	$\gamma_{LV} < 40$ dyn cm^{-1} (0.040 N m^{-1})	$\gamma_{LV} > 40$ dyn cm^{-1} (0.040 N m^{-1})
Obtained law	$\cos\theta = 1 + b(\gamma_c - \gamma_{LV})$	$\cos\theta = \dfrac{2\gamma_O - \gamma_{LV} - \pi_e}{\gamma_{LV}}$
Explanation of why some experimental points do not verify the law	They are points corresponding to hydrogen bonds ($\gamma_{LV} > 40$ dyn cm^{-1})	All experimental points verify the law
Explanation for the dispersion of experimental points having a low surface tension	No explanation	This slight dispersion is a consequence of the fact that π_e depends on the liquid's chemical nature
Expression of W_{AO}	$W_{AO} = (2 + b\gamma_c)\gamma_{LV} - b\gamma_{LV}^2$	$W_{AO} = 2\gamma_O$
Expression of W_A	$W_A = (2 + b\gamma_c)\gamma_{LV} - b\gamma_{LV}^2$	$W_A = 2\gamma_O - \pi_e$
Expression of γ_{SL}	$\gamma_{SL} = \gamma_{SO} - \gamma_{LV}(1 + b\gamma_c) + b\gamma_{LV}^2$	$\gamma_{SL} = \gamma_{SO} + \gamma_{LV} - 2\gamma_O$

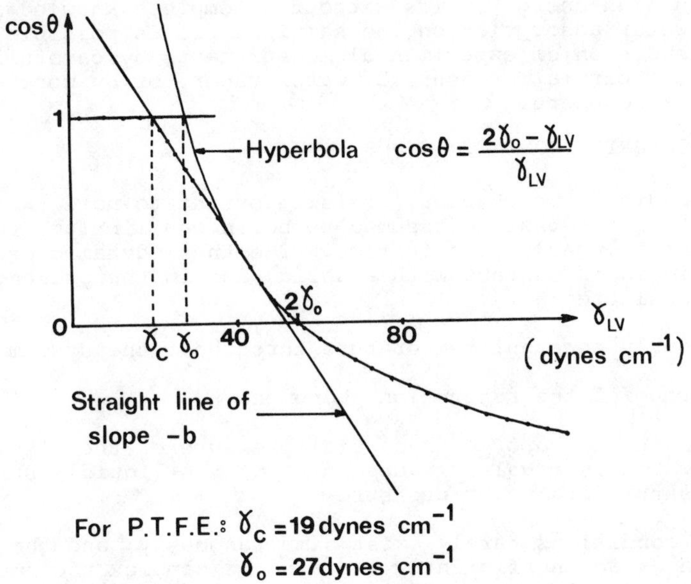

Fig. 74. Surface tension versus cos θ for miscellaneous liquids on PTFE.

to mercury for drawing the straight line $W_A - W_C = 2(\gamma_O - \gamma_{LV})$ and obtain γ_O.

In the case of solids with a high superficial tension, the conclusions that we have drawn for PTFE should also remain valid. However, experimental curves corresponding to measurements taken in an open enclosure may present some difficulties. In fact, on solids with a high tension, water should produce a nonnegligible spreading pressure π_e. Since water is found in the state of partial pressure in all nondried atmospheres, the presence of an adsorbed water film on the solid will be important at the moment of contact angle measurement of any liquid on the solid. It would be the same for any other compound in the state of partial pressure in the enclosure. In such a case the contact angle would verify

$$\gamma_{LV} \cos \theta = \frac{2\,\gamma_O - \gamma_{LV} - \pi_e - \pi_e'}{\gamma_{LV}} \qquad (152)$$

in which π_e represents the film pressure due to the saturing vapor of the liquid constituting the drop, and π_e' represents the film pressure due to vapors other than the liquid's.

The contact angles of liquids on solids having a high surface energy depend essentially on the adsorbed layers at the solid's surface, and they are particularly sensitive to the adsorption of water vapor. The result was already noted experimentally (see Sec. 6.6). It is thus impossible to interpret the contact angles of liquids on high-energy solids without a complete knowledge of the laws of water adsorption on the samples. It is possible to avoid the dispersion of experimental measurements by carefully controlling the partial pressure of water vapor, or by working in an anhydrous atmosphere.

9 NONEQUILIBRIUM CONTACT ANGLES

Before ending this chapter, it is important to note that we have studied systems that we assumed to be in equilibrium, that is, systems that no longer evolve in time. The thermodynamic equilibrium of a liquid in contact with a solid, and in the presence of a vapor, necessitates:

A rigorously constant temperature throughout the system

Equilibrium of the adsorption phenomena

Presence in the vapor of a partial pressure of the liquid's vapor, which is equal at every point of the liquid's surface to the saturating vapor pressure

All of these conditions rarely exist simultaneously, and the theories we have shown do not sufficiently explain the complex phenomena that appear at the time of wetting of solid surfaces under nonequilibrium conditions.

For example, a temperature gradient at the solid surface causes the migration of a liquid on a solid surface (Fote et al., 1976). Movements of liquid were also observed on the inner surfaces of a recipient containing wine or other alcoholic liquid (Thomson, 1855). In both cases, there are surface tension variations at different points on the liquid's surface. These variations are due, respectively, to differences in temperature or differences in concentration (following evaporation of the alcohol). These phenomena were first explained by Marangoni (1871), and were called "Marangoni effects." Scriven and Sternling (1960) made a bibliographic study of them. The cases in which there is a movement of the liquids at the solid surfaces are very numerous. These "surface-driven phenomena" have been reviewed by Levich and Krylov (1969).

In some cases (Chappuis et al., 1976), it is even possible to obtain continuous periodic movements of the line of junction between liquid and solid. The most likely explanations bring into play adsorption-desorption processes (Martin et al., 1977).

Let us note, finally, that a sessile drop on a perfectly plane solid surface cannot be a system in equilibrium. In fact, the saturating vapor pressure p of the liquid, must be, according to the curvature of the liquid's surface, greater than the saturating vapor pressure p_0 of a plane liquid surface (see Sec. 2.8). In the vicinity of the three-phase line, there must then be a condensation of that vapor on the solid surface (see Sec. 4.2). Local equilibrium conditions could then be obtained by slight temperature differences of the system's phases. In recent developments, reviewed by Good (1979), the microscopic configuration of the interface could be modified in the vicinity of the solid (up to 10 or 20 Å), because the molecules of the liquid are under the influence of the solid.

NOMENCLATURE

a capillary constant (Eq. 28)

a_s activity of a solution (Sec. 2.4)

A area of an interface (Sec. 2.1)

A_{SV} area of a solid-vapor interface (Sec. 5.1)

A_a apparent area of a solid surface (Sec. 6.5)

A_r real area of a solid surface (Sec. 6.5)

A_o quantity indicating the affinity of a liquid for a solid (Sec. 5.7)

$A_1, A_{12}, A_{345...}, A_{456...}, \ldots$ forces of attraction coming from the molecules of layers 1, 1 and 2, etc. (Sec. 2.2)

b absolute value of the slope of straight lines $\cos\theta = f(\gamma)$ in Zisman's interpretation (Sec. 8.4)

b' parameter characterizing the size of the meniscus; radius of curvature at its apex (Sec. 2.7)

B constant in Langmuir's adsorption isotherm (Sec. 4.2)

c concentration of a solution (Sec. 4.3)

C constant

d algebraic distance from the free liquid surface to the lower plane of the sample; $d = \overline{OM}$ on Fig. 52 (Sec. 7.2)

D distance between two molecules (Sec. 2.2)

e_s surface specific energy (Eqs. 5, 71)

E energy of desorption per mole (Sec. 4.1)

f resultant of forces exerted on the balance scale less the sample's weight in the air (Sec. 7.2)

f' force necessary to increase the extent of a surface reversibly (Sec. 2.1)

F force (Sec. 2.2)

F_B resultant of buoyancy forces (Sec. 7.2)

F_G gravitational force on the sample; $F_G = Mg$ (Sec. 7.2)

F_S resultant of surface forces (Sec. 7.2)

g gravity (Sec. 2.5)

g_s surface specific free energy (Eq. 72)

G free energy of a system (Sec. 2.1)

h algebraic distance from the free liquid surface to the solid-liquid junction line; $h = \overline{OH}$ on Fig. 52 (Sec. 7.2)

h' distance separating the bottom of the hollows and the summit of the asperities of a solid surface (Sec. 3.5)

H height of the sample (Sec. 7.2)

k Boltzmann's constant (Sec. 4.1)

k' ratio of the real area to the apparent area of a solid surface (Sec. 6.3)

K constant in van der Waals attractive law (Sec. 2.2)

K' constant in Freundlich's isotherm (Sec. 4.2)

ℓ perimeter of a plate or of a sample (Sec. 5.5)

$d\ell$ element belonging to a circumference (Sec. 2.5)

$d\ell'$ displacement of the line of discontinuity of a surface (Sec. 2.1)

L length of a plate in contact with the liquid (Sec. 5.6)

L' length of the line of discontinuity of a surface (Sec. 2.1)

L'' depth of immersion of a vertical plate in a liquid (Sec. 5.5)

L_S length of a solid plate (Sec. 3.1)

m mass of adsorbed substance (Sec. 4.3)

m_a mass of adsorbant (Sec. 4.3)

M mass of the sample (Sec. 7.2)

M' necessary mass for the equilibrium of a mechanical system (Sec. 5.5)

n constant in Freundlich isotherm (Sec. 4.2)

$n_1, n_2, \ldots, n_n$ molecular densities of molecular layers 1, 2, . . . ; n (Sec. 2.2)

N Avagadro's number (Sec. 4.1)

p pressure of the saturating vapor of a curved liquid surface (Eq. 59)

p_0 pressure of the saturating vapor of a plane liquid surface (Sec. 2.8)

P pressure (Sec. 2.5)

ΔP difference of pressure across an interface (Sec. 2.5)

P_A atmospheric pressure (Sec. 5.2)

P_I pressure inside a spherical drop (Sec. 5.2)

$P_{M'}$, $P_{M''}$ pressure at point M', M'' (Sec. 2.6)

P_0 pressure at the level of the plane portion of the interface (Sec. 2.6)

r radius of a spherical surface (Sec. 2.8)

r_1 radius of a spherical liquid drop (Sec. 5.2)

r_1', r_2' radii of curvature of a surface in two orthogonal planes containing a normal line to the surface (Sec. 2.5)

R	universal gas constant (Sec. 2.4)
$R_1, R_2, R_3, \ldots, R_6$	forces of repulsion coming from the molecules of layers 1, 2, 3, . . . , 6 (Sec. 2.2)
R'	radius of curvature situated in the plane of the figure (Sec. 2.6)
R_1', R_2'	principal radii of curvature of a surface (Sec. 2.5)
R_S	solid reaction exerted on each unit of length of the three-phase line (Sec. 5.2)
s_s	surface specific entropy (Sec. 5.1)
ds	element of a curved line (Sec. 2.6)
S	spreading coefficient (Eqs. 100, 101, 102)
S'	section of a sample (Sec. 3.1)
T	absolute temperature (Sec. 2.1)
$v_1, v_2, v_3, \ldots$	variables defining the system (Sec. 2.1)
V	volume of a meniscus (Sec. 2.6)
V_i	volume of phase i (Sec. 5.1)
V_m	liquid's molar volume (Sec. 2.8)
w	width of a section of meniscus (Sec. 2.6)
w_a	apparent weight of the drop (Sec. 5.2)
w'	width of a plate (Sec. 5.6)
W_A	work of adhesion of a liquid to a solid (Eqs. 92, 93)
W_{AO}	work of adhesion of a liquid to a bare solid (solid in vacuo) (Eq. 95)
W_C	work of cohesion of a liquid (Eq. 94)
W_M	weight of a cylindrical meniscus (Sec. 7.1)
W_f	total work of cylindrical meniscus formation (Eqs. 52, 55)
dW	work of wetting or dewetting of an area dA of a solid surface (Eq. 86)
dW'	work furnished by the operator when raising a plate by dh (Eq. 103)
dW_s	elementary work of surface extension (Eq. 46)
dW_g	elementary work against gravity (Eq. 45)
x	variable along axis Ox
x'	radius of a circular solid-liquid interface (Sec. 5.2)
y	variable along axis Oy
z	variable along axis Oz
z_1	height between two points (Sec. 2.5)

Greek Letters

a	angle containing the liquid, and limited on one hand by

the tangent at the point of junction to the profile of the liquid, and on the other hand by the vertical (Sec. 5.9)

α' fraction of a surface with a low surface tension (Sec. 6.2)

α'' percentage of zones of low superficial tension on a line of type (a) in Fig. 44 (Sec. 6.2)

α_1 angle between the vertical and stripes turned downward (Sec. 6.3)

β parameter characterizing the shape of a meniscus (Sec. 2.7)

β_1 angle between the vertical and stripes turned upward (Sec. 6.3)

γ superficial tension, or work of formation of a surface (Sec. 2.1)

γ_0 abscissa of the point of intersection of the axis $W_A - W_C = 0$ and of the straight line on which there are the experimental points of high surface tension, in the curves $W_A - W_C$ versus γ_{LV} (Sec. 8.4, Fig. 68)

γ_1, γ_2 liquid-vapor interfacial tensions of liquid 1, liquid 2 (Sec. 2.3)

$\gamma_{1(2)}$ interfacial tension of liquid 1, modified by liquid 2 (Sec. 2.3)

$\gamma_{2(1)}$ interfacial tension of liquid 2, modified by liquid 1 (Sec. 2.3)

γ_{12} interfacial tension of the interface separating liquids 1 and 2 (Sec. 2.3)

γ_{100}, γ_{110}, γ_{111} superficial tensions of a solid, the surfaces of which are located by Miller's crystallographic indexes (Sec. 3.2)

γ_{LV} interfacial tension of a liquid-vapor interface (Sec. 5.1)

γ_{LL} null interfacial tension of a surface located in the mass of a liquid (Sec. 5.6)

γ_{SL} interfacial tension of a solid-liquid interface (Sec. 3.2)

γ_{SV} interfacial tension of a solid-vapor interface (Sec. 3.3)

γ_{SO} surface tension of a bare solid (without any adsorbed molecule) (Sec. 4.4)

γ_c critical surface tension of Zisman (Sec. 8.4)

γ^d (γ_1^d, γ_2^d) part due to London dispersion forces in the superficial tension of a liquid (of liquid 1, liquid 2) (Sec. 2.3)

γ^h (γ_1^h) part due to molecular forces other than dispersion ones in the superficial tension of a liquid (of liquid 1) (Sec. 2.3)

Γ	adsorption: algebraic surface excess per unit area (Sec. 2.4)
$\Gamma(p)$	quantity adsorbed per unit area of a solid, at pressure p (Sec. 4.2)
ε	interaction energy of an adsorbed molecule with the solid (Sec. 4.1)
θ	contact angle of a liquid in contact with a solid; we so call the true angle, at the microscopic scale, containing the liquid and limited on one hand by the solid surface and, on the other hand, by the tangent at the point of junction to the profile of the liquid (Sec. 1.1)
θ_0	Young's intrinsic angle (Sec. 6.5)
θ_1	contact angle on a surface labeled 1 (of high surface tension) (Sec. 6.2)
θ_2	contact angle on a surface labeled 2 (of low surface tension) (Sec. 6.2)
θ_A	advancing contact angle (Sec. 1.1)
θ_E	angle of junction upon emersion of a vertical plate (receding angle) (Sec. 7.2)
θ_I	angle of junction upon immersion of a vertical plate (advancing angle) (Sec. 7.2)
θ_R	receding contact angle (Sec. 1.1)
θ_W	Wenzel's contact angle (Sec. 6.5)
θ_a	fraction of a solid surface covered by an adsorbed layer (Sec. 4.2)
θ'	apparent (macroscopic) contact angle (Sec. 6.3)
μ_i	chemical potential of phase i (Sec. 5.1)
ν	pseudo-frequency of vibration of a molecule in the adsorbed state (Sec. 4.1)
π_e	spreading film pressure (Eg. 68)
ρ (ρ_1, ρ_2)	volumic mass of a fluid (of fluid 1, 2) (Secs. 2.5 2.6)
ρ_A	volumic mass of air (Sec. 7.2)
ρ_L	volumic mass of a liquid (Sec. 5.2)
ρ_V	volumic mass of a vapor (Sec. 5.2)
ρ'	distance between point P and any point of the curve of Fig. 9 (Sec. 2.5)
$\Delta\rho$	difference between two volumic masses (Sec. 2.6)
τ	adhesion tension (Eq. 87)
τ_a	adsorption time; mean duration of adsorption of a molecule (Eq. 64)
τ_0	pseudo-period of vibration of a molecule in the adsorbed state (Sec. 4.1)
$d\tau'$	work necessary to increase the extent of a fluid surface by dA (Sec. 2.1)

φ	angle between the horizontal and the tangent at point *M* to the profile of the meniscus (Sec. 2.6)
ϕ	Good's interaction parameter (Sec. 2.3)
ϕ'	angle between the normal line to the meniscus at *S* and the axis of symmetry (Sec. 2.7)
ϕ_1	angle $\widehat{PNA}$ on Fig. 9 (Sec. 2.5)

REFERENCES

Adam, N. K. 1941. *The Physics and Chemistry of Surfaces*, 3rd ed. London: Oxford University Press.

Adam, V. K., and G. Jessop 1863. *J. Chem. Soc.*, p. 1925.

Antonow, G. 1907. *J. Chim. Phys.* vol. 5, p. 372.

Bangham, D. H., and R. I. Razouk 1937. Adsorption and the Wettability of Solid Surfaces. *Trans. Faraday Soc.* vol. 33, pp. 1459–1463.

Bashforth, F., and J. C. Adams 1892. *An Attempt to Test the Theory of Capillary Action.* Cambridge: Cambridge University Press and Deighton, Bell & Co.

Bigelow, W. C., D. L. Pickett, and W. A. Zisman 1946. Oleophobic Monolayers Films Adsorbed From Solutions in Non-polar Liquids. *J. Colloid Sci.* vol. 1, pp. 513–538.

Bikerman, J. J. 1941. *Ind. Eng. Chem.* vol. 13, p. 443.

Bikerman, J. J. 1967. Solid Surfaces. Contact Angles, Spreading and Wetting. *Second International Congress of Surface Activity, Vol. 3.* New York: Academic Press, pp. 125–135.

Blaisdell, B. E. 1940. *J. Math. Phys.* vol. 19, pp. 186, 217, and 220.

Brockway, L. O., and R. L. Jones 1964. Electron Microscopic Investigation of the Adsorption of Long-Chain Fatty Acid Monolayers on Glass. *Advan. Chem. Ser. 43*, pp. 275–294.

Brunauer, S. 1945. *The Adsorption of Gases and Vapors, Vol. 1.* Princeton, N.J: Princeton University Press.

Brunauer, S., P. H. Emmet, and E. Teller 1938. *J. Am. Chem. Soc.* vol. 60, p. 309.

Cassie, A. B. D., and S. Baxter 1944. Wettability of Porous Surfaces. *Trans. Faraday Soc.* vol. 40, pp. 546–551.

Chappuis, J. 1974. Contribution à l'Etude du Mouillage, Application aux Problèmes de Lubrification. Thèse de Doctorat d'Etat, Lyon, no. 264.

Chappuis, J. 1977. Vectorial Justification of Young's Equation: Experimental Evidence and Theoretical Model. *51st Colloid and Surface Science Symposium, Buffalo, New York*, p. 85.

Chappuis, J., and J. M. Georges 1974. Contribution à l'Etude du Mouillage; Analyse d'une Methode de Mesure. *J. Chim. Phys.* vol. 71, pp. 567–575.

Chappuis, J., M. Jacquet, and J. M. Georges 1974. Influence de la Vapeur d'Eau sur un Film Monomoléculaire d'Acide Stéarique. *C.R. Acad. Sci. Paris* vol. C278, pp. 1215–1218.

Chappuis, J., J. M. Martin, and J. M. Georges 1976. Mise en Evidence d'un Phénomène de Pulsations verticales d'une Goutte d'Huile Placée dans un Espace Capillaire avec Gradient Thermique. *C.R. Acad. Sci. Paris* vol. C282, pp. 775–778.

Chappuis, J., H. Montes, and J. M. Georges 1977. Description d'une Technique de Mesure de la Mouillabilité de Billes de Petits Diamètres. *J. Chim. Phys.* vol. 74, pp. 234–237.

Cooper, W. A. and W. A. Nuttall 1915. The Theory of Wetting, and the Determination of the Wetting Power of Dipping and Straying Fluids Containing a Soap Basis. *J. Agr. Sci.* vol. 7, pp. 219–239.

De Boer, J. H. 1953. *The Dynamic Character of Adsorption.* Oxford: Clarendon Press, p. 44.

Defay, R., and I. Prigogine 1951. *Tension Superficielle et Adsorption.* Liège: Desoer, p. 10.

Defay, R., I. Prigogine, A. Bellemans, and D. H. Everett 1966. *Surface Tension and Adsorption.* London: Longmans, Green

De Laplace, P. S. 1806. *Mécanique Céleste*, Suppl. to Book 10. Paris: Coureier.

Doss, K. S. and B. S. Rao 1938. *Proc. Indian Acad. Sci.* vol. 7A p. 117.

Dupré, A. 1869. *Théorie Mécanique de la Chaleur.* Paris: Gauthier Villars, p. 369.

Fordham, S. 1948. *Proc. Roy. Soc. (London)* vol. 194A, p. 1.

Fort, T. Jr., and H. T. Patterson 1963. A Simple Method for Measuring Solid-Liquid Contact Angles. *J. Colloid Sci.* vol. 18, pp. 217–222.

Fote, A. A., L. M. Dormant, and S. Feuerstein 1976. Migration of Hydrocarbon Oil on Metal Substrates Under the Influence of Temperature Gradients. *Lubrication Eng. ASLE* vol. 32, pp. 542–545.

Fowkes, F. M. 1962. Determination of Interfacial Tensions, Contact Angles, and Dispersion Forces in Surfaces by Assuming Additivity of Intermolecular Interactions in Surfaces. *J. Phys. Chem.* vol. 66, p. 382. See also Fowkes, F. M. 1964. *Advan. Chem. Ser. no. 43*, p. 99.

Fowkes, F. M. 1964. *Advan. Chem. Ser. 43*, p. 99.

Fowkes, F. M. 1967, Attractive Forces at Solid-Liquid Interfaces. *Wetting*. Soc. Chem. Ind. Monograph No. 25, p. 3.

Fox, H. W., and W. A. Zisman 1950. The Spreading of Liquids on Low-Energy Surfaces 1. Polytetrafluorethylene. *J. Colloid Sci.* vol. 5, pp. 514–531.

Fox, H. W., E. F. Hare, and W. A. Zisman 1953. The Spreading of Liquids on Low-Energy Surfaces; 6. Branched-Chain Monolayers, Aromatic Surfaces, and Thin Liquid Films. *J. Colloid Sci.* vol. 8, pp. 194–203.

Fox, H. W., and W. A. Zisman 1952a. The Spreading of Liquids on Low-Energy Surfaces; 2. Modified Tetrafluoroethylene Polymers. *J. Colloid Sci.* vol. 7, pp. 109–121.

Fox, H. W., and W. A. Zisman 1952b. The Spreading of Liquids on Low-Energy Surfaces; 3. Hydrocarbon Surfaces. *J. Colloid Sci.* vol. 7, pp. 428–442.

Freundlich, H. 1926. *Colloid and Capillary Chemistry*. London: Methuen.

Gibbs, J. W. 1961. On the Equilibrium of Heterogeneous Substances. *The Scientific Papers of J. W. Gibbs, Vol. 1.* New York: Dover Publications, pp. 55–353.

Girifalco, L. A., and R. J. Good 1957. A Theory for the Estimation of Surface and Interfacial Energies. 1. Derivation and Application to Interfacial Tension. *J. Phys. Chem.* vol. 61, pp. 904–909.

Good, R. J. 1952. *J. Am. Chem. Soc.* vol. 74, p. 504.

Good, R. J. 1977. Surface Free Energy of Solids and Liquids: Thermodynamics, Molecular Forces, and Structure. *J. Colloid Interface Sci.* vol. 59, pp. 398–419.

Good, R. J. 1979. Contact Angles and the Surface Free Energy of Solids. *Surface and Colloid Sci.* , *Vol. 11*, R. J. Good and R. R. Stromberg, Eds. New York: Plenum Press, pp. 1–29.

Guastalla, J. 1956. Recent Work on Surface Activity, Wetting and Dewetting. *J. Colloid Sci.* vol. 11, pp. 623–636.

Harkins, W. D., and H. K. Livingstone 1952. *J. Chem. Phys.* vol. 10, p. 348.

Harkins, W. D., and E. H. Loeser 1950. *J. Chem. Phys.* vol. 18, p. 556.

Herring, C. 1951. Surface Tension as a Motivation for Sintering. *The Physics of Powder Metallurgy*, W. E. Kingston, Ed. New York: McGraw-Hill, pp. 143–179.

Johnson, R. E., Jr. 1959. Conflicts Between Gibbsian Thermodynamics and Recent Treatments of Interfacial Energies in Solid-Liquid-Vapor Systems. *J. Phys. Chem.* vol. 63, pp. 1655–1658.

Johnson, R. E., Jr., and R. H. Dettre 1969. Wettability and Contact Angles. *Surface and Colloid Science*, vol. 2, E. Matijevic, Ed. New York: Wiley-Interscience, pp. 85–153.

Langmuir, I. 1918. *J. Am. Chem. Soc.* vol. 40, p. 1361.

Lester, G. R. 1961. Contact Angles of Liquids at Deformable Solid Surfaces. *J. Colloid Sci.* vol. 16, pp. 315–326.

Levich, V. G., and V. S. Krylov 1969. Surface-Driven Phenomena. *Ann. Rev. Fluid Mech.* vol. 1, pp. 293–316.

Linford, R. G. 1973. Surface Thermodynamics of Solids. *Solid State Surface Science*, M. Green, Ed. New York: Marcel Dekker.

Marangoni, C. G. M. 1871. *Ann. Phys. (Poggendorf)* vol. 143, p. 337.

Martin, J. M., J. Chappuis, and J. M. Georges 1977. Role des Additifs Anti-Usure des Huiles Lubrifiantes dans un Test de Frottement Simulant le Contact Segment-Chemise. *Rev. Inst. Fr. Pet.* vol. 23, pp. 113–130.

Mills, O. S. 1953. *J. Appl. Phys.* vol. 4, p. 24.

Mullins, W. W. 1963. *Metal Surfaces: Structure, Energetics and Kinetics.* Metals Park, Ohio: American Society for Metals, p. 17.

Neumann, A. W. 1974. Contact Angles and Their Temperature Dependence: Thermodynamic Status, Measurement, Interpretation and Application. *Advan. Colloid Interface Sci.* vol. 4, pp. 105–191.

Neumann, A. W., and R. J. Good 1972. The Thermodynamics of Contact Angles; 1. Heterogeneous Solid Surfaces. *J. Colloid Interface Sci.* vol. 38, pp. 341–358.

Neumann, A. W., and R. J. Good 1979. Techniques of Measuring Contact Angles. *Surface and Colloid Science, Vol. 11*, R. J. Good and R. R. Stromberg, Eds. New York: Plenum Press, pp. 31–91.

Neumann, A. W., and P. J. Sell 1964. *Z. Phys. Chem. (Leipzig)* vol. 227, p. 187.

Nicholas, J. F. 1968. *Austral. J. Phys.* vol. 21, p. 21.

Padday, J. F. 1969. Theory of Surface Tension. *Surface and Colloid Science*, vol. 1, E. Matijevic, Ed. New York: Wiley-Interscience, pp. 39–252.

Pethica, B. A., and T. J. P. Pethica 1967. The Contact Angle Equilibrium. *Second International Congress of Surface Activity, Vol. 3.* New York: Academic Press, p. 31.

Princen, H. M. 1969. Shape of Interfaces, Drops and Bubbles. *Surface and Colloid Science*, E. Matijevic, Ed. New York: Wiley-Interscience, pp. 1–84.

Lord Rayleigh 1890. *Phil. Mag.* vol. 30, p. 397.

Ries, H. R., Jr., and D. C. Walker 1961. Films of Mixed Horizontally and Vertically Oriented Compounds. *J. Colloid Sci.* vol. 16, pp. 361–374.

Scriven, L. E., and C. V. Sternling 1960. The Marangoni Effects. *Nature* vol. 187, pp. 186–188.

Schulman, F., and W. A. Zisman 1952. The Spreading of Liquids on Low-Energy Surfaces; 5. Perfluorodecanoic Acid Monolayers. *J. Colloid Sci.* vol. 7, pp. 465–481.

Shafrin, E. G., and W. A. Zisman 1952. The Spreading of Liquids on Low-Energy Surfaces; 4. Monolayer Coatings on Platinum. *J. Colloid Sci.* vol. 7, pp. 166–177.

Shuttleworth, R. 1950. The Surface Tension of Solids. *Proc. Phys. Soc. (London)* vol. A 63, pp. 444–457.

Shuttleworth, R., and G. L. J. Bailey 1948. The Spreading of a Liquid over a Rough Surface. *Discussions Faraday Soc.* vol. 3, pp. 16–22.

Tabor, D. 1969. *Gases, Liquids and Solids.* Penguin Library of Physical Science. Glasgow: Bell and Bain.

Tawde, N. R., and K. G. Parvatikar 1958. *Ind. J. Phys.* vol. 32, p. 174.

Thomson, J. 1855. On Certain Curious Motion Observable at the Surface of Wine and Other Alcoholic Liquids. *Phil. Mag.* vol. 10 (4), p. 330.

Thomson, W. (Lord Kelvin). 1871. *Phil. Mag.* vol. 42 (4), p. 448.

Udin, H., A. J. Shaler, and J. Wulff 1949. The Surface Tension of Solid Copper Metals. *Trans. AIME* vol. 1 (2), pp. 186–190.

Van Oss, C. J., B. H. Park, J. M. Bernstein, and C. F. Gillman 1977. Diminished Hydrophilicity of Granulocytes in Children Who Are Prone to Bacterial Infections. Influence of Serum Factors. *51st Colloid and Surface Science Symposium, Buffalo, New York.*

Wenzel, R. N. 1936. Resistance of Solid Surfaces to Wetting by Water. *Ind. Eng. Chem.* vol. 28, pp. 988–994.

Whalen, J. W., and W. H. Wade 1967. *J. Colloid Interface Sci.* vol. 24, p. 372.

Young, T. 1805. An Essay on the Cohesion of Fluids. *Phil. Trans. Roy. Soc. (London)* vol. 95, pp. 65–87.

Young, T. 1855. *Miscellaneous Works, Vol. 1*, G. Peacock, Ed. London: J. Murray, p. 418.

Zisman, W. A. 1959. Friction, Durability and Wettability Properties of Monomolecular Films on Solids. *Friction and Wear*, R. Davies, Ed. Amsterdam: Elsevier, pp. 110–148.

Zisman, W. A. 1964. Relation of the Equilibrium Contact Angle to Liquid and Solid Constitution. *Advan. in Chem. Ser. 43*, pp. 1–51.

Subject Index